Electrochemical Energy Conversion via Small Molecule Catalysis

Electrochemical Energy Conversion via Small Molecule Catalysis

Edited By

Zhicheng Zhang
Yuanmiao Sun

WILEY VCH

Authors

Professor Zhicheng Zhang
Department of Chemistry
School of Science
Tianjin University
No. 92 Weijin Road
Nankai District
People's Republic of China

Professor Yuanmiao Sun
Shenzhen Institute of Advanced Technology
1068 Xueyuan Avenue
Shenzhen University Town
Shenzhen, 518055
People's Republic of China

Cover Design: Wiley
Cover Image: © Love Employee/Shutterstock

Typesetting: Lumina Datamatics, Madurai, India
Printing and Binding: CPI Group (UK) Ltd, Croydon, CR0 4YY

Library of Congress Card No.: applied for

British Library Cataloguing-in-Publication Data
A catalogue record for this book is available from the British Library.

Bibliographic information published by the Deutsche Nationalbibliothek

The Deutsche Nationalbibliothek lists this publication in the Deutsche Nationalbibliografie; detailed bibliographic data are available on the Internet at http://dnb.d-nb.de.

Print ISBN 9783527354405
ePDF ISBN 9783527849284
ePub ISBN 9783527849277
oBook ISBN 9783527849291

C9783527354405_120925

Contents

Preface

With the ongoing increase in global energy demand and the diminishing availability of fossil fuels, there is a growing need for sustainable, efficient, and cost-effective energy conversion and storage systems. This challenge has become one of the most pressing issues in modern society. Electrocatalysis has emerged as a key technology in enabling the efficient transformation of energy, particularly in systems such as fuel cells, metal-air batteries, CO_2 conversion, and water splitting. Electrocatalysts enhance the rate of electrochemical reactions, playing a pivotal role in improving the efficiency of energy conversion processes.

This book provides a comprehensive examination of electrocatalytic materials and their applications in energy conversion, with a focus on electrochemical reactions such as the hydrogen evolution reaction, oxygen evolution reaction, CO_2 reduction, nitrogen fixation, and biomass conversion. It is divided into several sections, each exploring critical aspects of the field. The first section covers fundamental concepts and theories of electrocatalysis, detailing the mechanisms and pathways involved in various electrochemical reactions. The second section delves into the design principles, structural characteristics, and structure–activity relationships of state-of-the-art electrocatalysts, offering insights into how geometric and electronic factors influence catalytic performance. The final section addresses the challenges and opportunities for advancing electrocatalysis, proposing future directions for research and development.

Through this book, we aim to provide readers with a deeper understanding of the role of electrocatalysts in energy conversion, while stimulating further interest and investment in this rapidly evolving field. We hope that this book will inspire ongoing research and innovation, ultimately driving the development of practical, scalable solutions for sustainable energy production and utilization.

Zhicheng Zhang, Tianjin University
Yuanmiao Sun, Shenzhen Institute of Advanced Technology, Chinese Academy of Sciences

Acknowledgments

This book was supported by Innovation Funding Project of Science and Technology, China National Petroleum Corporation (2022DQ02-0408). We would like to thank editors Elisha Benjamin and Alice Qian from Wiley for their great efforts in editing this book.

1

Introduction

*Yiran Zhang, Zhicheng Zhang**

Tianjin Key Laboratory of Molecular Optoelectronic Sciences, Department of Chemistry, School of Science, Tianjin University, Tianjin, China

*Corresponding author: zczhang19@tju.edu.cn

1.1 Origins and Evolution of Small-molecule Catalysis for Electrochemical Energy Conversion

The pursuit of sustainable energy sources has led to significant interest in electrochemical energy conversion (EEC) technologies, which offer an efficient pathway for transforming chemical energy into electrical energy and vice versa. One key component in these technologies is catalysts, specifically small-molecule catalysts, which facilitate the electrochemical conversion processes. The roots of small-molecule catalysis trace back to early studies in electrochemistry and surface chemistry, where fundamental reactions like hydrogen evolution, oxygen evolution, and oxygen reduction were identified as critical steps in energy conversion processes [1].

Early discoveries focused on platinum and other noble metals due to their exceptional catalytic activity. However, the high cost and limited availability of these metals raised concerns about their scalability for large-scale applications. This led to a surge of interest in alternative catalysts, particularly small molecules, which could mimic the catalytic properties of noble metals. As the field of EEC expanded, researchers recognized that the key to unlocking sustainable energy solutions lies in developing cost-effective, efficient, and stable catalysts [2].

In recent decades, advancements in materials science, surface chemistry, and catalytic design have brought forth a wide array of small-molecule catalysts, such as molecular catalysts, non-precious metal-based catalysts, and hybrid materials. These catalysts are capable of promoting a variety of electrochemical reactions, including hydrogen oxidation, CO_2 reduction, and nitrogen fixation, which are central to the development of green energy technologies. Understanding the principles of small-molecule catalysis and developing efficient, sustainable catalysts remains one of the most active areas of research in EEC.

Electrochemical Energy Conversion via Small Molecule Catalysis, First Edition.
Edited by Zhicheng Zhang and Yuanmiao Sun.

1.2 Current State of Small-molecule Catalysis in Electrochemical Energy Conversion

Small-molecule catalysis has gained significant traction as a promising approach for advancing EEC technologies. The development of non-precious metal catalysts, molecular catalysts, and atomically dispersed catalysts has revolutionized the field, offering higher efficiency and sustainability compared to traditional platinum-based catalysts. These small-molecule catalysts are crucial for key electrochemical reactions such as hydrogen evolution, oxygen reduction, carbon dioxide reduction, and nitrogen fixation.

1.2.1 Hydrogen Evolution and Oxygen Evolution Reactions

Hydrogen evolution reaction (HER) and oxygen evolution reaction (OER) are the cornerstone reactions in water electrolysis, which splits water into hydrogen and oxygen. The development of efficient HER and OER catalysts has been a major focus of electrochemical energy research. Traditionally, platinum-based catalysts have been the benchmark for HER due to their high activity, but the scarcity and cost of platinum have led to extensive efforts in discovering non-precious metal alternatives. Transition metal-based catalysts, including phosphides, sulfides, and nitrides, have demonstrated remarkable activity for HER [3]. Similarly, non-precious metal-based OER catalysts such as cobalt-based oxides have shown high performance in alkaline media [4].

Furthermore, molecular catalysts that mimic natural enzymes, such as hydrogenases and oxygenases, have been designed for both HER and OER. These molecular catalysts offer significant advantages in terms of structural tunability, enabling researchers to optimize their catalytic properties for energy conversion [5]. This has led to the development of catalysts with improved efficiency and stability, which are essential for practical large-scale applications in green hydrogen production.

1.2.2 Oxygen Reduction Reaction

The oxygen reduction reaction (ORR) is another key process, particularly for fuel cell technologies. ORR involves the reduction of O_2 to H_2O or H_2O_2, depending on the reaction pathway, and is considered the rate-determining step in fuel cells [6]. While platinum remains the most efficient catalyst for ORR, research into small-molecule catalysts has made significant strides in developing alternatives. Additionally, other non-precious metal catalysts, such as cobalt-based complexes, are being actively studied for their potential to replace platinum in ORR, offering both cost and performance advantages [7].

1.2.3 CO_2 Reduction

Electrochemical CO_2 reduction is another exciting area in the field of EEC. It involves the conversion of CO_2 into valuable products such as carbon monoxide, methane, and formic acid, using renewable electricity. Small-molecule catalysts, including copper-based catalysts and metal–organic frameworks (MOFs), have shown promising results for CO_2 reduction [8]. Copper-based catalysts, in particular, are unique in their ability to selectively reduce CO_2 to a wide range of carbon-based products, making them central to the development of carbon-neutral fuels and chemicals. This area of research has gained significant attention due to its potential to mitigate climate change and provide alternative pathways for chemical production [9].

1.2.4 Nitrogen Fixation

Nitrogen fixation, the process of converting atmospheric nitrogen (N_2) into ammonia (NH_3), is crucial for sustainable agriculture. The traditional Haber–Bosch process for nitrogen fixation is energy-intensive and relies on fossil fuels. Electrochemical nitrogen fixation, on the other hand, offers a cleaner, more energy-efficient method for synthesizing ammonia from atmospheric nitrogen. Small-molecule catalysts, such as molybdenum and tungsten-based catalysts, have shown promise in facilitating nitrogen reduction reactions under milder conditions [10]. This represents a transformative approach to ammonia synthesis, with the potential to revolutionize the agricultural industry while reducing the environmental impact of fertilizer production.

1.3 Challenges in Small-molecule Catalysis for Electrochemical Energy Conversion

Despite the promising advances in small-molecule catalysis for EEC, several challenges remain to be addressed. One of the main challenges is the stability of these catalysts under the harsh operating conditions of electrochemical reactions. Many catalysts, especially non-precious metal and molecular catalysts, degrade over time, losing their activity and leading to reduced performance. The development of more stable catalysts, particularly those that can withstand corrosion and degradation in the long term, is critical for enabling their widespread application [11].

Another challenge is the optimization of catalyst selectivity. The electrochemical environment is complex, with competing reactions that can lead to unwanted side products. Achieving high selectivity for a specific reaction, while minimizing the formation of undesired products, is crucial for improving catalytic efficiency and reaction yields [12]. Researchers are focusing on fine-tuning the electronic structure of small-molecule catalysts to enhance their selectivity, but this remains a challenging task.

Furthermore, scaling up the use of small-molecule catalysts from laboratory studies to industrial applications presents significant hurdles. While many small-molecule catalysts exhibit impressive performance in controlled environments, scaling these processes to larger systems often reveals issues related to cost, catalyst stability, and system integration [13]. To overcome these barriers, a multidisciplinary approach is required, combining advances in catalyst design, reaction mechanism understanding, and process engineering.

1.4 Opportunities and Future Outlook

The future of small-molecule catalysis in EEC holds significant promise, with continued research focusing on overcoming current challenges and unlocking the full potential of these catalysts. Advances in computational chemistry, high-throughput screening, and in-situ characterization techniques are providing new insights into catalyst design, reaction mechanisms, and performance [14]. These tools will play a crucial role in accelerating the development of efficient and stable small-molecule catalysts for key electrochemical reactions.

Additionally, hybrid catalysts that combine the benefits of small-molecule catalysts with solid-state materials, such as MOFs and conductive polymers, are an exciting area of research. These hybrid systems could offer enhanced performance, stability, and scalability, paving the way for more efficient energy conversion systems [15]. The integration of small-molecule catalysis with renewable energy sources, such as solar and wind power, is also a key area of focus, enabling energy storage and conversion solutions that can help address the global energy crisis.

In conclusion, small-molecule catalysis represents a promising approach for advancing EEC technologies. Continued progress in catalyst design, performance optimization, and system integration will be essential for overcoming the current challenges and realizing the full potential of these catalysts in practical applications, including energy storage, green hydrogen production, and sustainable chemical synthesis [16].

Acknowledgment

This book was supported by the Innovation Funding Project of Science and Technology, China National Petroleum Corporation (2022DQ02-0408).

References

1 Yan, Y., Chen, G., She, P. et al. (2020). Mesoporous nanoarchitectures for electrochemical energy conversion and storage. *Adv. Mater.* 32 (44): 2004654.
2 Qin, Y., Wang, Y., Jin, G. et al. (2024). Construction and progress of small molecule-based coupled electrolyzers. *Adv. Energy Mater.* 14 (42): 2402429.
3 Yin, J., Fan, Q., Li, Y., Cheng, F. et al. (2016). Ni–C–N nanosheets as catalyst for hydrogen evolution reaction. *J. Am Chem. Soc.* 138 (44): 14546–14549.
4 Magnier, L., Cossard, G., Martin, V. et al. (2024). Fe–Ni-based alloys as highly active and low-cost oxygen evolution reaction catalyst in alkaline media. *Nat. Mater.* 21: 252–261.
5 Zhao, Y., Zhou, H., Zhu, X. et al. (2021). Simultaneous oxidative and reductive reactions in one system by atomic design. *Nat. Catal.* 4: 134–143.
6 Li, Q., Zhang, L., Dai, J. et al. (2018). Polyoxometalate-based materials for advanced electrochemical energy conversion and storage. *Chem. Eng. J.* 352: 446–458.

7 Liu, J., Wan, X., Liu, S. et al. (2021). Hydrogen passivation of M–N–C (M = Fe, Co) catalysts for storage stability and ORR activity improvements. *Adv. Mater.* 33 (38): 2103600.

8 Dutta, S., Gurumoorthi, A., Lee, S. et al. (2023). Sculpting in-plane fractal porous patterns in two-dimensional MOF nanocrystals for photoelectrocatalytic CO_2 reduction. *Angew. Chem. Int. Ed.* 62 (28): e202303890.

9 Chou, T.C., Chang, C.C., Yu, H.L. et al. (2020). Controlling the oxidation state of Cu electrode and reaction intermediates for electrochemical CO_2 reduction to ethylene. *J. Am. Chem. Soc.* 142 (6): 2857–2867.

10 Wang, J., Jiang, Z., Peng, G. et al. (2022). Surface valence state effect of MoO_{2+x} on electrochemical nitrogen reduction. *Adv. Sci.* 9 (12): 2104857.

11 Shao, M., Chang, Q., Dodelet, J.P. et al. (2016). Recent advances in electrocatalysts for oxygen reduction reaction. *Chem. Rev.* 116 (6): 3594–3657.

12 Zhang, Y., Nie, K., Yi, L. et al. (2023). Recent advances in engineering of 2D materials-based heterostructures for electrochemical energy conversion. *Adv. Sci.* 10 (31): 2302301.

13 Wang, J., Wang, J., Kong, Z. et al. (2017). Conducting-polymer-based materials for electrochemical energy conversion and storage. *Adv. Mater.* 29 (45): 1703044.

14 Zhao, Y., Liu, Y., Zhang, H. et al. (2020). 3D nanostructures for the next generation of high-performance nanodevices for electrochemical energy conversion and storage. *Adv. Energy Mater.* 10 (28): 2001460.

15 Wang, J., Dou, S., and Wang, X. (2021). Structural tuning of heterogeneous molecular catalysts for electrochemical energy conversion. *Sci. Adv.* 7 (13): eabf3989.

16 Xu, R., Du, L., Adekoya, D. et al. (2020). Well-defined nanostructures for electrochemical energy conversion and storage. *Adv. Energy Mater.* 11 (15): 2001537.

2

Electrochemical Energy Conversion Techniques

Zhongxin Chen, Huaiguang Li**

School of Science and Engineering, The Chinese University of Hong Kong, Shenzhen, Guangdong, China

*Corresponding authors: chenzhongxin@cuhk.edu.cn, lihuaiguang@cuhk.edu.cn

2.1 Introduction: The Future Is Electrifying

Climate change and renewable energy innovations are the twin drivers in accelerating the global energy transition from fossil fuels to clean electricity from wind, solar, hydro, and other renewable sources. To this end, direct and indirect electrification, two primary options in electrifying buildings, the industrial and transportation sectors, will become pervasive as end uses of energy switch from fossil fuels to electricity. The former uses electricity directly as an energy source, such as heat pumps or battery electric vehicles (BEVs). Meanwhile, indirect electrification involves converting renewable energy into hydrogen and synthetic fuels through electrolysis, serving as a replacement in the hard-to-abate sectors where direct electrification is not feasible [1].

Electrochemical energy conversion technologies are arguably the most critical pillar of the next sweeping energy transition. As shown in Figure 2.1, important fuels and feedstock chemicals, such as hydrogen, hydrocarbons, ammonia, and biofuels, can be produced in sustainable pathways by converting the abundant feedstock of water, carbon dioxide, and nitrogen in the Earth's atmosphere through electrochemical processes [2]. For example, green hydrogen can be produced from renewable and nuclear resources by electrochemical water splitting. This is driven by passing the electrical current through the water, where the electrical energy is converted to chemical energy at the electrode–solution interface through charge transfer reactions in an electrolyzer. Hydrogen gas is generated in the cathodic compartment of the electrolyzer by the hydrogen evolution reaction (HER, Chapter 6); oxygen gas is simultaneously produced in the anode by the oxygen evolution reaction (OER, Chapter 7). As an attractive energy carrier, green hydrogen can be used as fuel to produce clean electricity through hydrogen oxidation reaction (HOR, Chapter 8) and oxygen reduction reaction (ORR, Chapter 9) in fuel cells. Such a

Electrochemical Energy Conversion via Small Molecule Catalysis, First Edition.
Edited by Zhicheng Zhang and Yuanmiao Sun.

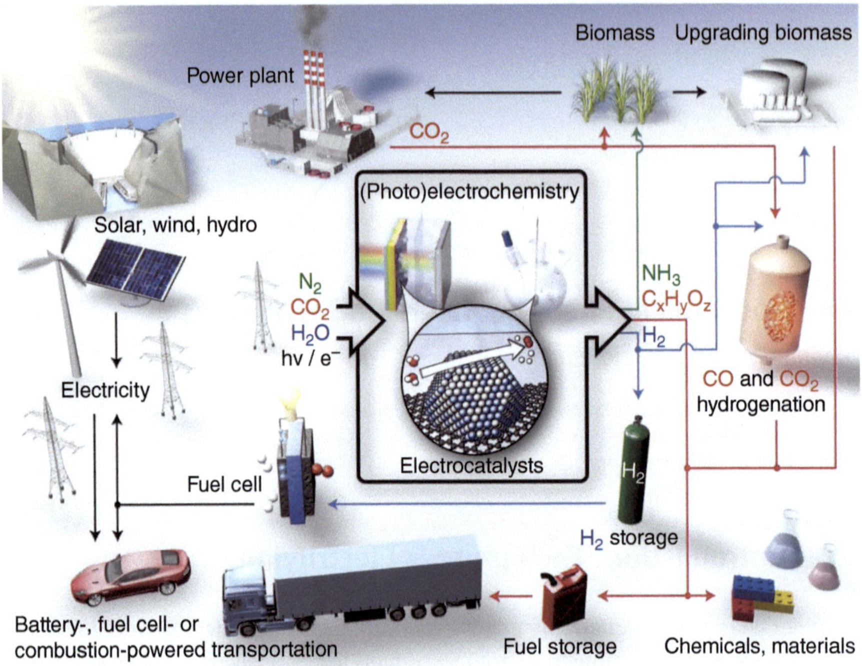

Figure 2.1 Schematic of a sustainable energy landscape based on electrocatalysis. *Source:* Reproduced with permission from [2] / American Association for the Advancement of Science - AAAS.

re-electrification pathway (electricity storage and regeneration) is essential in the long term to match the seasonality of demand and generation [3]. Additionally, fuel-cell electric vehicles (FCEVs) are complementary to BEVs in overcoming the existing limitations of batteries (weight, driving range, and refueling time), where commercial deployment has started in several regions of the world.

As the downstream of the "hydrogen economy," the electrification of chemical synthesis, or so-called electrosynthesis, is a promising route to promote sustainability without compromising economic competitiveness [4]. Hydrogen peroxide, a nexus chemical for paper-making and water treatment industries, can be synthesized through the two-electron-transfer pathway in ORR. Using electrons as the energy source and green reagent to drive the reaction holds tremendous potential in energy consumption and waste production against the energy-intensive anthraquinone process [5]. Likewise, the electrocatalytic CO_2 reduction reaction (CO_2RR, Chapter 11) is a sustainable route for converting CO_2 from the atmosphere or directly from point sources into value-added fuels and feedstocks, advancing a carbon-neutral economy. Ultrahigh Faradaic efficiencies (FEs) toward a specific product with high production rates have been reported for C_1 products, such as methane, carbon monoxide, formate, and methanol, in an industrially viable flow electrolyzer configuration. Ethylene and ethanol, the two most valuable CO_2RR products because of their high energy density, market demand, and industrial value, can be selectively synthesized, albeit of lower current density and selectivity [6].

Apart from the conversion of atmospheric CO_2, the electrochemical biomass valorization (Chapter 10) also represents a sustainable pathway to produce a variety of liquid and gaseous fuels and other carbon-based chemicals and intermediates. Typical derivatives, such as glycerol, glucose, furfural, and 5-hydroxymethylfurfural (5-HMF), can be selectively upgraded through electrochemical oxidation, coupling, and hydrogenation reactions [7].

Regarding the nitrogen cycle, the electrochemical fixation of nitrogen to ammonia (NRR, Chapter 12) would allow for the production of fertilizers sustainably and locally, eliminating distribution costs stemming from the inflexibly large-scale, centralized Haber–Bosch process. Electrochemical nitrate reduction reaction (NO_3RR), by converting excessive nitrate in water into environmentally benign dinitrogen gas or value-added chemicals such as ammonia under mild conditions, has also received much attention owing to its much higher selectivity and yield rate. This represents an eco-friendly and energy-efficient route for recirculating nitrogen species into the nitrogen cycle and economy when powered by renewable electricity [8]. Meanwhile, electrochemical carbon–nitrogen coupling reactions using CO_2 and inorganic nitrogenous species (N_2, nitrate, nitrite, and ammonia) as raw materials in Chapter 13 allow the production of valuable organonitrogen compounds (e.g., urea, amides, amines, and amino acids), opening a green synthetic strategy in organic synthesis, materials chemistry, and biochemistry [9]. More complicated molecular structures can be accessed through electrochemical fluorination (Chapter 14) and electrochemical polymerization (Chapter 15). In summary, as the world's desire to achieve decarbonization intensifies, our future will be electrified with innovative energy conversion technologies by interconverting small molecules.

2.2 Fuel Cells and Electrolyzers

In any electrochemical cell, electrons flow from one place to another, leading to an oxidative reaction at the anode and a reductive reaction at the cathode (Figure 2.2a). From a thermodynamic viewpoint, the electrochemical cells can be further classified into two types: the galvanic cell (fuel cell) uses energy released during a spontaneous redox reaction to generate electricity, and the electrolytic cell (electrolyzer) consumes electrical energy from an external source to drive a nonspontaneous redox reaction. Consequently, the reversible or equilibrium cell voltage is defined as the minimum voltage necessary to be applied between the anode and cathode for electrochemical reactions to happen, which can be calculated by the sum of their standard electrode potentials (Figure 2.2b). In the following, we will briefly analyze the electrochemical conversion techniques from a reaction-orientated viewpoint.

2.2.1 Water Splitting Reaction (Water Electrolyzer)

The electrolysis of water is a simple way to produce hydrogen. Unfortunately, splitting pure water into hydrogen and oxygen is not thermodynamically favorable at standard conditions. As shown in Figure 2.2b, the standard potential of the water-electrolysis cell is 1.23 V at 25 °C at pH = 0. Since pure water has an inferior

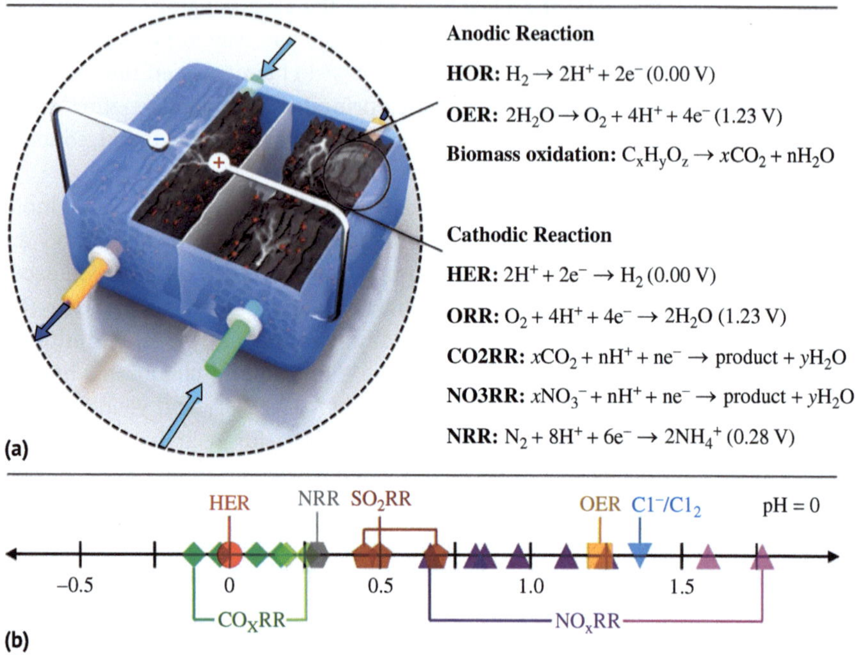

Figure 2.2 (a) Representative anodic and cathodic reactions in the electrochemical conversion devices; (b) visualization of standard potentials for various electrochemical reactions relative to the reversible hydrogen electrode (RHE) at pH = 0.

conductivity of 0.055 μS cm^{-1}, the electrolysis of pure water proceeds very slowly or not at all at the reversible cell voltage, requiring excess energy in the form of overpotential to overcome various activation barriers. In general, the efficiency of water electrolysis could be promoted by adding an electrolyte such as strong acids (H_2SO_4) and strong bases (KOH). The cell voltage can be as low as 1.8 V when using a solid polymer electrolyte, such as anion-exchange or cation-exchange membranes, along with a specific catalyst on each side of the membrane [10, 11].

Water electrolyzers are therefore categorized into alkaline water electrolyzers (AWE), anion-exchange membrane water electrolyzers (AEMWE), proton-exchange membrane water electrolyzers (PEMWE), and solid oxide electrolyzers (SOEC) in Figure 2.3. Among them, alkaline water electrolysis is the most mature hydrogen production technology and has been implemented for industrial hydrogen production for decades [10]. The AWE cells consist of two metallic electrodes separated by a porous separator and operate in an aqueous electrolyte of 25–40 wt.% KOH or NaOH at 70–90 °C to maximize the electrical conductivity. The AWE technology presents several advantages, mostly related to the alkaline electrolyte, allowing the use of non-platinum group metal (PGM) catalysts without compromising the performance and durability of the operation. The main drawback of AWE cells is linked to the intermixing of H_2 and O_2 bubbles in the porous separator, which has adverse consequences regarding operational safety and gas purity. AWE cells also

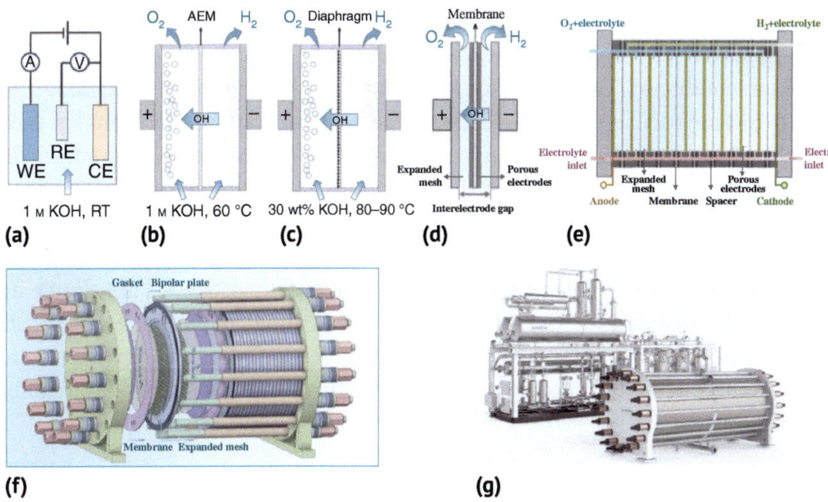

Figure 2.3 Comparison of electrocatalytic setups from laboratory to industry: (a) three-electrode electrochemical cell, (b) anion-exchange membrane (AEM) water electrolyzer, (c) alkaline water electrolyzer (AWE) with a diaphragm, (d) zero-gap electrolyzer, and (e) bipolar-type AWE stack. (f) Schematic of an industrial AWE stack and (g) an SHME-series AWE with an H_2 production rate of (SUNGROW). *Source:* Reproduced from [11] / wih permission of John Wiley & Sons.

need several hours to reach their steady state in practice, making their coupling with intermittent renewable energy (solar/wind) difficult [10].

Meanwhile, AEMWEs are an emerging technology for efficient hydrogen production through water electrolysis. These systems utilize an AEM that selectively conducts hydroxide ions while preventing the passage of gases and other impurities. Like AWE, this unique feature allows for using non-precious metal catalysts, reducing costs, and enhancing sustainability. AEMWEs operate at lower temperatures and pressures than AWEs, making them attractive for various applications, including renewable energy integration and decentralized hydrogen production. However, the lifetime and stability of the AEMWE system still need to be further improved to meet the requirements of commercial applications.

PEMWEs are the most effective water electrolysis technology. Modern PEMWEs contain perfluorinated sulfonic acid copolymer membranes such as Nafion because of their high ionic conductivity, mechanical strength, and relatively strong chemical stability. Since liquid electrolytes are absent in PEMWEs, the electrodes can be pressed tightly against the membrane in a zero-gap configuration, thus maximizing the electrical contact between the electrodes [10]. The catalysts used in PEMWEs are generally PGM catalysts. Therefore, PEMWEs usually operate at exceptionally high current densities and high energy efficiency, generating high-purity hydrogen gas. Moreover, PEM electrolysis features rapid ramp-up and ramp-down capabilities and a wide dynamic operating range of 0–100%. This makes it particularly well-suited for generating hydrogen using excess renewable energy [12]. The main drawbacks are high initial capital investment and complicated zero-gap electrolyzer designs.

In sharp contrast, SOECs operate at a very high temperature of 800–1000 °C for oxygen ion (O^{2-}) migration in either a fuel cell or electrolyzer configuration. Oxide-ion conducting ceramics such as yttrium and scandium oxide-stabilized zirconia (YSZ) are used as the solid electrolyte and cell separator for efficient ion transport. SOEC technologies are driven by the possibility to operate at ultrahigh current densities and efficiencies. Challenges include understanding and controlling electrochemical degradation and thermomechanical stability to meet the hydrogen production demands from intermittent renewable electricity [10].

In summary, the technology and equipment of water electrolysis are simple and can meet the demand for high-purity hydrogen without pollution. However, the energy consumption and H_2 production costs are relatively high compared to other industrial production technologies, which urges further investigation at the catalyst and reactor level.

2.2.2 Chlor-alkali Process

The chlor-alkali process is an industrial process for the electrolysis of sodium chloride solutions to produce chlorine, sodium hydroxide (caustic soda), and hydrogen (around 4% of global supply). Chlorine (Cl_2) is a crucial precursor for many industrial processes, including polymer synthesis, disinfection goods production, pharmaceutical manufacturing, and wastewater treatment. Up to 2020, the global production of Cl_2 was more than 88 million tons per year [11, 13]. Analogous to water splitting, the reaction is driven by electricity to promote the HER (cathodic reaction) and chloride oxidation reaction or so-called chlorine evolution reaction (anodic reaction), as follows:

$$\text{Anode: } 2Cl^- \rightarrow Cl_2 + 2e^- \quad (1.36\ \text{V vs. RHE})$$

$$\text{Cathode: } 2H_2O + 2e^- \rightarrow H_2 + 2OH^- \quad (0.00\ \text{V vs. RHE})$$

$$\text{Overall: } 2NaCl + 2H_2O \rightarrow Cl_2 + H_2 + 2NaOH$$

Apparently, the chloride oxidation reaction will compete with OER due to its higher standard electrode potential, as shown in Figure 2.2b. Such chlor-alkali process is also highly energy-intensive, where around 2500 kWh (9000 MJ) of electricity is consumed per ton of sodium hydroxide produced. To prevent unwanted side reactions, the chlor-alkali process is usually performed in a saturated brine solution so that the oxidation of chloride ions is preferred. Three major processes based on mercury, diaphragm, and membrane cells were developed (Figure 2.3) [11, 13]. The diaphragm is a key component in separating product gases and preventing the mixing of hydrogen and chlorine. Porous diaphragms based on asbestos were used for decades until the mid-1970s when they were banned because of their toxicity and high gas permeability. The mercury cell was the next commercial technology. It dominated the chlor-alkali industry for a long period as it requires a lower concentration of raw brine and has a comparatively large production capacity. Nevertheless, the growing concerns about the environmental and health hazards of mercury poisoning have severely hindered its use and further development. Due to its eco-friendliness, high efficiency, and product purity, membrane-based electrolysis is currently the most attractive approach. Roughly 83% of the chlorine production in 2020 was met by membrane technology, followed by 13% by diaphragm cells, 2% by mercury cells, and 2% by others [13].

As the anodic chloride oxidation is highly energy-intensive, the electrocatalyst design significantly reduces the energy consumption of the integrated chlor-alkali process. Various metal/carbon-based materials have been reported in the past century. Among them, mixed metal oxides (MMOs) based on precious metals Ru or Ir, such as the dimensionally stable anodes (DSAs, IrO_2/RuO_2-modified titanium supports), have been predominantly employed. However, MMO-based catalysts are also superior to OER, where the OER and chloride oxidation are intimately coupled [13]. Developing new catalysts with high activity, excellent selectivity, and long-term stability is highly desirable. Meanwhile, the industrial technical revolution is also crucial for promoting the efficiency of membrane reactors. The pH difference between the anodic and cathodic chambers is one of the major challenges in this field owing to the use of cation exchange membranes to maintain the acidic environment in the anodic chamber. Hydroxide ions will accumulate in the cathodic chamber and eventually lead to membrane corrosion and rapid degradation in activity. New reactor designs, such as integrating anion/cation exchange membranes, may prevent performance degradation caused by pH differences. It is also of great promise to use the generated chlorine as a redox mediator for combining with other electrocatalytic reactions, such as CO_2 and N_2 reduction, for value-added chemical production [13].

2.2.3 CO_2 Electrolyzers

Electrochemical CO_2RR is a sustainable pathway to produce low-carbon-intensity chemicals and fuels with only water and renewable electricity as inputs. If the CO_2 is sourced from the air, CO_2RR can close the carbon cycle for fuels and consumed chemicals. Pioneer research on CO_2RR was conducted in the 1980s. The field has progressed significantly over the last 10 years from a narrow focus on electrocatalysts to a much broader combination of system engineering and catalyst design. Electrolyzer system designs have proliferated and are motivated by achieving energy and carbon efficiency. Nevertheless, significant challenges remain before CO_2RR becomes commercially viable [6, 14].

During the CO_2RR, CO_2 is fed into an electrolyzer and reacts on a catalyst surface under an applied potential. Since electrolysis is a paired operation, OER is usually chosen as the counter-reaction of CO_2RR when the electrolysis is conducted in an aqueous solution. The CO_2RR/OER system is a nonspontaneous process, and a voltage must be applied to drive the reaction [14]. CO_2RR catalysts decrease the activation energy barriers so that the applied potential can overcome these barriers and make the reaction kinetically favorable at low pressures and temperatures. It should be mentioned that CO_2RR is not a single-product reaction. Instead, more than 16 products (such as C_1: carbon monoxide, formate, methanol, methane; C_2:acetate,ethylene,ethanol;C_3:n-propanol,etc.)havebeenreportedfromCO_2RR.The thermodynamic potential for each CO_2RR is similar to the thermodynamic potential for HER, which means the product selectivity cannot be tuned by changing the applied potential alone. More importantly, unwanted HER must be suppressed by careful designs of the CO_2RR system due to similar standard potentials [10, 12].

Depending on their behavior and product selectivity, the CO_2RR catalysts can be categorized into three groups: the formate-producing catalysts, CO-producing catalysts, and finally, hydrocarbon and oxygenate-producing catalysts. Formate-producing catalysts (such as tin, indium, bismuth, and lead) exhibit the ability to adsorb *HCOO, a critical intermediate in the formation of formate. For CO-producing catalysts, the CO_2RR proceeds through the initial binding of the *COOH species on their surfaces to form a weakly bound *CO intermediate, followed by the desorption of *CO to generate CO. Common catalysts in this category include silver, gold, zinc, and nickel. The reaction pathway for hydrocarbons and oxygenates is much more complicated. The binding strength of *CO has to be strong enough to facilitate subsequent reactions (such as C—C coupling). Copper is the only monometallic catalyst of this type, where exploring advanced catalysts for C_2 and C_{2+} products represents new frontiers in this field [6,14]. Since CO is a crucial intermediate on the pathway to further reduce hydrocarbons and oxygenates, it can be fed directly as a reactant, which is known as the CO reduction reaction (CORR). The coupling of CORR with an upstream CO_2-to-CO electrolysis has been pursued as an alternative to the direct reduction of CO_2 in generating multi-carbon products. Such a tandem system enjoys the advantage of energy-efficient CO_2-to-CO conversion in an SOEC to reduce the energy intensity of the overall process. SOECs can achieve CO production at low voltage with high FEs in the absence of water, allowing an energy efficiency of >70% even at current densities of 700–800 mA cm^{-2}. Hence, SOEC technology for CO_2-to-CO is currently being commercialized by Topsoe with a technology readiness level (TRL) of 7–8 [12]. With the combined advantages of high-efficiency SOEC and carbonate-free alkaline CORR, there is an opportunity to increase energy efficiency in producing C_{2+} chemicals.

2.2.4 Fuel Cells

Fuel cells are compact electricity producers that consume fuels (chemical energy) through fuel oxidation (e.g., hydrogen, methanol, and hydrocarbon) at the anode and ORR as the counter-reaction at the cathode, as shown in Figure 2.4. Fuel cell efficiency is not limited by the Carnot cycle, and it converts the chemical energy in the fuel directly to electrical energy with efficiencies capable of exceeding 60%, which is a revolutionary advantage over other energy-producing devices (such as conventional internal combustion engines). As no combustion reactions are involved, they do not produce harmful emissions (such as CO_2, NO_x, SO_2, and particulate matter) that are typically associated with the use of fossil fuels in conventional devices. Therefore, fuel cells are environmentally friendly [3, 15]. Fuel cells also stand out for their reliability and longevity. Unlike batteries that lose charge over time, hydrogen or other fuels can be stored as a gas or liquid ready for use, ensuring a dependable energy source for critical applications. Fuel cells have notable operational durability, with some designs demonstrating lifespans exceeding 20 000 hours. This

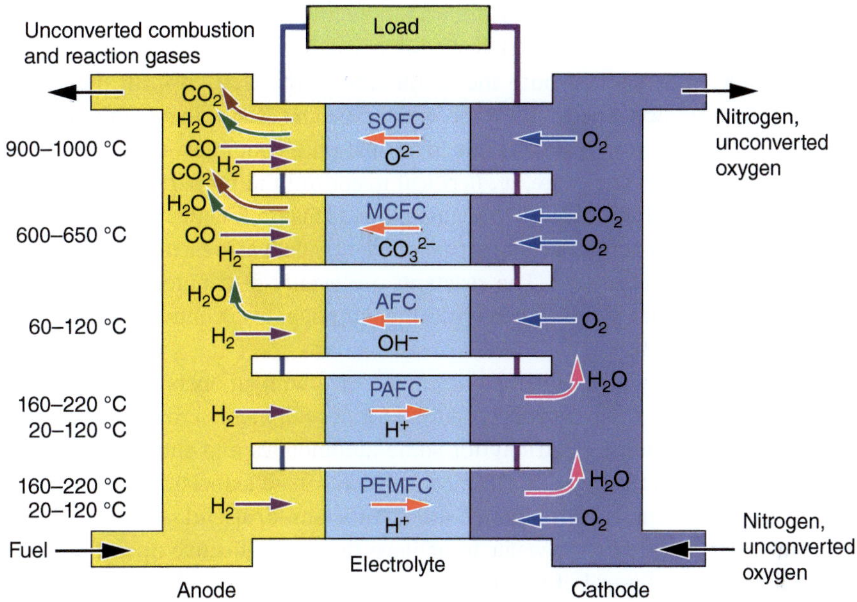

Figure 2.4 An overview of modern fuel cell technologies. *Source:* Reproduced from [15] / with permission of Elsevier.

reliability, coupled with the minimal maintenance needs of fuel cell systems, positions fuel cells as a sustainable solution for long-term energy needs [16, 17].

2.2.4.1 Proton Exchange Membrane Fuel Cells

Proton exchange membrane fuel cell (PEMFC) is one of the most widely used fuel cells (Figure 2.4), which employs hydrogen as the fuel and oxygen as the oxidant. The process occurring in a PEMFC is basically the reverse of water electrolysis, that is, the recombination of oxygen and hydrogen to form water through HOR and ORR reactions to produce a direct electric current. In an ideal $H_2 - O_2$ PEMFC, the cell voltage is 1.23 V and is independent of the current drawn. However, this is not realized even at the open circuit (zero current) due to the irreversibilities that arise during fuel cell operation. It is essential to prevent the crossover of the fuel and oxidant to the other side; otherwise, it will reduce the efficiency and lifespan of fuel cells and result in safety issues. Hence, a proton-conducting, thin, permeable polymer membrane (such as Nafion) is used as the solid electrolyte to separate the anode and cathode sides. Hydrogen is catalytically oxidized at the anode to give protons (H^+) and electrons. The electrolyte is unique in that it allows only the passage of ions, e.g., protons, but not electrons. The resulting electrons flow away from the anode through the external circuit toward the cathode. The protons pass through the electrolyte to the cathode, where oxygen reacts with the incoming electrons from the external circuit and protons from the electrolyte to generate heat and water [17].

The membrane electrode assembly (MEA) is the core component of a PEMFC and comprises four key components: the polymer electrolyte membrane, catalyst layer, gas diffusion layers (GDLs), and bipolar plates. The polymer electrolyte membrane, a solid acidic polymer, allows for the hydrogen to be dissociated as protons through the membrane [16, 17]. An ink of catalyst and ionomer (Nafion® solution) is sprayed or painted onto the membrane or the GDL to form the catalyst layer, which interacts directly with the membrane and GDLs. Inside the MEA, the electrolyte, catalyst, and reactants mix in a triple-phase boundary (TPB), where the cell reactions occur. Therefore, it is critical to control the properties of TPB, such as porosity and hydrophobicity, to regulate the diffusion kinetics of the reactants. The GDL features a porous structure that efficiently delivers reactants and removes products from the electrode. The reactants are evenly distributed to the catalyst layer through bipolar plates. Meanwhile, bipolar plates also conduct electrical currents for the cell [18].

PEMFCs have a typical cell efficiency of 40–70% with an operating temperature of 50–120 °C. They show a strong potential for development across diverse energy scales and can produce electricity for some automobiles and stationary and portable applications. However, a significant disadvantage associated with PEMFCs is low operation temperatures, which are insufficient for useful cogeneration. The cell must be saturated with water to achieve the most effective operation of the electrolyte, where control of the moisture level at the anode and cathode becomes an important consideration. Water evaporates either too slowly (the MEA will be flooded) or too fast (the polymer electrolyte membrane will be dry), causing damage to stability and power output. It is also challenging to lower the platinum loading of the MEA without loss of performance and improve the cell stability concerning platinum dissolution and carbon-support corrosion [17].

2.2.4.2 Alkaline Fuel Cells

A key motivative in developing alkaline fuel cells (AFCs) lies in the much-enhanced reaction kinetics of ORR in the alkaline environment (such as concentrated KOH electrolyte), which allows using non-PGM catalysts as the electrocatalysts and reducing the cost of the whole system. In traditional PEMFCs, a much higher Pt loading in the cathode, along with more active and durable ORR electrocatalysts than pure Pt catalysts, is needed since the kinetics of ORR is around five orders of magnitude slower than HOR. This problem is addressed in AFCs due to higher ORR kinetics and low dissolution rates of base metals (Fe, Co, Ni, etc.). Furthermore, AFCs feature competitive advantages on wider selections of materials and components that are stable at high pH and inexpensive solid electrolytes that do not need fluorinated ionomers (i.e., Nafion) [19]. AFCs have proven to provide high power densities with considerable lifetimes, thus making them a worthy competitor to PEMFCs.

Despite its compelling merits, traditional AFCs working on concentrated 30–45% KOH electrolytes gained little attention for decades, mainly due to their high sensitivity to atmospheric CO_2. The formation of carbonate by the side reaction between hydroxide ions and CO_2 will reduce the concentration of hydroxide ions and potentially block the pores of the porous electrode through the precipitation of the metal carbonate. Even a small amount of CO_2 in the air can dramatically affect cell performance and durability due to carbonate formation [19, 20]. This issue is partially addressed by replacing the KOH solution with a solid anion exchange membrane (AEM) without mobile cations in Figure 2.4, which has a lower susceptibility to CO_2 poisoning and can minimize the electrode flooding and drying problems associated with imbalance in the management of liquid electrolyte. Such AEMFCs are economically qualified for providing a power of 10–100 kW and work optimally at 80 °C with a cell efficiency of 60–70%. Nonetheless, CO_2 still affects the cell performance and durability compared to PEMFCs. The main challenges for AEMFCs include tolerance to carbon dioxide, membrane conductivity and durability, higher temperature operation, water management, power density, and anode electrocatalysis [19, 20].

2.2.4.3 Phosphoric Acid Fuel Cells

Phosphoric acid fuel cells (PAFCs) are considered the "first generation" of modern fuel cells and are typically used for stationary power generation. PAFCs use phosphoric acid and its derivatives as an electrolyte (Figure 2.4); for instance, pyrophosphoric acid ($H_4P_2O_7$), a polymeric phosphoric acid with higher conductivity than the parent acid, to form an ionic solution. The charge carrier of PAFCs is protons, which is similar to the PEMFCs. PAFCs are more than 85% efficient when used for the cogeneration of electricity and heat to produce <100 MW power at a working temperature of 150–220 °C and a cost of US$4–4.5 per W. They are also more tolerant of impurities in fossil fuels that have been reformed into hydrogen than PEMFCs. However, PAFCs are less efficient at generating electricity alone (37–42%), which is only slightly higher than that of combustion-based power plants (around 33%). PAFCs are also less powerful than other fuel cells, given the same weight and volume. As a result, these fuel cells are typically large and heavy [21].

2.2.4.4 Molten Carbonate Fuel Cells

Molten carbonate fuel cells (MCFCs) are high-temperature fuel cells that use a molten carbonate salt mixture (e.g., potassium carbonate, K_2CO_3) as the electrolyte. They operate at temperatures of 600–650 °C to allow the transportation of carbonate ions (rather than protons) in a porous and chemically inert ceramic matrix (Figure 2.4), giving a typical energy efficiency of 50–60% and an output power of 100 MW. By integrating waste heat recovery, the overall efficiency of MCFCs could reach as high as 85%, which is a notable advantage over PAFCs. Additionally, more energy-dense fuels (such as hydrocarbons) can be directly fed to MCFCs owing to the internal reforming of fuels to hydrogen at such high temperatures, thus preventing the use of external reformers in low-temperature fuel cells (PEMFCs, AFCs, etc.). Likewise, non-PGM catalysts can be used as catalysts at the anode and cathode for cost reduction. MCFCs are not prone to carbon monoxide or carbon dioxide poisoning and are highly resistant to impurities compared to other fuel cells. Indeed, they can even use carbon oxides as fuel, making them more attractive for fueling with gases made from coal in natural gas and coal-based power plants for electrical utility, industrial, and military applications. The primary disadvantage of MCFCs is low durability due to high operational temperature and the corrosive nature of electrolytes, urging for corrosion-resistant materials and better fuel cell designs to double the life span to 10 years without performance degradation [16, 17].

2.2.4.5 Solid Oxide Fuel Cells

Solid oxide fuel cells (SOFCs) are another type of high-temperature fuel cells, which operate at very high temperatures (>600 °C) and use a solid oxide electrolyte to conduct negative oxygen ions from the cathode to the anode. The electrochemical oxidation of fuels by oxygen ions occurs on the anode side to give an energy efficiency of 50–80% toward electricity production or 85% if waste heat is also utilized. Similar to MCFCs, high-temperature operation removes the need for PGM catalysts and external reformers, thereby enabling wider fuel and material selection and reducing cost. SOFCs are also more resistant to sulfur and carbon monoxide. To this end, light hydrocarbon fuels, such as methane, propane, and butane, can be directly fed to SOFCs; heavier hydrocarbons, such as gasoline, diesel, jet fuels, or biofuels, can be used when coupled with external reformers to generate mixtures of hydrogen, carbon monoxide, carbon dioxide, steam, and methane [16, 17]. Nevertheless, thermal expansion demands a uniform and well-regulated heating process at startup, necessitating a long heating process in the order of an hour to operating temperature and significant thermal shielding to retain heat and protect personnel. The high operating temperatures also place stringent durability requirements on materials. It remains a key challenge in developing low-temperature SOFCs operating at 700 °C or below for this technology.

2.3 Batteries and Supercapacitors

Although it is not our prime focus, batteries and supercapacitors are among the most important electrochemical energy devices for a sustainable future. Due to the large fluctuation in generation, electricity produced from renewable energies must be efficiently stored to supply the world with energy on demand. Batteries and supercapacitors represent the two leading technologies for energy storage, as illustrated in a graph of energy density in versus power density in $W\,kg^{-1}$ (the Ragone plot, Figure 2.5) [22, 23].

A battery is a device that stores and converts chemical energy into electricity. The chemical reactions in a battery involve the flow of electrons from one material (electrode) to another through an external circuit. The flow of electrons provides an electric current that can be used to do work. To balance the flow of electrons, charged ions also flow through an electrolyte solution in contact with both electrodes. Different electrodes and electrolytes produce different chemical reactions that affect the basic attributes of a battery. Some common batteries are single-use only (also known as primary batteries), as the electrode materials are irreversibly changed during discharge. A typical example is the alkaline battery for flashlights and portable electronics. Secondary (rechargeable) batteries can be discharged and recharged multiple times using an applied electric current; the original composition of the electrodes can be restored by reverse current. Examples include the lead–acid batteries used in vehicles and lithium–ion batteries (LIBs) for portable electronics such as laptops and mobile phones. Batteries come in many shapes and sizes, from miniature cells used to power hearing aids and wristwatches to huge battery banks for standby or emergency power. As shown in Figure 2.5, batteries

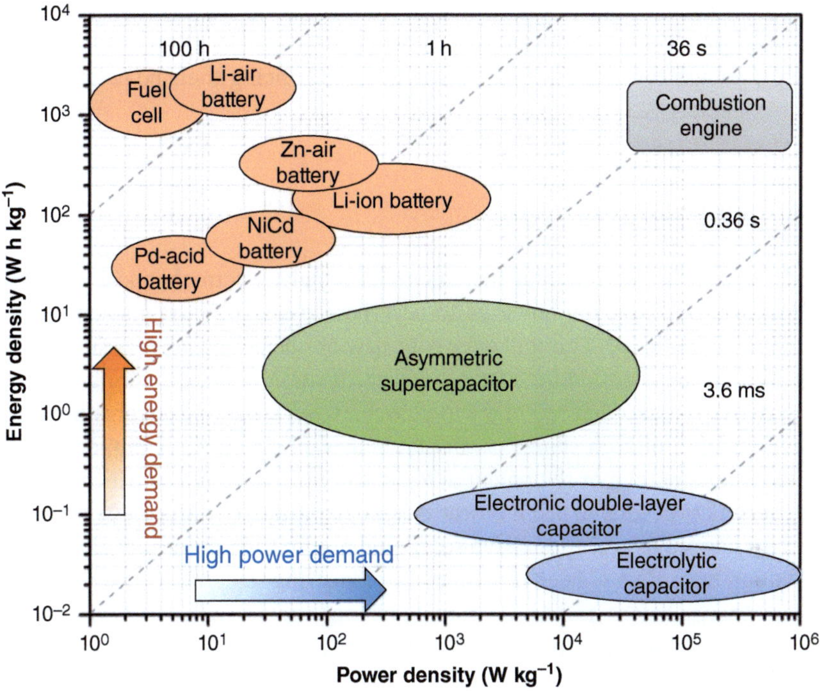

Figure 2.5 Ragone plot illustrating the performances of specific power versus specific energy for different electrical energy-storage technologies. Times shown in the plot are the discharge time, obtained by dividing the energy density by the power density. *Source:* Reproduced from [22] / with permission of American Chemical Society.

have much lower specific energy (energy per unit mass) than common fuels such as gasoline. In automobiles, this can be offset by the higher efficiency of electric motors in converting electrical energy to mechanical work compared to combustion engines. In particular, LIBs are currently widely used in consumer electronics owing to their high energy density (approaching 180 W h kg^{-1}). However, due to a variety of resistive losses from sluggish electron and ion transport, batteries give rise to heat generation and dendrite formation when operated at high power, which can lead to serious safety issues [22, 24].

Supercapacitors, also known as electrochemical capacitors, can complement or even replace batteries in particular applications due to their ability to supply high power and rapid charging while maintaining an extremely long cycle life (>100 000 cycles). They are classified into three types, namely electric double-layer capacitors, pseudocapacitors, and hybrid capacitors. Supercapacitors have sparked a surge of interest in applications that demand fast charging, exceptional cycling stability, and high power density. For example, supercapacitors are now used in heavy-duty vehicles, hybrid platforms for trucks and buses, load-leveling systems for intermittent renewable energy sources, and the regenerative braking energy of electric vehicles and light rail. Although commercially available supercapacitors can deliver much higher energy density (~5 W h kg^{-1}) than traditional solid-state electrolytic capacitors, this is still significantly lower than batteries (up to 200 W h kg^{-1}) and fuel cells (up to 350 $Wh\,kg^{-1}$) in Figure 2.5. Thus, the widespread use of supercapacitors has been limited, and extensive research efforts are underway to achieve the energy-storage capability of batteries without sacrificing their power density and cycling stability [22].

2.4 Summary

With renewables dominating the electricity supply, electrochemical conversions (electrification) have the potential to decarbonize a significant portion of energy demand in the future. This will offer both new obstacles and new possibilities. Two specific acts are required for successful electrification. First, delivering smart electrification strategies that leverage synergies between clean power consumption and the integration of high shares of variable renewable energy; second, developing a better understanding of the potential technical limitations of electrification in various sectors over time to avoid power system disruption or unbearable costs. Much of the knowledge and innovative solutions required for the smart electrification transition have already been developed by pioneering research. However, to ensure secure energy access and tackle climate change, uptake must be accelerated and broadened well beyond the pioneers and tailored to local needs.

References

1 International Renewable Energy Agency (IRENA) (2022). *Smart Electrification with Renewables: Driving the Transformation of Energy Services.* International Renewable Energy Agency.

2 Seh, Z.W., Kibsgaard, J., Dickens, C.F. et al. (2017). Combining theory and experiment in electrocatalysis: insights into materials design. *Science* 355 (6321): eaad4998.

3 International Renewable Energy Agency (IRENA) (2018). *Hydrogen from Renewable Power: Technology Outlook for the Energy Transition.* International Renewable Energy Agency.

4 Chen, S.S., Glezakou, V.-A., Holewinski, A. et al. (2024). Advances in electrosynthesis for a greener chemical industry. *Green Chem* 26 (8): 4240–4241.

5 Xia, C., Xia, Y., Zhu, P. et al. (2019). Direct electrosynthesis of pure aqueous H_2O_2 solutions up to 20% by weight using a solid electrolyte. *Science* 366 (6462): 226–231.

6 Xu, D., Li, K., Jia, B. et al. (2022). Electrocatalytic CO_2 reduction towards industrial applications. *Carbon Energy* 5 (1): e230.

7 Lucas, F.W.S., Grim, R.G., Tacey, S.A. et al. (2021). Electrochemical routes for the valorization of biomass-derived feedstocks: from chemistry to application. *ACS Energy Lett.* 6 (4): 1205–1270.

8 Min, B., Gao, Q., Yan, Z. et al. (2021). Powering the remediation of the nitrogen cycle: progress and perspectives of electrochemical nitrate reduction. *Ind. Eng. Chem. Res.* 60 (41): 14635–14650.

9 Peng, X., Zeng, L., Wang, D. et al. (2023). Electrochemical C-N coupling of CO_2 and nitrogenous small molecules for the electrosynthesis of organonitrogen compounds. *Chem. Soc. Rev.* 52 (6): 2193–2237.

10 Chatenet, M., Pollet, B.G., Dekel, D.R. et al. (2022). Water electrolysis: from textbook knowledge to the latest scientific strategies and industrial developments. *Chem. Soc. Rev.* 51 (11): 4583–4762.

11 Wang, N., Song, S., Wu, W. et al. (2024). Bridging laboratory electrocatalysts with industrially relevant alkaline water electrolyzers. *Adv. Energy Mater.* 14 (16): 2303451.

12 Badgett, A., Brauch, J., Buchheit, K. et al. (2022). *Water electrolyzers and fuel cells supply chain.* US Department of Energy, 26. https://www.energy.gov/sites/default/files/2022-02/FuelCells.

13 Wang, Y., Liu, Y., Wiley, D. et al. (2021). Recent advances in electrocatalytic chloride oxidation for chlorine gas production. *J. Mater. Chem. A* 9 (35): 18974–18993.

14 O'Brien, C.P., Miao, R.K., Shayesteh Zeraati, A. et al. (2024). CO_2 electrolyzers. *Chem. Rev.* 124 (7): 3648–3693.

15 Tanç, B., Arat, H.T., Baltacıoğlu, E. et al. (2019). Overview of the next quarter century vision of hydrogen fuel cell electric vehicles. *Int. J. Hydrogen Energy* 44 (20): 10120–10128.

16 Rolo, I., Costa, V.A.F., Brito, F.P. et al. (2023). Hydrogen-based energy systems: current technology development status, opportunities and challenges. *Energies* 17 (1): 180.

17 Aminudin, M.A., Kamarudin, S.K., Lim, B.H. et al. (2023). An overview: current progress on hydrogen fuel cell vehicles. *Int. J. Hydrogen Energy* 48 (11): 4371–4388.

18 Bhosale, A.C., Ghosh, P.C., and Assaud, L. (2020). Preparation methods of membrane electrode assemblies for proton exchange membrane fuel cells and unitized regenerative fuel cells: a review. *Renew. Sustain. Energy Rev.* 133 (2020): 110286.

19 Xiao, F., Wang, Y.C., Wu, Z.P. et al. (2021). Recent advances in electrocatalysts for proton exchange membrane fuel cells and alkaline membrane fuel cells. *Adv. Mater.* 33 (50): e2006292.

20 Pan, Z.F., An, L., Zhao, T.S. et al. (2018). Advances and challenges in alkaline anion exchange membrane fuel cells. *Prog. Energy Combust. Sci.* 66: 141–175.

21 Zucconi, A., Hack, J., Stocker, R. et al. (2024). Challenges and opportunities for characterisation of high-temperature polymer electrolyte membrane fuel cells: a review. *J. Mater. Chem. A* 12 (14): 8014–8064.

22 Shao, Y., El-Kady, M.F., Sun, J. et al. (2018). Design and mechanisms of asymmetric supercapacitors. *Chem. Rev.* 118 (18): 9233–9280.

23 Zhu, Z., Jiang, T., Ali, M. et al. (2022). Rechargeable batteries for grid-scale energy storage. *Chem. Rev.* 122 (22): 16610–16751.

24 Fan, E., Li, L., Wang, Z. et al. (2020). Sustainable recycling technology for Li-ion batteries and beyond: challenges and future prospects. *Chem. Rev.* 120 (14): 7020–7063.

3

Constant-potential Modeling in Electrochemical CO_2 Reduction

*Yaqiong Su**

School of Chemistry, Engineering Research Center of Energy Storage Materials and Devices of Ministry of Education, National Innovation Platform (Center) for Industry-Education Integration of Energy Storage Technology, Xi'an Jiaotong University, Xi'an, Shanxi, China

*Corresponding author: yqsu1989@xjtu.edu.cn

3.1 Introduction

At present, the primary energy source of the entire globe is highly reliant on fossil fuels. However, the continuous utilization of fossil fuels leads to the growing accumulation of carbon dioxide (CO_2) in the atmosphere, resulting in serious environmental problems and energy crisis [1–3]. To alleviate the environmental problems and effectively reuse the abundant CO_2 resources, great research efforts have been devoted to CO_2 conversion and utilization. Recently, electrochemical CO_2 reduction reaction (CO_2RR) to high-value-added chemicals has been attracting much attention, with the advantages of good controllability, mild reaction conditions, and high energy density of the delivered products [4–7].

Among the electrocatalysts of CO_2RR, copper (Cu) has been the leading candidate for the electrochemical CO_2RR to highly valued chemicals [8–10]. Various Cu-based electrocatalysts have been developed, including metallic Cu, Cu alloy, Cu oxides, and supported Cu-based catalysts (single-atom, clusters, or nanoparticles) [11–16]. However, multiple factors, such as crystal facets, surface charge states (Cu^0, Cu^+, Cu^{2+}), electrolytes (cations, anions, pH values, etc.), and grain boundaries, have been generally reported to be responsible for influencing the catalytic activity and duration of Cu-based electrocatalysts [16–22]. Owing to potential synergistic effects among various factors at the electrode–electrolyte interfaces, it is difficult to understand the origin of catalytic activity and product selectivity control in CO_2RR on Cu-based electrocatalysts.

A momentous breakthrough in understanding the catalytic performance has been proposed by some theoretical studies of Cu-based electrocatalysts toward CO_2RR, such as the structural evolution and oxidation states of the Cu-based catalysts and their impact on the CO_2RR performance could be explored at the atomic level [23–27].

Electrochemical Energy Conversion via Small Molecule Catalysis, First Edition.
Edited by Zhicheng Zhang and Yuanmiao Sun.

Earlier, Asthagiri and Janik for the first time performed the density functional theory (DFT) calculations to identify the activation barriers of elementary steps of CO_2 electroreduction to methane and ethylene on Cu(111) surface [28, 29]. Goddard III's group started the quantum molecular dynamics simulations that included approximately five layers of explicit water at the water/electrode interface using enhanced sampling methodology to obtain the free energies of CO_2RR on Cu(100) surface [30] and found that chemisorbed hydroxyl-methylene (CH—OH) is the key intermediate determining the selectivity for methane over methanol. Then, they developed good methodologies and conducted a series of great works to investigate the CO_2RR mechanism and the structural evolution of Cu electrodes under different work conditions [31–40]. Sautet et al. conducted the first-principles calculations to model the structural dynamics of a Cu(111) electrode under reduction conditions and suggested that restructuring in electroreduction conditions creates highly active Cu adatom sites that account for the CO_2RR performance [41–43]. Most studies do not consider the change of charge states of Cu electrodes under CO_2RR conditions and cannot give an enough accurate description. Therefore, the constant-potential methods considering the external potential effect are required and developed recently.

3.2 Principle of Constant-potential Modeling

Theoretical modeling of electrochemical reactions is very difficult due to the complicated heterogeneous electrode–electrolyte interfaces. Besides, an accurate description of electrochemical potential is significant for a realistic model of the electrochemical interface. Commonly, the computational hydrogen electrode (CHE) model is adopted to calculate the grand canonical free energy of electrochemical reactions that involve coupled proton–electron transfers calculated from [44]:

$$\Delta G_n(U) = \Delta G_n(U = 0) + neU \tag{3.1}$$

where n is the number of transferred protons and electrons and U is the external potential versus the reversible hydrogen electrode (RHE). $\Delta G_n(U = 0)$ is the free energy at 0 V versus RHE and can be calculated from DFT electronic energies by considering zero-point energy and entropic corrections.

Electrochemical steps include the transfer of a charge-neutral $(H^+ + e^-)$ couple. This has the advantage of maintaining charge neutrality and the reaction $H^+ + e^- \rightarrow ½H_2$ can be used to calculate grand canonical potential-dependent free energies. Only reaction enthalpies that are electron transfer steps are roughly affected by the potential through the energy correction of neU. Typically, the electrode–electrolyte interfaces are modeled using a fixed charge method, which seriously ignores the electrode response to local charge fluctuations in the electrolyte solution under work conditions. In the charging process, ions in the electrolytes aggregate at the electrode–electrolyte interface to form an electric double layer (EDL), which in turn induces charges on the electrode surfaces.

To explicitly take such charge fluctuations into account, the constant-potential method (CPM) was developed for atomic simulations of electrochemical systems [45]. In the CPM, the applied potential Ψ_i on electrodes is constrained at each intermediate state to a given external potential V, which is constant over a given electrode. This constraint leads to the following equation for the charge, q_i, on each electrode atom:

$$V = \Psi_i = \frac{\partial U}{\partial q_i} \tag{3.2}$$

where U is the total Coulomb energy of the system.

In this method, the total number of electrons n is considered as an additional degree of freedom, which is added on top of the Cartesian coordinates, r. Thus, the corresponding enthalpy landscape on (r, n) can be written as follows:

$$H(r,n) = E(r,n) - (n - n_0)\phi \tag{3.3}$$

where $E(r, n)$ is the internal potential energy, n_0 is the total number of nuclear charges, and ϕ is the effective energy level of the counter electrode.

Taking the standard hydrogen electrode (SHE) with the Fermi level of $E_{fs} = -4.6$ eV as the reference electrode, we have

$$\phi = -U + \frac{E_{fs}}{e}A \tag{3.4}$$

where U is the applied potential with respect to SHE. Because the work function of SHE measurement is scattered from 4.4 through 4.8 eV, the average value of 4.6 eV is generally taken for such simulations.

The electron chemical potential is then

$$\mu = \frac{\partial E}{\partial N} - \phi = E_f - \phi \tag{3.5}$$

where E_f is the Fermi level of the working electrode with adsorbates.

Herein, E_f can be obtained from the output of Vienna Ab-initio Simulation Package (VASP) calculations and then referenced to vacuum in post-processing by subtracting the surface electrostatic potential. The full gradient of is then $(\partial H/\partial r,\ \partial H/\partial n) = (-\mathbf{f},\ \mu)$, where $\mathbf{f}$ is the atomic force vector. Then, the local optimization can be performed on $H(r,\ n)$ with the generalized force $(\mathbf{f},\ -\mu)$. This direct optimization approach simultaneously relaxes the atomic positions and the number of electrons. The details are shown in Figure 3.1.

3.3 Constant-potential Modeling of Electrochemical CO_2RR

3.3.1 Single-atom Catalysis for Electrochemical CO_2RR

Electrocatalysis of the CO_2RR is a promising strategy to mitigate the severe CO_2 concentration in the atmosphere and achieve the conversion of gaseous CO_2 into renewable feedstock and sustainable energies. To date, numerous electrocatalysts have been extensively investigated for highly efficient electrochemical CO_2 hydrogenation. The single-atom catalysts (SACs) containing atomically dispersed metal atoms on suitable substrates, such as graphene and porphyrin, have become

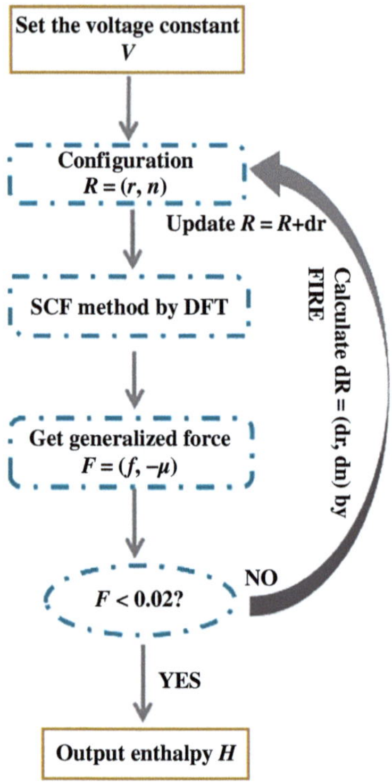

Figure 3.1 Process for saddle search and local optimization with continuous potential. The same loop updates both *r* and *n*, reducing computing costs. *Source:* Reproduced from [46] / with permission of John Wiley & Sons.

cutting-edge catalysts since they were first proposed in 2011 by Chinese Scientists [47]. The precise simulations of SACs during electrochemical CO_2 reduction have also been studied, and the constant-potential methods play an important role during the accurate description of electrode–electrolyte interfaces that catalyze CO_2 conversion into carbon monoxide, formic acid, methanol, methane, ethylene, etc. Recently, Wang's group developed a methodology of Langmuir adsorption model–derived potential-dependent kinetics (LPD-K) to probe the potential-dependent kinetics of the CO_2RR on single-nickel atom nitrogen–graphene catalysts [48], as shown in Figure 3.2. Moreover, they also investigated the reaction mechanisms and energetics of CO_2RR to CO at the electrochemical interface using *ab initio* molecular dynamics (AIMD) simulations with fully explicit solvation, in conjunction with thermodynamic integration methods and electrode potential analysis. They found that both reaction energies and barriers for CO_2 adsorption and further protonation are approximately linearly correlated with the applied electrode potentials.

Xia's group employed the grand canonical density functional theory (GC-DFT) combining the CANDLE (Charge-Asymmetric Nonlocally Determined Local-Electric) continuum solvation model to precisely simulate CO_2RR upon cobalt-doped phthalocyanine monolayer (Co-Pc) catalyst [49]. The calculation results

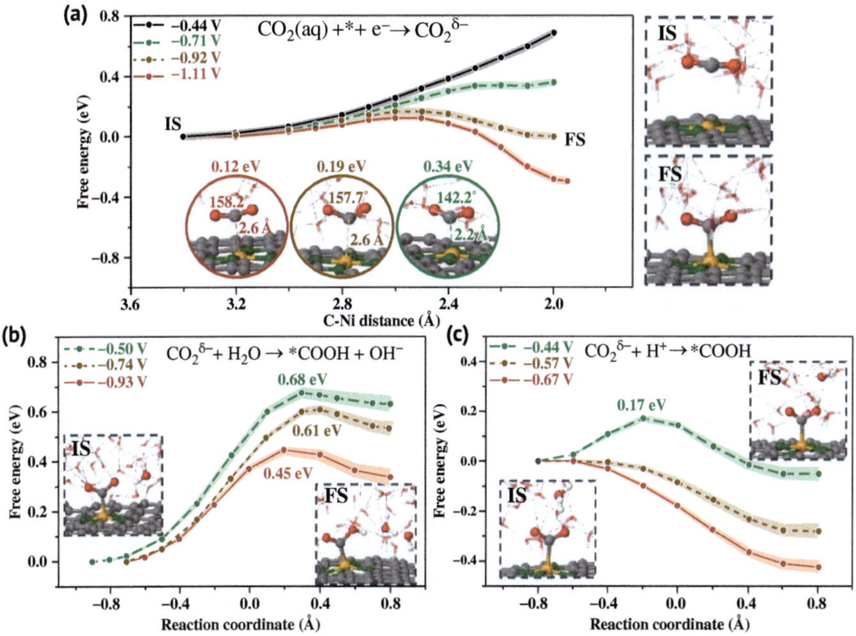

Figure 3.2 Reaction-free energy profiles with error bars for (a) CO_2 adsorption and (b, c) *COOH formation under basic and acidic environment NiN_4-Gra sites under different electrode potentials (vs. RHE). The shaded regions indicate 95% confidence intervals based on standard errors obtained using the method of block averaging. The propagation of uncertainty within the thermodynamic integration has been accounted for. The insets depict representative snapshots of the structural evolution during electrochemical steps, such as IS and FS. The corresponding bond distances and barriers of the transition state (TS) change with the applied potential are labeled. *Source:* Reproduced from [48] / with permission of American Chemical Society.

reasonably explained the activation of CO_2 and hydrogen evolution reaction (HER) competition and predicted potential determined step and onset potential, which all are excellently consistent with experimental values. They also well explained the potential-dependent competition among physical and chemical adsorption of CO_2 and H formation for high CO_2RR selectivity at a specific potential range and suggested that the charge of the reaction intermediate is the key to potential-dependent behaviors and larger electron transfer in H adsorption than CO_2 adsorption is the primary reason for HER suppression of CO_2RR at high potential.

Luo and Goddard III recently applied the developed grand canonical potential kinetics (GCP-K) formulation of quantum mechanics to predict the kinetics as a function of applied potential (U) to determine Faradic efficiency, turn-over frequency, and Tafel slope for CO and H_2 production for all the Ni sites ($Ni—N_2C_2$, $Ni—N_3C_1$, $Ni—N_4$) in graphene-supported Ni-single atom catalysts (Ni—SACs) [50]. As shown in Figure 3.3, they successfully predicted an onset potential of −0.84 V (vs. RHE) at 10 mA cm^{-2} for $Ni—N_2C_2$ site and −0.92 V for $Ni—N_3C_1$ site in agreement with experiments and also predicted that the highest current is for $Ni—N_4$, leading to 700 mA cm^{-2} at $U = -1.12$ V.

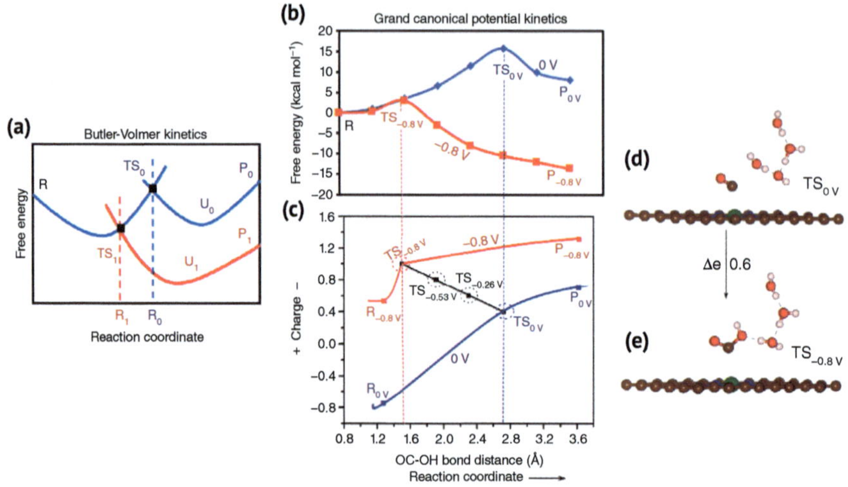

Figure 3.3 (a) The traditional Butler–Volmer reaction kinetics. Starting from the product (blue curves), at zero applied potential, (U0) the reaction between reactant, R, and product, P0, shows a transition state (TS0) with R0 spatial distance along the reaction path. The transition state (TS1) shifts toward the reactant with lower free energy giving R1 spatial distance at applied potential U1 between reactant R and product P1 (red curves). The product forms with a single electron transfer from electrode to product, leading to an electron transfer jump along the reaction pathway. (b, c) Grand canonical potential kinetics methodology, illustrating the relationship between the TS geometry and charge as the reaction changes continuously with applied potential, leading to a continuous potential-dependent reaction path. The blue curve represents the reaction pathways for cis-COOH to CO conversion on Ni—N_4 site at zero (0 V) potential while the red curve is for −0.8 V applied potential. At 0 V, the reaction between the reactant, R or R0V, and product, P0V, produces a transition state (TS0) with OC—OH = 2.77 Å (blue dotted line) and an energy barrier of 16.89 kcal mol^{-1}. At −0.8 V applied potential, the energy barrier for the transition state decreases to 2.98 kcal mol^{-1} with OC—OH = 1.46 Å (red dotted line). The black line connecting the transition states at 0 V (TS0) and −0.8 V (TS = −0.8 V) shows the linear continuous relation between the applied potential and species charge within the system. The dotted circles indicate the transition states at different applied potentials. (d, e) The geometry of the transition state at zero (TS0) and −0.8 V (TS = −0.8 V) applied potential, resulting in a charge difference of 0.6 e. The arrow represents the TS direction on the applied potential (brown: carbon; blue: nitrogen; green: nickel; red: oxygen; off-white: hydrogen atom). *Source:* [50] / Springer Nature / CC BY 4.0.

Jing's group recently used constant-potential calculations with explicit consideration of electrode potentials and solvent effects to systematically explore the role of tensile strain in determining the CO_2RR selectivity of graphyne-supported metal SACs [51]. They revealed the change in potential of zero charges of *COOH to be more sensitive to tensile strain than that of *H species on Co-GY SACs, resulting in a slower change of its adsorption energy than that of *H species under working potentials and consequently enhanced CO_2RR selectivity toward CO production.

Moreover, Wang and Ling successfully established the structure–stability relation of N-doped carbon-supported SACs (MN_4 SACs) under working conditions of CO_2 reduction, by using advanced constant-potential density functional theory calculations [52]. They further constructed a simple descriptor consisting of electron number and electronegativity, realizing accurate and rapid prediction of the stability of SACs. They accordingly proposed strategies by modifying the local geometric structure to improve stability without changing the active centers, which are supported by related experiments.

The above studies mainly focus on the design of SACs and the corresponding CO_2RR mechanisms by means of the constant-potential simulations. However, the anchored single atoms are generally unstable under reaction conditions and experience spillover and sintering to form metal clusters for the original activity of electrochemical CO_2RR. The applied potential can adjust the charge states of the electrode and accordingly influence the stability and catalytic performance of supported single metal atoms, while the traditional methods without constant-potential correction cannot give good modeling for the structural evolution under work conditions of CO_2RR, as the dynamic adjustment of charge states of electrode surfaces is neglected.

Recently, Su's group employed constant-potential density functional theory calculations to investigate the sintering process of Cu single atoms of Cu–N–C single-atom catalysts into clusters under eCO_2RR conditions [27]. They conducted systematic constant-potential AIMD simulations and revealed that the supported single Cu atoms can agglomerate into clusters and increase the Cu^{2+} concentration or the applied potential can efficiently suppress Cu sintering, as shown in Figure 3.4. The sintered Cu clusters provide the original active sites for generating C_2 products during eCO_2RR. In their other work, they revealed that copper atoms in the $Cu^{\delta+}$ NCN complex can activate CO_2 and the delocalized electrons around copper sites facilitate carbon–carbon coupling by reducing the Gibbs free energy for *CHC formation by means of constant-potential DFT calculations [53]. Meanwhile, Li's group also performed extensive constant-potential calculations with explicit solutions to investigate the underlying mechanism of Cu SACs and disclosed the absence of C_{2+} activity in Cu_1 sites regardless of the different substrates [54]. The original Cu_1 sites cannot facilitate *CO hydrogenation and CO—CO coupling due to the lack of synergistic active sites. These single-atom sites are also unstable under work conditions and lead to leaching and aggregation to form small Cu clusters. They further revealed that the derived Cu clusters composed of at least three Cu atoms can efficiently promote CO—CO coupling for C_2 products. Previously, Liu and Wang took a single copper atom embedded in N-doped graphene as an example and adopted the "constant-potential hybrid-solvation dynamic model" to evaluate the reversible transformation between copper single atoms and clusters under realistic reaction conditions [55]. They also suggested that the collision of the Cu atoms forms a transient Cu cluster structure as the active sites to promote ethanol formation.

3.3.2 Cu-based Surface Catalysis for Electrochemical CO_2RR

It is generally believed that Cu catalysts have the best ability to generate C_{2+} chemicals with significant Faradaic efficiency (FE), attributed to their ideal *CO binding strength in accordance with the Sabatier principle. However, the morphological transformations or surface reconstruction in Cu-based catalysts during CO_2RR seriously make the investigation of active sites complicated. This simultaneous

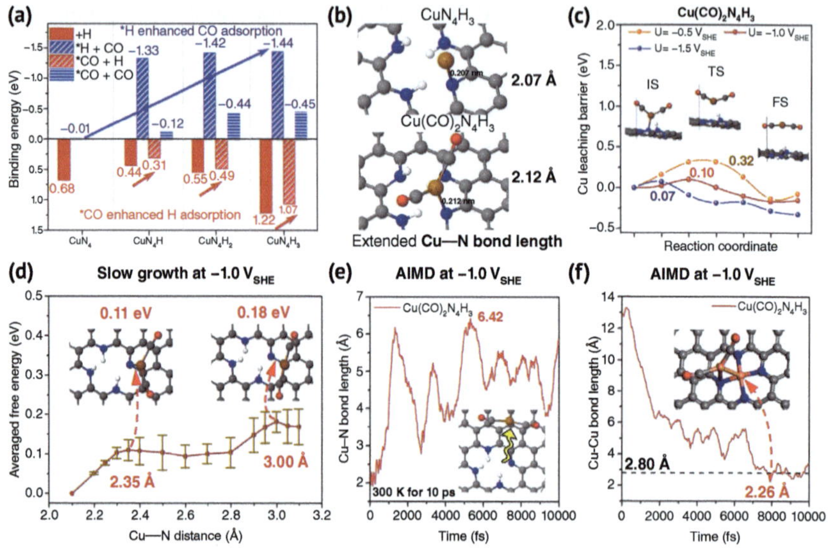

Figure 3.4 (a) The binding energies of CO and H on different substrates, where "*H+CO" denotes the CO adsorption energy on an H-adsorbed substrate, "*CO+H" denotes the H adsorption energy on a CO adsorbed substrate, and "*CO+CO" denotes the adsorption energy of one additional CO. (b) The optimized structures of CuN_4H_3 and $Cu(CO)_2N_4H_3$, where the Cu—N bond lengths are measured to be 2.07 Å and 2.12 Å, respectively. (c) The constant-potential Cu leaching barriers on $Cu(CO)_2N_4H_3$ at U = −0.5, −1.0, and −1.5 V_{SHE}; the corresponding initial state (IS), transition state (TS), and final state (FS) structures are shown in the insets. (d) The grand free energy profile during the Cu—$(CO)_2$ moiety leaching process in $Cu(CO)_2N_4H_3$ at − 1.0 V versus SHE calculated by the "slow-growth" approach. The results represent the average of multiple calculations, with corresponding standard deviations indicated as error bars. (e) Cu—N bond lengths in $Cu(CO)_2N_4H_3$ during constant-potential AIMD simulation at U = −1.0 VSHE for 10 ps. (f) Cu—Cu bond lengths in $Cu(CO)_2N_4H_3$ during the additional 10 ps of the AIMD simulation after Cu—$(CO)_2$ moiety leaching. *Source:* Reproduced from Qin et al. [27], with permission of Wiley.

Cu reconstruction during electrochemical CO_2RR is highly interrelated to the reaction conditions [56]. Typically, the Cu surfaces are sensitive to electrolyte additives, modified molecules containing small molecules and polymers, cations, and anions, as well as the applied potentials and pH values. It is mainly due to the weak cohesive energy and high surface mobility of metallic Cu catalysts, and the micro-environmental factors at the electrode–electrolyte interfaces significantly bring about atomic re-arrangement and compositional change during electrochemical CO_2RR [57]. The reconstruction behavior of Cu is a double-edged sword for the CO_2RR. Although the initial well-defined morphology and highly active sites tend to be lost during electrocatalysis, surface reconstruction generally can generate unique active sites for promising catalytic performance toward the desired products. Understanding the reconstruction of Cu electrodes under experimental conditions plays an important role in the optimization and design of highly efficient Cu-based catalysts. The constant-potential modeling can provide a precise description of interactions between Cu electrode and electrolyte at the atomic level and give the panorama of Cu reconstruction under work conditions.

Recently, Valery and coworkers performed grand-canonical density functional calculations adopting an implicit solvation model with potential control to investigate the influence of adsorbates and applied potential on the stability of different facets of metallic Cu electrocatalysts [59]. They found that CO intermediates exhibit higher kinetically and thermodynamically accessible coverages on (100) than (111) facets, and the (100) facet gets more stable during CO_2RR. He's group conducted constant-potential slow-growth simulations to investigate a Cu dissolution trajectory under work conditions and proposed that the potential is indeed an important factor for the dissolution of copper [58]. As shown in Figure 3.5, they also inferred that at an electrode potential more negative than the onset potential (i.e., about −0.4 VRHE at our conditions) and with the presence of alkali cations

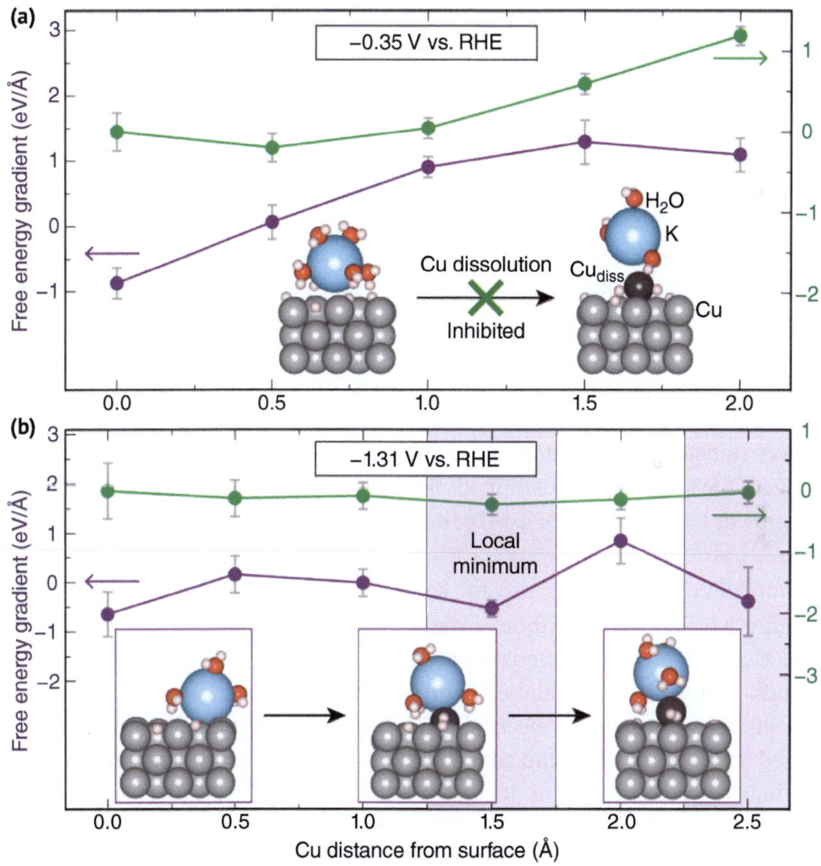

Figure 3.5 Free energy gradient (purple) and free energy (green) as functions of Cu distance from the surface along the coordinate of the dissolution reaction. The gradient shows an overall monotonically increasing trend at −0.35 V versus RHE (a) while it shows negative local minima at −1.31 V versus RHE (b). Inset schematic in (a) shows that copper dissolution is inhibited at the less negative voltage of −0.35 V versus RHE due to the increasing gradient and high free energy barrier. Inset in (b) shows the atomic structures (with Cu distance from the surface of 0, 1.5, and 2.5 Å) of the copper dissolution reaction that happens at the more negative voltage of −1.31VRHE. *Source:* Reproduced from Ref. [58], Springer Nature, CC BY 4.0.

in the solution, Cu forms soluble intermediate ternary hydride species, which may react quickly with adsorbed hydrogen species or water from the solution and get deposited back onto the electrode, leading to the observed reconstruction of the Cu catalyst and the formation of smaller Cu particles and the reconstruction of Cu catalysts.

3.3.3 Cation Effect for Electrochemical CO_2RR

It is a growing recognition that alkali metal cations in the electrolyte can significantly impact the CO_2RR performance. However, the molecular underpinning of this cation effect remains an important topic of discussion. Resasco et al. performed the systematic DFT investigations to show that the alkali metal cations can influence the distribution of CO_2RR products formed as a consequence of electrostatic interactions between solvated cations present at the outer Helmholtz plane (OHP) and adsorbed species having large dipole moments [60]. The observed trends in activity with cation size are attributed to an increase in the concentration of cations at the OHP with increasing cation size. Lu and Xu experimentally suggested that larger cations at the OHP have also been shown to affect several near-electrode properties such as the water structure around surface intermediates [61], while Malkani et al. found that the proposed enhanced buffering capacity of larger cations is unlikely to account for the cation effect in the CO_2RR and suggested that the mechanisms through which cations impact the CO_2RR remain unresolved [62]. To this end, several researchers also conducted thoroughly constant-potential modeling to unravel the cation effect on CO_2RR.

Melander and coworkers studied cation effects by simulating both outer-sphere electron transfer (OS-ET) and inner-sphere electron transfer (IS-ET) pathways during CO_2RR through constrained DFT molecular dynamics (cDFT-MD) and slow-growth DFT-MD (SG-DFT-MD), respectively [63]. They found that cations can effectively promote CO_2 activation through IS-ET with the barrier of only 0.61 eV (K^+) and 0.91 eV (Li^+), generating the key intermediate (adsorbed ${CO_2}^{\delta-}$), as shown in Figure 3.6. Moreover, without cations, CO_2-to-${CO_2}^{\delta-}$ conversion cannot take place. Using AIMD simulations with explicit solvation and enhanced sampling methods, Zhang et al. systematically investigated alkaline cation effect on CO_2RR at the interface between Cu electrode and electrolytes [64]. They found that H+ is repelled from the interface and constrained by alkaline cations, and the competing HER must be considered simultaneously to understand the crucial role of alkali metal cations in CO_2RR. Hansen's group provided a systematic mechanism study toward CO_2RR through AIMD simulations integrated with the slow-growth sampling (SG-AIMD) method [65]. They unraveled that K^+ cations can promote CO_2 activation with a free energy barrier of only 0.66 eV and as a result enhance CO_2RR performance at the inner-sphere interface. Furthermore, they also found that competitive HER can be inhibited by interfacial cations with the induced kinetic blockage effect.

Recently, Pan's group systematically investigated the combined effect of interfacial water and alkali metal cations on the C—C coupling at the interface of Cu(100)

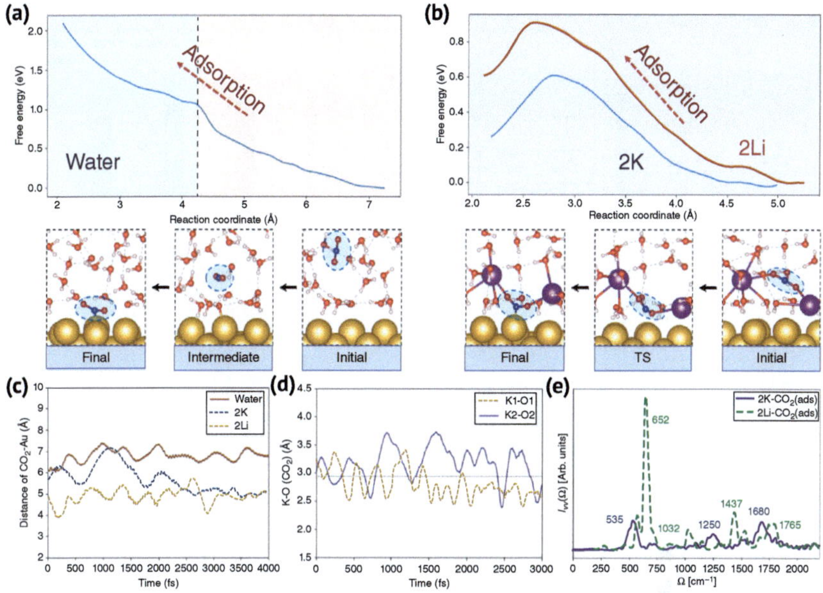

Figure 3.6 Free energy profile (top) and key structures (bottom) of IS-ET at Au–water (a) and Au–water–2K (b) interfaces, with the free energy curve at Au–water–2Li interfaces inserted for comparison. (c) Distance variations between CO_2 and Au surface during 4ps DFT-MD simulations for water and AM^+-containing electrolytes. (d) Distance variations of K-O (CO_2) during SG-DFT-MD at Au–water–2K interfaces. (e) Simulated vibration spectra of CO_2(ads) at Au–water–2K and Au–water–2Li interfaces. *Source:* Reproduced from Ref. [63], Springer Nature, CC BY 4.0.

electrode and electrolytc using AIMD simulations [66]. As shown in Figure 3.7, they observed a linear correlation between the water-adsorbate stabilization effect and the corresponding alleviation in the C—C coupling free energy. Large cations can create a local hydrophobic environment with a scarcity of hydrogen bonds owing to their great ionic radius, limiting the hydrogen of surrounding interfacial water to approach the adsorbed *CO. Moreover, it circumvents the further hydrogenation of *CO toward the C_1 pathway, serving as the predominant factor through which a larger cation facilitates C—C coupling. Xu and Xiao determined the impact of cations on the enthalpy and entropy of CO adsorption on Cu under electrochemical conditions combined with experimental and constant-potential modeling investigations [67]. They suggested that CO adsorption becomes increasingly unfavorable in the sequence $Li^+ > Na^+ > K^+ > Cs^+$ with a substantial enthalpy–entropy compensation effect.

Su's group recently performed constant-potential AIMD simulations to study the dynamic behavior of interfacial K^+ ions over Cu surfaces during C—C coupling and the origin of the cation effect [68]. As shown in Figure 3.8, they found that the specific adsorption of K^+ cation readily occurs at the surface sites adjacent to CO intermediates on the Cu surfaces. Furthermore, this specific adsorption of K^+ cation is more important than quasi-specific adsorption for enhancing CO—CO

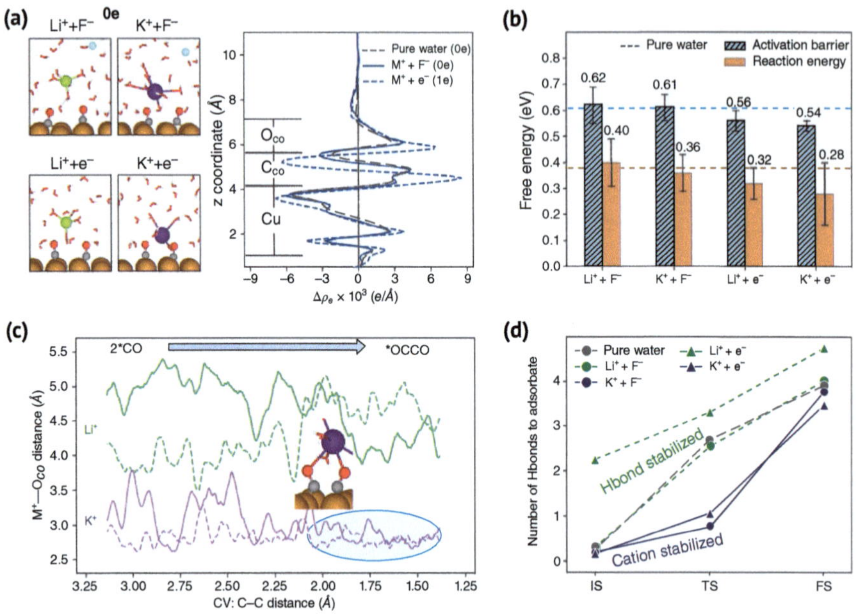

Figure 3.7 Influences of cations on the CO dimerization step. (a) Snapshots of M^+ + F^- and $M^+ + e^-$ (M = Li/K) models and their planar-averaged differential charge density. $\Delta\rho e = \rho_{Cu+ads+wat+ion} - \rho_{Cu+wat+ion} - \rho_{ads}$. For clarity, $Li^+ + F^-$ and $Li^+ + e^-$ models are presented, while $K^+ + F^-$ and $K^+ + e^-$ models exhibit similarity to $Li^+ + F^-$ and Li^+ + e^-. To clearly show the local structure of cations and adsorbates, we extracted only a portion of the entire model. (b) Free energy profiles of the CO dimerization under different conditions. (c) Distance of M^+ – OCO along the collective variable. For clarity, $Li^+ + e^-$ and $K^+ + e^-$ models serve as illustrative examples, with $Li^+ + F^-$ and $K^+ + F^-$ models having similar behavior. The solid line and dashed line, respectively, represent the distances between the cation and each of the two *CO molecules. (d) Variations in the number of hydrogen bonds during CO dimerization. Color code: Cu, brown; O, red; H, white; C, gray; Li, green; K, purple; F, cyan. *Source:* Reproduced from Ref. [66], with permission of American Chemical Society.

coupling kinetics, reducing the coupling barriers by approximately 0.20 eV, due to the charge redistribution occurring between the specifically adsorbed K^+ cation, *CO, and Cu sites.

3.3.4 Surface-modified Species for Electrochemical CO_2RR

The surface modification of molecules and polymers can also have an important influence on electrochemical CO_2RR. In 2016, Wang's group reported the earliest attempt to modify the Cu catalyst surface with organic moieties to improve CO_2RR activity by using various amino acids as modifiers [69]. Early studies mainly focused on the local molecular interactions of CO_2 or its reduction intermediates on Cu surfaces with introduced chemical moieties [70]. Understanding the nature of surface modification and this influence on electrochemical CO_2RR requires the assistance of constant-potential modeling.

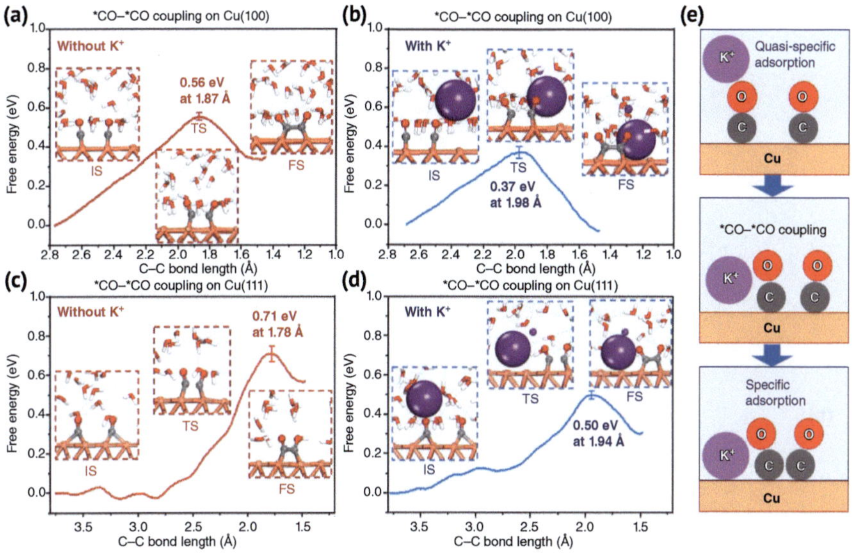

Figure 3.8 Without K^+ ions in the electrolyte, the grand free energy profile of *CO—*CO coupling on (a) Cu(100) and (b) Cu(111) surfaces at −1.5 VSHE calculated by the "slow-growth" approach. With K+ ions in the electrolyte, the grand free energy profile of *CO—*CO coupling on (c) Cu(100) and (d) Cu(111) surfaces at −1.5 VSHE. The CV is defined as the distance between the two involved C atoms. (e) Illustration of the K+ -specific adsorption-assisted *CO—*CO coupling process. *Source:* Reproduced from Ref. [68], with permission of American Chemical Society.

In 2019, Toma and Toste conducted the ReaxFF reactive molecular dynamics simulations and disclosed that the hydrophilic/hydrophobic modifiers influence the formation of surface hydrides for producing formic acid or H_2 under CO_2RR conditions [71]. As shown in Figure 3.9, they also demonstrated the universal effect of various polymers and molecules on the Cu CO_2RR, revealing that organic species affect catalysis by regulating the behavior of water on the Cu catalyst surface. Grubbs and Goddard III employed molecular dynamics simulations to show that the components of a polymer modifier on Cu surfaces can cooperatively improve the C_{2+} production selectivity. They also suggested that the modifier is likely to induce a combination of nonspecific effects such as gathered local CO_2 concentration, increased porosity for gas diffusion, and a local electric field effect rather than the specific molecular interactions of the components affecting CO_2RR activity. Combining experimental and theoretical investigations, Wang and Su reported that the surface-modified tannic acid (TA) molecules can adaptively regulate the reconstruction of a Cu-based material and facilitate CO_2 reduction to C_2H_4 products [72]. The TA-induced $Cu^{\delta+}/Cu^0$ interfaces can significantly lower the activation energy for CO-CO coupling, thereby promoting C_2H_4 production. Wang and co-workers conducted experimental characterization and DFT calculations to investigate the influence of the 2-position alkyl chain length of benzimidazole-based Cu coordination polymer on their acidic CO_2RR performance and unveiled that longer substituents slow down H^+ diffusion and reduce the coordination number of exposed Cu atoms of Cu nanoparticles, creating interface conditions beneficial for CO_2RR to C_{2+} products [73].

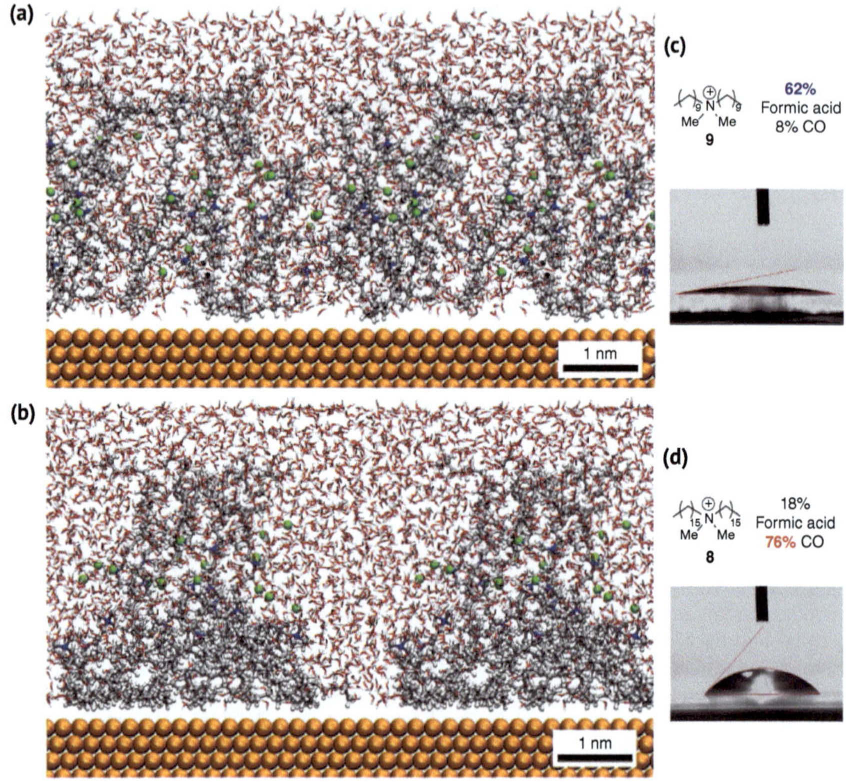

Figure 3.9 Study of Cu surfaces modified with 8 and 9. Last snapshot of the interface between Cu, water, and hydrophilic species 9 (a) and hydrophobic species 8 (b) extracted from 5 ns of reactive molecular dynamic canonical ensemble (NVT) simulations at 298 K to equilibrate the interface. The colors are N in blue, C in gray, O in red, H in white, and Br in green. The dimension of the simulation is 4.77 nm × 7.87 nm. The contact angle between a 3 μL drop of water and Cu surfaces modified with 9 (c) and 8 (d). *Source:* Reproduced from [71] / with permission of American Chemical Society.

Che and Gu recently performed DFT calculations to comprehensively understand CO_2RR to C_{2+} products over aminothiolate-coated Cu catalysts [74]. As shown in Figure 3.10, they proposed that the CO_2RR performance is strongly related to the alkyl chain length, ligand coverage, ligand configuration, and Cu facets. The aminothiolate–Cu interface can significantly promote initial CO_2 activation and lower the activation barrier of C—C coupling over the interfacial active sites. Su and coworkers adopted constant-potential molecular dynamics simulations to probe the dynamic change of interfacial water structure on an electrified hydrophobic electrode surface. They revealed that interfacial water near the hydrophobic electrode surface exhibits rigid intermolecular hydrogen bonding interaction. Moreover, the strong intermolecular hydrogen bond makes water reorientation hard and leads to a long metal–H distance, which inhibits water dissociation and decreases *H coverage. Moderate *H coverage not only impedes hydrogen evolution but also ensures the hydrogenation of CO_2RR intermediates.

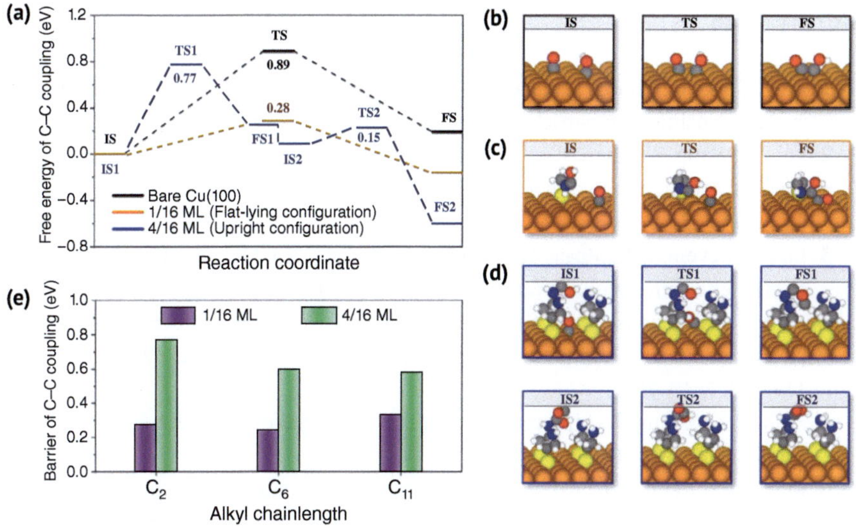

Figure 3.10 C—C coupling ($CO^* + COH^* \rightarrow OCCOH^* + ^*$) reaction-free energies and activation barriers on Cu(100). (a) Free-energy diagram of C—C coupling with and without C_2 ligands. The activation barriers for all the examined cases are indicated in the plot. The corresponding geometries of IS, TS, and FS are shown in (b–d). (e) Activation barrier of C—C coupling over the ligand–metal interface, including C_2, C_6, and C_{11} ligands at 1/16 and 4/16 ML. *Source:* Reproduced from [74] / with permission of American Chemical Society.

3.4 Conclusion and Outlook

The modeling of elementary steps in electrocatalysis is not only significant but also technologically difficult due to the complicated interface between electrodes and electrolytes. With the increase in computer power and the development of appropriate methods, first-principles simulations of electrocatalytic processes get good accuracy, especially within the framework of constant-potential modeling. The understanding of electrochemical CO_2RR gets clear and precise through the constant-potential modeling at the atomic or molecular scales. It is extremely important for the rational design of electrochemical catalysts for energy storage and conversion. However, it demands massive computational power. With the consideration of the complexity of electrochemical electrolyte/electrode interfaces, the acceleration of constant-potential modeling requires the development of corresponding algorithms at the prerequisite of enough simulation accuracy. Thus, there are still substantial theoretical and numerical challenges associated with the modeling of electrocatalytic processes.

Acknowledgment

This work was supported by the National Key R&D Program of China (No. 2023YFB4203900).

References

1 Zhu, J., Hu, L., Zhao, P. et al. (2019). Recent advances in electrocatalytic hydrogen evolution using nanoparticles. *Chem. Rev.* 120 (2): 851–918.

2 Shi, R., Wang, Z., Zhao, Y. et al. (2021). Room-temperature electrochemical acetylene reduction to ethylene with high conversion and selectivity. *Nat. Catal.* 4 (7): 565–574.

3 Li, F., Thevenon, A., Rosas-Hernández, A. et al. (2020). Molecular tuning of CO_2-to-ethylene conversion. *Nature* 577 (7791): 509–513.

4 Nitopi, S., Bertheussen, E., Scott S.B. et al. (2019). Progress and perspectives of electrochemical CO_2 reduction on copper in aqueous electrolyte. *Chem. Rev.* 119 (12): 7610–7672.

5 Shih, C.F., Zhang, T., Li, J. et al. (2018). Powering the future with liquid sunshine. *Joule* 2 (10): 1925–1949.

6 Ma, S. and Kenis, P.J. (2013). Electrochemical conversion of CO_2 to useful chemicals: current status, remaining challenges, and future opportunities. *Curr. Opin. Chem. Eng.* 2 (2): 191–199.

7 Pan, F. and Yang, Y. (2020). Designing CO_2 reduction electrode materials by morphology and interface engineering. *Energy Environ. Sci.* 13 (8): 2275–2309.

8 Zhang, H., Li, J., and Cheng, M.J. (2018). CO electroreduction: current development and understanding of Cu-based catalysts. *ACS Catal.* 9 (1): 49–65.

9 Tomboc, G.M., Choi, S., Kwon, T. et al. (2020). Potential link between Cu surface and selective CO_2 electroreduction: perspective on future electrocatalyst designs. *Adv. Mater.* 32 (17): 1908398.

10 Woldu, A.R., Huang, Z., Zhao, P. et al. (2022). Electrochemical CO_2 reduction (CO_2RR) to multi-carbon products over copper-based catalysts. *Coordin. Chem. Rev.* 454: 214340.

11 Gawande, M.B., Goswami, A., Felpin, F.X. et al. (2016). Cu and Cu-based nanoparticles: synthesis and applications in catalysis. *Chem. Rev.* 116 (6): 3722–3811.

12 Loiudice, A., Lobaccaro, P., Kamali, E.A. et al. (2016). Tailoring copper nanocrystals towards C_2 products in electrochemical CO_2 reduction. *Angew. Chem. Int. Ed.* 55 (19): 5789–5792.

13 Li, H., Liu, T., Wei, P. et al. (2021). High-rate CO_2 electroreduction to C_{2+} products over a copper-copper iodide catalyst. *Angew. Chem.* 133 (26): 14450–14454.

14 Zhang, W., Huang, C., Xiao, Q. et al. (2020). Atypical oxygen-bearing copper boosts ethylene selectivity toward electrocatalytic CO_2 reduction. *J. Am. Chem. Soc.* 142 (26): 11417–11427.

15 Lum, Y., Kwon, Y., Lobaccaro, P. et al. (2016). Trace levels of copper in carbon materials show significant electrochemical CO_2 reduction activity. *ACS Catal.* 6 (1):202–209.

16 Lee, S.Y., Jung, H., Kim, N.K. et al. (2018). Mixed copper states in anodized Cu electrocatalyst for stable and selective ethylene production from CO_2 reduction. *J. Am. Chem. Soc.* 140 (28): 8681–8689.

17 Ma, D., Zhi, C., Zhang, Y. et al. (2024). A review on the influence of crystal facets on the product selectivity of CO_2RR over Cu metal catalysts. *ACS Nano* 18 (33): 21714–21746.

18 Choi, W., Chae, Y., Liu, E. et al. (2024). Exploring the influence of cell configurations on Cu catalyst reconstruction during CO_2 electroreduction. *Nat. Commun.* 15 (1): 8345.

19 Wang, X., Guo, C., Zhu, B. et al. (2023). Reaction-induced iodine adsorption on Cu surfaces facilitates electrocatalytic CO_2 reduction. *J. Chem. Phys.* 158 (20): 204701.

20 Shi, H., Luo, L., Li, C. et al. (2024). Stabilizing Cu^+ species in Cu_2O/CuO catalyst via carbon intermediate confinement for selective CO_2RR. *Adv. Funct. Mater.* 34 (11): 2310913.

21 Li, Z., Wang, L., Sun, L. et al. (2024). Dynamic cation enrichment during pulsed CO_2 electrolysis and the cation-promoted multicarbon formation. *J. Am. Chem. Soc.* 146 (34): 23901–23908.

22 Wang, J., Qin, Y., Jin, S. et al. (2023). Customizing CO_2 electroreduction by pulse-induced anion enrichment. *J. Am. Chem. Soc.* 145 (48): 26213–26221.

23 Birdja, Y.Y., Pérez-Gallent, E., Figueiredo, M.C. et al. (2019). Advances and challenges in understanding the electrocatalytic conversion of carbon dioxide to fuels. *Nat. Energy* 4 (9): 732–745.

24 Chou, T.C., Chang, C.C., Yu, H.L. et al. (2020). Controlling the oxidation state of the Cu electrode and reaction intermediates for electrochemical CO_2 reduction to ethylene. *J. Am. Chem. Soc.* 142 (6): 2857–2867.

25 Zhu, P., Qin, Y.C., Cai, X.H. et al. (2025). Understanding oxidation state of Cu-based catalysts for electrocatalytic CO_2 reduction. *J. Mater. Sci. Technol.* 218: 1–24.

26 Lei, Q., Huang, L., Yin, J. et al. (2022). Structural evolution and strain generation of derived-Cu catalysts during CO_2 electroreduction. *Nat. Commun.* 13 (1): 4857.

27 Qin, Y., Zhao, W., Xia, C. et al. (2024). CO intermediate-assisted dynamic Cu sintering during electrocatalytic CO_2 reduction on Cu-N-C catalysts. *Angew. Chem. Int. Ed.* 63 (23): e202404763.

28 Nie, X., Esopi, M.R., Janik, M.J. et al. (2013). Selectivity of CO_2 reduction on copper electrodes: the role of the kinetics of elementary steps. *Angew. Chem. Int. Ed.* 52 (9).

29 Nie, X., Luo, W., Janik, M.J. et al. (2014). Reaction mechanisms of CO_2 electrochemical reduction on Cu(111) determined with density functional theory. *J. Catal.* 312: 108–122.

30 Cheng, T., Xiao, H., and Goddard III, W.A. (2015). Free-energy barriers and reaction mechanisms for the electrochemical reduction of CO on the Cu(100) surface, including multiple layers of explicit solvent at pH = 0. *J. Phys. Chem. Lett.* 6 (23): 4767–4773.

31 Xiao, H., Goddard III, W.A., Cheng, T. et al. (2017). Cu metal embedded in oxidized matrix catalyst to promote CO_2 activation and CO dimerization for electrochemical reduction of CO_2. *P. Natl. Acad. Sci. USA* 114 (26): 6685–6688.

32 Cheng, T., Xiao, H., and Goddard III, W.A. (2016). Reaction mechanisms for the electrochemical reduction of CO_2 to CO and formate on the Cu (100) surface at 298 K from quantum mechanics free energy calculations with explicit water. *J. Am. Chem. Soc.* 138 (42): 13802–13805.

33 Xiao, H., Cheng, T., and Goddard III, W.A. (2017). Atomistic mechanisms underlying selectivities in C_1 and C_2 products from electrochemical reduction of CO on Cu (111). *J. Am. Chem. Soc.* 139 (1): 130–136.

34 Cheng, T., Xiao, H., and Goddard III, W.A. (2017). Full atomistic reaction mechanism with kinetics for CO reduction on Cu (100) from *ab initio* molecular dynamics free-energy calculations at 298 K. *P. Natl. Acad. Sci. USA* 114 (8): 1795–1800.

35 Cheng, T., Fortunelli, A., and Goddard III, W.A. (2019). Reaction intermediates during operando electrocatalysis identified from full solvent quantum mechanics molecular dynamics. *P. Natl. Acad. Sci. USA* 116 (16): 7718–7722.

36 Cheng, T., Xiao, H., and Goddard, W.A. (2017). Nature of the active sites for CO reduction on copper nanoparticles; suggestions for optimizing performance. *J. Am. Chem. Soc.* 139 (34): 11642–11645.

37 Xiao, H., Cheng, T., and Goddard III, W.A. et al. (2016). Mechanistic explanation of the pH dependence and onset potentials for hydrocarbon products from electrochemical reduction of CO on Cu (111). *J. Am. Chem. Soc.* 138 (2):483–486.

38 Liu, G., Lee, M., Kwon, S. et al. (2021). CO_2 reduction on pure Cu produces only H_2 after subsurface O is depleted: theory and experiment. *P. Natl. Acad. Sci. USA* 118 (23): e2012649118.

39 Osella, S. and Goddard III, W.A. (2023). CO_2 reduction to methane and ethylene on a single-atom catalyst: a grand canonical quantum mechanics study. *J. Am. Chem. Soc.* 145 (39): 21319–21329.

40 Naserifar, S., Chen, Y., Kwon, S. et al. (2021). Artificial intelligence and QM/MM with a polarizable reactive force field for next-generation electrocatalysts. *Matter* 4 (1): 195–216.

41 Cheng, D., Wei, Z., Zhang, Z. et al. (2023). Restructuring and activation of Cu(111) under electrocatalytic reduction conditions. *Angew. Chem.* 135 (20): e202218575.

42 Zhang, Z., Gee, W., Sautet, P. et al. (2024). H and CO co-induced roughening of Cu surface in CO_2 electroreduction conditions. *J. Am. Chem. Soc.* 146 (23): 16119–16127.

43 Wei, Z. and Sautet, P. (2022). Improving the accuracy of modelling CO_2 electroreduction on copper using many-body perturbation theory. *Angew. Chem.* 134 (43): e202210060.

44 Nørskov, J.K., Rossmeisl, J., Logadottir, A. et al. (2004). Origin of the overpotential for oxygen reduction at a fuel-cell cathode. *J. Phys. Chem. B* 108 (46): 17886–17892.

45 Wang, Z., Yang, Y., Olmsted, D.L. et al. (2014). Evaluation of the constant potential method in simulating electric double-layer capacitors. *J. Chem. Phys.* 141 (18): 184102.

46 Wang, Z., Qin, Y., Wu, T. et al. (2023). Theoretical study on B-doped FeN_4 catalyst for potential-dependent oxygen reduction reaction. *Chem. Phys. Chem.* 24 (16): e202300152.

47 Qiao, B., Liu, J., Allard, L. et al. (2011). Single-atom catalysis of CO oxidation using Pt_1/FeO_x. *Nature Chem.* 3 (8): 634–641.

48 Zhao, H., Cao, H., Zhang, Z. et al. (2022). Modeling the potential-dependent kinetics of CO_2 electroreduction on single-nickel atom catalysts with explicit solvation. *ACS Catal.* 12 (18): 11380–11390.

49 Wang, T., Zhou, L., Xia, S. et al. (2023). Theoretical understanding of potential-dependent electrocatalytic CO_2RR and competition with HER upon cobalt single atom supported by phthalocyanine monolayer. *J. Phys. Chem. C* 127 (6): 2963–2973.

50 Hossain, M.D., Huang, Y., Yu, T.H. et al. (2020). Reaction mechanism and kinetics for CO_2 reduction on nickel single atom catalysts from quantum mechanics. *Nat. Commun.* 11 (1): 2256.

51 Liu, T., Xu, T., Li, T. et al. (2024). Selective CO_2 reduction over γ-graphyne supported single-atom catalysts: crucial role of strain regulation. *J. Am. Chem. Soc.* 146 (34): 24133–24140.

52 Cui, Y., Ren, C., Wu, M. et al. (2024). Structure–stability relation of single-atom catalysts under operating conditions of CO_2 reduction. *J. Am. Chem. Soc.* 146 (42): 29169–29176.

53 Yue, K., Qin, Y., Huang, H. et al. (2024). Stabilized Cu^0-Cu^{1+} dual sites in a cyanamide framework for selective CO_2 electroreduction to ethylene. *Nat. Commun.* 15 (1):7820.

54 Zhang, J., Wang, Y., and Li, Y. (2024). Not one, not two, but at least three: activity origin of copper single-atom catalysts toward CO_2/CO electroreduction to C_{2+} products. *J. Am. Chem. Soc.* 146 (22): 14954–14958.

55 Bai, X., Zhao, X., Zhang, Y. et al. (2022). Dynamic stability of copper single-atom catalysts under working conditions. *J. Am. Chem. Soc.* 144 (37): 17140–17148.

56 Choi, C., Kwon, S., Cheng, T. et al. (2020). Highly active and stable stepped Cu surface for enhanced electrochemical CO_2 reduction to C_2H_4. *Nature Catal.* 3 (10): 804–812.

57 Han, Z., Han, D., Chen, Z. et al. (2022). Steering surface reconstruction of copper with electrolyte additives for CO_2 electroreduction. *Nat. Commun.* 13 (1): 3158.

58 Liu, S., Li, Y., Wang, D. et al. (2024). Alkali cation-induced cathodic corrosion in Cu electrocatalysts. *Nat. Commun.* 15 (1): 5080.

59 Yu, H., Govindarajan, N., Weitzner, S.E. et al. (2024). Theoretical investigation of the adsorbate and potential-induced stability of Cu facets during electrochemical CO_2 and CO reduction. *ChemPhysChem* 25 (10): e202300959.

60 Resasco, J., Chen, L.D., Clark, E. et al. (2017). Promoter effects of alkali metal cations on the electrochemical reduction of carbon dioxide. *J. Am. Chem. Soc.* 139 (32): 11277–11287.

61 Li, J., Wu, D., Malkani, A.S. et al. (2020). Hydroxide is not a promoter of C_{2+} product formation in the electrochemical reduction of CO on copper. *Angew. Chem.* 132 (11): 4494–4499.

62 Malkani, A.S., Anibal, J., and Xu, B. (2020). Cation effect on interfacial CO_2 concentration in the electrochemical CO_2 reduction reaction. *ACS Catal.* 10 (24): 14871–14876.

63 Qin, X., Hansen, H.A., Honkala, K. et al. (2023). Cation-induced changes in the inner-and outer-sphere mechanisms of electrocatalytic CO_2 reduction. *Nat. Commun.* 14 (1): 7607.

64 Zhang, Z., Li, H., Shao, Y. et al. (2024). Molecular understanding of the critical role of alkali metal cations in initiating CO_2 electroreduction on Cu(100) surface. *Nat. Commun.* 15 (1): 612.

65 Qin, X., Vegge, T., and Hansen, H.A. (2023). Cation-coordinated inner-sphere CO_2 electroreduction at Au–water interfaces. *J. Am. Chem. Soc.* 145 (3): 1897–1905.

66 Yang, X., Ding, H., Li, S. et al. (2024). Cation-induced interfacial hydrophobic microenvironment promotes the C–C coupling in electrochemical CO_2 reduction. *J. Am. Chem. Soc.* 146 (8): 5532–5542.

67 Xu, Y., Xia, Z., Gao, W. et al. (2024). Cation effect on the elementary steps of the electrochemical CO reduction reaction on Cu. *Nature Catal.* 7: 1–10.

68 Qin, Y., Xia, C., Wu, T. et al. (2024). Specific adsorption of alkaline cations enhances CO–CO coupling in CO_2 electroreduction. *J. Am. Chem. Soc.* 146 (47): 32539–32549.

69 Xie, M.S., Xia, B.Y., Li, Y. et al. (2016). Amino acid modified copper electrodes for the enhanced selective electroreduction of carbon dioxide towards hydrocarbons. *Energy Environ. Sci.* 9 (5): 1687–1695.

70 Ahn, S., Klyukin, K., Wakeham, R.J. et al. (2018). Poly-amide modified copper foam electrodes for enhanced electrochemical reduction of carbon dioxide. *ACS Catal.* 8 (5): 4132–4142.

71 Buckley, A.K., Lee, M., Cheng, T. et al. (2019). Electrocatalysis at organic–metal interfaces: identification of structure–reactivity relationships for CO_2 reduction at modified Cu surfaces. *J. Am. Chem. Soc.* 141 (18): 7355–7364.

72 Chen, S., Ye, C., Wang, Z. et al. (2023). Selective CO_2 reduction to ethylene mediated by adaptive small-molecule engineering of copper-based electrocatalysts. *Angew. Chem.* 135 (50): e202315621.

73 Deng, H., Liu, T., Zhao, W. et al. (2024). Substituent tuning of Cu coordination polymers enables carbon-efficient CO_2 electroreduction to multi-carbon products. *Nat. Commun.* 15 (1): 9706.

74 Wan, M., Yang, Z., Morgan, H. et al. (2023). Enhanced CO_2 reactive capture and conversion using aminothiolate ligand–metal interface. *J. Am. Chem. Soc.* 145 (48): 26038–26051.

4

Advanced In Situ Characterization Techniques for Direct Observation of Gas-involved Electrochemical Reactions

Yue Zhai[#], Jiamin Zhu[#], ShuHui Li[#], Luohua Liu[#], Li An[*]

State Key Laboratory of Applied Organic Chemistry, Frontiers Science Center for Rare Isotopes, Key Laboratory of Nonferrous Metal Chemistry and Resources Utilization of Gansu Province, College of Chemistry and Chemical Engineering, Lanzhou University, Lanzhou, China

[*]Corresponding author: anli@lzu.edu.cn
[#]These authors contributed equally to this work.

4.1 Introduction

This study delves into advanced in situ characterization techniques, emphasizing their applications in catalysis and energy storage systems. In situ infrared spectroscopy enables the direct observation of functional group dynamics and adsorbed intermediates during reactions, exemplified by CO_2 reduction processes on catalytic surfaces. Techniques like EQCM provide real-time insights into ion transport, mass changes, and charge storage dynamics in supercapacitors, linking material structure to EC performance.

In situ XRD unravels structural evolution and phase transitions in battery electrodes, crucial for understanding active materials in lithium–sulfur and zinc ion systems. Similarly, in situ Raman spectroscopy captures dynamic structural changes and active intermediate formation during oxygen evolution reactions, with enhanced sensitivity achieved through surface-enhanced Raman techniques. These insights are vital for tuning catalytic and material performance in energy applications.

Emerging methods, such as DEMS, further enhance mechanistic understanding by quantifying gaseous byproducts during catalytic and EC processes. Collectively, these state-of-the-art techniques provide a comprehensive framework for studying material transformations and reaction mechanisms, enabling breakthroughs in sustainable energy technologies and catalysis.

Electrochemical Energy Conversion via Small Molecule Catalysis, First Edition.
Edited by Zhicheng Zhang and Yuanmiao Sun.

4.2 In Situ Infrared Technique

In situ infrared spectroscopy is an analytical method used to determine the changes in a sample or reaction system with time, temperature, pressure, and environmental changes and to study the chemical and physical properties of materials. It directly performs infrared analysis of the reaction products by recording the absorption or transmission spectrum generated when the material interacts with infrared light. This technique can reveal details at the molecular level, such as the presence of functional groups and changes in molecular structure. In spectral analysis, when infrared light irradiates the sample, it will be absorbed by specific chemical bonds, causing a change in the molecular vibration energy level and forming a characteristic absorption peak. Figure 4.1 shows infrared spectroscopic instrument.

In situ infrared technology allows direct monitoring of microscopic changes in the sample or reaction system, including structural changes in functional groups, under actual reaction conditions. This direct measurement method can simulate the experimental process more accurately, which is particularly important for the analysis of the reaction mechanism and the catalytic principle of the catalyst. In situ infrared analysis is widely used in the field of catalyst characterization (such as adsorption state, surface acidity of solid, and active center), reaction kinetics, and polymer reaction, crystallization, solidification kinetics, and thermal stability. This technology is not only applicable to the field of chemistry, but also widely used in materials science, environmental science, and resource science and technology. It is often used in the study of the catalytic mechanisms of heterogeneous reactions and the phase transition of substances caused by temperature change.

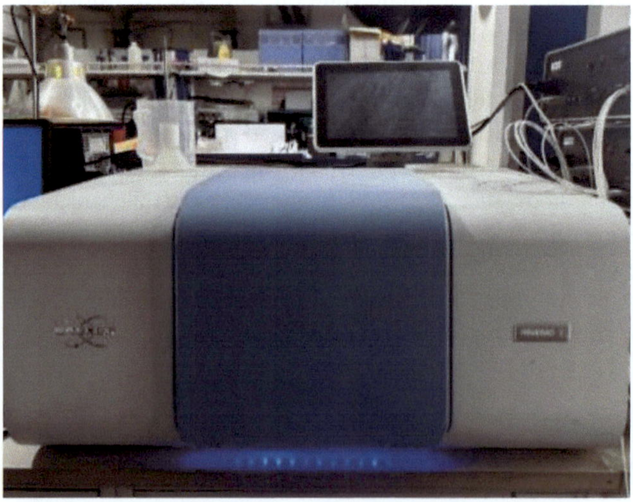

Figure 4.1 Infrared spectroscopic instrument.

4.2.1 Example Study

It is suitable for the study of surface adhesion dynamics of photocatalytic catalysts. The signal of catalyzed intermediates was observed and the catalytic process path was studied. It explores the composition of materials as they are heated, pressurized, and exposed to different atmospheres. For example, in catalyst research, researchers use in situ infrared technology to monitor changes in intermediates and reactants during catalytic reactions to understand the catalytic mechanism. For example, in the study of catalytic conversion of carbon dioxide, the adsorption state of carbon dioxide on the catalyst surface and the formation of intermediates in the reaction process can be observed by in situ infrared technology.

4.2.2 In Situ Characterization of Catalyst Reaction Mechanism

Research team and achievements: The DRM photocatalyst $Rh/CeXWO_3$ designed by Professor Xu Dongsheng's team uses photothermal and photoelectric processes to overcome the thermodynamic limitations under conventional conditions. The researchers propose that photoinduced metal-to-metal charge transfer plays a crucial role in DRM reactions, i.e., the induced REDOX cycle between Ce and W species can reduce the activation energy. In situ irradiated X-ray photoelectron spectroscopy (ISIXPS) reveals a photoinduced metal-to-metal charge transfer process, which further reduces the activation energy barrier for DRM reactions.

Photocatalyst performance: The catalyst maintains the stability and coking resistance of the light-driven DRM without the need for external heating and demonstrates a high photochemical energy efficiency of 4.65% at moderate light intensity. *Mechanism exploration*: Photoinduced metal-to-metal charge transfer plays a crucial role in reducing the activation energy of DRM reactions. ISIXPS analysis revealed the photoinduced metal-to-metal charge transfer process, further reducing the activation energy barrier of DRM reactions.

Quantum mechanical studies show that the high oxygen mobility of Ce_XWO_3 leads to the elimination of a large number of carbon substances deposited during the reaction process, which provides the possibility of energy-saving industrial processes for DRM. *Catalyst design implications*: This design strategy paves the way for the development of low-temperature catalysts for DRM. *Test results – Test setup and conditions*: In situ diffuse Fourier transform spectroscopy (DRIFTS) was measured using a Fourier transform infrared (FT-IR) spectrometer equipped with a reaction chamber. *Sample treatment*: The Rh/Ce_XWO_3 sample was ground and diluted with KBr to 5 wt% and then treated in an Ar atmosphere at 300 °C to remove surface adsorbent.

The results show that the active species and electronic states of Rh/Ce_XWO_3 change under light irradiation, which helps to understand the DRM reaction pathway. The study revealed the active species and electronic state changes of Rh/Ce_XWO_3 in DRM reactions and proved that the REDOX cycle between Ce and W atoms promotes the catalytic performance, providing important information for understanding and optimizing the DRM pathway of Rh/Ce_XWO_3 catalysts.

4.2.3 In Situ Infrared Spectrum Analysis

In situ IR spectra of Rh/CeXWO$_3$ were measured under the conditions of He flow and CH_4/CO_2 mixture after CO_2 saturation adsorption. Carbonate-like species, including single-toothed carbonates (mCO_3^{2-}) and double-toothed carbonates (bCO_3^{2-}), as well as bicarbonate (HCO_3^-), were observed, and these species adsorbed to the OH— groups of the catalyst. Above 100 °C, HCO_3^- is completely desorbed, while mCO_3^{2-} and bCO_3^{2-} remain on the surface even at 300 °C, indicating that they are the main active substances as shown in Figure 4.2.

4.2.4 In Situ FT-IR Characterization of the Electrode Reaction Process of Calcium Ion Batteries

The team of Tang Yongbing, Shenzhen Institute of Advanced Technology, Chinese Academy of Sciences, developed a new type of calcium-based dual-ion battery, which uses 3.5 m Ca(FSI)$_2$ salt completely dissolved in carbonate solvent as the electrolyte, using environmentally friendly, low-cost 3,4,9,10 perylene tetracarboxylic dianhydride (PTCDA) as the negative electrode and graphite as the positive electrode.

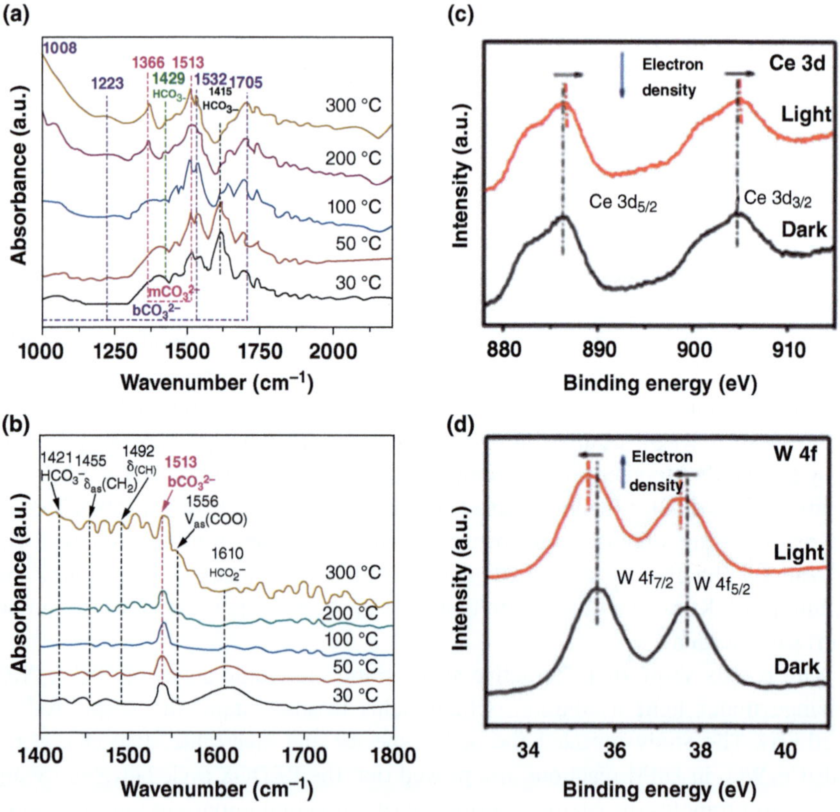

Figure 4.2 In situ infrared spectroscopy. (a) He flow after CO_2 saturated adsorption and in (b) CH_4/CO_2 mixture under different temperature. ISI-XPS spectra of Ce 3d (c) and W 4f (d) of Rh/Ce$_x$WO$_3$ with and without illumination.

Research team and findings: The team of Professor He Chi from Xi'an Jiaotong University proposed that by regulating the metal carrier electron interaction in the Pt_1CuO monoatomic catalyst, its activity in acetone oxidation can be significantly improved. *Electron interaction and catalytic performance*: Studies have shown that the strength of electron interaction on metal support positively correlates with catalytic performance. This interaction enhances the adsorption and activation of reactants by promoting charge redistribution and regulating the d-band structure of the atomic Pt site. *Optimization of the catalytic process*: It was found that positively charged Pt atoms can effectively activate acetone at low temperatures, while stretched CuO bonds promote the activation of lattice oxygen and further participate in the oxidation process.

Catalyst performance: The Pt_1CuO monoatomic catalyst with a unified bonding mode exhibits the strongest metal-supported electron interaction and can completely convert 800 ppm acetone to CO_2 at 210 °C, showing a low activation energy (54.32 kJ mol^{-1}). *Experimental methods*: In situ infrared spectrum data of acetone oxidation were collected using a Tensor 37 infrared spectrometer with MCT detector. Samples were pretreated and tested in an in situ reaction cell covered by KBr window and gas flow control system. The surface mechanism of the acetone oxidation reaction was revealed by in situ infrared spectroscopy, which showed that the electron interaction of the metal carrier had an important effect on the catalytic process.

Analysis of test results: characterization of PTCDA as a calcium ion battery negative electrode: (a) battery devices for in situ FT-IR testing; (b) charge and discharge curves of PTCDA| graphite positive battery during in situ test; and (c) in situ FT-IR results of PTCDA negative electrode.

Figure 4.3a shows the battery device diagram of the in situ FT-IR test, and Figure 4.3b shows charge and discharge curves of the PTCDA| graphite cathode battery during the in situ test. To clearly show the spectral changes, the background and peaks of the original PTCDA are subtracted, as shown in Figure 4.3c. During the charging process, the transmission of carbonyl group (C=O) in PTCDA gradually became positive at 1700 cm^{-1} and negative at 1306 cm^{-1} (CO's stretching mode)

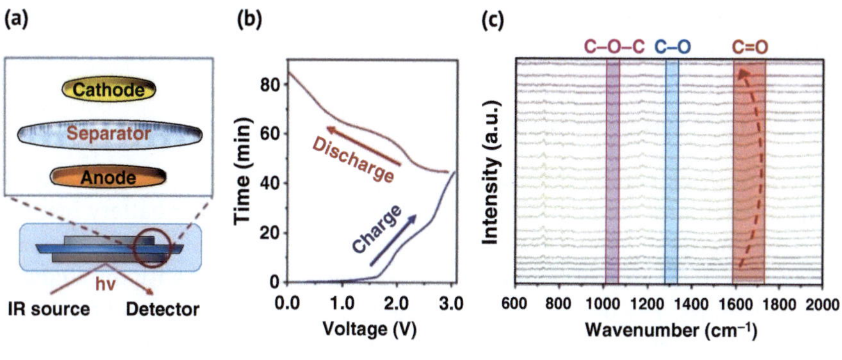

Figure 4.3 (a) Cell configuration of in situ FT-IR setup. (b) Charge/discharge curves of PTCDA || graphite cathode cell during in situ test. (c) In situ FT-IR results for PTCDA anode.

and 1029 cm^{-1} (COC's stretching mode). This indicates the EC conversion of C=O to CO after Ca^{2+} insertion. When discharged, CO and C=O can be restored to the initial state, confirming the reversible EC reaction of PTCDA during Ca^{2+} insertion and deem bedding.

Mechanism analysis: In situ infrared spectroscopy test showed that the acetone oxidation process of the Pt_1CuO catalyst was mainly carried out by the Mars–van Krevelen mechanism, while the process on the Pt_1CuO catalyst was mainly carried out by the Langmuir–Hinshelwood mechanism, which was affected by the adsorption of oxygen substances.

Conclusion: The reversible intercalation process of the anion (FSI) in graphite layers and the reversible EC reaction of PTCDA in Ca^{2+} insertion and deintercalation were demonstrated by in situ XRD and in situ FT-IR characterization. A novel calcium-based dual-ion battery was successfully constructed by combining high-concentration electrolyte, environmentally friendly graphite positive electrode, and low-cost PTCDA organic negative electrode. This research is of great significance for the development of more economical and high-performance calcium ion batteries [1, 2].

4.3 Electrochemical Quartz Crystal Microbalance

In recent years, advanced characterization techniques have greatly promoted the study of EC phenomena and mechanisms in batteries. Of all the emerging technologies, EQCM is a relatively inexpensive and nondestructive characterization method that can provide in situ information on mass changes and reflect the structural evolution of the battery electrode body and surface. In addition, the sample preparation of EQCM is quite simple, and the sample can simulate the behavior of real electrodes to a large extent. Therefore, EQCM is a powerful tool for studying EC phenomena and mechanisms in batteries.

EQCM provides real-time EC and gravimetric measurement information by combining conventional quartz crystal microbalances (QCM) with EC technology in a solid/liquid system. QCM is an ultrasensitive weight detector with a sensitivity of nanograms, which works on the principle of the inverse piezoelectric effect of the quartz crystal and the resonant frequency change due to the change in the sample mass.

4.3.1 EQCM System Diagram

There are three core components in a typical EQCM setup: a quartz crystal chip, an oscillator, and an EC workstation (potentiostat). For battery research, electrode material can be coated onto the Au layer. Due to the occurrence of EC process, the resonant frequency of the electrode changes with the change in electrode mass. The oscillator records the change in frequency and the impedance of the quartz crystal. EC workstations are used to perform EC tests such as cyclic voltammetry (CV) and electrostatic charge–discharge, as shown in Figure 4.4 [3–6].

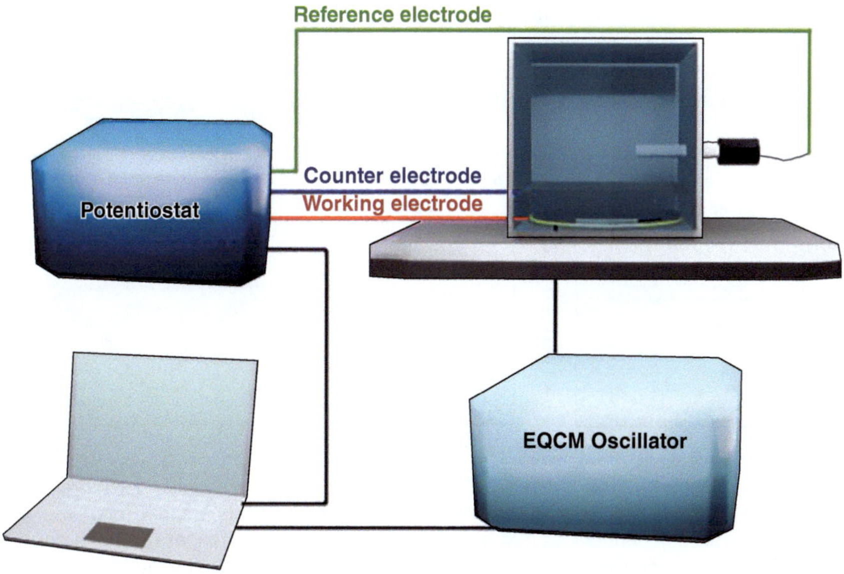

Figure 4.4 Schematic diagram of EQCM system.

4.3.2 Development History of EQCM

The most important milestone in the development of EQCM was the advent of the Sauerbrey formula in 1959, which successfully correlated the frequency change of a quartz chip with the mass change of its surface. In 2009, EQCM was first used to monitor the flow of ions in porous materials. In 2011, the emergence of the EQCA model extended the application of EQCM to monitor the change in viscoelasticity of materials. In 2016, the combination of EQCA and EQCM-D greatly expanded the practical application range of EQCM [7–9].

4.3.3 Monitoring the Ion Transport Mode of Supercapacitors

The results of EQCM measurement of supercapacitors show that electrolyte concentration, ion size, electrode aperture, and surface functional groups have important effects on ion transport mode. In YP-17 microporous-activated carbon, when the concentration of CsCl aqueous solution increased from 0.025 to 0.1 M, the increase in the number of ions led to the enhancement of co-ion and counter-ion exchange effects, making ion exchange dominate charge storage in a wider voltage range. In terms of ion size, compared with NH^{4+} ions, the ion-exchange region of the larger TAA^+ cation in YP-17 expands to the higher charge density region, possibly because the larger ion size leads to the decrease in desolvation energy, which is conducive to ion exchange. In terms of electrode aperture, in 2 M EMI-TFSI acetonitrile solution, the electrode mass of CDC-0.65 nm electrode hardly changed when positive voltage was applied, because TFSI-anions could not enter the hole, and the increased mass of anions adsorbed on the electrode surface was offset by the discharged acetonitrile molecules. The CDC-1 nm

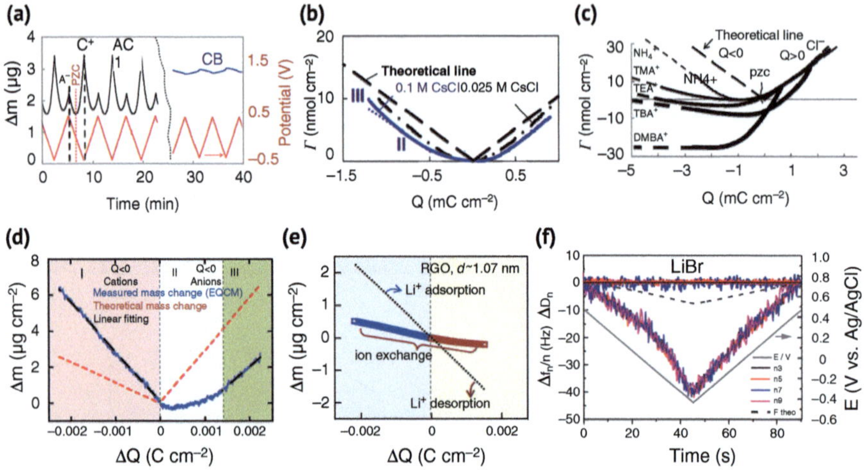

Figure 4.5 EQCM is used to monitor the ion transport mode of supercapacitors. (a) Potential–time dependencies (red line) and the related mass changes (Δm) of activated carbon (black line) and carbon black (blue line) electrodes. The dashed straight lines mark the mass change peaks of the cation and the anion. The position of the point of zero charges (PZC) is shown by the dotted red line (b) related molar amounts of ions (Γ) during charge–discharge vs. electrode charge (Q) (c) the amounts of adsorbed ions (Γ) as a function of the electrode charge density (d) electrode mass change (Δm) versus charge (ΔQ) during the polarization of CDC-1 nm in 2 M EMIM-TFSI+ acetonitrile (ACN) (e) the simultaneous mass change (Δm) of reduced graphene oxide (RGO) film during the cyclic voltammetry (CV) cycle. The blue and yellow backgrounds refer to the electrodes under negative and positive polarizations, respectively. The dotted line represents the theoretical mass change of pure solvated Li+ adsorptionanddesorptio (f) frequency (Δf) and dissipation changes (ΔD) as a function of time for third through ninth overtone orders. The theoretical frequency changes are denoted as black dashed lines.

electrode with larger pore size showed obvious mass change, and charge storage was mainly carried out by canal-cation exchange. In single-layer graphene, the mass does not change significantly under negative voltage, and the mass decreases linearly with the amount of charge under positive voltage. Combined with the presence of capacitance in CV curves under negative voltage, the EQCM results mentioned above indicate that ionic charge storage occurs through ion rearrangement when negative voltage is applied, while cations are expelled when positive voltage is applied. This different behavior is thought to be related to the strong π–π interaction between EMI^+ and the graphene surface, as shown in Figure 4.5 [10, 11].

4.3.4 Monitor the Ion Solvation Number of Supercapacitors

Using 2 M EMI-TFSI acetonitrile solution as an electrolyte, the solvation number of EMI^+ decreased from 3.7 to 1.6 when the pore size of CDC decreased from 1 to 0.65 nm. Similar to the CDC porous electrode, by adjusting the layer spacing of two-dimensional (2D) graphene, it is found that the solvation number decreases with the decrease in the layer spacing, and the solvation number of Li^+ in 1 M

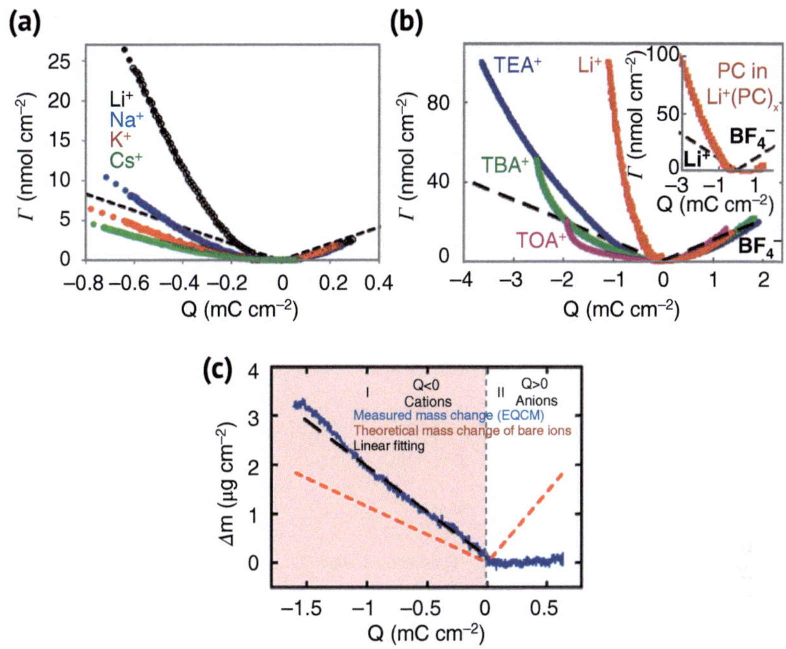

Figure 4.6 EQCM is used to monitor the ion solvation number of supercapacitors. (a) $\Delta\Gamma$ vs electrode charge (Q) obtained with BP-880 carbon in 0.025 M solutions of alkaline metal chlorides (b) amount of cations and anions ($\Delta\Gamma$) as a function of the charge density (Q) for the different salts dissolved in propylene carbonate (PC). The inset compares the number of solvent molecules (the solid red curve) (c) electrode mass change (Δm) vs. charge (Q) (solid blue lines) during the polarization of CDC-0.65 nm in 2 M EMIM-TFSI + ACN. The red dashed lines are the theoretical mass change of neat ions calculated from Faraday's law. The black dashed line shows the linear fitting of measured mass change.

Li_2SO_4 aqueous solution at ~0.4 nm is 1.4. These results support the view that partial desolvation occurs when cations enter small pore sizes, and the degree of desolvation increases with smaller pore sizes. However, in both alkali metal cations and halide anions, the solvation number of BP-880 (containing 16 nm macropores) is smaller than that of microporous carbon YP-17 (with an average pore size of about 1 nm). This phenomenon is explained by the fact that the water molecules around the ions are more easily replaced and conducive to desolvation under the condition of weak confinement in mesoporous carbon, as shown in Figure 4.6 [8, 12, 13].

4.3.5 Combination of EQCM and Other Detection Methods

Both charge and mass variables can be obtained in EQCM testing; however, the electrolyte is usually composed of cations, anions, and solvent molecules, making the quantitative characterization of ion transport difficult. Undoubtedly, the current qualitative results have greatly helped researchers to understand the ion transport behavior in supercapacitors, but further elucidating it requires more accurate

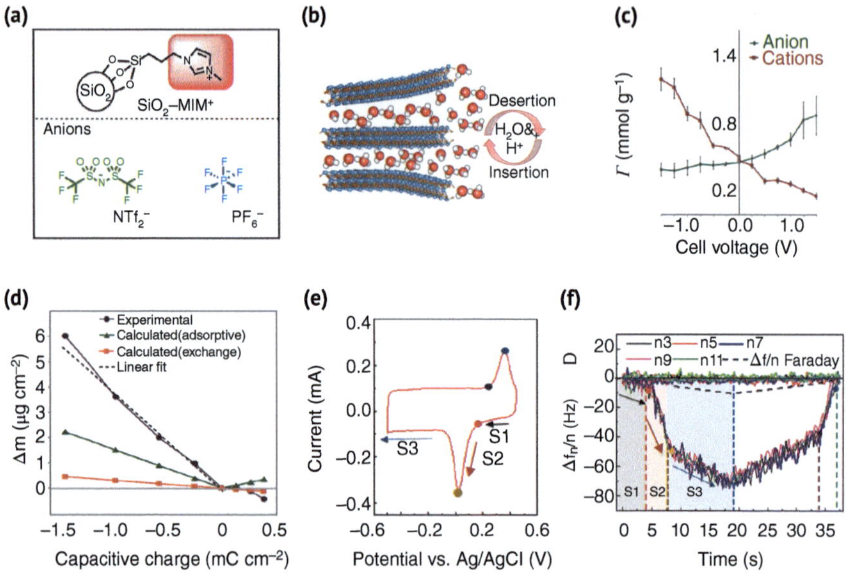

Figure 4.7 Combination of EQCM and other detection methods. (a) Structure of anions, and silica nanoparticle-grafted cation (b) scheme of MXene electrode interaction with 1 M H_2SO_4 aqueous electrode (c) in-pore ion populations (Γ) for YP-50F intherange −1.5 V to +1.5 V, deconvoluted from the resonance intensities of NMR (d) experimental and calculated electrode mass changes (Δm) for a YP-50 F electrode (e) CV recorded of $Ti_3C_2T_x$ in water-in-salt electrolytes during EQCM measurements (f) frequency (Δf) and dissipation (D) changes as a function of time obtained in saturated LiBr for MXene electrode.

quantitative characterization. Attempts to achieve this can be broadly divided into two directions (as shown in Figure 4.7) [14–16]:

1. Changes to the test object itself, such as using an ionic liquid containing only two ions or altering the electrode/electrolyte to ensure only a single ion participates in ion transport.
2. The combination of multiple experimental and simulation analysis methods, such as EQCM and in situ NMR, EQCM, and molecular dynamics (MD) simulation.

4.3.6 Summary and Prospect

Although EQCM technology for supercapacitors has been rapidly developed, its application in supercapacitors is still limited by the following points: in systems containing three or more ions, since there are only two types of data, the information obtained from EQCM data using current analytical methods is still qualitative, hindering the in-depth study of supercapacitors. Current approaches to EQCM testing and data analysis require further innovation to gather more information.

The combination of EQCM with other simulation methods can provide more molecular-level information. However, researchers hope for a more comprehensive and quantitative combination between experiments and simulations. Among different simulation methods, MD simulation can capture detailed information such as mass change, ion exchange, and distribution. Therefore, the in-depth combination of EQCM and MD simulation can greatly improve the understanding of the energy storage process of supercapacitors and will be an important development direction in the future.

Although the combination of EQCM and different experimental techniques has achieved remarkable results, various test methods cannot be carried out in the same system, which will affect the quality of the obtained data, and the synchronous measurement of EQCM and other experimental techniques in the same system would greatly improve the accuracy of the experiment.

4.4 X-ray Powder Diffraction

XRD is a primary method utilized for investigating the phase and crystal structure of materials. When materials are subjected to XRD analysis, they undergo irradiation, resulting in varying degrees of diffraction phenomena that provide valuable information regarding the composition, crystal form, intramolecular bonding mode, molecular configuration, conformation, and other properties of the material. These factors collectively determine the specific diffraction patterns observed. In EC processes, electrode materials often experience phase transitions, morphological changes, or volume expansion/contraction which directly impact battery performance. In situ XRD techniques offer direct evidence of these changes, including structural modifications and morphology as well as volume alterations in electrodes, interface reactions, and solid electrolyte interphase (SEI) layer formation, along with establishing relationships between EC properties and structure. This enables a deeper understanding of the complex mechanisms involved in EC processes [17].

Based on available literature, the first in situ XRD study was conducted by Rao et al., in 1978 to observe Li^+ intercalation into the Li/TiSz system which was referred to as "dynamic X-ray diffraction" [18]. Ex situ XRD involves analyzing disassembled batteries using scattered X-rays produced by periodic arrangement of atoms inside crystals to qualitatively analyze crystalline phases, lattice parameters, and crystal defects. The obtained diffraction pattern is as unique as a human fingerprint. In situ XRD operates similarly to ex situ XRD but detects electrode material during testing without interruption, as shown in Figure 4.8.

Rechargeable lithium–sulfur (Li–S) batteries show great potential for energy storage due to higher energy density than LIBs. In situ XRD is often employed to understand the reaction mechanism and the influence of electrolyte on the dissolution of sulfur and polysulfides. Chen et al. constructed a twin-borne ultrathin 2D graphene-based mesoporous $SnO_2/SnSe_2$ hybrid as a polysulfide immobilizer and lithium regulator for Li–S chemistry. The in situ XRD test was implemented to gain an in-depth understanding of the phase transformation of $SnSe_2$, by using the G-m$SnO_2/SnSe_2$ electrode, and its results manifest the multistep conversion process of sulfur, which is accompanied with the intercalations of Li_1SnSe_2 to Li_3SnSe_2, as shown in Figure 4.9 [19].

Nowadays, aqueous system batteries are attracting extensive attention for energy storage, which can benefit from their high safety, low cost, convenient

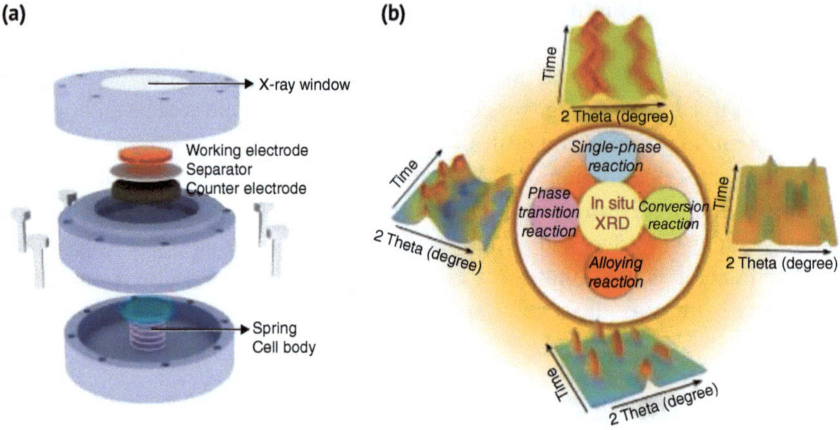

Figure 4.8 (a) Schematic diagram of in situ cell. (b) Schematic diagram of in situ XRD patterns about four typical reaction mechanisms: single-phase reaction, phase-transition reaction, alloying reaction, and conversion reaction.

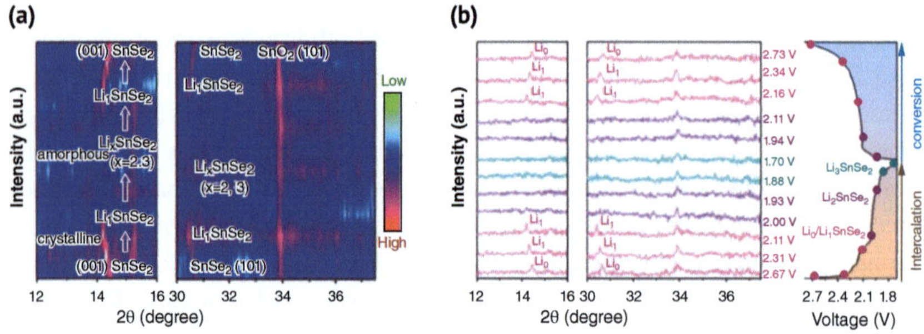

Figure 4.9 (a, b) In situ XRD contour plots of a pristine G-mSnO_2/$SnSe_2$ electrode in the first discharge/charge process within the voltage window of 1.7–2.8 V.

manufacturing, and environmentally friendly electrolyte. Nevertheless, unlike the nonaqueous system, the choice of electrode materials in aqueous electrolyte is seriously limited by several problems, such as the unstable EC window of water, instability of electrode materials, and undesired side reactions. In situ XRD technique is increasingly applied to explore the storage mechanism in aqueous system.

Ma et al. employed transmission-mode operando XRD to accurately document the real-time phasic evolution of the layered cathode and synchronized strain change induced by Zn deposition, which present a feasible solution for health monitoring by the miniaturized electronics. To further probe the real-time phasic transition upon the dynamic discharge–charge cycle, they conducted an operando XRD test for the Zn || V_2O_5 cells equipped with various electrolytes. In this test, they compared the influence of quasi-solid-state PVA-gel and PSC-gel electrolytes with lower mobile H_2O molecule contents with a dilute liquid electrolyte (2 M $ZnCl_2$) on the cathode lattice breathing at 0 °C. Based on the peak shifts observed from the operando XRD patterns, the phase transitions during the intercalation process can be identified by the changes in the lattice parameters at different voltage states, as shown in Figure 4.10 [20].

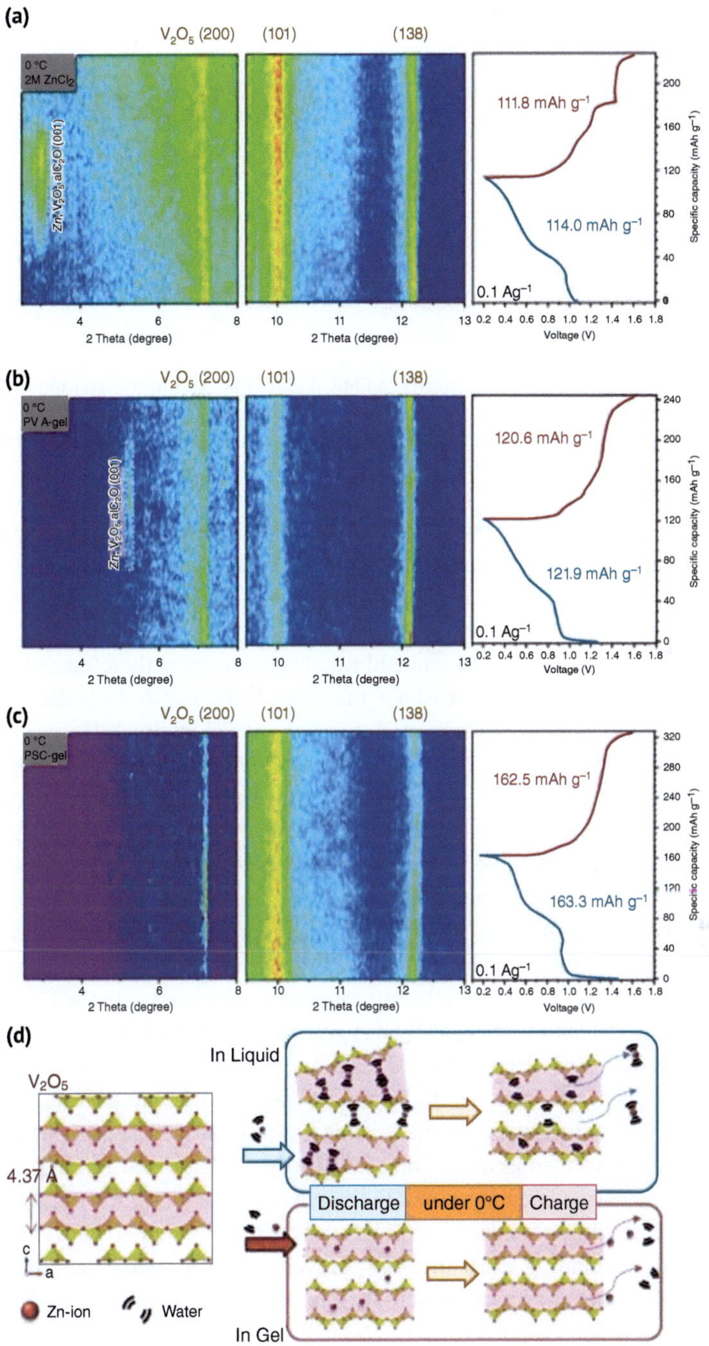

Figure 4.10 Transmission-mode operando XRD tracking of the V_2O_5 cathode with (a) 2 M $ZnCl_2$ electrolyte, (b) PVA gel, and (c) PSC gel for the initial cycle at 0 °C with corresponding galvanostatic charge–discharge curves at 0.1 A g^{-1}. (d) Schematic illustration of the dynamic structural evolution for the V_2O_5 cathode in both the liquid and PSC electrolytes.

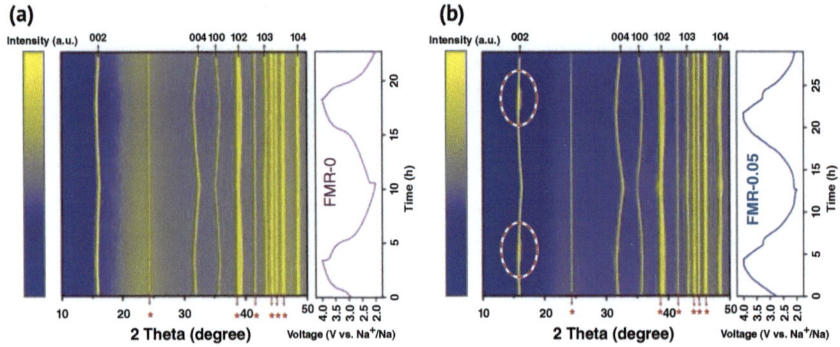

Figure 4.11 (a, b) In situ XRD patterns of FMR-0 and FMR-0.05 in the 2θ range of 10–50° with the charge/discharge curves during the first and second cycles. Asterisks denote diffraction peaks of inactive impurities including Be and Al, which do not interfere with the evolution of the cathode material during the cycle.

A new type of battery known as metal-ion batteries promises better performance than existing batteries. In terms of energy storage, they could prove useful and eliminate some of the problems existing batteries face. During the actual charging and discharging process, the structure of cathode materials will undergo a series of evolutions with the releasing and embedding of sodium ions; therefore, further operando XRD characterization of cathode materials is the key to study the mechanism of cell structure evolution. Chen et al. carried out EC in situ XRD experiments at lower current densities (<0.2 C) and obtained XRD patterns and complete in situ structure evolution maps every fixed time. Based on the in situ X-ray experimental structure, it can be concluded that the cathode material undergoes a classical simple solid solution reaction in the voltage range of 2–4 V, as shown in Figure 4.11 [21].

EC co-reduction of CO_2 and nitrogenous species (N_2, NO, NO^{3-}, NO^{2-}, etc.) is regarded as a green and sustainable urea synthesis process under ambient conditions, providing an alternative to the traditional urea synthesis route. However, owing to the high energy barrier between the intermediates associated with the C—N coupling reaction, it is crucial to strategically design catalysts that can enhance the activity and selectivity of the electrocatalytic reaction. Wu's study focused on the dynamic reconstruction of ultrathin 2D Bi_2Se_3 nanosheets into Bi nanosheets through EC CO_2 and NO^{3-} reduction reactions. In this work, in situ XRD analysis along with Raman spectra were employed to monitor the dynamic reconstitution process from Bi_2Se_3 nanosheets to Bi catalysts and their active species formation during reduction processes. The study revealed how Bi nanosheet reconstruction contributes to urea synthesis by combining experimental observations with theoretical analysis while providing insights into structure–activity interactions, as shown in Figure 4.12 [22].

4.5 In Situ Differential Electrochemical Mass Spectrometer

EC methods are widely employed in sensing, reaction monitoring, and mechanism studies across various fields such as chemistry, biology, energy, and environment due to their inherent advantages of high sensitivity, accuracy, and cost-effectiveness.

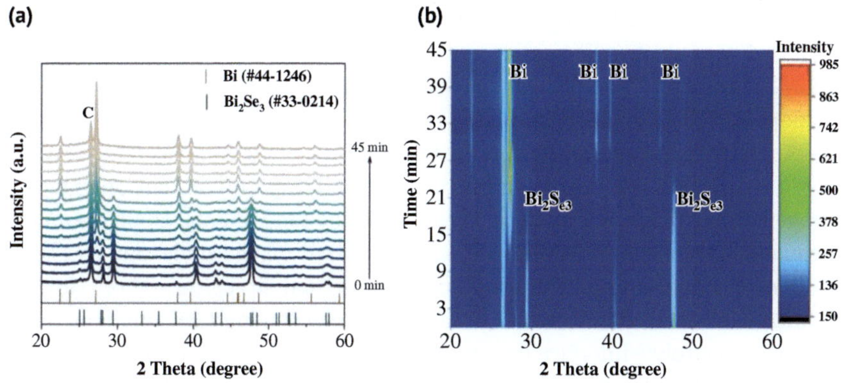

Figure 4.12 (a, b) In situ XRD patterns at −0.4 V for various time points.

However, traditional EC techniques, including voltammetry, potentiometry, and coulometry, solely relying on the measurement of potential, current, and conductance fail to provide comprehensive compositional and structural information at EC interfaces under reaction conditions. Mass spectrometry (MS), another powerful analytical technique characterized by rapid response time, exceptional sensitivity, and excellent selectivity, is also utilized for reaction monitoring and mechanism studies. On the one hand, MS enables precise determination of molecular weight for low-abundance intermediates and products; on the other hand, tandem MS can furnish detailed structural insights into analytes. These two advantages compensate for the limitations of traditional EC methods, as shown in Figure 4.13 [23].

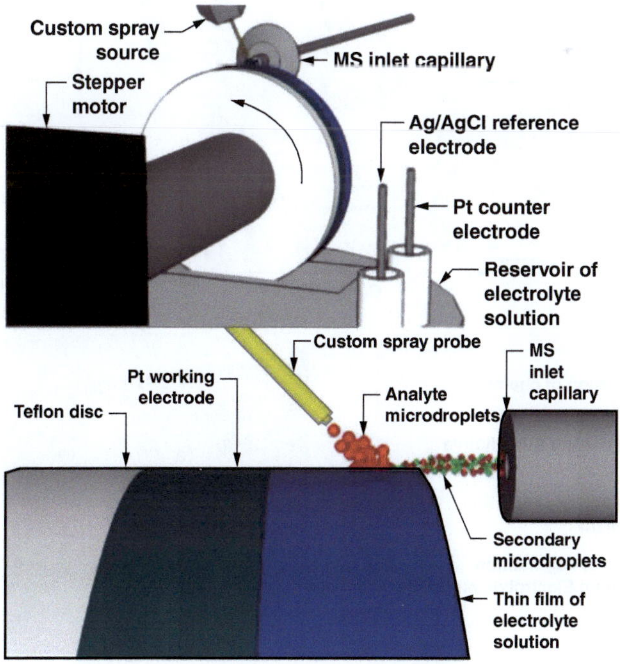

Figure 4.13 Apparatus used to detect electrochemical reaction intermediates, employing a rotating waterwheel electrode with DESI-MS.

In 1971, the in situ EC-MS was realized for the first time by a porous electrode interfaced with electron ionization mass spectrometry (EI-MS) [24]. Subsequently, a DEMS method was established to monitor the gaseous and volatile products of EC reactions in 1984 [25]. Electrocatalysis is a kind of catalysis that controls and accelerates the electron transfer reaction at the interface of electrode and electrolyte, which plays a more and more important role in water splitting, carbon dioxide reduction, nitrogen cycle, fuel cells, and lithium batteries. The fundamental challenge in electrocatalysis is the rational design of catalysts based on the understanding of the reaction pathways and mechanisms to improve catalytic performance. However, traditional ex situ characterization cannot provide enough information on the electrode–electrolyte interface under working conditions. The most common in situ EC-MS applied in electrocatalysis is DEMS, as shown in Figure 4.14 [26].

Comprehensively understanding the gas-generating mechanisms in various cathodes, anodes, and electrolytes is key to achieving safe, high-performance battery operation. As a suitable analytical technique for real-time detection of the various gaseous reaction products of lithium-ion batteries (LIBs), DEMS offers two distinct advantages: real-time EC analysis that can facilitate gas observation and highly accurate quantitative gas measurements, because the ionization-based gas molecule detection permits quantitative analysis in moles [26].

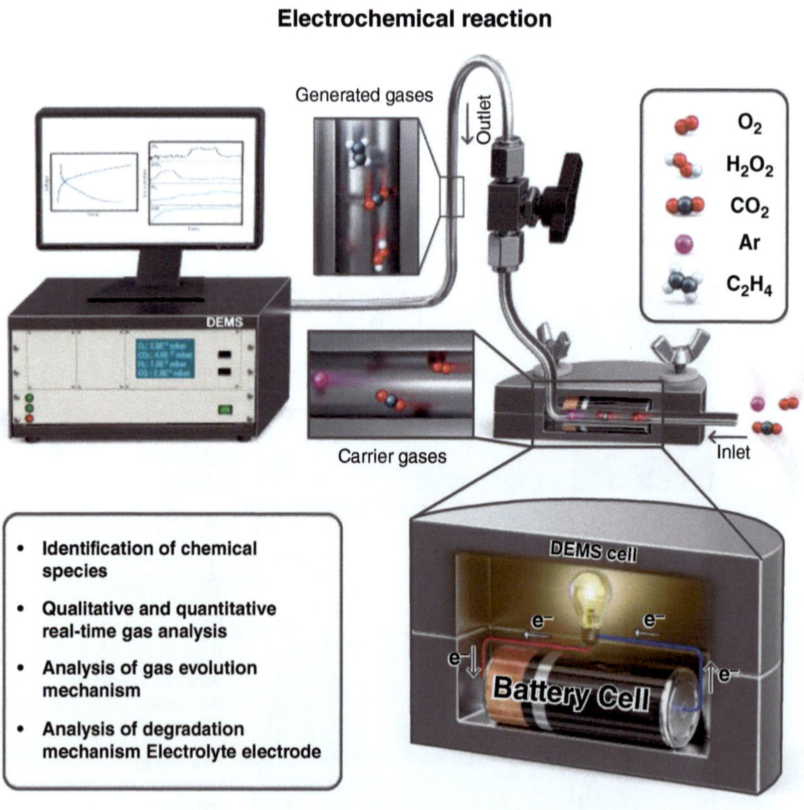

Figure 4.14 General features of a DEMS apparatus used for analyzing ARBs: inlet/outlet gas lines, a DEMS cell, a gas-detecting spectrometer, and a data-processing computer. *Source:* [26] / John Wiley & Sons.

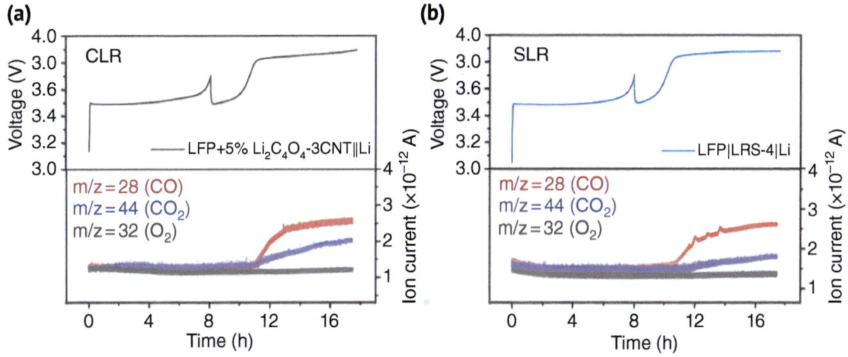

Figure 4.15 Comparison of the electrochemical performance of LFP||Gr full cells with CLR (LFP + 5% $Li_2C_4O_4$−3CNT||Gr) and SLR (LFP|LRS-4|Gr) strategies. (a) and (b) In situ DEMS tests of the LFP||Li half cells. *Source:* [27] / The Royal Society of Chemistry / CC BY 3.0

Wang et al. employed DEMS measurements to investigate the gas evolution during the decomposition of $Li_2C_4O_4$–3CNT, which presented direct evidence to support the long-held hypothesis, as shown in Figure 4.15 [27].

Min et al. used DEMS to investigate gas evolution of $LiCoO_2$/Li during first charge process at 50 °C as well, and a high reactivity between LiDFOB-PC and $LiCoO_2$ is indicated by significant generation of CO_2, as shown in Figure 4.16 [28].

The design of stable and efficient OER catalyst is the key point of the long-term stable operation of seawater electrolysis. Shao et al. established a general design principle of OER electrodes for seawater electrolysis, a universal structure–activity-durability relationship of non-noble metal catalysts was elucidated at the (Ni, Fe)O(OH)@NiCoS NAs via the operando EC-DEMS analysis. The catalysts were carried out isotope-labeling by reacting in the 18O-labeled 1 M KOH aqueous solution, then washed with the 16O aqueous solution several times, and processed under 60 °C vacuum drying over 2 hours. Ultimately, polarization potentials were applied to (Ni, Fe)O(OH)@NiCoS NAs in the 16O-labeled 1 M KOH electrolyte during operando DEMS measurement for comparison with the results without isotope-labeling, and the proportion of 18O and 16O products can demonstrate the mechanism proportion of AEM and LOM. These results demonstrate that the

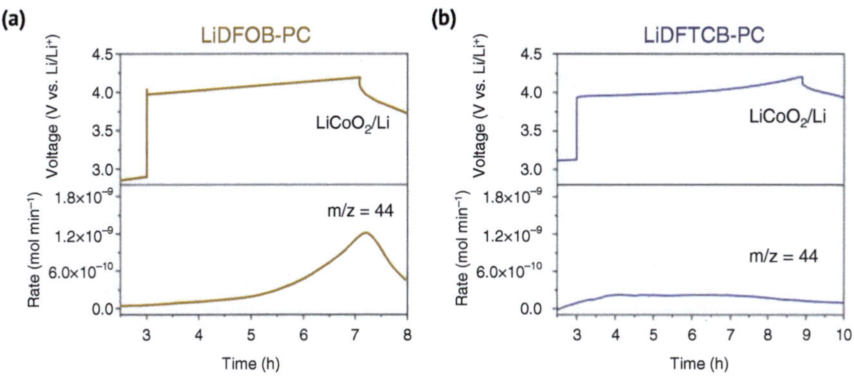

Figure 4.16 In situ DEMS signals of $LiCoO_2$/Li cells using (a) LiDFOB-PC and (b) LiDFTCB-PC at 50 °C.

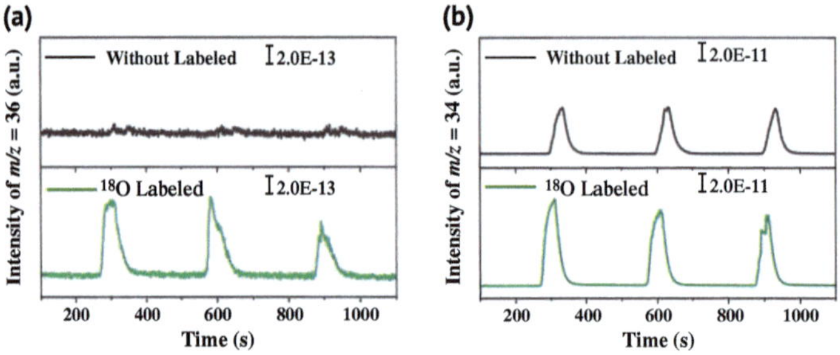

Figure 4.17 DEMS signals of the (a) $^{18}O^{18}O$ (m/z = 36) products and (b) $^{16}O^{18}O$ (m/z = 34) products of ^{18}O-labeled (Ni, Fe)O(OH)@NiCoS NAs during three times of LSV measurements at the potential range of 1.00–1.75 V vs. RHE (scan rate: 5 mV s^{-1}) in the 1 M KOH ^{16}O aqueous solution.

oxygen intermediates are generated both at the metal-bonded lattice defects and the surface of metal sites during the lattice-oxygen-participated mechanism (LOM) proceeds of (Ni, Fe)O(OH)@NiCoS NAs, as shown in Figure 4.17 [29].

4.6 In Situ Raman Spectroscopy

Raman was discovered by Indian scientist C.V. Raman in 1928, and it is a light scattering technique. Raman spectra technique studies the frequency of scattered light resulting from the interactions of molecules and light. In general, a certain amount of Raman peaks exists in a Raman spectrum. Every Raman peak usually corresponds to specific molecular bond vibrations, which can used to analyze the bond information of object. Raman spectral region ranges from 40 to 4000 cm^{-1}. The instrument consists of a light source, external optical path system, sample pool, monochromator, signal processing, and output system. Typical laser can be from a laser device with ultraviolet, visible, and wavelengths. Raman spectrum is a nondestructive analysis technique that provides information about chemical construction, phase, crystallinity, and molecular interaction. Therefore, simple instrument design and large wavelength range provide the feasibility of an application. However, the low sensitivity of Raman spectroscopy promotes the development of surface-enhanced Raman spectroscopy (SERS) and tip-enhanced Raman spectroscopy.

With the rapid development of characterization technology, the in situ Raman characterization technique has been used to reflect different reaction mechanisms, such as EC synthesis, EC catalytic processes, and chemical deposition processes. In situ Raman spectrum adds the design of detecting reactions based on ex situ Raman spectroscopy. It can detect the change of objects and reaction intermediates with the change in reaction conditions. EC in situ Raman spectroscopy is one of the fastest developed technology due to its convenient design. Moreover, in situ Raman spectroscopy has been studied for use in high-temperature and high-pressure reaction detection processes in recent years. EC in situ Raman spectroscopy equipment consists of a Raman spectrometer and an in situ EC Raman cell. The EC reaction occurs in the cell. Meanwhile, the change in object via Raman spectroscopy equipment can be detected. From the above combination, the Raman peak can be analyzed to get those reaction mechanisms. Nowadays, the research detecting reaction process by in situ Raman spectroscopy is very common. Especially applied in EC catalytic reactions such as OER, HER, CO_2RR, and ORR. This part will illustrate the application of in situ Raman spectroscopy following the EC reaction classification.

Because the characteristic peak of metal sites (Fe, Ni, Co, Cu, Ir, Ru, etc.) can be detected by Raman spectroscopy, Raman spectroscopy is one of the methods to study OER process at present. Regarding the restructuring process of the catalyst during the OER process, the Xi group studied the OER catalytic process using in situ Raman spectroscopy technology. Raman spectra of the CuS OER process revealed that CuS converted to $Cu(OH)_2$, and S was oxidized to oxysulfide within the OER voltage range [30]. They also found that a little doped Fe in NiO/NiS could alleviate the corrosion of sulfur to some extent. The peak of γ-FeOOH faded away as the potential increased, meaning that FeOOH is not scaled with catalytic activity [31]. At the same time, in situ Raman spectra were used to track the transformation during EC reconstruction. The half-width of the Ag peak in CeO_2-$CoS_{1.97}$ and $CoS_{1.97}$-CeO_2 broadened 50% and 30% compared with $CoS_{1.97}$, respectively, suggesting the smaller domain sizes and a higher degree of lattice disorder, as shown in Figure 4.18 [32].

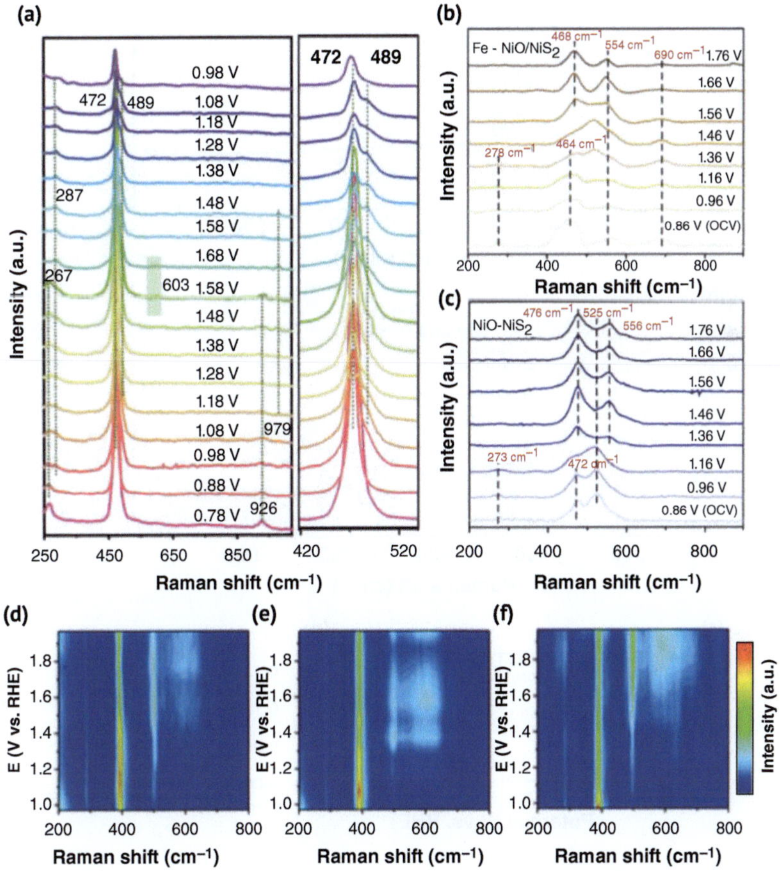

Figure 4.18 (a) In situ Raman spectrogram of CuS; potential-dependent in situ Raman spectra of (b) Fe-NiO/NiS_2 and (c) NiO/NiS; potential-dependent in situ Raman spectra of (d) $CoS_{1.97}$, (e) CeO_2-$CoS_{1.97}$, and (f) $CoS_{1.97}$-CeO_2 from 0.97 to 1.97 V (vs. RHE).

Zhang et al. performed the in situ Raman experiment in different catalytic environments for mechanistic investigations. They describe the phenomenon in which water is successively oxidized to form Mn—OH, Mn—O, and Mn—OOH species and finally produces oxygen. These results show that HOO* is the dominant species on the $Mn_{7.5}O_{10}Br_3$ surface and that $Mn_{7.5}O_{10}Br_3$ possesses better stability than $Mn_8O_{10}Cl_3$ and [33]. Sun et al. noticed that defects can promote the surface restructuring during OER electrocatalysis via quantification of Raman characteristic peaks [34]. They consider that the intensity ratio of the characteristic peaks at 528 and 457 cm^{-1} (I_{528}/I_{457}) can be used as an index to reflect the fraction of defective or disordered Ni(OH)x species. Prior to the onset of oxygen evolution, the value of I_{528}/I_{457} for both electrocatalysts increases gradually with increasing potential, showing increasing local conversion of crystalline surface Ni(OH)x to more defective or disordered Ni(OH)x species. The result showed the occurrence of more extensive surface restructuring due to the presence of cationic defects, as shown in Figure 4.19.

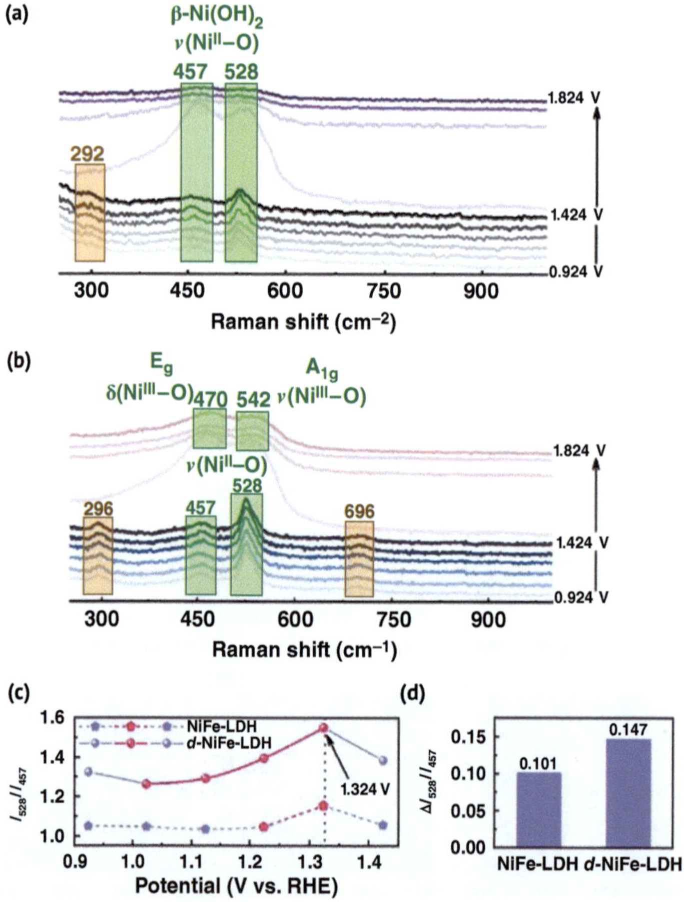

Figure 4.19 In situ Raman spectroscopy measurements of (a) the pristine NiFe-LDH and (b) d-NiFe-LDH in OER electrocatalysis. (c) The evolution of I_{528}/I_{457} vs. potential. (d) $\Delta I_{528}/I_{457}$ of NiFe-LDH and d-NiFe-LDH.

Zou group further researched the role of different types of defects in Co_3O_4 structural evolution and active motifs during the oxygen evolution reaction [35]. They further explored the effects of defects on the driving force and dynamics of structure reconstruction during the OER process using in situ Raman spectra. For *Co_3O_4—Vo*, the signal of Co—O E_g peak appears at a relatively lower potential (0.95 V vs. RHE) owing to the thermodynamic advantage for *OH adsorption induced by O vacancies. This implies that *Co_3O_4—VCo* has high OER process completeness, rather than staying in the first or second step (RDS of OER), which leads to the accumulation of intermediates and irreversible transformation, as shown in Figure 4.20.

Fu group combined in situ Raman spectroscopy and theoretical calculations to reveal the consistent Ce protection and the successful construction of a stable basic unit of OER-active Ce—O—Co [36], in which Ce protects the Co—O site and provides a stable and efficient unit for OER through Ce-induced electronic restructuring, as shown in Figure 4.21.

Furthermore, Zheng et al. confirmed the promoting effect of CeO_2 modification on the formation and transformation of intermediates in the OER process by in situ Raman. They proved that the Co^{3+} site at the interface is primarily responsible for the observed catalysis. The high-valence Co^{3+} is more conducive to the formation of Co—OOH intermediate, thereby enabling a quicker OER at a lower potential.

In situ Raman spectroscopy plays a key role in dynamic detection of catalytic processes. In recent years, the combination of surface-enhanced Raman spectroscopy

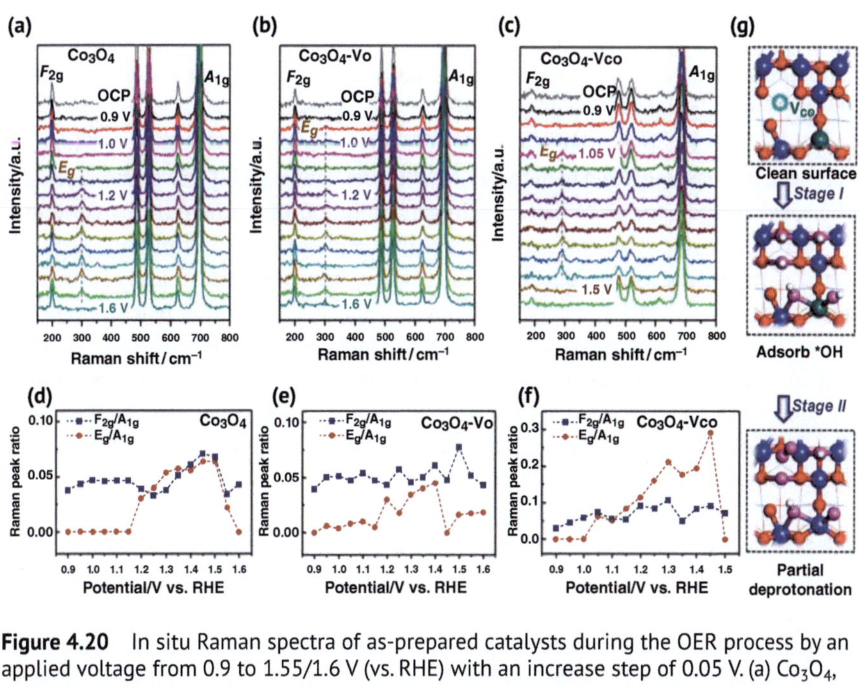

Figure 4.20 In situ Raman spectra of as-prepared catalysts during the OER process by an applied voltage from 0.9 to 1.55/1.6 V (vs. RHE) with an increase step of 0.05 V. (a) Co_3O_4, (b) Co_3O_4-V_O, and (c) Co_3O_4-V_{Co}. (d–f) Raman peak area ratio of F_{2g}/A_{1g} ($[CoO_4]/[CoO_6]$) and E_g/A_{1g}($[Co(OH)_6]/[CoO_6]$). (g) Scheme of adsorption and deprotonation on (110) surface of Co_3O_4-V_{Co}. The blue, cyan, red, pink, and white balls represent the Co^{3+}, Co^{2+}, lattice oxygen, adsorbed oxygen, and hydrogen atoms, respectively.

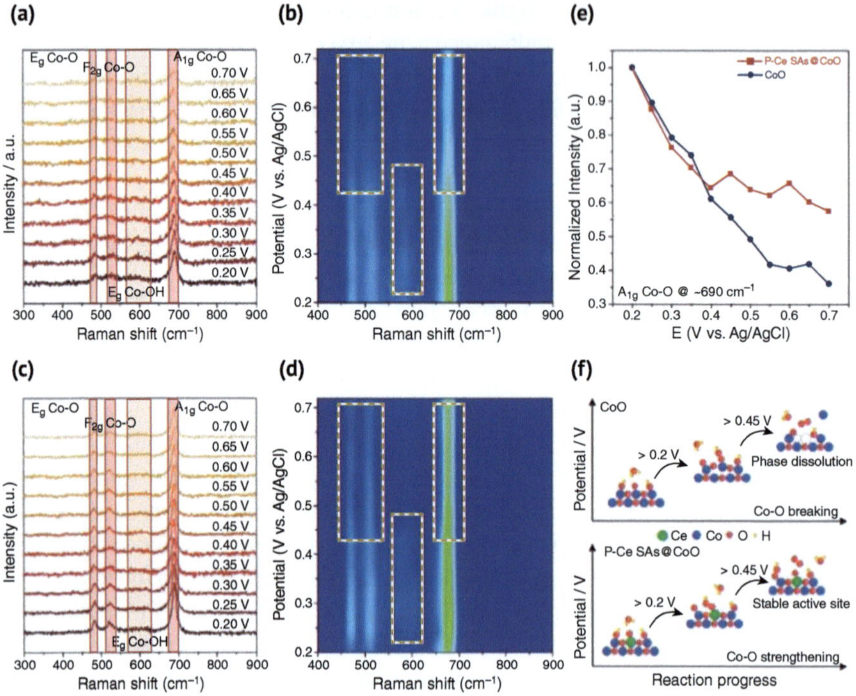

Figure 4.21 In situ Raman spectroscopy for investigating the OER mechanism of P-CeSAs@CoO. (a) In situ Raman spectrum and (b) the corresponding contour plot of CoO. (c) In situ Raman spectra and (d) the corresponding contour plot of P-CeSAs@CoO. (e) Normalized Raman peak at 690 cm^{-1}. (f) Schematic comparison of the surface evolution process of OER for P-Ce SAs@CoO and CoO.

has been increasingly used to detect H_2O molecule changes [37]. Li's group elucidated the mechanistic aspects of better hydrogen evolution of Ruthenium by combining in situ Raman spectroscopy and theoretical calculations. They simultaneously captured dynamic spectral evidence of Ru surfaces, interfacial water, and *H and *OH intermediates. Ru surfaces exist in different valence states in the reaction potential range, dissociating interfacial water differently and generating two distinct *H, resulting in different activities. Future development directions of Raman spectroscopy include more accurate signals, more convenience, and a wider range of applications. In addition, high space time, high resolution, and fast sampling Raman spectroscopy are also the main orientation of the scientist [22, 38, 39].

In situ fluorescence spectroscopy, a nondestructive analysis technology, can be used to study the electronic structure and fluorescence properties. In situ fluorescence spectroscopy can be measured with the materials in their reaction environment, so it is highly practical and operable. This section will elaborate the principle, application, and development of in situ fluorescence spectroscopy. The principle of in situ fluorescence spectroscopy focused on fundamental fluorescence phenomenon and measures principle of in situ fluorescence spectroscopy. Fluorescence is the light phenomenon emitted by substances after being excited by energy. When a substance is excited by light whose energy is greater than its ground state level, some electrons will be excited to a higher energy level and then quickly return from the excited state to the ground state [40]. At that time, the fluorescence produced and can be measured by a detector. The wavelength and intensity of the emitted light in the fluorescence spectra can be analyzed to study the electronic structure and chemical properties. Moreover, the measurement principle of in situ fluorescence spectrum will be elaborated as follows.

4.7 In Situ Fluorescence Spectrum

In situ fluorescence spectroscopy includes three processes [41]. Initially, the fluorescent probe is in direct contact with the sample, then the fluorescence is excited by the exciter to act on the sample, and finally the fluorescence change of the sample emission is detected. The measurement system of in situ fluorescence spectra includes a laser, a fluorescence probe, an optical system, and a photodetector. Lasers are used to excite fluorescence. The fluorescent probe is responsible for contacting and receiving the excited light. Optical systems are used to collect and transform fluorescence spectra. Photodetectors are used to measure the intensity and wavelength of a spectrum. In situ fluorescence spectroscopy is widely used, such as material science, environmental monitoring, life science, and medicine. When applied in the field of environmental monitoring, it can be used to monitor and assess the environment, such as the concentration of organics, metal ions, and hazardous substances in daily water. Identification and quantitative studies of fluorescent signals in human tissues and cells can be realized by in situ fluorescence spectroscopy. In situ fluorescence spectroscopy is also used to study the structure and transformation of biological macromolecules and the interaction of biological interfaces.

The application of in situ fluorescence spectra in material science has attracted more and more attention. For example, it can be used to study the band structure, bandgap width, and electronic structure of materials by in situ fluorescence spectroscopy. At the same time, in situ fluorescence spectroscopy can also be used to study the surface properties and structural changes of materials as well as the defects and impurities of materials. In situ fluorescence spectroscopy is widely used to detect reaction intermediates during electrochemical reactions (such as OER, HER, and ORR). However, research in this area remains limited. Zhang et al. [42] developed the technique of in situ fluorescence EC spectroscopy (FES) to monitor the $[CO_2]_i$ during CO_2RR. Time-dependent in situ fluorescence electrochemical spectroscopy (FES) is developed to quantitatively analyze the interfacial CO_2 transportation process during CO_2RR. Fluorescence probes show very high sensitivity to trace amounts of CO_2, high spatial resolution, fast time responses, and specific CO_2 molecular target recognition in complex environments, though they have never previously been used to study EC systems for CO_2RR. They used 8-hydroxypyrene-1,3,6-trisulfonic acid (HPTS) as a fluorescence probe to detect local CO_2 concentrations. The quantification of interfacial CO_2 transport behavior through in situ fluorescence EC spectroscopy further revealed the fundamental role of interfacial CO_2 transportation in determining the stability of CO_2 equilibrium concentration during EC reaction, as shown in Figure 4.22.

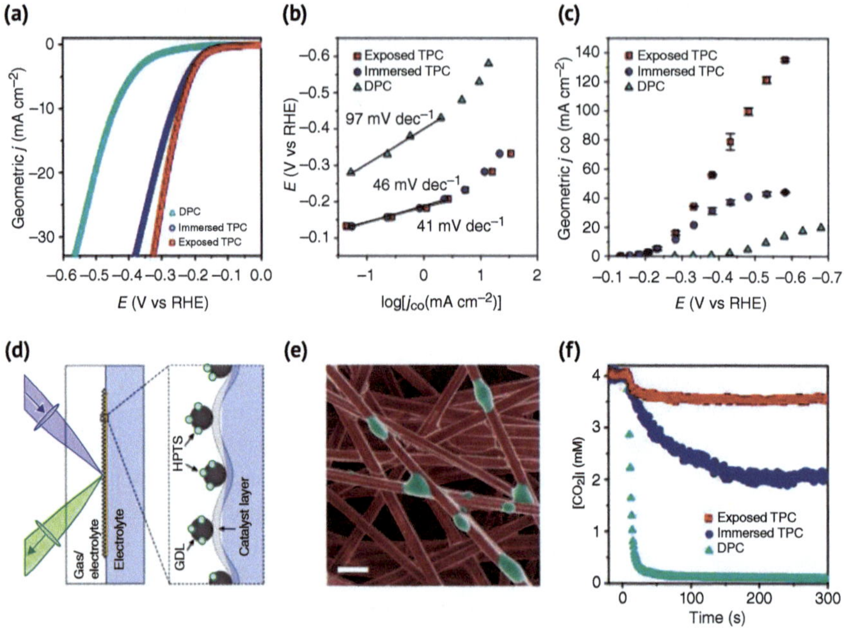

Figure 4.22 Electrochemical CO_2RR over different interfacial structures. (a) LSV of Au/C-P-0.5 electrodes in 1 M KOH for the TPC systems and in CO_2-saturated 1 M $KHCO_3$ for the DPC system. (b) CO production Tafel plot of Au/C-P-0.5 electrodes over exposed TPC and immersed TPC and DPC systems. (c) Geometric j_{CO} of Au/C-P-0.5 over the three systems vs. applied cathodic potentials. Error bars represent the standard deviation of three independent experiments. (d) Schematic illustration of in situ fluorescence electrochemical spectroscopy measurements. (e) Color-modified SEM image of HPTS-labeled gas diffusion layer. Bright green is the CO_2-sensitive HPTS gel immobilized on the surface of carbon fibers, scale bar: 20 μm. (f) Time-dependent $[CO_2]_l$ during CO_2RR at 50 mA cm^{-2}. Time = 0 s represents the start of electrolysis. *Source:* Run Shi et al. (2020) / Springer Nature / CC BY 4.0.

With the development of science and technology and the increase in the demand for applications, the research and application of in situ fluorescence spectroscopy in various fields are also expanding and deepening. In situ fluorescence spectroscopy will increasingly be combined with other analytical techniques to achieve multifunctional applications. For example, in materials research, in situ fluorescence spectroscopy can be combined with in situ Raman spectroscopy to achieve simultaneous measurement of structural and chemical information of materials. At the same time, in environmental monitoring, in situ fluorescence spectroscopy and in situ EC technology can be combined to achieve multiparameter detection of environmental pollutants. In situ fluorescence spectra will increasingly be miniaturized and remote to achieve real-time monitoring under non-localized environmental conditions [43]. For example, a miniaturized in situ fluorescence spectroscopy system can be used for real-time monitoring of pollutants and organisms in the atmosphere through optical fiber transmission and remote control. In situ fluorescence spectroscopy will enable the development of in situ imaging technology. For example, in the life sciences, in situ fluorescence spectroscopy can be used to achieve in situ imaging of tissues and cells for the diagnosis and treatment of diseases. Meanwhile, in materials science, in situ fluorescence spectroscopy can be used for surface imaging and structural analysis of materials.

In conclusion, as a nondestructive analytical technique, in situ fluorescence spectroscopy and in situ Raman characterization techniques have high practicability and operability. Its research and application in the fields of materials science, environmental monitoring, life science, and medicine are also becoming more and more extensive.

4.8 X-Ray Photoelectron Spectrometer

4.8.1 Theory of XPS

The energy levels of the inner electrons in the molecule are minimally affected by the environment of the molecule. The electron binding energy of the inner layer of the same atom is very little different in different molecules, so the photoelectron energy obtained by stimulating the inner layer electron is only related to the energy of the incident photon (i.e., the radiation source energy) and the atomic orbital binding energy. Thus, for a specific monochromatic excitation source and a specific atomic orbit, its photoelectron energy is characteristic. When the excitation source energy is fixed, its photoelectron energy is only related to the type of elements and the atomic orbit of the ionized excitation.

The principle of XPS is to use X-ray to deradiate samples, and its photon energy is between 1000 and 1500 eV, which can not only ionize the valence electrons of molecules but also excite the inner electron or valence electrons of atoms or molecules. To measure the energy of the photoelectron, the optoelectronic energy spectrum can be made with the kinetic energy/binding energy of the photoelectron and the relative intensity of the ordinate to obtain information about the sample. The X-ray photoelectron energy spectrum is also known as the electron energy spectrum for chemical analysis because it is the most useful for chemical analysis, as shown in Figure 4.23 [44].

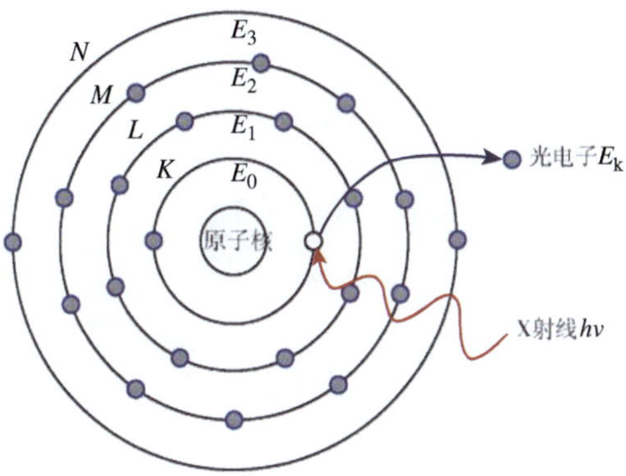

Figure 4.23 XPS principle.

4.8.2 X-ray Photoelectron Spectrometer

The XPS instrument design has made very obvious progress compared with the earliest experimental instruments, but all modern XPS instruments are based on the same structure: a vacuum chamber and its connected pumping system; sample introduction and operating system; X-ray source; electronic energy analyzer and its connected input (or transmission) electronic optical lens system; electronic detection system and service data processing system based on a PC or workstation. These components are enclosed in an ultra-high vacuum envelope and are usually made of stainless steel. With the continuous improvement of XPS characterization technology, a variety of functions, including large area XPS, small area XPS, imaging XPS, ion beam sputtering XPS, and angular resolution depth analysis XPS, and the overall trend is developing toward high spatial resolution, high energy resolution, and image analysis. Currently, the XPS spectrometers used in the laboratory are mostly from two brands: Kratos (United Kingdom) and Shimadzu (China), as shown in Figure 4.24.

4.8.3 Characteristic

As a surface analysis method, X-ray photoelectron energy spectroscopy can provide information about the elemental content and morphology of the sample surface. Its advantages lie in the simple sample processing, wide applicability, the rich chemical information obtained which is not destructive to the sample, the sampling depth less than 10 nm, and the detection limit of atomic concentration of 0.1–1.0%. Furthermore, XPS can be identified in addition to H, He all elements, widely used in metals, alloys, organic matter, inorganic, film, nanomaterials, including a variety of organic and inorganic solid material surface elements and valence for qualitative and quantitative analysis, and atomic and molecular valence band structure research surface elements of chemical imaging and depth analysis, etc. [45].

The main feature of XPS is that it can conduct surface analysis under low vacuum, which is impossible for other methods. In addition, the chemical shift effect is another feature of the XPS method different from other methods, and we can now make a mature analysis of its influence.

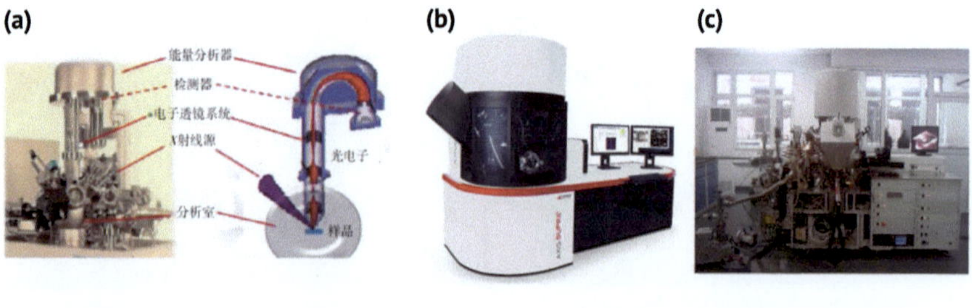

Figure 4.24 Diagram of (a) an X-ray photoelectron spectrometer, (b) AXIS SUPRA+, and (c) AXIS ULTRA DLD. *Source:* [49] / Elsevier.

The surface sampling depth of XPS is related to the properties of the material, the energy of the optoelectronics, and also the surface and the angle of the analyzer. With its high spatial resolution, it can also carry out micro-area spot selection analysis, line distribution scanning analysis, and element surface distribution analysis [46].

4.8.4 In Situ X-ray Photoelectron Spectroscopy

Of course, in situ representation is not easy. Generally speaking, the design goal of instruments and equipment is to meet detection environment and detection conditions, while there are no special design requirements for the reaction environment and reaction conditions. Therefore, if the reaction conditions interfere with the detection conditions, issues arise. For example, the most familiar measurement method, X-ray photoelectron spectroscopy (XPS), is a familiar analysis tool for surface analysis and catalytic researchers, which can give the element information and electronic structure information on the surface of the material [47]. However, due to the strong interaction of X-ray-excited photoelectrons with matter, if there is gas in the environment, the photoelectrons will scatter with the gas molecules, resulting in a signal reduction and unable to obtain effective detection. Therefore, XPS usually works in an ultra-high vacuum environment. If we had a large enough XPS spectrometer that we wanted to see the Earth's surface, we would need to remove everything from the Earth's surface that causes electron scattering, such as the atmosphere, gaseous and liquid water, the carbon dioxide and nitrogen that plants need, and the oxygen that animals need to survive. Under such test conditions, the Earth may only show her slightly barren "traces." But we all know that she is actually a beautiful blue water balloon, on which everything is full of vitality. Therefore, the problem of "how to enter gases or even liquids in XPS to measure the state of matter in the real environment" has become a technical obstacle to the development of in situ XPS.

XPS surface element qualitative analysis, quantitative analysis, surface chemical structure analysis on the basis can be widely used in the surface science and engineering in the field of analysis and research, such as surface oxidation (silicon oxide layer thickness determination, etc.), surface coating, surface catalytic mechanism, surface band structure analysis (semiconductor band structure determination, etc.) and high polymer friction charged phenomenon analysis, etc.

The energy spectrum analysis platform of Tsinghua University conducted XPS under light conditions, directly observed the internal charge separation and transfer, and revealed the photocatalytic mechanism of 3D C_3N_4@C. Details of synchronous illumination X-ray photoelectron spectroscopy (SI-XPS) experiments can be found in Ref. [48].

Synchronized illumination with SI-XPS was performed on a XPS Thermo Fisher 250Xi$^+$. The radiation source was a laser from Wilton Technology Ltd. with a wavelength of 450 nm. The XPS analysis was first performed in complete darkness. Then, the laser was turned on for in situ irradiation, and 2 minutes later for XPS analysis, irradiation, and sampling. Finally, data processing and

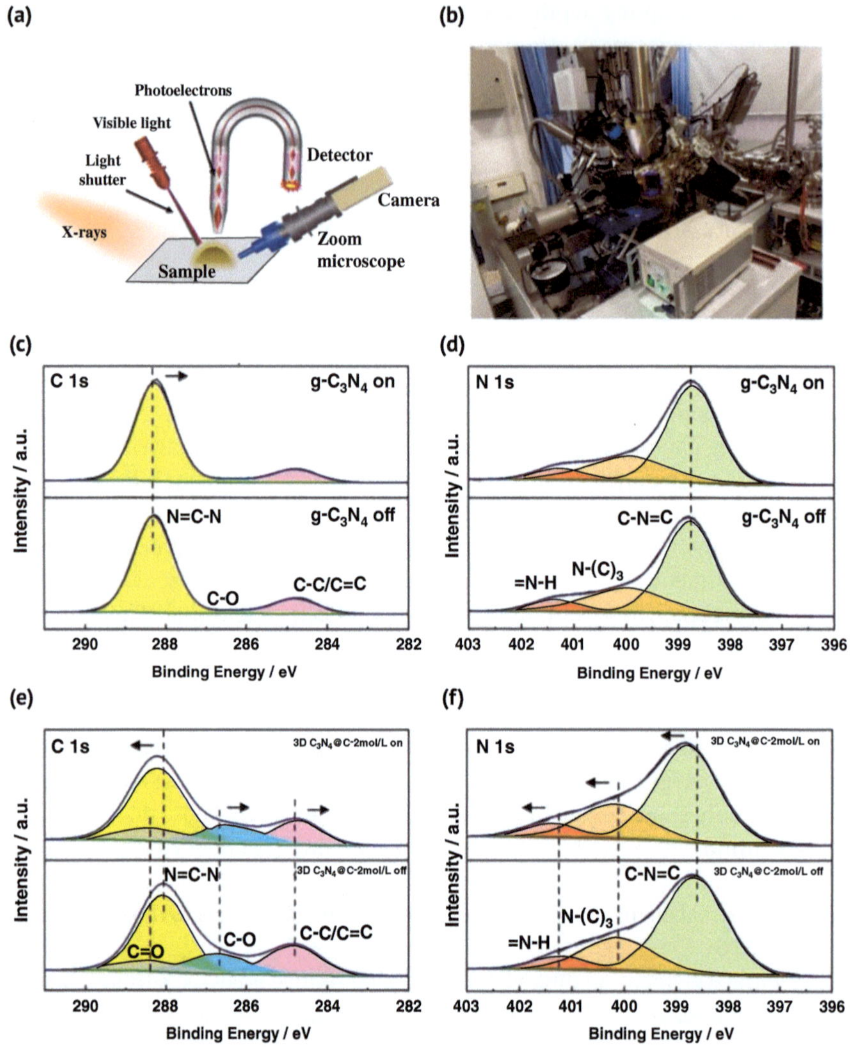

Figure 4.25 (a) Schematic diagram of SI-XPS test. (b) Photo of SI-XPS test; XPS spectra of g-C_3N_4 and 3D C_3N_4@C-2 mol L^{-1} before and after light irradiation. (c) C 1s peaks of g-C_3N_4. (d) N 1s peaks of g-C_3N_4. (e) C 1s peaks of 3D C_3N_4@C-2 mol L^{-1}. (f) N 1s peaks of 3D C_3N_4@C-2 mol L^{-1}. *Source:* [50] / Elsevier.

analysis were performed using Thermo Fisher's Avantage software, as shown in Figure 4.25.

An in situ XPS system was used in situ to represent the EC deposition process of lithium metal on the surface of a typical sulfide solid electrolyte $Li_{10}GeP_2S_{12}$ (LGPS) to study the interfacial stability of LGPS and lithium metal, as shown in Figure 4.26 [49].

According to the position of the known related component spectrum peaks, the data can be estimated in comparison by fitting the in situ XPS spectrum peaks.

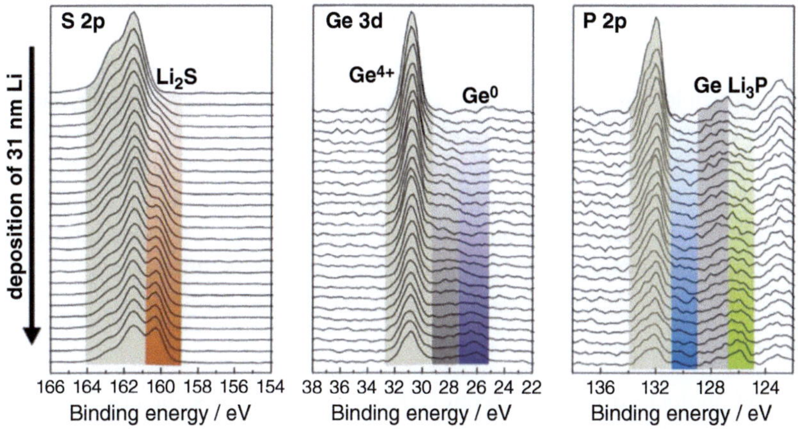

Figure 4.26 X-ray photoelectron (XPS) spectra recorded during deposition of 31 nm Li metal on $Li_{10}GeP_2S_{12}$. S 2p, Ge 3d, and P 2p/Ge 3p detail spectra are shown for different deposition states. With increasing Li deposition time, LGPS decomposes. The identified species are marked and labeled in the spectra.

4.9 Ultraviolet Photoelectron Spectrometer

4.9.1 Theory of Ultraviolet Photoelectron Spectrometer

Ultraviolet photoelectron spectrometer (UPS) uses ultraviolet light to excite the surface, sending emitted electrons to the surface. The excitation light source is a He lamp, and the synchrotron radiation light source can also be used as a vacuum ultraviolet light source due to its continuous adjustable energy. Compared with XPS, the ultraviolet light energy is low, so most of the outgoing photoelectrons come from valence electrons and are rarely used for quantitative analysis. However, valence electrons are generally involved in chemical bonding interactions, so the UPS is particularly suitable for studying the bonding interactions indicated. At the same time, UPS can also provide the solid work function, band structure, and other information, as shown in Figure 4.27 [50].

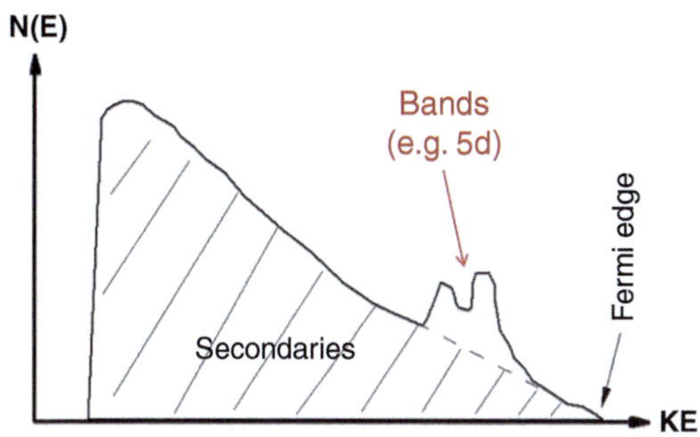

Figure 4.27 Typical UPS spectrum. The secondary electron cutoff edge is on the left and the high kinetic energy region on the right is the Fermi edge. The photoelectron-excited electrons are inelastic scattering near the surface layer, and the excited secondary electrons also have enough energy to escape from the surface, forming the secondary electron background in the spectrogram.

4.9.2 Ultraviolet Photoelectron Spectroscopy

UV photoelectron energy spectrometer includes the following main parts: monochrome ultraviolet light source (hν = 21.21 eV), electronic energy analyzer, vacuum system, sputtering ion gun source or electron source, sample chamber, information amplification, recording, and data processing system, as shown in Figure 4.28 [51].

The photoelectron-excited electrons are inelastic scattering near the surface layer, and the excited secondary electrons also have enough energy to escape from the surface, forming the secondary electron background in the spectrogram.

4.9.3 Characteristic

Ultraviolet photoelectron spectroscopy (UPS) operates on the same principle as the XPS, with the only difference being that the UPS uses ionizing radiation of dozens of eV to induce the photoelectric effect. In the laboratory, gas discharge lamps are used to generate UV photons. Gas discharge lamps are usually filled with helium gas, but they can also be filled with other gases, such as argon gas and neon gas. Due to the use of lower energy photons, the UPS cannot acquire the photoemission of most core levels, so the spectral acquisition is limited to the valence band region [52]. Two types of experiments are performed using UPS: valence band acquisition and electron detachment work measurement.

4.9.4 In Situ UPS

UPS, like XPS, is mainly used for material surface analysis [52]. Through the analysis project "Application of ultraviolet electron spectroscopy and XPS in metal material corrosion," we can obviously feel the difference between UPS and XPS in practical application:

Combined with the valence electronic structure of metal materials and different chemical components, and the interaction between different chemical components, the energy density distribution, the binding energy and its weight, Fermi level and surface work function can be obtained through UPS. Through XPS, we can obtain the chemical composition of the oxide membrane and the

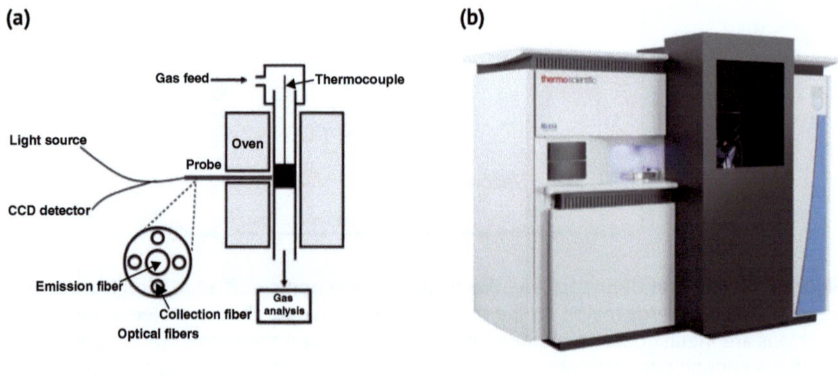

Figure 4.28 Diagram of (a) an ultraviolet photoelectron spectroscopy and (b) Thermo Fisher (Nexsa).

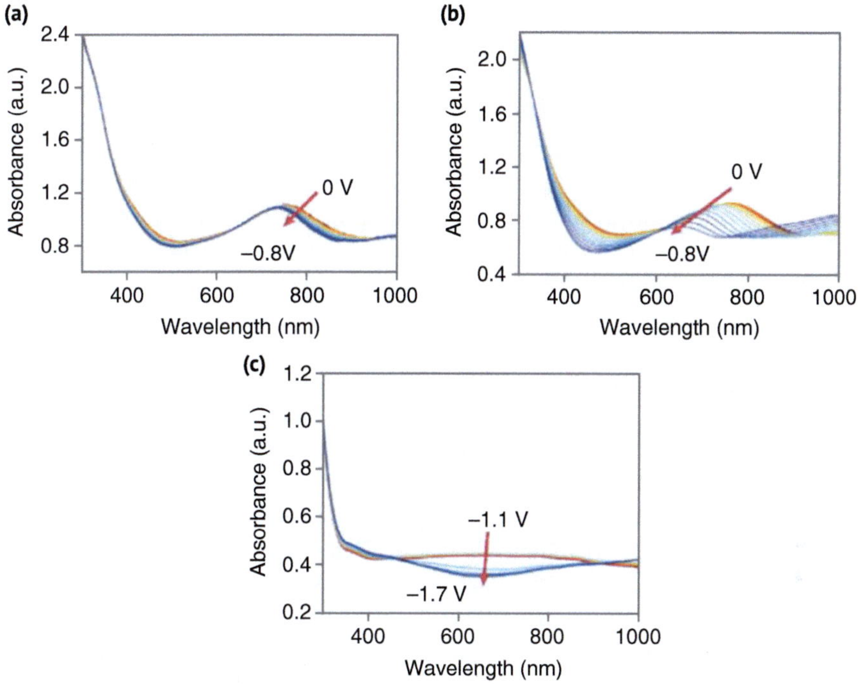

Figure 4.29 In situ UV-Vis absorption profiles.

atomic fraction of different elements, atoms, or the binding energy of ions of different valence states.

The team proposed a method using in situ UV-visible absorption spectroscopy (UV-Vis) to distinguish the EC energy storage mechanism and quantify the number of electrons transferred in the acidic electrolyte of their selected system and found that it was very similar to the results of in situ XAS. In this work, the research group first selected three EC systems, lithium titanate in 1M lithium perchlorate/acetonitrile (battery energy storage), $Ti_3C_2T_x$ in 1M sulfuric acid electrolyte (pseudocapacitive energy storage), and $Ti_3C_2T_x$ in 1M lithium sulfate solution (double-layer capacitance energy storage) [53]. The three systems separately tested the cyclic voltammograms for in situ UV-Vis devices and the cyclic voltammogram reconstructed using the MUSCA method. The cyclic voltammogram of the device is very consistent with the three systems reported in the literature, which proves the EC test. Finally distinguished between the three EC mechanisms, as shown in Figure 4.29.

In situ XRD reflects the buffering effect of the activated carbon during the reaction. In situ UV-visible absorption spectrum confirmed the high reversibility of iodine redox reactions occurring in the cocrystal electrolytes compared to aqueous electrolytes.

References

1 Yan, Y., Chai, Z., Qin, X. et al. (2022). Light-induced redox looping of a rhodium/Ce_xWO_3 photocatalyst for highly active and robust dry reforming of methane. *Angew. Chem. Int. Ed.* 61: e202200567.

2 Li, J., Han, C.J., Ou, X.W. et al. (2022). Concentrated electrolyte for high-performance Ca-Ion battery based on organic anode and graphite cathode. *Angew. Chem. Int. Ed.* 61: e202116668.

3 Forse, A.C., Merlet, C., Griffin, J.M. et al. (2016). New perspectives on the charging mechanisms of supercapacitors. *J. Am. Chem. Soc.*138: 5731–5744.

4 Pal, B., Yasin, A., Kaur, R. et al. (2021). Understanding electrochemical capacitors with in-situ techniques. *Renew. Sustain. Energy Rev.* 149: 111418.

5 Patra, A., Namsheer, K., Jose, J.R. et al. (2021). Understanding the charge storage mechanism of supercapacitors: in situ/operando spectroscopic approaches and theoretical investigations. *J. Mater. Chem. A* 9: 25852–25891.

6 Wang, H., Koster, T.K., Trease, N.M. et al. (2011). Real-time NMR studies of electrochemical double-layer capacitors. *J. Am. Chem. Soc.* 133: 19270–19273.

7 Levi, M.D., Salitra, G., Levy, N. et al. (2009). Application of a quartz-crystal microbalance to measure ionic fluxes in microporous carbons for energy storage. *Nat. Mater.* 8: 872–875.

8 Levi, M.D., Levy, N., Sigalov, S. et al. (2010). Electrochemical quartz crystal microbalance (EQCM) studies of ions and solvents insertion into highly porous activated carbons. *J. Am. Chem. Soc.* 132: 13220–13222.

9 Griffin, J.M., Forse, A.C., Tsai, W.Y. et al. (2015). In situ NMR and electrochemical quartz crystal microbalance techniques reveal the structure of the electrical double layer in supercapacitors. *Nat. Mater.* 14: 812–819.

10 Sigalov, S., Levi, M.D., Salitra, G. et al. (2010). EQCM as a unique tool for determination of ionic fluxes in microporous carbons as a function of surface charge distribution. *Electrochem. Commun.* 12: 1718–1721.

11 Shpigel, N., Chakraborty, A., Malchik, F. et al. (2021). Can anions be inserted into Mxene? *J. Am. Chem. Soc.* 143: 12552–12559.

12 Levi, M.D., Sigalov, S., Aurbach, D. et al. (2013). In situ electrochemical quartz crystal admittance methodology for tracking compositional and mechanical changes in porous carbon electrodes. *J. Phys. Chem. C* 117: 14876–14889.

13 Tsai, W.Y., Taberna, P.L., Simon, P. et al. (2014). Electrochemical quartz crystal microbalance (EQCM) study of ion dynamics in nanoporous carbons. *J. Am. Chem. Soc.* 136: 8722–8728.

14 Wang, X., Mathis, T.S., Sun, Y. et al. (2021). Titanium carbide MXene shows an electrochemical anomaly in water-in-salt electrolytes. *ACS Nano* 15: 15274–15284.

15 Dou, Q., Liu, L., Yang, B. et al. (2017). Silica-grafted ionic liquids for revealing the respective charging behaviors of cations and anions in supercapacitors. *Nat. Commun.* 8: 1–9.

16 Zheng, K., Xian, Y.Q., Lin, Z.F. et al. (2022). A method for deconvoluting and quantifying the real-time species fluxes and ionic currents using in situ electrochemical quartz crystal microbalance. *Adv. Mater. Interf.* 9: 2200112.

17 Xia, M., Liu, T., Peng, N. et al. (2019). Lab-Scale in situ X-ray diffraction technique for different battery systems: designs, applications, and perspectives. *Small Methods* 3: 1900119.

18 Chianelli, R., Scanlon, J.C., and Rao, B.M.L. (1978). Dynamic X-ray diffraction. *J. Electrochem. Soc.* 125: 1563.

19 Sushmita, C., Prabir, D., and Kanishka, B. (2022). High-performance thermoelectrics based on solution-grown SnSe nanostructures. *ACS Nano* 16 (7): 10783–10797.

20 Wang, Z., Xue, R., Zhang, H. et al. (2024). A hydrogel electrolyte toward a flexible zinc-ion battery and multifunctional health monitoring electronics. *ACS Nano* 18 (10): 7596–7609.

21 Chen, Z., Yang, M., Chen, G. et al. (2022). Triggering anionic redox activity in Fe/Mn-based layered oxide for high-performance sodium-ion batteries. *Nano Energy* 94: 106958, 2211–2855.

22 Wang, Y., Xia, S., Cai, R. et al. (2024). Dynamic reconstruction of two-dimensional defective Bi nanosheets for efficient electrocatalytic urea synthesis. *Angew. Chem. Int. Ed.* 63: e202318589.

23 Zhang, X., Zhan, J., Yu, Z., et al. (2023). Recent advances in real-time analysis of electrochemical reactions by electrochemical mass spectrometry. *Chin. J. Chem.* 41: 214–224.

24 Bruckenstein, S. and Gadde, R.R. (1971). Use of a porous electrode for in situ mass spectrometric determination of volatile electrode reaction products. *J. Am. Chem. Soc.* 93: 793–794.

25 Wolter, O. and Heitbaum, J. (1984). Differential electrochemical mass spectroscopy (DEMS) – a new method for the study of electrode processes. *Ber. Bunsenges. Phys. Chem.* 88: 2–6.

26 Kim, S., Kim, H.-S., Kim, B. et al. (2023). In situ gas analysis by differential electrochemical mass spectrometry for advanced rechargeable batteries: a review. *Adv. Energy Mater.* 13: 2301983.

27 Liu, G., Wang, W., Nie, Q. et al. (2024). Controllable long-term lithium replenishment for enhancing energy density and cycle life of lithium-ion batteries. *Energy Environm. Sci.* 17: 1163–1174.

28 Min, X., Han, C., Zhang, S. et al. (2023). Highly oxidative-resistant cyano-functionalized lithium borate salt for enhanced cycling performance of practical lithium-ion batteries. *Angew. Chem. Int. Ed.* 62: e202302664.

29 Na, J., Yu, H., Jia, S. et al. (2024). Electrochemical reconstruction of non-noble metal-based heterostructure nanorod arrays electrodes for highly stable anion exchange membrane seawater electrolysis. *J. Energy Chem.* 91: 370–382, 2095–4956.

30 Zhu, J., Zi, S.J., Zhang, N. et al. (2023). Surface reconstruction of covellite CuS nanocrystals for enhanced OER catalytic performance in alkaline solution. *Small* 19: 2301762.

31 Zhang, N., Hu, Y., An, L. et al. (2022). Surface activation and Ni-S stabilization in NiO/NiS_2 for efficient oxygen evolution reaction. *Angew. Chem. Int. Ed.* 134: e202207217.

32 Dai, Y., Zhang, X., Sun, M. et al. (2021). Uncovering the promotion of $CeO_2/CoS_{1.97}$ heterostructure with specific spatial architectures on oxygen evolution reaction. *Adv. Mater.* 33: 2102593.

33 Pan, S., Liu, H., Liu, D. et al. (2022). Efficient and stable noble-metal-free catalyst for acidic water oxidation. *Nat. Commun.* 13: 2294.

34 Wu, Y.-J., Yang, J., Tu, T.-X. et al. (2021). Evolution of cationic vacancy defects: a motif for surface restructuration of OER precatalyst. *Angew. Chem. Int. Ed.* 60: 26829.

35 Zhang, R., Pan, L., Guo, B. et al. (2023). Tracking the role of defect types in Co_3O_4 structural evolution and active motifs during oxygen evolution reaction. *J. Am. Chem. Soc.* 145: 2271.

36 Li, M., Wang, X., Liu, K. et al. (2023). Reinforcing Co-O covalency via Ce(4f)-O(2p)-Co(3d) gradient orbital coupling for high-efficiency oxygen evolution. *Adv. Mater.* 35: 2302462.

37 Li, Y., Zhang, X., Zheng, Z. et al. (2022). CeO_2 functionalized cobalt layered double hydroxide for efficient catalytic oxygen-evolving reaction. *Small* 18: 2107594.

38 Chen, X., Wang, X.-T., Le, J.-B. et al. (2023). Revealing the role of interfacial water and key intermediates at ruthenium surfaces in the alkaline hydrogen evolution reaction. *Nat. Commun.* 14: 5289.

39 Dai, X., Li, T., Wang, B. et al. (2023). Tailoring active Cu_2O/Copper interface sites for N-Formylation of aliphatic primary amines with CO_2/H_2. *Angew. Chem. Int. Ed.* 62: e202217380.

40 Amao, Y. and Nakamura, N. (2004). Optical CO_2 sensor with the combination of colorimetric change of α-naphtholphthalein and internal reference fluorescent porphyrin dye. *Sensor Actuat. B-Chem.* 100: 347.

41 Nivens, D.A., Schiza, M.V., and Angel, S.M. (2002). Multilayer sol–gel membranes for optical sensing applications: single layer pH and dual layer CO_2 and NH_3 sensors. *Talanta* 58: 543.

42 Shi, R., Guo, J., Zhang, X.R. et al. (2020). Efficient wettability-controlled electroreduction of CO_2 to CO at Au/C interfaces. *Nat. Commun.* 11: 3028.

43 Wolfbeis, O.S., Weis, L.J., Leiner, M.J.P. et al. (2002). Fiber-optic fluorosensor for oxygen and carbon dioxide. *Anal. Chem.* 60: 202.

44 Dong, H. and Cheng, B. (2022). X-ray photoelectron analysis. In: *Spectroscopy in Instrumental*, ch 14. Beijing: Chemical Industry Press. (in Chinese).

45 Isaacs, M.A., Jones, J.D., Davies, P.R. et al. (2021). Advanced XPS characterization: XPS-based multi-technique analyses for comprehensive understanding of functional materials. *Mater. Chem. Front.* 5: 7931–7963.

46 Moulder, J.F., Stickle, W.F., Sobol, P.E. et al. (1992). *Handbook of X-ray Photoelectron Spectroscopy. Eden Prairie*, MN: Perkin-Elmer Corp.

47 Wang, L., Guan, E., Wang, Y. et al. (2018). Silica accelerates the selective hydrogenation of CO_2 to methanol on cobalt catalysts. *Nat. Commun.* 9: 1362.

48 Wang, Y., Liu, M., Fan, F. et al. (2022). Enhanced full-spectrum photocatalytic activity of 3D carbon-coated C_3N_4 nanowires via giant interfacial electric field. *Appl. Catal. B: Environm. Energy* 318: 121829.

49 Wenzel, S., Randau, S., Leichtweiß, T. et al. (2016). Direct observation of the interfacial instability of the fast ionic conductor $Li_{10}GeP_2S_{12}$ at the lithium metal anode. *Chem. Mater.* 28: 2400–2407.

50 Liu, Y., Han, N., Zhang, Y. et al. (2015). *Louis Scudiero, Ultraviolet Photoelectron Spectroscopy (UPS)-1ppT.3.*

51 Meunier, F.C. (2010). The design and testing of kinetically-appropriate operando spectroscopic cells for investigating heterogeneous catalytic reactions. *Chem. Soc. Rev.* 39: 4602–4614.

52 Bols, M.L., Ma, J., Rammal, F. et al. (2024). Schoonheydt in situ UV–Vis–NIR absorption spectroscopy and catalysis. *Chem. Rev.* 124: 2352–2418.

53 Zhang, D., Wang, R, Wang, X., et al. (2023). In situ monitoring redox processes in energy storage using UV–Vis spectroscopy. *Nat. Energy* 8: 567–576.

5

Dynamic Structural Evolution Identification via X-ray Absorption Fine Structure

Chengyi Liu[1,2], *Shibo Xi*[2,*]

[1] *College of Safety and Environmental Engineering, Shandong University of Science and Technology, Qingdao, China*
[2] *Institute of Sustainability for Chemicals, Energy and Environment (ISCE2), Agency for Science, Technology and Research (A*STAR), Singapore, Republic of Singapore*

*Corresponding author: Xi_shibo@isce2.a-star.edu.sg

5.1 Introduction

Electrochemical energy conversion processes, such as water splitting, carbon dioxide reduction, and nitrogen fixation, are pivotal in addressing global energy and environmental challenges. Central to these processes are catalysts that facilitate the transformation of small molecules into value-added products. The efficiency and selectivity of these catalysts are intrinsically linked to their structural and electronic properties, which can undergo significant changes under operational conditions. For instance, the synergy between high temperature and gas environments during reactions can lead to structural reconstruction, degradation, and surface segregation, resulting in uncertain active sites [1, 2]. Understanding the dynamic structural evolution of catalysts during electrochemical reactions is essential for the rational design of more robust and efficient catalytic systems.

X-ray absorption fine structure (XAFS) spectroscopy has emerged as a powerful tool for probing the local structural and electronic environments of catalytic materials [3, 4], offering advantages such as element specificity, short-range sensitivity, and high sensitivity. Since XAFS was recognized in the 1970s [5, 6] as a novel method for studying local structures, early studies primarily used it to measure the coordination numbers of supported single-metal and bimetallic nanoparticles to determine their average size. With the development of multiple scattering theory in the 1990s [7], more advanced analytical methods were employed. XAFS itself is not surface-sensitive and can penetrate through materials. By tuning the X-ray energy to the absorption edge, it is possible to focus on specific elements or metal centers, effectively filtering out structural correlations arising from other spectator species, and independently studying the local environment of each atom.

Electrochemical Energy Conversion via Small Molecule Catalysis, First Edition.
Edited by Zhicheng Zhang and Yuanmiao Sun.

X-ray spectroscopy is not limited by sample state and is sensitive only to the local site structure [8]. By analyzing the absorption of X-rays as a function of energy, XAFS provides detailed information on oxidation states, coordination numbers, and interatomic distances. This technique encompasses two primary regions: X-ray absorption near edge structure (XANES), which is sensitive to the electronic structure and oxidation state of the absorbing atom, and extended X-ray absorption fine structure (EXAFS), which offers insights into the local geometric arrangement of atoms surrounding the absorber.

The application of XAFS, with its superior spatial, temporal, and energy resolution, in an in situ or operando mode – where measurements are conducted under actual reaction conditions – enables real-time monitoring of structural changes in catalysts. This capability is crucial for capturing transient species and intermediate states that occur during electrochemical reactions, providing a deeper understanding of the mechanisms that govern catalytic performance. For instance, in situ XAFS studies have elucidated the oxidation state changes and local structural rearrangements in transition metal catalysts during the oxygen evolution reaction, shedding light on the active sites responsible for catalysis [9–11].

Advancements in XAFS methodologies, including high-energy resolution fluorescence-detected XAFS (HERFD-XAFS) and time-resolved XAFS, have further enhanced the ability to investigate dynamic processes with greater precision. These techniques allow for the discrimination of subtle electronic and structural variations, facilitating a more comprehensive understanding of catalyst behavior under operational conditions.

In this chapter, we will explore the principles and applications of XAFS in studying the dynamic structural evolution of catalysts involved in electrochemical energy conversion. We will discuss the fundamentals of XAFS spectroscopy, the methodologies for data analysis and interpretation, and the design of in situ and operando experimental setups. Through case studies, we will illustrate how XAFS has been employed to unravel the structure–performance relationships in various catalytic systems, providing insights that drive the development of more efficient and durable catalysts for sustainable energy applications.

5.2 Fundamentals of XAFS

XAFS spectroscopy is an advanced and versatile technique used to explore the local atomic and electronic structure of materials [12, 13]. XAFS is the modulation of the X-ray absorption coefficient $\mu(E)$ near or above the absorption edge, caused by the chemical and physical states of atoms, as well as the oscillations of the photoelectron wave function at the absorbing atom due to scattering from neighboring atoms. By measuring the absorption of X-rays as a function of energy, XAFS provides detailed insights into the environment surrounding specific atoms in a sample. This information includes bond lengths, coordination numbers, oxidation states, and electronic configurations. Divided into two primary components, XANES and EXAFS, XAFS has become indispensable in various scientific disciplines.

5.2.1 Principles of XAFS

XAFS is essentially a quantum mechanical phenomenon based on the X-ray photoelectric effect, which arises from the photoelectric effect where an incident X-ray photon excites a core electron from a bound state in an atom into the continuum [14]. It is important to note that when the energy of the incident X-ray is lower than the binding energy of electrons in the element's orbitals, the electrons will not be excited to the highest unoccupied states or to the vacuum. When X-ray photons strike an atom in the sample, they are absorbed by the electrons in the tightly bound quantum core energy levels (such as 1s or 2p levels) of the atom, and a photoelectron with a wavevector k is ejected from an inner atomic orbital. The excess X-ray energy is transferred to the ejected photoelectron. The ejected photoelectron interacts with neighboring atoms, creating scattering effects that modulate the absorption coefficient. This modulation carries information about the local atomic structure and electronic properties of the absorbing atom. Since core electrons are located near the nucleus of an element, XAFS spectra provide information about the local density of unoccupied states around the photon-absorbing atom and can also offer chemical information about surface species by detecting these unoccupied states.

The initial state in XAFS is the bound state, and the final state is a continuum state with higher kinetic energy, where a core hole exists. Atoms with core holes are in an excited, unstable state, and the lifetime of this excited state is limited by the core-hole lifetime (i.e., the vacancy in the initial 1s state). As a result, electrons in higher energy levels rapidly transition to lower energy levels to fill the core hole. XAFS can measure transitions between the bound initial state and both the bound and continuum states. Below the absorption edge, allowed dipole or quadrupole transitions may occur into unfilled states with appropriate symmetry, resulting in the formation of local peaks, known as the pre-edge (Figure 5.1). In X-ray absorption spectra, the initial state (such as the 1s state for the K edge or the 2s or 2p states for

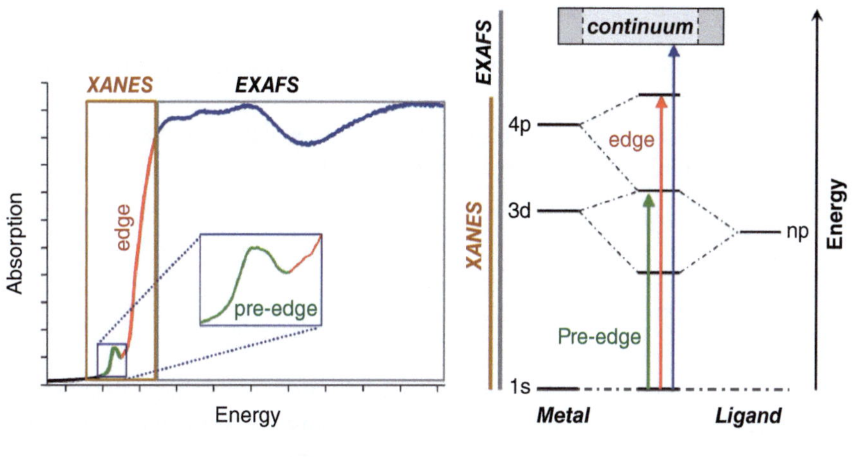

Figure 5.1 Illustration of the XAFS: On the left is an example XAFS spectrum, highlighting its division into the XANES and EXAFS regions. On the right is the corresponding energy level diagram, depicting the associated transitions. *Source:* Reproduced from [16] / with permission of John Wiley & Sons.

the L edge) has very simple symmetry and is highly localized near the absorbing atom center. XAFS spectra depend solely on the value of the final state wavefunction within the small region where the initial state is localized [15, 16].

5.2.1.1 XANES

XANES covers the region near the absorption edge, typically within 50 eV of it, transitioning to unfilled bound states, close to the bound state and continuum state. This portion of the spectrum is highly sensitive to the oxidation state, coordination geometry, and local symmetry of the absorbing atom, making it useful as a phase fingerprint [17]. The sharp peaks and features in the XANES region arise from electronic transitions into unoccupied states, providing a wealth of information about electronic structure and chemical bonding. A key characteristic of XANES is the multiple scattering resonance of low-energy photoelectrons, which has a large scattering cross-section and is dominated by multiple scattering. To resolve XANES features, the energy bandwidth of the X-ray beam should be less than or equal to 1 eV, or at most a few eV. XANES analysis is crucial for determining valence states and understanding the chemical environment of catalysts and other complex systems.

5.2.1.2 EXAFS

EXAFS extends several hundred eV above the absorption edge, transitioning to the continuum state, with high sensitivity to local structure and spatial resolution. The oscillatory patterns generated arise from the interference between the outgoing photoelectron wave and the waves scattered by surrounding atoms. These oscillations provide quantitative structural information, such as bond distances, coordination numbers, and atomic species. The EXAFS equation incorporates several parameters, including the photoelectron mean free path, phase shifts, and Debye–Waller factors, to model these oscillations accurately. EXAFS oscillations in k-space are approximately periodic [18], given by

$$k = \sqrt{2m(E - E_0)/\hbar^2} \tag{5.1}$$

where E is the photon energy, E_0 is the initial photon energy, and m is the electron mass.

EXAFS is calculated by the Debye–Waller factor, which suppresses each path at high k [19]. The EXAFS equation can be expressed as follows:

$$\begin{aligned} \chi(k) &= \frac{Nf(k)e^{-2k^2\sigma^2}}{kR^2}\sin[2kR + \delta(k)] \\ &= \sum_j \frac{N_j S_0^2 f_j(k) e^{-2R_j/\lambda(k)} e^{-2k^2\sigma_j^2}}{kR_j^2}\sin\left[2kR_j + \delta_j(k)\right] \end{aligned} \tag{5.2}$$

where f is the frequency, which depends on the distance (scattering path length) between the absorbing atom and the neighboring atoms; $\lambda(k)$ is the mean free path of the photoelectron, which diverges at low k, rendering the EXAFS equation invalid at that point; S_0^2 is the amplitude reduction factor; $N = \frac{2\Delta k \Delta R}{\pi}$, and Δk and ΔR are the effective ranges of EXAFS in k-space and R-space, respectively. The amplitude depends on the atom types and their distribution.

For EXAFS, even for disordered samples like powders and solutions, local structural information around the studied atoms can be obtained. EXAFS is highly sensitive to the first-shell distance and, to a lesser extent, to the coordination number. A common issue with EXAFS is the difficulty in distinguishing scattering atoms with small differences in atomic number, and it is often challenging to determine whether a Fourier peak should correspond to a distance with a large disorder parameter or to two distances with smaller disorder parameters [20, 21].

5.2.2 Key Concepts in XAFS

5.2.2.1 Absorption Coefficient μ

The X-ray absorption coefficient μ represents the likelihood of X-ray absorption in a material and exhibits sharp increases known as absorption edges. These edges correspond to the binding energy of core electrons and are unique to each element, making XAFS an element-specific technique. This specificity enables researchers to study individual elements in complex materials with high precision.

The X-ray absorption coefficient $\mu(E)$ describes the relationship between the absorbed X-ray intensity and the X-ray energy E [22, 23], and it follows Beer's law:

$$I = I_0 e^{-\mu t} \tag{5.3}$$

where I_0 is the incident X-ray intensity, which is energy-dependent based on the experimental setup. I is the transmitted X-ray intensity, which is energy-dependent based on the material's absorption characteristics. t is the sample thickness.

In most X-ray energy ranges, the absorption coefficient μ is a smooth function of energy, which largely depends on the X-ray energy E and the atomic number Z:

$$\mu \approx \frac{\rho Z^4}{AE^3} \tag{5.4}$$

and can be considered as the probability of photon absorption or the probability of electron presence at the origin, or the wavevector at the origin. Photoelectrons can be scattered from electrons in neighboring atoms, and the scattered photoelectrons can return to the absorbing atom. The absorption coefficient depends on the availability of an electronic state, i.e., whether there is an electron at the atom's position with the appropriate energy and momentum, and photoelectrons scattered from neighboring atoms can alter the absorption coefficient.

According to Fermi's golden rule,

$$\mu(E) = |\langle f|\mathcal{H}|i\rangle|^2 \tag{5.5}$$

where the initial state $|i\rangle$ contains X-ray photons and core-level electrons, which are not affected by neighboring atoms, and the initial state does not change; the final state $|f\rangle$ contains photoelectrons in the core–shell and the continuum, without photons, and affected by neighboring atoms; $\mathcal{H}$ is the interaction term, $\mathcal{H} = e^{ikr} \approx 1$.

5.2.2.2 Local Probing Capability

Unlike diffraction techniques, XAFS does not require crystalline order and is therefore suitable for investigating amorphous materials, liquids, and disordered systems. Its sensitivity to short-range order makes it an ideal method for characterizing heterogeneous samples, including biological tissues and nanostructures. This feature allows XAFS to complement other structural characterization techniques.

5.2.2.3 Element-specific Nature

XAFS is inherently element-specific, targeting individual elements within a sample. This capability is particularly valuable in studying complex mixtures or systems with trace elements, such as catalysts where active sites often contain low-concentration species. By isolating signals from specific elements, XAFS provides unparalleled insights into the local environment of target atoms.

5.2.3 Experimental Setup for XAFS

5.2.3.1 Synchrotron Radiation

Modern XAFS experiments are conducted at synchrotron radiation facilities, where the primary radiation source is bending magnets. These magnets utilize Lorentz force to bend the electron beam along a curved path, thereby emitting synchrotron radiation in the tangential direction of the horizontal plane. These facilities provide intense and tenable X-ray beams. Synchrotron radiation is generated when charged particles are accelerated to relativistic speeds, meaning their velocity approaches the speed of light, c. When these charged particles move along a curved path under the influence of a magnetic field, they emit electromagnetic radiation. As the particles travel at speeds close to c, the electromagnetic radiation's spectrum is shifted to higher frequencies (and energies) due to relativistic length contraction and time dilation, and the radiation pattern becomes highly directional [24]. X-ray beams from synchrotron radiation sources are almost always plane-polarized. A typical method for selecting specific energies is using a double-crystal monochromator, which consists of two nearly perfect parallel crystals, with silicon being a typical choice (Figure 5.2). The first crystal rotates to a specific angle using the incoming X-ray beam from the light source as a reference, reflecting X-rays at a particular energy following Bragg's law. The second crystal, having the same lattice spacing, rotates together with the first crystal and is positioned to intercept the diffracted beam and re-diffuse it, making it parallel to the original X-ray beam. This monochromator allows the selection of monochromatic X-rays over a wide energy range. Synchrotron radiation sources include both storage rings and linear accelerators. Synchrotron sources enable high-energy resolution and the ability to focus X-ray beams onto small samples, allowing for precise measurements even under challenging experimental conditions. The tunability of synchrotron radiation is critical for achieving the fine energy resolution required for XAFS.

5.2.3.2 Data Acquisition

XAFS data can be collected in transmission, fluorescence, or electron yield modes, depending on the sample's nature.

Transmission mode measures the intensity of X-rays before and after passing through a sample. During testing, it is essential to ensure that the beam is well aligned with the sample, the sample thickness is appropriate, and the sample is uniform without pinholes [26]. This mode is typically used for samples where the target element is the main component. For powder samples, it is necessary to

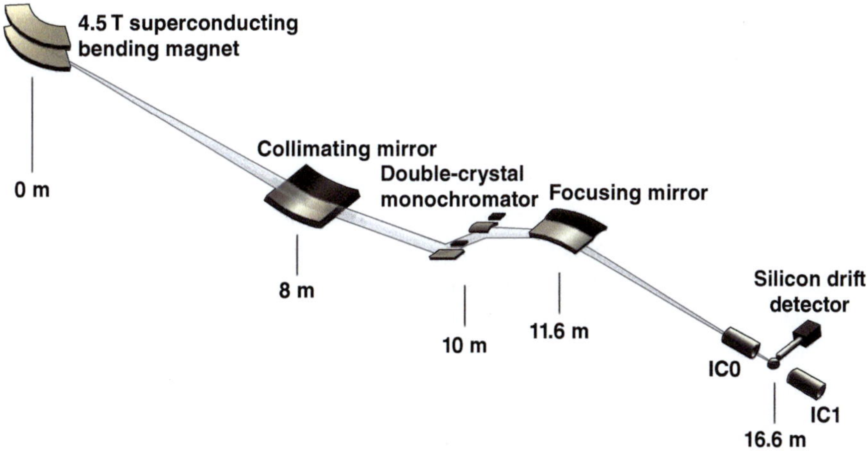

Figure 5.2 Schematic view of the XAFCA beamline. The collimating mirror aligns the beam vertically parallel before it reaches the double-crystal monochromator (DCM), while the focusing mirror refocuses the beam at the sample position. The DCM features two sets of crystals: Si (111) and KTP (011). The second Si (111) crystal provides sagittal (horizontal) focusing. *Source:* Reproduced from [25] / with permission of International Union of Crystallography.

ensure that their grain size is not significantly larger than the absorption length. This testing method is suitable for model compounds and elements with concentrations greater than 10%. During the test, two ionization chambers are arranged in a line, with the sample placed between them. The first ionization chamber measures the intensity of the incident X-rays (I_0), while the second ionization chamber measures the intensity of the X-rays passing through the sample (I). A reference sample, typically a standard metal foil, is often used for calibration. When the sample is placed in front of the entrance slit, rather than in its usual position, no absorption edge should appear in the XAFS spectrum. If a noticeable "edge" is observed in the spectrum, it may be due to nonlinearity of the detection system, improper offset adjustment of the detection system, or significant higher order content in the incident X-ray beam.

Fluorescence mode is used to detect X-ray fluorescence emitted by the sample. Fluorescence is particularly useful for dilute or thick samples where transmission measurements may be impractical. In fluorescence XAFS testing, the X-rays emitted from the sample will include fluorescence lines from the target element, fluorescence lines from other elements in the sample, and elastic and inelastic (Compton effect) X-rays [27]. Two factors need to be considered during testing: solid angle and energy resolution. Since the X-rays from the synchrotron source are polarized in the synchrotron radiation plane, the fluorescence detector is placed at a 90° angle to the plane of the incident beam, with the sample at a 45° angle to the incident beam [15]. Due to detector saturation, not all fluorescence can be counted, resulting in "dead time." Additionally, since all materials attenuate X-rays, the fluorescence intensity and XAFS oscillations may be suppressed by the self-absorption effect. Fluorescence ionization chambers are typically used with filters and slit components and generally require a lower electric field. The filter absorbs scattered X-rays, with a $Z - 1$ filter (where Z is the atomic number) commonly used [28]. The slit component is used to reduce the fluorescence X-rays and prevent them from being detected by the ionization chamber. When scattered X-rays are stronger than fluorescence X-rays, a semiconductor detector can be used, which absorbs X-ray photons and generates electron–hole pairs. However, its energy resolution is sometimes lower.

Electron yield mode is often employed for surface-sensitive studies. It indirectly measures absorption events by detecting the decay products generated when core holes are refilled, specifically measuring the electrons emitted from the sample surface. Due to its surface sensitivity, this mode avoids the "self-absorption" effect seen in fluorescence mode.

5.2.4 Applications of XAFS

The versatility of XAFS makes it indispensable in numerous scientific and industrial applications:

- *Catalysis*: XAFS is extensively used to study heterogeneous catalysts, providing insights into active site structures, oxidation states, and reaction mechanisms. Real-time operando studies reveal dynamic changes in catalyst structures under working conditions.
- *Energy Materials*: In energy storage and conversion, XAFS is employed to investigate battery electrodes, fuel cell catalysts, and photovoltaic materials. By understanding the structural and electronic changes during operation, researchers can optimize material performance and durability.
- *Environmental Science*: XAFS is crucial for studying heavy metal speciation, pollutant interactions with soils, and the fate of contaminants in aquatic systems. Its element-specific sensitivity allows for the precise identification of trace elements in complex matrices.
- *Biological Systems*: Metalloproteins and enzymes often contain metal centers essential for their function. XAFS provides direct insights into the coordination environment and oxidation states of these metal centers, aiding in the design of biomimetic catalysts and therapeutic agents.

XAFS remains a cornerstone technique for elucidating the local structure and electronic properties of materials across various scientific fields. Its ability to probe noncrystalline and complex systems with element specificity and short-range sensitivity sets it apart from other characterization methods. With ongoing advancements in synchrotron technology and analytical methodologies, XAFS is poised to unlock even greater insights into the fundamental properties of materials, paving the way for breakthroughs in catalysis, energy, biology, and beyond.

5.3 XAFS Data Analysis and Interpretation

Data analysis in XAFS spectroscopy is critical for extracting meaningful structural and electronic information about materials. XAFS analysis involves several steps, including preprocessing raw data, background subtraction, normalization, and fitting of the oscillatory signal to obtain quantitative structural parameters such as interatomic distances, coordination numbers, and disorder factors. Advanced analytical techniques and computational tools have further enhanced the precision and scope of XAFS data interpretation, paving the way for deeper insights into complex material systems. This section provides a comprehensive overview of standard and advanced techniques for XAFS data analysis, emphasizing their applications, challenges, and integration with modern computational methodologies.

5.3.1 Key Components of XAFS Analysis

5.3.1.1 Data Preprocessing

Preprocessing is the foundation of reliable XAFS analysis. The process begins with energy calibration, specifically determining the absorption edge or threshold energy E_0 (the energy level of the continuous state at k=0), to ensure alignment of absorption edges across multiple datasets. The most common approach is to select the maximum value of the first derivative of $\mu(E)$, a method that is easy to reproduce and applicable to any spectrum, though it may require further refinement during the fitting process (Figure 5.3b). Another approximation is half of the edge step. For transition metal complexes with pre-edge transitions, the threshold is typically located at the first inflection point of the edge. Noise removal and artifact correction are crucial steps, often performed using smoothing algorithms or wavelet transforms. EXAFS measurements require a signal-to-noise ratio (S/N) greater than 10^3 to accurately resolve the spectrum in the region approximately 600–1000 eV above the absorption edge. Typically, this range must be measured above the edge to achieve sufficient spatial resolution. Preprocessing also involves energy range selection to optimize data quality for both XANES and EXAFS regions. Effective preprocessing ensures the integrity of subsequent analyses and enhances reproducibility across experiments.

5.3.1.2 Background Subtraction

Background subtraction is essential for isolating the oscillatory fine structure from the raw absorption spectrum, as this step removes low-frequency components before performing the Fourier transform, preventing data contamination. Algorithms for background subtraction like AUTOBK and spline fitting are commonly used to model and remove the atomic absorption background [29]. These methods determine an empirical background spline based on the difference between the data and the background in the Fourier components. In data analysis, the background caused by pre-edge absorption or scattering is first fitted and subtracted, followed by the removal of the smoothed atomic background μ_0, which corresponds to the absorption of the atom in the absence of neighboring atoms. The data are then converted into the $\chi(k)$ function. The oscillatory part, $\chi(k)$, contains the sum of all contributions from neighboring atomic groups $\chi_i(k)$ that are approximately at the same distance from the absorbing atom (within the ith shell). It provides a correction to the X-ray absorption coefficient due to the presence of neighboring atoms. Theoretically,

$$\chi(E) = \frac{[\mu(E) - \mu_0(E)]}{\Delta\mu_0} \tag{5.6}$$

where $\mu(E)$ is the measured absorption coefficient and $\mu_0(E)$ is the absorption coefficient of the central atom, excluding scattering from neighboring atoms. Experimentally,

$$\chi = \frac{(\mu_{exp} - \mu_b)}{\Delta\mu_{exp}} \tag{5.7}$$

where μ_{exp} is the experimentally measured absorption or fluorescence, $\Delta\mu_{exp}$ is the edge step, and μ_b is the smoothed background curve fitted to the data. This step enables accurate extraction of the extended fine structure, which contains information about the local atomic environment. Advanced background subtraction techniques, including polynomial fitting and Fourier-based methods, have been developed to handle complex datasets and improve accuracy in noisy spectra.

5.3.1.3 Normalization

Normalization ensures that XAFS spectra are independent of sample thickness, concentration, or detector efficiency. This process involves scaling the post-edge region of the absorption spectrum to unity, i.e., normalizing $\mu(E)$ so that it is typically close to 0 below the absorption edge and approximately 1 above the edge, representing the absorption of 1 X-ray, enabling direct comparison of spectra from different samples or experimental conditions. Normalized data are typically presented after subtracting the regression of the quadratic curve and adjusting for the slope differences between the pre- and post-edge polynomials. Good normalization requires careful handling, and it is important to assess the effectiveness and consistency of this process for a specific dataset. Normalization also facilitates the identification of subtle spectral features, such as weak pre-edge peaks or fine structural variations.

5.3.2 EXAFS Analysis

EXAFS analysis focuses on the oscillatory fine structure beyond the absorption edge, providing quantitative information about the local atomic environment around the absorbing atom. The core of EXAFS analysis lies in extracting information about atomic distances, coordination numbers, and mean square relative displacements from the data in k-space. This process involves several steps, including Fourier transform, curve fitting, and wavefunction transformations.

5.3.2.1 Fourier Transform

The Fourier transform (FT) converts the EXAFS signal from k-space (photoelectron wavenumber) to R-space (real-space radial distances) (Figure 5.3f), yielding a radial distribution function that highlights contributions from neighboring atoms [31]. The Fourier transform is actually a complex function (with both real and imaginary parts), a complex quantity with phase. Half of the information is contained in the phase, while the modulus is most commonly displayed. The relative positions of the maximum values of the imaginary part and the modulus are influenced by the scattering phase shift constant. Peaks in R-space correspond to atomic groups at different distances, and coordination shells consist of atomic groups with similar distances. Roughly speaking, the height of the Fourier transform peaks represents the

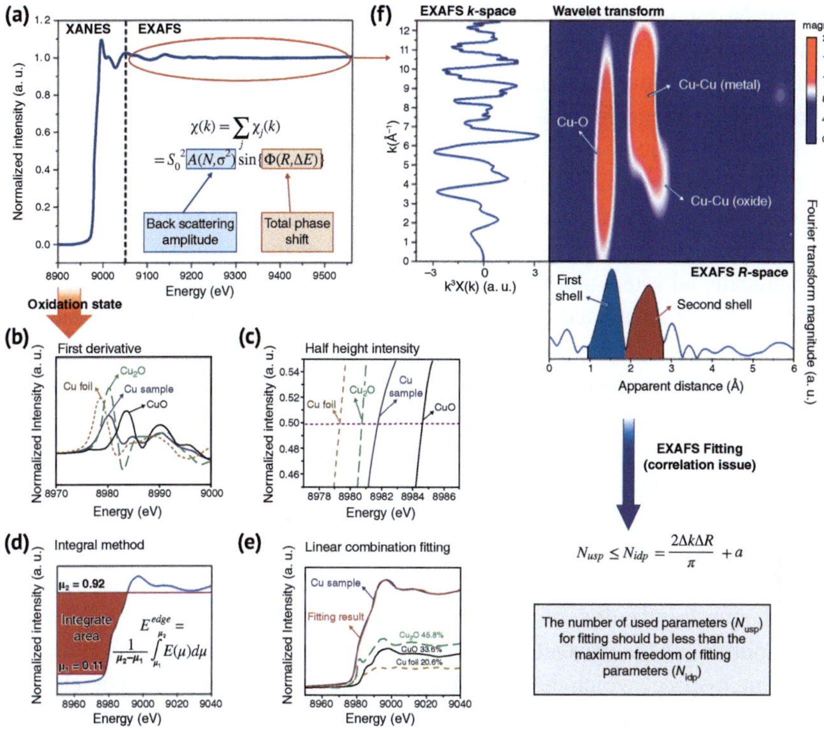

Figure 5.3 (a) A typical XAS spectrum consisting of XANES and EXAFS regions, with an inset showing the quantitative parameterization of the EXAFS spectra. (b–e) Methods for determining the oxidation state from the XANES spectra of a Cu-based mixture sample: (b) first derivative; (c) half-height intensity; (d) integral method; and (e) linear combination fitting. An equation for defining the average edge energy is included in (d), where the inverse function $E(\mu)$ of the spectrum is integrated between μ_1 and μ_2. (f) Cu K-edge EXAFS spectrum for a Cu-based mixture sample, along with its Fourier and wavelet transforms. The Nyquist criterion is used to estimate the upper bound of the number of independent variables for EXAFS fitting. *Source:* [30] / Springer Nature / CC BY 4.0.

average k^n-weighted EXAFS amplitude of the shell signal within the k-space transformation range. The position of the peaks in R-space indicates the average slope of the phase in the k-space transformation range, which is largely determined by the atomic distance term $2kr$. The position of the peaks is generally shifted towards lower R, usually by around 0.5 Å, due to scattering and phase shifts (Figure 5.3a). The width of the peaks is primarily determined by the data transformation range, i.e., the width of the k window. Advanced FT techniques, such as phase-corrected transforms, improve resolution and accuracy in identifying atomic shells.

5.3.2.2 Curve Fitting

Curve fitting involves modelling the EXAFS signal using theoretical spectra generated by programs like FEFF [32]. This process extracts structural parameters such as bond lengths, coordination numbers, and Debye–Waller factors, which can be determined by comparing with known standard samples, without the need for shell-to-shell fitting. Statistical tools, including the F-test and Akaike information criterion (AIC), are used to evaluate the quality of the fit and the reliability of extracted parameters [33]. Common data processing and analysis methods include nonlinear least squares fitting, linear combination analysis (Figure 5.3e), and principal component analysis [34]. Currently, two methods are used to enhance EXAFS fitting through theoretical modeling: ab initio calculations of the Debye–Waller factor system and EXAFS spectral simulations for systems with large dynamic disorder. When fitting XAFS data, a theoretical model containing unknown parameters is typically constructed. The assumed values of the unknown parameters are systematically adjusted, and the square differences between the data and the theoretical model are calculated. This process helps to refine the parameter values and achieve the best fit between the experimental data and the theoretical predictions. Multi-shell fitting techniques enable the analysis of complex structures, such as alloys or multicomponent systems (Figure 5.3f).

5.3.2.3 Wavelet Transform

Wavelet transform (WT) is a powerful technique that provides simultaneous localization of EXAFS signals in both k-space and R-space (Figure 5.3f). Unlike Fourier Transform, it performs an integral transform by localizing wave trains, providing simultaneous localization of signals in both time and frequency domains [35]. The amplitude of the wavelet transform offers radial distance resolution and resolution in k-space, which helps distinguish the coordination of light elements [36, 37]. WT is particularly effective in resolving overlapping contributions from different scattering paths and distinguishing single- and multiple-scattering events. Applications of WT include the analysis of nanostructured materials, where traditional FT methods may fail to resolve complex structural features.

5.3.3 XANES Analysis

XANES analysis provides critical insights into the electronic structure, oxidation states, and local symmetry of materials. This information is vital for understanding the chemical and functional properties of materials in catalysis, energy storage, and environmental remediation.

Since XANES cannot provide a numerical parameterization in terms of oxidation states or coordination chemistry, simple spectral analysis methods are often used to determine these properties. For XANES analysis, spectra are typically compared by amplitude, so any error in the edge step of a spectrum will directly affect its relative weight. The fitting intensity of peaks can be compared to spectral curves of known substances for calibration. Another approach is to treat the spectrum as a characteristic or fingerprint of the material and determine the proportions of the components in the sample based on spectra from known materials, often using statistical methods like principal component analysis (Figure 5.3c–e). In addition, XANES can provide qualitative analysis based on coordination chemistry (e.g., regular, distorted octahedral, or tetrahedral coordination), molecular orbitals (e.g., p-d orbital hybridization, crystal field theory), band structure (e.g., density of available electronic states), and multiple scattering (e.g., photoelectron multiple scattering events) [38].

5.3.3.1 Pre-edge Analysis

Pre-edge features are characteristics caused by electronic transitions to empty bound states, with transition probabilities governed by dipole selection rules. They offer valuable information about the electronic configuration, ligand field effects, and local symmetry of the absorbing atom. Pre-edge intensity is proportional to the deviation from centrosymmetry, and its form depends on spin state, oxidation state, ligand field splitting, and multiplet effects. The weighted average energy of the pre-edge intensity is modulated by the strength of the ligand field. Deconvolution techniques are employed to separate overlapping pre-edge peaks, enabling detailed analysis of oxidation states and coordination geometries.

5.3.3.2 Edge Position and Shape

The position and shape of the absorption edge are indicative of the oxidation state and electronic environment. For example, an upward shift in edge energy often corresponds to higher oxidation states. Advanced edge-fitting techniques allow for the precise determination of oxidation states in mixed-valence systems.

5.3.3.3 Multiple Scattering Contributions

XANES spectra are highly sensitive to multiple-scattering effects, which provide three-dimensional structural information. Computational modelling using programs like FEFF and MXAN helps interpret these contributions, revealing complex geometries and bonding interactions.

5.3.4 Integration with Computational Methods

The integration of XAFS data with computational approaches, such as Density Functional Theory (DFT) and molecular dynamics (MD), has significantly advanced the field. These methods enable the prediction and validation of structural and electronic properties, enhancing the reliability of experimental interpretations. Machine learning algorithms are increasingly being used to analyze large XAFS datasets, offering automated feature extraction and pattern recognition.

5.3.5 Challenges and Future Directions

5.3.5.1 Noise and Artifacts

Noise and artifacts remain significant challenges, particularly in dilute systems or low-concentration samples. Advanced denoising techniques, including wavelet-based methods and neural network algorithms, are being developed to address these issues.

5.3.5.2 Model Dependency

EXAFS fitting relies heavily on theoretical models, which may introduce biases if the models do not accurately represent the sample. Efforts are underway to develop more robust simulation algorithms and comprehensive databases to improve model reliability and applicability.

5.3.5.3 High-throughput and Automation

The increasing volume of XAFS data necessitates the development of high-throughput analysis pipelines. Automated workflows, incorporating preprocessing, fitting, and statistical validation, are being implemented to streamline data analysis and reduce human error.

Data analysis and interpretation are central to the application of XAFS spectroscopy, transforming raw experimental data into actionable insights about material properties. Advanced analytical tools, computational methodologies, and machine learning techniques continue to expand the capabilities of XAFS, enabling the study of increasingly complex systems with greater precision and efficiency. As these methods evolve, XAFS will remain a cornerstone of materials characterization, driving innovation across catalysis, energy storage, and nanotechnology.

5.4 In Situ and Operando XAFS Techniques

In the context of catalysis, "in situ" refers to structural characterization of the catalyst under realistic or model reaction conditions, but it does not necessarily involve real-time data collection during the reaction process. In contrast, "operando" refers to studies conducted under the actual working conditions of a catalyst, typically involving real reaction environments. This approach enables the direct acquisition of data during the reaction process to analyze structural changes in the catalyst. Operando studies often include simultaneous online monitoring of reaction performance, utilizing techniques such as mass spectrometry (MS) or gas chromatography (GC) to assess the catalyst's activity and selectivity in real time. In situ and operando XAFS techniques represent a transformative approach for investigating the behavior of materials under realistic operational conditions [39]. These methodologies enable real-time monitoring of structural and electronic changes in catalysts and other functional materials during active processes, bridging the gap between static characterization and real-world applications. The ability to probe materials in their functional states provides critical insights into reaction mechanisms, catalyst dynamics, and structural stability. Such insights are indispensable for designing *advanced* materials tailored for applications in catalysis, energy storage, and environmental remediation.

In recent years, the demand for in situ and operando studies has grown significantly due to the need for understanding the dynamic nature of materials under operational conditions. Traditional characterization techniques, while invaluable, often fail to capture the transient phenomena that occur during material functionality. By contrast, in situ and operando XAFS provide time-resolved data that reveal how materials evolve at the atomic level, shedding light on active sites, structural changes, and reaction intermediates in real time.

5.4.1 Key Concepts

5.4.1.1 In Situ versus Operando XAFS

In situ XAFS involves studying materials under conditions that simulate operational environments, such as specific temperatures, pressures, and chemical atmospheres

[40, 41]. However, it does not necessarily measure the material's performance concurrently. This makes it an excellent tool for pre-screening material behavior under controlled settings before testing real-world performance. For instance, Matos et al. employed in situ EXAFS to investigate the effects of different pretreatment methods and thermal annealing environments on the sintering behavior of nanoparticles, providing an effective strategy for enhancing thermal stability (Figure 5.4) [42].

- *Operando XAFS*, on the other hand, takes the concept further by simultaneously monitoring the material's structural changes and functional performance metrics, such as catalytic activity, electrical conductivity, or electrochemical efficiency. This dual approach enables researchers to directly correlate structural dynamics with functional properties, providing a comprehensive understanding of material behavior. Operando XAFS has become a cornerstone for understanding how real-time changes influence overall performance [43]. It should be noted that the operando method often probes the resting state of the catalyst or the equilibrium between different catalyst states. As a result, a transient intermediate state may not be detected in standard XAFS experiments.

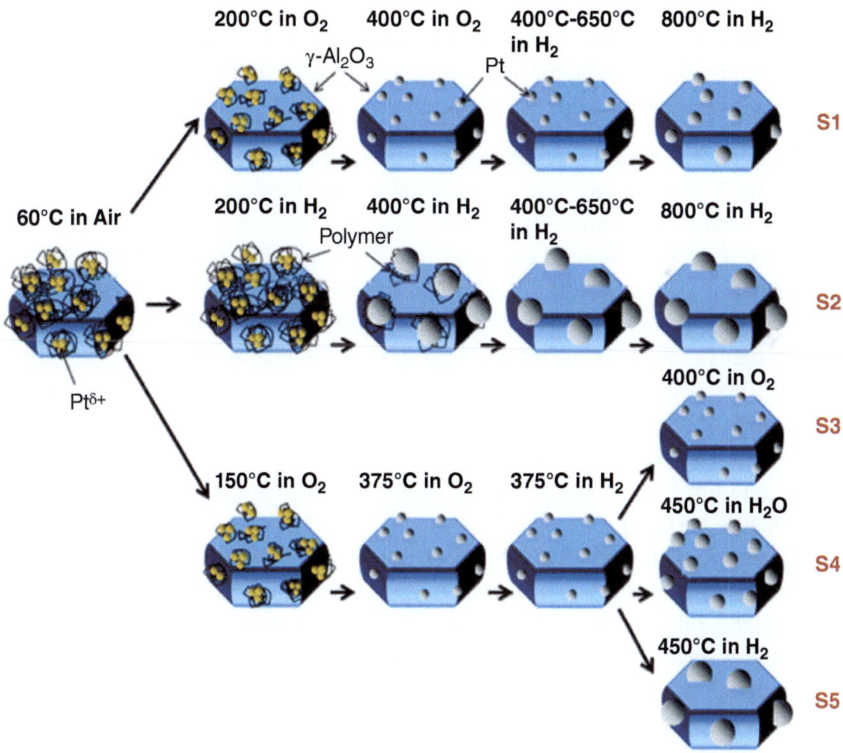

Figure 5.4 Schematic representation of the evolution of the structure and chemical state of our inverse micelle prepared Pt/g-Al_2O_3 samples (S1–S5) under various thermal and chemical treatments inferred from the analysis of in situ EXAFS and ex situ STEM data. *Source:* Reproduced from [42] / with permission of Royal Society of Chemistry.

5.4.2 Applications of In Situ and Operando XAFS

5.4.2.1 Catalysis

Most catalytic reactions involve chemical adsorption and electron transfer, leading to changes in oxidation states and local structures. Catalysis research has greatly benefited from in situ and operando XAFS techniques. These methods allow researchers to observe how catalysts transform during chemical reactions, providing insights into their active states and stability. For instance,

- Studies have demonstrated that copper-based catalysts undergo reversible structural reconstructions during CO_2 reduction reactions, forming highly active nanoclusters that enhance reaction efficiency.
- Operando investigations into platinum and palladium catalysts used in fuel cells have revealed oxidation state changes and local coordination shifts, enabling optimization of their durability and catalytic efficiency.
- Transition metal oxides, such as manganese and cobalt oxides, have been studied for oxygen evolution reactions (OER), with operando XAFS identifying the redox-active species responsible for catalytic activity.

5.4.2.2 Energy Conversion and Storage

In energy storage applications, understanding the structural and chemical evolution of materials is crucial for improving performance and longevity. In situ XAFS provides invaluable data on the following:

- *Battery Materials*: Li-ion battery cathodes and anodes have been investigated to track phase transitions, ion diffusion, and degradation mechanisms during charge–discharge cycles [44]. For example, studies on layered oxides have revealed structural collapses that limit cycle life, paving the way for new material designs. Additionally, considering that electrochemical reactions typically occur at the electrode–electrolyte interface, it is often necessary to design in situ electrochemical cell devices to ensure that the measurement process accurately reflects the dynamic changes within the battery. For instance, Wang et al. used an in situ electrochemical cell with an O-ring sealing design to reveal the evolution of the oxidation state and local structure of vanadium ions in the g-LiV_2O_5 battery during charge and discharge cycles (Figure 5.5) [45].
- *Fuel Cells*: Operando XAFS has been used to study the oxidation state dynamics of iridium oxide (IrO_2) during oxygen evolution reactions (OER), offering insights into catalyst activation and deactivation pathways. In addition, in situ time-resolved XAFS has been employed to reveal the redox reactions and degradation mechanisms of cathode electrocatalysts in polymer electrolyte fuel cells (PEFCs), enabling real-time visualization of dynamic structural changes (Figure 5.6).
- *Supercapacitors*: For example, Ti_3C_2-Co and Ti_3C_2-Cu_1Co_8 supercapacitors based on MXenes (Figure 5.7), by mapping the structural changes in electrode materials during rapid charging and discharging cycles, have provided guidance for designing materials with enhanced energy densities and lifespans.

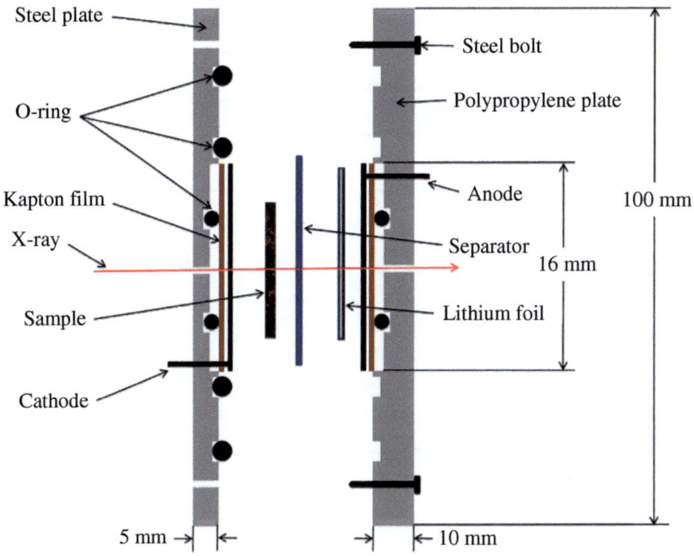

Figure 5.5 Schematic sketch of the in situ cell. The battery cell with a diameter of 16 mm is formed in between a steel and a polypropylene plate. These outside plates have an X-ray window in the center, which is made of a Kapton film instead of a toxic Beryllium plate. The sample space, shielded by O-rings, includes a cathode and an anode, separated by a polyolefin film. This in situ electrochemical cell is re-usable and portable, and the O-ring sealing allows the experiments to run for several days without significant degradation of the electrodes. *Source:* Reproduced from [45] / with permission of Elsevier.

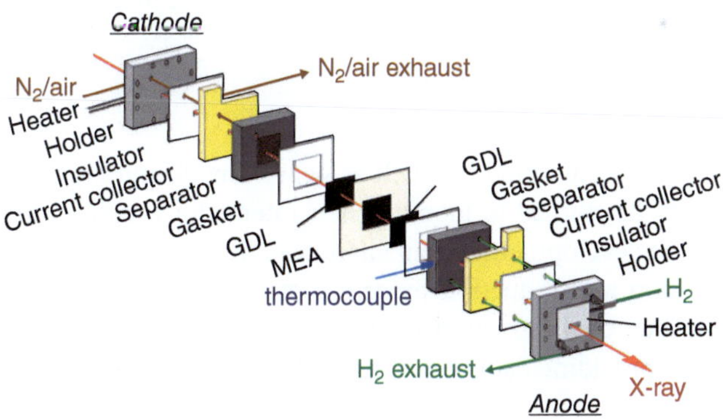

Figure 5.6 Schematic of a laboratory-made PEFC cell designed for operando XAFS measurements. *Source:* Reproduced from [46] / with permission of Springer Nature.

5.4.2.3 Environmental Studies

In situ and operando XAFS have proven essential in environmental science for tracking the speciation and transformation of hazardous elements in complex systems. For instance,

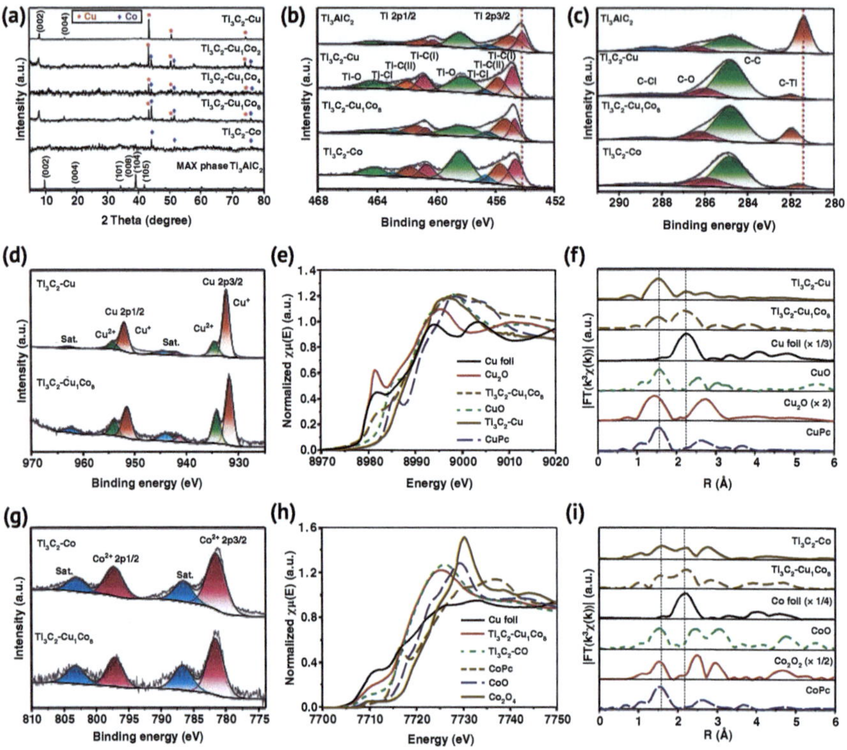

Figure 5.7 (a) XRD patterns of Ti_3AlC_2 and Ti_3C_2-Cu/Co hybrids. (b–d, g) High-resolution Ti 2p (b), C 1s (c), Cu 2p (d), Co 2p (g) XPS spectra of Ti_3AlC_2 and Ti_3C_2-Cu/Co hybrids. (e, f) Normalized Cu K-edge XANES (e) and Fourier transform of EXAFS (f) spectra of Ti_3C_2-Cu, Ti_3C_2 Cu_1Co_8 and references. (h, i) Normalized Co K-edge XANES (h) and Fourier transform of EXAFS (i) spectra of Ti_3C_2-Co, Ti_3C_2-Cu_1Co_8 and references. *Source:* Reproduced from [47] / with permission of John Wiley & Sons.

- Soil and water contaminants, such as heavy metals and organometallic compounds, can be studied to monitor their interactions with natural substrates, aiding in the development of remediation strategies.
- The oxidation and reduction processes of arsenic, mercury, and chromium under various environmental conditions have been elucidated using these techniques, providing data critical for pollution control efforts.

5.4.3 Methodologies and Experimental Considerations

Specially designed reaction cells are essential for simulating operational environments during XAFS measurements. These cells are typically equipped with liquid nitrogen cooling and resistive heating functions to enable a wide range of temperature control. These cells must accommodate the following:

- Controlled atmospheres, including reactive gases like CO, H_2, and O_2.
- Precise temperature and pressure conditions, often mimicking industrial settings.

- Electrochemical configurations for studying batteries, fuel cells, and electrocatalysts. For instance, advanced spectroelectrochemical cells have enabled real-time monitoring of electrode processes in water-splitting reactions.

5.4.4 Challenges and Future Directions

5.4.4.1 Technical Challenges

- *Cell Design*: Creating reaction cells that replicate real-world conditions without interfering with X-ray transmission is challenging. Materials must be X-ray transparent while maintaining mechanical and chemical stability [48].
- *Data Analysis*: The dynamic and often noisy nature of in situ and operando data requires advanced computational tools for accurate analysis and modelling. Machine learning and AI are increasingly being employed to interpret complex datasets.
- *Temporal Resolution*: Capturing ultrafast processes, such as electron transfer or bond cleavage, demands cutting-edge detection systems and synchrotron configurations.

5.4.4.2 Future Directions

The future of in situ and operando XAFS lies in the following:

- *Integration with Complementary Techniques*: Combining XAFS with X-ray diffraction (XRD), Raman spectroscopy, infrared spectroscopy, and electron microscopy will provide multidimensional insights into material behavior under operational conditions [49].
- *Advances in Beamline Technology*: Emerging beamlines with higher energy resolution and faster data acquisition rates will enable studies of transient phenomena at unprecedented temporal scales.
- *Increased Accessibility*: Efforts to develop laboratory-based XAFS setups will make these powerful techniques accessible to a broader range of researchers, reducing reliance on synchrotron facilities.

In situ and operando XAFS techniques have revolutionized the study of functional materials, offering a detailed understanding of their dynamic behavior under realistic conditions. These methods have provided critical insights into reaction mechanisms, structural evolution, and performance optimization across a wide range of applications, from catalysis and energy storage to environmental science. As experimental setups and computational tools continue to evolve, the scope and impact of in situ and operando XAFS are expected to expand, unlocking new frontiers in material science and technology.

5.5 Advanced XAFS Methods

Advanced XAFS methods represent the forefront of spectroscopic innovation for characterizing atomic and electronic structures in materials. These methods extend traditional XAFS capabilities, providing enhanced resolution, sensitivity, and time-resolved data collection to probe the intricacies of material behavior under complex reaction environments. By combining advanced analytical techniques with state-of-the-art instrumentation, these methods have become indispensable for understanding the nuanced dynamics of catalysts, single-atom systems, and multi-component materials in diverse applications, including electrocatalysis, environmental science, and energy storage. Techniques such as High-Energy Resolution Fluorescence Detected XAFS (HERFD-XAFS), Grazing-Incidence XAFS (GI-XAFS), Quick-Scanning XAFS (Q-XAFS), and in situ/operando XAFS enable researchers to achieve unparalleled precision in studying local atomic environments and reaction mechanisms.

5.5.1 Key Techniques

5.5.1.1 High-energy Resolution Fluorescence Detected XAFS (HERFD-XAFS)

The first HERFD experiment was conducted by Hämäläinen and coworkers [50], who recorded the Dy L_3-edge by detecting 2p3d fluorescence. Its energy resolution surpassed the broadening caused by core-hole lifetime effects. HERFD spectroscopy is applicable to the K-edges of 3d transition metals and the L_2 and L_3 edges of rare earth and actinide elements. HERFD-XAFS delivers exceptional energy resolution by employing a crystal analyzer to selectively capture fluorescence emissions, allowing fluorescence yields to be recorded within an energy window of a few electron volts. This approach mitigates the lifetime broadening of the core hole in XAS spectra, resulting in clearer spectral features and achieving resolutions as fine as ~1 eV. This heightened sensitivity is particularly valuable for resolving subtle electronic and structural changes in the XANES region, which are often obscured in conventional fluorescence–detection setups. For instance, HERFD-XAFS has revealed the oxidation state dynamics of Cu nanoparticles during CO_2 reduction, where transitions between Cu(I) and Cu(0) states were correlated with enhanced selectivity toward multicarbon products such as ethylene and ethanol [51]. Furthermore, this technique has proven essential in investigating the electronic interactions within single-atom catalysts, where fine details of oxidation and spin states critically influence catalytic performance [52, 53].

5.5.1.2 Grazing-Incidence XAFS (GI-XAFS)

GI-XAFS is designed for surface-sensitive analysis, with a near-surface detection depth at the nanoscale, achieved by utilizing shallow incidence angles to probe atomic interactions in the topmost layers of materials. This technique is particularly effective for studying thin films, nanoparticles, and surface-anchored catalysts under operational conditions. Applications include analyzing the behavior of single-atom catalysts on oxide supports, revealing how anchoring mechanisms and metal–support interactions influence their physical and chemical properties [54]. GI-XAFS has also been employed to monitor the dynamic restructuring of metal clusters during thermal and electrochemical reactions, providing insights into surface dynamics critical to optimizing catalytic efficiency [55, 56].

5.5.1.3 Quick-scanning XAFS

Quick-Scanning XAFS (Q-XAFS) usually refers to facilitates high-speed data acquisition by continuously oscillating the channel-cut monochromator. Its time resolution depends on hardware-related factors, such as the scanning speed of the monochromator, the response time of the detector and measurement system, as well as sample-related factors. Influenced by the monochromator scanning speed and the response time of ionization chambers, the minimum time resolution of Q-XAFS is approximately 1 ms [57]. The millisecond time resolution (Figure 5.8a) [58, 59] is instrumental for capturing transient phenomena and reaction kinetics in real time. For example, Q-XAFS has been applied to monitor intermediate species during water splitting and CO_2 reduction reactions, offering time-resolved insights into the formation and stabilization of active sites [60]. Such data are invaluable for elucidating reaction mechanisms and refining catalytic designs. Q-XAFS is suitable for spatially nonuniform samples and samples surrounded by highly scattering materials. It can be performed in the same mode as step-scan XAFS. Additionally, Q-XAFS can be combined with time-resolved synchrotron radiation (SR) X-ray methods as well as non-X-ray-based methods. However, it requires specially designed operando cells and customized setups for the sample cells and detectors. Time-resolved XAFS methods also include energy-dispersive XAFS (DXAFS,

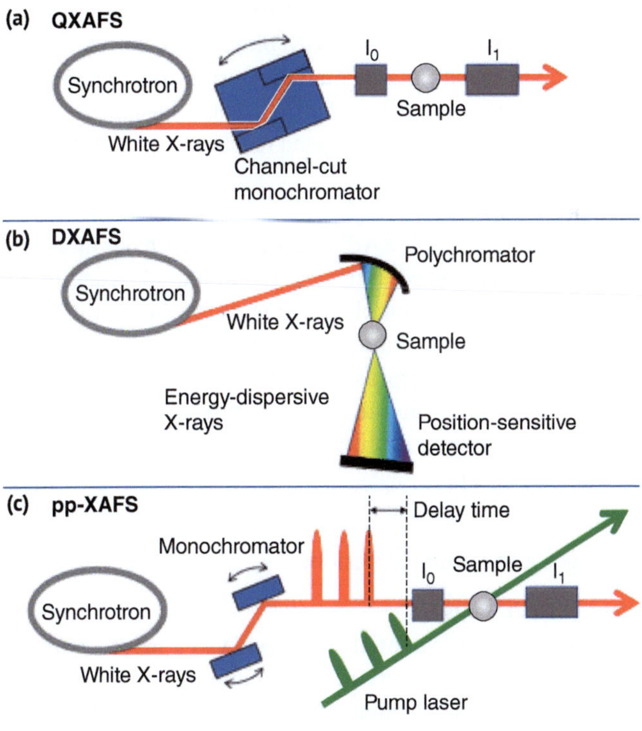

Figure 5.8 Schematic diagrams of (a) quick-XAFS (QXAFS), (b) energy-dispersive XAFS (DXAFS), and (c) pump–probe XAFS (pp-XAFS). *Source:* Reproduced from [46] / with permission of Springer Nature.

Figure 5.8b) [61] and pump-probe-XAFS (pp-XAFS, Figure 5.8c) [62–64], with time resolutions of several microsecond and around 100 picosecond, respectively.

5.5.2 Applications in Electrocatalysis

5.5.2.1 Single-atom Catalysts

Advanced XAFS techniques have been pivotal in elucidating the structural and electronic properties of single-atom catalysts (SACs), which feature isolated metal atoms dispersed on supports. HERFD-XAFS and GI-XAFS are particularly effective for characterizing SACs, revealing details about local coordination environments and oxidation states that govern catalytic activity [65]. For instance, studies on Pt atoms anchored on ceria supports using HERFD-XAFS demonstrated enhanced stability and unique electronic interactions that contribute to superior CO oxidation activity [66, 67]. These findings underscore the role of advanced characterization in the rational design of SACs.

5.5.2.2 Multi-component Systems

Bimetallic and alloy catalysts present unique challenges in characterization due to the overlap of spectral features from different elements. Advanced XAFS techniques address this complexity by enabling precise deconvolution of overlapping signals, thereby providing a clearer understanding of structural and electronic interactions. For example, operando XAFS studies of Pt-Ni bimetallic catalysts during the oxygen reduction reaction (ORR) have highlighted the critical role of metal–support interactions in enhancing activity and stability [68]. These insights are instrumental in optimizing catalyst compositions for targeted reactions.

5.5.2.3 Reaction Mechanisms

The combination of advanced XAFS techniques with computational modelling has significantly advanced our understanding of reaction mechanisms. Q-XAFS, for instance, has been employed to monitor the evolution of active sites and intermediates during CO_2 reduction, capturing transient species and structural rearrangements that underpin catalytic performance [69, 70]. Additionally, HERFD-XAFS has provided atomic-scale insights into the electronic changes occurring during the hydrogen evolution reaction, aiding in the identification of key catalytic pathways [71].

5.5.3 Challenges and Future Directions

5.5.3.1 Technical Limitations

Despite their transformative potential, advanced XAFS methods face several challenges. High instrumentation costs and the complexity of experimental setups limit their accessibility to a broader research community. Moreover, improving the temporal resolution of Q-XAFS and extending the energy range accessible to GI-XAFS remain critical goals for advancing these techniques.

5.5.3.2 Integration with Complementary Techniques

The integration of XAFS with complementary methods, such as X-ray diffraction (XRD), Raman spectroscopy, and electron microscopy, provides a multidimensional perspective on material behavior, including studies on disorder (mean square displacement of atomic distances) and particle size (based on the coordination number of higher shell neighboring atoms) [72]. Multimodal setups are expected to further enhance the ability to correlate local structural dynamics with macroscopic properties. For instance, combining XAFS with in situ XRD enables simultaneous monitoring of short- and long-range structural changes during phase transitions. This is because XAFS is not limited to crystalline phases and provides information on local disorder rather than long-range disorder.

5.5.3.3 Computational Approaches

Machine learning and advanced computational tools, such as Reverse Monte Carlo, neural networks, and MCR-ALS, offer promising avenues for accelerating the analysis of XAFS data. These approaches can facilitate automated spectral interpretation, analyze the local environment of active species, and identify correlations between structural features and material performance, which is beneficial for breakthroughs in heterogeneous and complex processes. Coupling computational modelling with experimental data can also provide predictive insights, guiding the design of next-generation catalysts.

Advanced XAFS methods have redefined the landscape of materials characterization, enabling unprecedented insights into the atomic-scale dynamics of catalysts and functional materials. By bridging the gap between experimental observation and theoretical modelling, these techniques offer powerful tools for addressing critical challenges in energy conversion, environmental remediation, and sustainable chemistry. As instrumentation and analytical methodologies continue to evolve, advanced XAFS is poised to play an increasingly vital role in the development of innovative materials and technologies.

5.6 Dynamic Structural Evolution in Catalysts

Catalysts are fundamental to a wide array of chemical transformations, ranging from industrial-scale reactions to renewable energy applications [73]. Their performance and efficiency are intricately linked to their structural and electronic properties. However, these properties are rarely static; they continuously evolve under operational conditions due to external stimuli such as temperature, pressure, and chemical environment. This dynamic nature often determines the reactivity, selectivity, and longevity of catalytic systems. Understanding the nuances of this structural evolution is critical for uncovering reaction mechanisms, identifying transient active sites, and designing next-generation catalysts that are more efficient, robust, and sustainable. Among the advanced techniques available for probing these changes, XAFS spectroscopy, especially in its in situ and operando modes, stands out as a powerful analytical tool [74, 75]. It provides unparalleled atomic-level insights into the structure–function relationships that govern catalytic behavior.

5.6.1 Role of Surface and Bulk Dynamics

Dynamic structural evolution can manifest at both the surface and bulk levels of catalysts, each with distinct implications for performance.

- *Surface Dynamics*: The surface of a catalyst, often considered the "active zone," undergoes significant transformations during catalytic reactions. Active sites, which are typically located at defects, edges, or specific crystal facets, may form, dissolve, or reorganize dynamically under reaction conditions. For instance, palladium catalysts employed in the aerobic oxidation of cinnamyl alcohol exhibit pronounced structural transformations [76]. In situ XAFS studies reveal that surface palladium oxides (PdO) are reduced to metallic palladium (Pd) as the reaction progresses [77]. This reduction correlates directly with a decline in catalytic activity, emphasizing the importance of maintaining an optimal oxidation state to achieve sustained catalytic performance.
- *Bulk Dynamics*: While surface transformations dominate catalytic activity, changes in the bulk structure also play a pivotal role. Bulk restructuring may involve phase transitions, crystallinity alterations, or atomic diffusion. Temperature-dependent XAFS studies on gold nanoparticles (AuNPs) have demonstrated that coordination numbers and bond lengths of gold atoms vary significantly with temperature and reaction conditions [78]. These bulk vibrational dynamics, particularly the softer vibrational modes of surface atoms, can profoundly influence the stability and reactivity of the catalyst. These findings underline the interplay between surface and bulk dynamics in determining overall catalytic behavior.

5.6.2 Dynamic Effects in Nanostructured Catalysts

Nanostructured catalysts, owing to their high surface-to-volume ratios and unique electronic properties, are especially susceptible to dynamic changes.

- *Size-dependent Effects*: Nanoparticles, particularly those with diameters below 10 nm, exhibit pronounced structural fluctuations. For instance, smaller gold clusters display lower Einstein temperatures and greater bond-length disorder compared to bulk gold [23]. These effects arise due to surface tension and interactions with ligands or support materials, which collectively influence the nanoparticle's reactivity and thermal stability.
- *Shape-dependent Dynamics*: The geometry of nanoparticles, including facets, edges, and corners, plays a crucial role in determining catalytic activity. Dynamic reshaping of nanoparticles under reaction conditions can lead to the exposure or occlusion of highly active sites, further affecting performance.

The dynamic structural evolution of catalysts under reaction conditions is a key determinant of their performance, stability, and durability. In situ and operando XAFS techniques, complemented by computational modelling and advanced data analysis, have provided profound insights into these dynamic processes. These methods have not only deepened our understanding of reaction mechanisms but have also paved the way for rational catalyst design. As experimental setups and analytical tools continue to evolve, the potential to develop catalysts with tailored properties for specific applications becomes increasingly attainable. Future advancements in synchrotron radiation sources and machine learning algorithms are poised to further revolutionize this field, enabling real-time monitoring and predictive modelling of catalytic systems for sustainable chemical processes.

5.7 Application of XAFS in Electrochemical Energy Conversion

Electrochemical energy conversion technologies, such as water splitting, fuel cells, and carbon dioxide reduction, are central to the global transition to renewable and sustainable energy. These technologies depend critically on advanced catalysts to lower energy barriers, enhance efficiency, and enable selectivity in complex electrochemical reactions. XAFS spectroscopy has emerged as a vital tool for characterizing these catalytic materials, providing unparalleled insights into their structural and electronic properties under operational conditions [79, 80]. By integrating real-time analysis with advanced computational techniques, XAFS has enabled researchers to elucidate reaction mechanisms, optimize catalyst design, and enhance performance across diverse energy conversion processes. This section explores the pivotal applications of XAFS in studying key reactions and developing advanced catalytic materials for electrochemical energy conversion.

5.7.1 Hydrogen Evolution Reaction

The hydrogen evolution reaction (HER) is a key half-reaction in water splitting, necessitating the design of efficient catalysts to minimize overpotential and maximize hydrogen production.

- *Catalyst Design*: Transition metal sulfides, phosphides, and carbides have been widely studied for HER. Operando XAFS has been instrumental in revealing the role of MoS_2 and the active Ni_2P phases formed during HER under reaction conditions [81]. These studies demonstrate how XAFS enables the identification of active sites and structural dynamics critical for HER optimization. Furthermore, XAFS has elucidated the formation of metastable phases and the evolution of coordination environments during electrochemical cycling, providing deeper insights into material durability.
- *Electronic Structure*: XANES analysis has provided insights into the changes in electronic density of states and oxidation states of catalytic centers, aiding in the design of non-precious metal-based HER catalysts. Advanced HER studies using XAFS have shown the importance of hybrid materials, such as MoS_2 nanosheet-supported transition metals, where strong metal–support interactions play a key role in enhancing catalytic performance [82].

5.7.2 Oxygen Evolution Reaction

The oxygen evolution reaction (OER), often considered a bottleneck in water splitting, is limited by slow reaction kinetics and high overpotentials.

- *Oxide Catalysts*: Advanced XAFS techniques, including HERFD-XAFS, have been employed to investigate transition metal oxides such as Ni-Fe-layered double hydroxides (LDHs) and their analogues, as well as Co-based spinels. These studies reveal the formation of high-valence states during OER, providing crucial insights into the active phases [83]. Operando experiments have further demonstrated the dynamic restructuring of catalyst surfaces under high oxidative potentials, highlighting the importance of maintaining structural integrity to sustain performance.
- *Deactivation Mechanisms*: Operando XAFS has been utilized to monitor the degradation of noble-metal-based catalysts such as RuO_2 under cycling conditions, helping to design more stable alternatives [84]. By identifying pathways of catalyst dissolution and passivation, these studies contribute to the development of long-lasting OER catalysts.

5.7.3 Oxygen Reduction Reaction

The oxygen reduction reaction (ORR) is critical in fuel cells and metal–air batteries, where the efficiency and stability of catalysts directly impact device performance.

- *Platinum-Based Catalysts*: XAFS has provided key insights into the role of alloying and surface modifications in enhancing the ORR activity of Pt-based catalysts. For example, HERFD-XAFS has revealed the lattice strain effects and electronic redistribution in Pt-Co alloys, which contribute to improved catalytic performance [85]. Operando XAFS studies have further shown how electrolyte composition influences the stability and activity of Pt surfaces during long-term operation.
- *Non-precious Catalysts*: XANES and EXAFS have been used to study Fe-N-C catalysts, elucidating the coordination environment of Fe centers and their role in ORR activity and stability [86]. These studies highlight the importance of nitrogen-doping configurations in stabilizing Fe centers and enhancing catalytic activity.

5.7.4 Carbon Dioxide Reduction Reaction (CO_2RR)

The electrochemical reduction of CO_2 offers a promising pathway for mitigating greenhouse gas emissions and producing valuable chemicals. Qiu et al. studied the MOF-transformed In_2O_{3-x}@C nanocorn electrocatalyst for efficient CO_2 reduction to HCOOH using operando XAFS (Figure 5.9) [43]. They identified the active site of In^{3+} for formate production, despite the possible reduction of In^{3+} to In at the applied negative potentials during the CO_2 reduction reaction.

- *Copper Catalysts*: Operando XAFS studies on Cu-based catalysts have revealed the dynamic formation of under-coordinated Cu sites and their role in

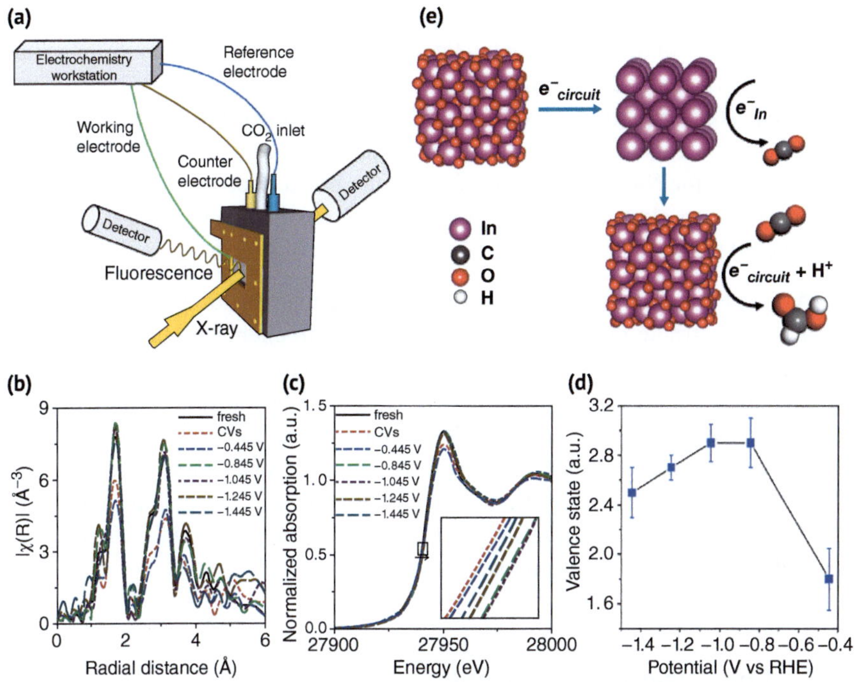

Figure 5.9 (a) Schematic of the experimental setup. (b) Fourier-transformed (FT) k^3-weighted *R*-space EXAFS spectra. (c) In K-edge XANES spectra of MIL-68-N_2 tested under different applied potentials. (d) Changes in the In valence state of MIL-68-N_2 under varying applied potentials. (e) The proposed reaction mechanism based on operando measurements. *Source:* [87] / Springer Nature / CC BY 4.0

stabilizing intermediates for multi carbon product formation [88]. These findings highlight the importance of real-time structural monitoring. Advanced studies using time-resolved XAFS have shown how reaction intermediates, such as CO and CHO, evolve during catalysis, providing direct evidence of the mechanisms driving selectivity toward ethylene and ethanol [89].

- *Single-atom Catalysts*: XAFS has been employed to investigate SACs such as Ni and Fe supported on N-doped carbon [90]. These studies demonstrate how single-atom configurations enable selective CO_2 reduction to desired products. Operando XANES has further revealed how electronic coupling between single-atom sites and the carbon matrix enhances stability and activity under varying potentials.

5.7.5 Nitrogen Reduction Reaction

The electrochemical nitrogen reduction reaction (NRR) represents a sustainable alternative to the Haber–Bosch process for ammonia synthesis.

- *Active Site Identification*: XAFS studies on Mo- and Fe-based catalysts have identified single-atom active sites as crucial for breaking the strong $N \equiv N$ bond [91, 92]. Operando experiments provide evidence of intermediate species such as N_2H_x, offering insights into reaction pathways. These studies emphasize the importance of optimizing metal–ligand coordination environments to facilitate electron transfer during NRR.

- *Support Effects*: Studies using XAFS have highlighted the influence of different support materials on the electronic structure and activity of single-atom catalysts for NRR. For instance, graphene-based supports have been shown to enhance electron mobility and stabilize intermediate species, improving overall efficiency [93].

XAFS spectroscopy provides unparalleled capabilities for probing the structural and electronic dynamics of catalysts in electrochemical energy conversion processes. Its ability to operate under real-world conditions and offer atomic-level insights makes it an indispensable tool for catalyst development. Future advancements in instrumentation, computational integration, and data analysis promise to further enhance its impact, driving innovations in sustainable energy technologies. As XAFS techniques continue to evolve, their role in addressing global energy challenges will become increasingly vital, bridging the gap between fundamental research and practical applications.

5.8 Conclusion and Outlook

XAFS spectroscopy, with its unparalleled ability to probe the atomic-scale structure and electronic properties of materials, has proven indispensable in this endeavor. Operando and in situ XAFS techniques have provided real-time insights into the dynamic evolution of catalysts during reactions, elucidating critical aspects such as active site transformations, intermediate formation, and degradation mechanisms. From uncovering the unique properties of single-atom catalysts to advancing multi-component systems and alloy designs, XAFS has become a fundamental tool in guiding catalyst innovation.

Beyond its capabilities in characterizing catalysts, the integration of XAFS with computational tools such as density functional theory (DFT) and machine learning is accelerating the pace of discovery. These methodologies enable predictive modelling and high-throughput screening, bridging the gap between fundamental research and practical applications. The synergy between experimental and computational approaches promises to unlock new frontiers in catalyst design, optimizing materials for specific reactions and operational conditions.

Despite its transformative impact, challenges remain. Enhancing temporal resolution to capture ultrafast processes, improving data analysis techniques to handle complex spectra, and developing scalable methods for high-throughput characterization are areas ripe for advancement. As instrumentation continues to evolve and analytical methodologies mature, the role of XAFS in shaping the future of electrochemical energy conversion will only grow more significant.

In conclusion, XAFS spectroscopy represents a critical pillar in the ongoing quest for efficient, sustainable, and scalable energy conversion technologies. By providing atomic-level insights and enabling the rational design of advanced materials, XAFS is not only advancing our understanding of electrochemical processes but also paving the way for a cleaner and more sustainable energy future.

Acknowledgment

This work is supported by Agency for Science, Technology and Research (A*STAR) through Low Carbon Energy Research Finding Initiative (LCERFI01-0033 | U2102d2006).

References

1 Gao, DF., Liu, TF., Wang, GX. et al. (2021). Structure sensitivity in single-atom catalysis toward CO_2 electroreduction. *ACS Energy Lett* 6 (2): 713–727.

2 Wang, K., Yu, B.Y., Yang, F. et al. (2022). Review-structural design and dynamic evolution of intermetallic compound nanocatalysts revealed by environmental TEM. *ECS J Solid State Sci Technol* 11: 101006.

3 Weckhuysen, B. M. (2003). Operando spectroscopy: fundamental and technical aspects of spectroscopy of catalysts under working conditions. *Phys Chem Chem Phys* 5 (20): vi–vi.

4 Weckhuysen, B.M. (2005). In-situ spectroscopy of catalysts. *ChemInform* 1–11.

5 Lytle, F.W. (1976). Determination of d-band occupancy in pure metals and supported catalysts by measurement of the L_{III} X-ray absorption threshold. *J Catal* 43 (1): 376–379.

6 Sinfelt, J.H., Via, G.H., and Lytle, F.W. (1978). Extended X-ray absorption fine structure (EXAFS) of supported platinum catalysts. *J Chem Phys* 68: 2009–2010.

7 Mustre, J., Yacoby, Y., Stern, E. A. et al. (1990). Analysis of experimental extended x-ray-absorption fine-structure (EXAFS) data using calculated curved-wave, multiple-scattering EXAFS spectra. *Phys Rev B* 42 (17): 10843–10851.

8 Iwasawa, Y. (2017). History and progress of X-ray absorption fine structure (XAFS). In: *XAFS Techniques for Catalysts, Nanomaterials, and Surfaces* (ed. Y. Iwasawa, K. Asakura, and M. Tada), 3–10. Cham: Springer.

9 Lim, A., Tesch, M.F., and Spanos, I. (2023). The power of operando analysis: Understanding the critical characteristics of OER catalysts from atomistic to systemic scales. *Curr Opin Electrochem* 39: 101272.

10 Rotonnelli, B., Lassalle-Kaiser, B., Bonnefont, A. et al. (2024). Instability of Cobalt-substituted polyoxometalates during the oxygen evolution reaction: an operando X-ray absorption spectroscopy study. *J Phys Chem C* 128 (19): 7968–7976.

11 Fabbri, E., Abbott, D.F., Nachtegaal, M. et al. (2017). Operando X-ray absorption spectroscopy: a powerful tool toward water splitting catalyst development. *Curr Opin Electrochem* 5(1): 20–26.

12 Newville, M. (2004). *Fundamentals of XAFS*, vol. 78. Chicago, IL: Consortium for Advanced Radiation Sources, University of Chicago (USA).

13 Rehr, J.J. and Albers, R.C. (2000). Theoretical approaches to x-ray absorption fine structure. *Rev Mod Phys* 72 (3): 621.

14 Song, H., Song, Z., Zhou, W. et al. (2024). XAFS method for the structural characterization of single atom catalysts. *Sci China Chem*.

15 Bunker, G. (2010). *Introduction to XAFS: A Practical guide to X-ray Absorption Fine Structure Spectroscopy*. Cambridge University Press.

16 Kowalska, J.K., Lima, F.A., Pollock, C.J. et al. (2016). A practical guide to high-resolution x-ray spectroscopic measurements and their applications in bioinorganic chemistry. *Isr J Chem* 56 (9–10): 803–815.

17 Guda, A.A., Guda, S.A., Martini, A. et al. (2021). Understanding X-ray absorption spectra by means of descriptors and machine learning algorithms. *NPJ Comput Mater* 7 (1): 203.

18 Sharpe, L.R., Heineman, W.R., and Elder, R.C. (1990). EXAFS spectroelectrochemistry. *Chem Rev* 90 (5): 705–722.

19 Ankudinov, A.L., Ravel, B., Rehr, J.J. et al. (1998). Real-space multiple-scattering calculation and interpretation of x-ray-absorption near-edge structure. *Phys Rev B* 58 (12): 7565–7576.

20 Frenkel, A.I. (2012). Applications of extended X-ray absorption fine-structure spectroscopy to studies of bimetallic nanoparticle catalysts. *Chem Soc Rev* 41 (24): 8163–8178.

21 Beale, A.M. and Weckhuysen, B. M. (2010). EXAFS as a tool to interrogate the size and shape of mono and bimetallic catalyst nanoparticles. *Phys Chem Chem Phys* 12 (21): 5562–5574.

22 Russell, A.E. and Rose, A. (2004). X-ray absorption spectroscopy of low temperature fuel cell catalysts. *Chem Rev* 104 (10): 4613–4636.

23 Katsikini, M. and Paloura, E.C. (2016). XAFS for characterization of nanomaterials. In: *X-ray and Neutron Techniques for Nanomaterials Characterization* (ed. C.S.S.R. Kumar), 157–246. Berlin, Heidelberg: Springer.

24 Newville, M. (2014). Fundamentals of XAFS. *Rev Mineral Geochem* 78 (1): 33–74.

25 Du, Y., Zhu, Y., Xi, S. et al. (2015). XAFCA: a new XAFS beamline for catalysis research. *J Synchrotron Radiat* 22 (3): 839–843.

26 Lu, K.-Q. and Stern, E.A. (1983). Size effect of powdered sample on EXAFS amplitude. *Nucl Instrum Methods Phys Res* 212 (1): 475–478.

27 Jaklevic, J., Kirby, J.A., Klein, M.P. et al. (1993). Fluorescence detection of EXAFS: sensitivity enhancement for dilute species and thin films. *Solid State Commun* 88 (11): 1105–1108.

28 Lytle, F.W., Greegor, R.B., Sandstrom, D.R. et al. (1984). Measurement of soft X-ray absorption spectra with a fluorescent ion chamber detector. *Nucl Instrum Methods Phys Res A* 226 (2): 542–548.

29 Ravel, B. and Newville, M. (2005). ATHENA, ARTEMIS, HEPHAESTUS: data analysis for X-ray absorption spectroscopy using IFEFFIT. *J Synchrotron Radiat* 12 (4): 537–541.

30 Wang, J., Hsu, C.-S., Wu, T.-S. et al. (2023). In situ X-ray spectroscopies beyond conventional X-ray absorption spectroscopy on deciphering dynamic configuration of electrocatalysts. *Nature Commun* 14 (1): 6576.

31 Sun, Z., Liu, Q., Yao, T. et al. (2015). X-ray absorption fine structure spectroscopy in nanomaterials. *Sci China Mater* 58 (4): 313–341.

32 Ankudinov, A.L., Ravel, B., Rehr, J.J., et al. (1998). Real-space multiple-scattering calculation and interpretation of x-ray-absorption near-edge structure. *Phys Rev B* 58 (12): 7565.

33 Downward, L., Booth, C.H., Lukens, W.W. et al. (2007). A variation of the F-test for determining statistical relevance of particular parameters in EXAFS fits. *AIP Conf Proc* 882 (1): 129–131.

34 Kuzmin, A. and Chaboy, J. (2014). EXAFS and XANES analysis of oxides at the nanoscale. *IUCrJ* 1 (6): 571–589.

35 Timoshenko, J. and Kuzmin, A. (2009). Wavelet data analysis of EXAFS spectra. *Comput Phys Commun* 180 (6): 920–925.

36 Wang, M.Y., Arnadóttir, L., Xu, Z.C.J. et al. (2019). In situ X-ray absorption spectroscopy studies of nanoscale electrocatalysts. *Nano-Micro Lett* 11 (1): 47.

37 Funke, H., Chukalina, M., and Scheinost, A.C. (2007). A new FEFF-based wavelet for EXAFS data analysis. *J Synchrotron Radiat* 14 (Pt 5): 426–432.

38 Fei, H., Dong, J., Chen, D. et al. (2019). Single atom electrocatalysts supported on graphene or graphene-like carbons. *Chem Soc Rev* 48(20): 5207–5241.

39 Timoshenko, J. and Roldan Cuenya, B. (2021). In situ/operando electrocatalyst characterization by X-ray absorption spectroscopy. *Chem Rev* 121 (2): 882–961.

40 Bordiga, S., Groppo, E., Agostini, G. et al. (2013). Reactivity of surface species in heterogeneous catalysts probed by in situ X-ray absorption techniques. *Chem Rev* 113 (3): 1736–1850.

41 Mesilov, V.V., Bergman, S.L., Dahlin, S. et al. (2021). Differences in oxidation-reduction kinetics and mobility of Cu species in fresh and SO_2-poisoned Cu-SSZ-13 catalysts. *Appl Catal B: Env* 284: 119756.

42 Matos, J., Ono, L.K., Behafarid, F. et al. (2012). In situ coarsening study of inverse micelle-prepared Pt nanoparticles supported on γ-Al_2O_3: pretreatment and environmental effects. *Phys Chem Chem Phys* 14 (32): 11457–11467.

43 Li, J. and Gong, J. (2020). Operando characterization techniques for electrocatalysis. *Energy Environ Sci* 13 (11): 3748–3779.

44 Aquilanti, G., Giorgetti, M., Dominko, R. et al. (2017). Operando characterization of batteries using X-ray absorption spectroscopy: advances at the beamline XAFS at synchrotron Elettra. *J Phys D Appl Phys* 50 (7): 074001

45 Wang, H., Isobe, J., Shimizu, T. et al. (2017). Preparation of γ-LiV_2O_5 from polyoxovanadate cluster $Li_7V_{15}O_{36}$ (CO_3) as a high-performance cathode material and its reaction mechanism revealed by operando XAFS. *J Power Sources* 360: 150–156.

46 Ishiguro, N. and Tada, M. (2018). Structural kinetics of cathode events on polymer electrolyte fuel cell catalysts studied by operando time-resolved XAFS. *Catal Letters* 148: 1597–1609.

47 Bai, Y., Liu, C., Chen, T. et al. (2021). MXene-Copper/Cobalt hybrids via Lewis acidic molten salts etching for high performance symmetric supercapacitors. *Angewandte Chemie* 60 (48): 25318–25322.

48 Doronkin, D.E., Lichtenberg, H., and Grunwaldt, J.-D. (2017). Cell designs for in situ and operando studies. *XAFS Tech Catal Nanomat Surf* 75–89.

49 Marinkovic, N.S., Wang, Q., Barrio, L. et al. (2011). Combined in situ X-ray absorption and diffuse reflectance infrared spectroscopy: an attractive tool for catalytic investigations. *Nucl Instrum Methods Phys Res A* 649(1): 204–206.

50 Hämäläinen, K., Siddons, D.P., Hastings, J.B. et al. (1991). Elimination of the inner-shell lifetime broadening in X-ray-absorption spectroscopy. *Phys Rev Let* 67 (20): 2850–2853.

51 Gao, D., Sinev, I., Scholten, F. et al. (2019). Selective CO_2 electroreduction to ethylene and multicarbon alcohols via electrolyte-driven nanostructuring. *Angewandte Chemie* 131 (47): 17203–17209.

52 Lu, Y., Wang, J., Yu, L. et al. (2019). Identification of the active complex for CO oxidation over single-atom Ir-on-$MgAl_2O_4$ catalysts. *Nat Catal* 2 (2): 149–156.

53 Martini, A., Timoshenko, J., Grosse, P. et al. (2024). Adsorbate configurations in Ni single-atom catalysts during CO_2 electrocatalytic reduction unveiled by operando XAS, XES, and machine learning. *Phys Rev Lett* 133 (22): 228001.

54 Zhang, M., He, G., and Pan, G. (2015). Binding mechanism of arsenate on rutile (1 1 0) and (0 0 1) planes studied using grazing-incidence EXAFS measurement and DFT calculation. *Chemosphere* 122: 199–205.

55 Winzely, M., Clark, A. H., Diercks, J. S. et al. (2024). Electrochemical cell for operando grazing-incidence X-ray absorption spectroscopic studies of low-loaded electrodes. *Anal Chem.*

56 Phadke, S., Coroa, J., Abbas, I. et al. (2024). High-pressure cell for in situ grazing incidence XAS characterization of model catalysts on planar supports. *Chem Meth* e202400014.

57 Iwasawa, Y., Asakura, K., and Tada, M. (eds.). (2017). *XAFS Techniques for Catalysts, Nanomaterials, and Surfaces.* Cham: Springer International Publishing.

58 Frahm, R., Richwin, M., and Lützenkirchen-Hecht, D. (2005). Recent advances and new applications of time-resolved X-ray absorption spectroscopy. *Phys Scr* T115: 974.

59 Stötzel, J., Lützenkirchen-Hecht, D., and Frahm, R. (2010). A new flexible monochromator setup for quick scanning x-ray absorption spectroscopy. *Rev Sci Instr* 81 (7): 073109.

60 Gorlin, Y., Lassalle-Kaiser, B., Benck, J. D. et al. (2013). In situ X-ray absorption spectroscopy investigation of a bifunctional manganese oxide catalyst with high activity for electrochemical water oxidation and oxygen reduction. *J Am Chem Soc* 135 (23): 8525–8534.

61 Pascarelli, S., Mathon, O., Mairs, T. et al. (2016). The Time-resolved and Extreme-conditions XAS (TEXAS) facility at the European Synchrotron Radiation Facility: the energy-dispersive X-ray absorption spectroscopy beamline ID_{24}. *J SyncRad* 23 (1): 353–368.

62 Chen, L.X., Zhang, X., and Shelby, M. L. (2014). Recent advances on ultrafast X-ray spectroscopy in the chemical sciences. *Chem Sci* 5 (11): 4136–4152.

63 Sato, T., Nozawa, S., Tomita, A. et al. (2012).Coordination and electronic structure of ruthenium (II)-tris-2, 2'-bipyridine in the triplet metal-to-ligand charge-transfer excited state observed by picosecond time-resolved Ru K-edge XAFS. *J Phys Chem C* 116 (27): 14232–14236.

64 Wang, H., Yu, C., Wei, X. et al. (2017). Development of picosecond time-resolved X-ray absorption spectroscopy by high-repetition-rate laser pump/X-ray probe at Beijing Synchrotron Radiation Facility. *J Sync Rad* 24 (3): 667–673.

65 Li, Y. and Frenkel, A.I. (2021). Deciphering the local environment of single-atom catalysts with X-ray absorption spectroscopy. *Acc Chem Res* 54 (11): 2660–2669.

66 Nagai, Y., Hirabayashi, T., Dohmae, K. et al. (2006). Sintering inhibition mechanism of platinum supported on ceria-based oxide and Pt-oxide–support interaction. *J Catal* 242 (1): 103–109.

67 Wang, H. and Yang, M. (2024). Two different atomically dispersed Pt atoms supported on ceria. *Inorg Chem* 63 (43): 20592–20599.

68 Liu, D., Li, Y., Kottwitz, M. et al. (2018). Identifying dynamic structural changes of active sites in Pt–Ni bimetallic catalysts using multimodal approaches. *ACS Catal* 8 (5): 4120–4131.

69 Lin, S.-C., Chang, C.-C., Chiu, S.-Y. et al. (2020). Operando time-resolved X-ray absorption spectroscopy reveals the chemical nature enabling highly selective CO_2 reduction. *Nat Commun* 11 (1): 3525.

70 Zhang, N. and Wang, J. (2024). Deciphering the local environment of electrocatalytic metal sites with X-ray absorption fine structure. *Small Struct* 5 (6): 2300555.

71 Mei, B., Liu, C., Li, J. et al. (2022). Operando HERFD-XANES and surface sensitive $\Delta\mu$ analyses identify the structural evolution of copper (II) phthalocyanine for electroreduction of CO_2. *J Energy Chem* 64: 1–7.

72 Cheng, W., Zhao, M., Lai, Y. et al. (2024). Recent advances in battery characterization using in situ XAFS, SAXS, XRD, and their combining techniques: From single scale to multiscale structure detection. *Exploration* 4 (1): 20230056.

73 Suchorski, Y., Zeininger, J., Buhr, S. et al. (2021). Resolving multifrequential oscillations and nanoscale interfacet communication in single-particle catalysis. *Science*. 372 (6548): 1314–1318.

74 Beale, A.M., Jacques, S.D.M., and Weckhuysen, B.M. (2010). Chemical imaging of catalytic solids with synchrotron radiation. *Chem Soc Rev* 39 (12): 4656–4672.

75 Timoshenko, J., Duan, Z., Henkelman, G. et al. (2019). Solving the structure and dynamics of metal nanoparticles by combining X-ray absorption fine structure spectroscopy and atomistic structure simulations. *Ann Rev Anal Chem* 12: 501–522.

76 Shang, Q., Tang, N., Qi, H. et al. (2020). A palladium single-atom catalyst toward efficient activation of molecular oxygen for cinnamyl alcohol oxidation. *Chinese J Catal* 41 (12): 1812–1817.

77 Lee, A. F. and Wilson, K. (2004). Structure–reactivity correlations in the selective aerobic oxidation of cinnamyl alcohol: in situ XAFS. *Green Chem* 6 (1): 37–42.

78 Duan, Z., Li, Y., Timoshenko, J. et al. (2016). A combined theoretical and experimental EXAFS study of the structure and dynamics of Au 147 nanoparticles. *Catal Sci Technol* 6 (18): 6879–6885.

79 Sheng, B., Chu, Y., Cao, D. et al. (2023). Application of X-ray absorption spectroscopy in carbon-supported electrocatalysts. *Nano Res* 16 (11): 12438–12452.

80 Braun, A., Shrout, S., Fowlks, A.C. et al. (2003). Electrochemical in situ reaction cell for X-ray scattering, diffraction and spectroscopy. *J Synch Rad* 10 (4): 320–325.

81 Liu, M., Wang, J.-A., Klysubun, W. et al. (2021). Interfacial electronic structure engineering on molybdenum sulfide for robust dual-pH hydrogen evolution. *Nat Commun* 12 (1): 5260.

82 Duan, H., Liu, W., Guo, P. et al. (2020). Identifying the single active site in MoS_2-based hydrogen evolution electrocatalyst by XAFS technique. *Rad Phys Chem* 175: 108151.

83 Bai, L., Hsu, C.-S., Alexander, D. T. L. et al. (2021). Double-atom catalysts as a molecular platform for heterogeneous oxygen evolution electrocatalysis. *Nat Energy* 6 (11): 1054–1066.

84 Chang, C.-J., Chu, Y.-C., Yan, H.-Y. et al. (2019). Revealing the structural transformation of rutile RuO_2 via in situ X-ray absorption spectroscopy during the oxygen evolution reaction. *Dalton Trans* 48 (21): 7122–7129.

85 Gao, Y., Thakur, N., Uchiyama, T. et al. (2023). Investigating degradation mechanisms in PtCo alloy catalysts: the role of Co content and a Pt-rich shell using operando high-energy resolution fluorescence detection X-ray absorption spectroscopy. *ACS Appl Mater Interfaces* 15 (45): 52473–52484.

86 Wartner, G., Müller-Hülstede, J., Trzesniowski, H. et al. (2024). Operando X-ray absorption spectroscopy of Fe–N–C catalysts based on carbon black and biomass-derived support materials for the ORR. *Sustain Energy Fuels* 8 (10): 2309–2320.

87 Qiu, C., Qian, K., Yu, J. et al. (2022). MOF-transformed In_2O_3-x@ C nanocorn electrocatalyst for efficient CO_2 reduction to HCOOH. *Nano-Micro Lett* 14 (1): 167.

88 Zhang, Y., Chen, Y., Wang, X. et al. (2024). Low-coordinated copper facilitates the $^*CH_2CO$ affinity at enhanced rectifying interface of Cu/Cu_2O for efficient CO_2-to-multicarbon alcohols conversion. *Nat Commun* 15 (1): 5172.

89 Zhang, X.Y., Lou, Z.X., Chen, J. et al. (2023). Direct OC-CHO coupling towards highly C^{2+} products selective electroreduction over stable Cu^0/Cu^{2+} interface. *Nat Commun* 14 (1): 7681.

90 Hossain, M.N., Zhang, L., Neagu, R. et al. (2024). Free-standing single-atom catalyst-based electrodes for CO_2 reduction. *Electrochem Energy Rev* 7 (1): 5.

91 Han, L., Liu, X., Chen, J. et al. (2019). Atomically dispersed molybdenum catalysts for efficient ambient nitrogen fixation. *Angewandte Chemie* 131 (8): 2343–2347.

92 Yang, H., Liu, Y., Luo, Y. et al. (2020). Achieving high activity and selectivity of nitrogen reduction via Fe–N_3 coordination on iron single-atom electrocatalysts at ambient conditions. *ACS Sustain Chem Eng* 8 (34): 12809–12816.

93 Matsoso, J.B., Antonatos, N., Dekanovský, L. et al. (2024). Enhancing nitrogen reduction reaction through formation of 2*D*/2D hybrid heterostructures of MoS_2@ rGO. *ACS Appl Mater Interfaces* 16 (19): 24514–24524.

6

Electrochemical Hydrogen Evolution Reaction

*Xinran Sun#, Boxin Xiao#, Jiaqing Liu#, Xue Feng Lu**

State Key Laboratory of Photocatalysis on Energy and Environment, College of Chemistry, Fuzhou University, Fuzhou, China

*Corresponding author: luxf@fzu.edu.cn
#These authors contributed equally to this work.

6.1 Introduction

The limited reserves of fossil fuels and high pollutant emissions have caused significant social and environmental problems, underscoring the urgent need for energy alternatives to reduce dependence on fossil fuels [1]. Hydrogen, an important energy carrier for facilitating the green and low-carbon transition in energy consumption, has been recognized for its high-combustion calorific value and zero carbon emissions [2]. Hydrogen has wide-ranging applications in agriculture, medicine, and transportation, going far beyond petrochemicals, ammonia production, metal smelting, and fuel desulfurization (Illustration 6.1). Hydrogen Council estimates that the total investment in the hydrogen energy sector worldwide is expected to reach $500 billion by 2030, with a demand for clean hydrogen reaching 75 million tons, highlighting the necessity of highly efficient hydrogen production technologies [3]. Currently, hydrogen is mainly produced through steam reforming of fossil fuels, a method that has low conversion efficiency and emits carbon dioxide. In contrast, electrochemical water splitting is considered a clean, efficient, and sustainable process for carbon-free hydrogen production, playing a pivotal role in shaping future renewable energy strategies.

Water electrolysis involves two distinct reactions occurring at separate electrodes: hydrogen evolution reaction (HER) at the cathode and oxygen evolution reaction (OER) at the anode [4]. The HER mechanism depends on the electrolyte's pH value and can be divided into acidic and alkaline systems (Figure 6.1a) [5]. In an acidic medium, the Volmer step occurs first, where protons gain electrons to generate adsorbed hydrogen atoms (H*) on the electrode surface ($H^+ + e^- \rightarrow H^*$). Next, H* reacts with H^+ in the solution to generate hydrogen through either the Heyrovsky step ($H^* + H^+ + e^- \rightarrow H_2$) or the Tafel step ($2H^* \rightarrow H_2$), or both. In an alkaline medium,

Electrochemical Energy Conversion via Small Molecule Catalysis, First Edition.
Edited by Zhicheng Zhang and Yuanmiao Sun.

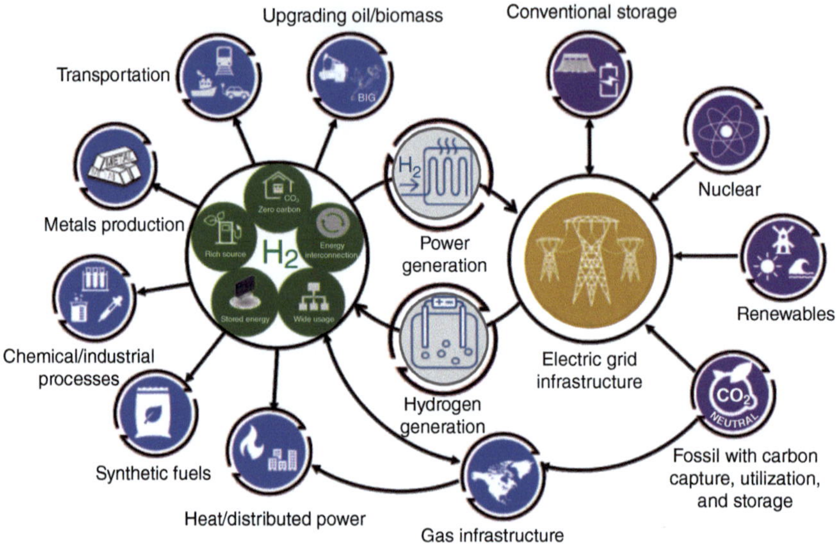

Illustration 6.1 The primary application areas of hydrogen energy.

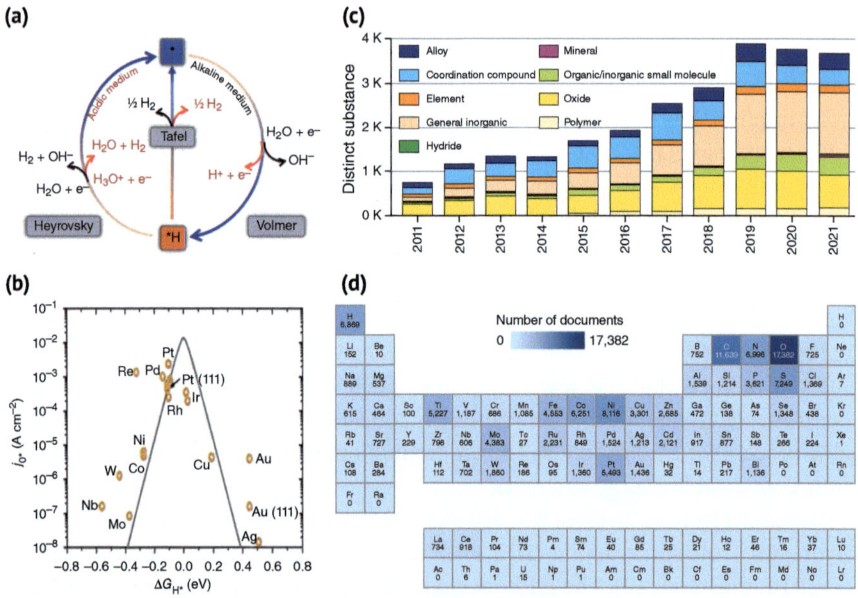

Figure 6.1 (a) The mechanism of the HER process, wherein * represents the active sites of the catalyst. (b) A volcano plot of experimentally measured exchange current density as a function of the DFT-calculated Gibbs free energy of adsorbed atomic hydrogen. *Source*: Adapted from [9]. (c) Annual distribution of the substance categories used in green hydrogen production from 2011 to 2021. (d) The number of documents related to elements in materials used as green hydrogen production catalysts from 2011 to 2021. *Source*: Leilani Lotti Diaz [12], with permission of American Chemical Society.

a water molecule acts as a proton source and splits into OH^- and H* (Volmer step, $H_2O + e^- \rightarrow H^* + OH^-$). Next, H* combines with another water molecule and an electron to form hydrogen (Heyrovsky step, $H * + H_2O + e^- \rightarrow H_2 + OH^-$), or two H* directly couple to generate hydrogen (Tafel step, $2H^* \rightarrow H_2$). Both H* and electron transfer play crucial roles in the HER process, regardless of the specific reaction pathway. The generation of H* in alkaline solutions poses greater challenges, so the HER kinetics in alkaline solutions are usually 2–3 orders of magnitude slower than those in acidic conditions [6]. In addition, different from acidic or alkaline HER, the reaction mechanism of neutral HER is more complex, and its reactants change with cathode potential [7]. To further elucidate the rate-determining step (RDS) of HER, Tafel plots are commonly constructed, with the apparent value of the Tafel slope serving as an indicator for RDS determination. Specifically, when the HER process is dominated by Volmer, Heyrovsky, or Tafel mechanisms, their corresponding Tafel slopes are approximately 118, 39, and 30 mV dec^{-1}, respectively [8].

As an important descriptor in the acidic pathway, the hydrogen adsorption-free energy (G_{H^*}) is widely used as a key indicator for evaluating HER catalysts, and a volcanic curve has been established for the relationship between exchange current density (j_0) and metal—hydrogen bond strength (Figure 6.1b) [9]. The platinum group metals (PGMs) are located at the peak of the volcano plot, with a value close to zero and an excellent j_0 value obtained in the experiment [10]. Research findings indicate that an optimal H* sorption, neither too strong nor too weak, benefits highly active HER catalysts (the Sabatier principle) [11]. Therefore, controlling the hydrogen binding energy on the surface is a research hotspot for designing high-performance electrocatalysts. However, solely focusing on changes in hydrogen binding energy cannot explain the loss of HER activity for Pt from acid to alkali. Moreover, catalysts like Pt-based alloys, iridium, 3d metal hydroxide, chalcogenide, and phosphide exhibit superior catalytic activity in alkaline media despite their less optimal hydrogen binding.

In the process of explaining the above experimental phenomena, the interaction between catalysts and electrolytes (or reactants) has gained significant attention. Unlike surface adsorbents, the substances involved in the interaction between catalysts and electrolytes are directly derived from electrolytes or reactants in the reaction process. Yan and colleagues propose that hydrogen binding energy is a unique descriptor for HER in all pH environments and the slow kinetics of HER in alkaline media can be attributed to the fact that hydrogen binding energy increases with increasing pH [13]. Markovic and colleagues suggest that adsorbing OH* may affect HER kinetics by competing with active intermediates at the same surface sites [14]. Koper et al. demonstrate that increasing electrolyte pH enhances surface cation concentration, stabilizing the Volmer step, while excessive cation concentration (at high pH) can reduce activity by blocking the electrode [15]. Chen et al. suggest that a significant water gap at the alkaline interface reduces the connectivity of the hydrogen bond network, thereby increasing the barrier for hydrogen transfer [16]. Employing a Pt–Ru alloy as the catalyst model, the OH* on the Ru sites modulates the water connectivity and improves the hydrogen bond networks in the electrical double layer, resulting in higher HER activity than Pt. Although there is still no consensus on the alkaline HER so far, strategies have been developed to enhance HER performance by utilizing the catalyst–electrolyte interplay, including using the influence of pH, designing adsorbents in the inner Helmholtz plane (IHP), designing the interaction between catalyst and interfacial water in the IHP, and so on [17, 18].

Since HER occurs at the gas–liquid–solid three-phase interface, the unfavorable transfer of electrons, ions, and molecules at catalytic active sites thus leads to the occurrence of polarization phenomena, which makes the actual potential of hydrogen evolution higher than the theoretical value. In this regard, high-quality electrocatalysts are highly desired to lower the reaction kinetic barriers and facilitate the catalytic efficiency. Noble metals, especially PGMs, are generally considered to be the most-effective electrocatalysts for HER [19, 20]. While taking into account cost and activity, a significant area of research focuses on reducing the usage of PGMs or increasing the activity of non-noble catalysts [21]. Figure 6.1c shows the annual distribution of the substance categories used in green hydrogen production from 2011 to 2021, a growing focus was on various inorganic compounds, small organic/inorganic molecules, polymers, and oxides [22]. In addition, the relative popularity of elements in catalysts employed for green hydrogen production is also assessed at the document level (Figure 6.1d). In general, an obvious emphasis on carbonaceous materials as well as transition metal oxides and sulfides exists. Significant interest has been observed in Co, Ni, and Pt, with the highest publication volume focused on the d^8 transition metals [23].

Currently, highly effective and stable acid HER electrocatalysts are still limited to PGMs, which are expensive and scarce, posing significant obstacles to large-scale commercialization. The dominant strategy in this field focuses on balancing the relationship between hydrogen production efficiency and material costs. For alkaline HER, numerous non-noble metal-based materials have been explored as high-performance electrocatalysts. Additionally, seawater is a universal neutral electrolyte, accounting for more than 96% of the Earth's water resources. Direct seawater electrolysis (SWE) has become a popular and promising research direction for synthesizing green hydrogen without relying on high-purity water. In this chapter, with a start of introducing performance evaluation criteria and methods, we present the research progress on HER electrocatalysts in acidic, alkaline, and neutral electrolytes. Finally, future opportunities and challenges for electrochemical HER are highlighted from the perspectives of both fundamental understandings and practical applications.

6.2 Performance Evaluation Criteria and Methods

Water electrolysis catalysts require key parameters to assess efficiency, stability, and applicability, enabling quantitative comparison, optimization, and understanding of the reaction mechanism [24]. Establishing unified performance evaluation criteria is essential for standardizing catalyst development and facilitating research comparisons, driving industrial applications in hydrogen production.

6.2.1 Overpotential

Overpotential (η) is defined as the potential difference between the applied potential (E_a) at a specific current density and the theoretical potential (E_0), which originates from the activation of the reaction, diffusion of charge carriers, and series resistance. In standard conditions, the E_0 for HER is 0 V versus (vs.) the reversible hydrogen electrode (RHE). Due to the energy barriers in actual systems, the E_a is more negative than E_0. So, the η can be calculated by the equation:

$$\eta = E_0 - (E_a - iR)$$

where iR is the ohmic correction that excludes the effect of resistance on the potential. However, for practical applications, η is obtained from the polarization curve without iR correction, especially at high current densities.

During testing, to minimize the non-Faradaic current, linear sweep voltammetry (LSV) at low scan rates (e.g., 2 mV s^{-1}, 5 mV s^{-1}) is commonly performed to obtain η [25]. Different reference electrodes are selected based on the pH of the electrolyte. The Hg/$HgSO_4$ reference electrode is typically used under acidic conditions, the Ag/AgCl reference electrode and saturated calomel electrodes (SCE) are typically used under neutral conditions [26], while the Hg/HgO reference electrode is commonly used under alkaline conditions. Taking the Hg/HgO reference electrode (Standard electrode potential is 0.098 V) as an example, the E_a can be calibrated to RHE according to the Nernst equation:

$$E_a \text{ (V vs. RHE)} = E \text{ (V vs. Hg/HgO)} + 0.098 + 0.05916 \times \text{pH}$$

Currently, researchers commonly use the η at 10 mA cm^{-2}, denoted as η_{10}, as a metric to compare the activity of different materials. The 10 mA $cm^{-2}{}_{geo}$ benchmark originates from the field of artificial photosynthesis, where 10% solar-to-hydrogen energy conversion efficiency is considered the standard for developing artificial photosynthesis devices, and this conversion efficiency corresponds to a current density of 10 mA $cm^{-2}{}_{geo}$. However, η_{10} does not necessarily represent the intrinsic activity of the catalyst, as it may be influenced by catalyst loading. Additionally, η_{10} is based on the current density normalized to the geometric surface area of the electrode, which overlooks the fact that electrochemical catalysis is fundamentally a surface process.

6.2.2 Onset Potential

Onset potential refers to the potential at which hydrogen begins to be generated on the surface of the electrode material [27]. It is usually determined by LSV (the potential at 1 mA cm^{-2}) or the inflection point method of the straight-line part of the Tafel curve at low potential (the low potential interval is selected in the Tafel curve to make a tangent to the curve, and the intersection with the overpotential of the vertical axis is the starting overpotential). It directly reflects the capacity of catalysts to initiate the electrochemical reaction, with performance improving as the onset potential approaches the theoretical value.

6.2.3 Electrochemical Impedance Spectroscopy

Electrochemical impedance spectroscopy (EIS) involves applying a small amplitude alternating current (AC) signal to the electrode to perturb the system and study the relationship between the AC impedance of the electrochemical system and the frequency [28]. The AC impedance tests typically cover a frequency range from 10 Hz to 100 kHz, with an amplitude of 5 mV. The starting voltage is chosen based on the specific testing conditions, generally corresponding to the open-circuit voltage of the system.

The equivalent circuit model can be obtained by fitting the impedance spectrum. In this model, the constant phase element (CPE) is the double-layer capacitance (C_{dl}) between the catalyst and the electrolyte, as well as the adsorption capacitance of intermediates on the catalyst surface. R_s and R_{ct} correspond to the solution resistance and the charge transfer resistance, respectively. R_p represents the gas diffusion resistance within the electrode material, reflecting the porosity of the electrode. Smaller values of R_s and R_{ct} indicate lower resistance during the reaction process, leading to a faster HER rate and better performance [29].

6.2.4 Tafel Slope and Exchange Current Density

The Tafel slope refers to the slope of the overpotential and the log |current density| in the linear region of the Tafel plot, which can be calculated by drawing the corresponding LSV curve [25]. The formula is as follows:

$$\eta = a + b \log j$$

where η is the overpotential, a is a constant related to temperature, electrolyte, and electrode; b is the Tafel slope; and j is the current density. The Tafel slope can be used to obtain information about the reaction kinetics and determine the RDS. Another approach is to determine R_{ct} through EIS and plot the linear relationship between log R_{ct} and η to derive the Tafel slope, offering a more accurate representation of charge transfer kinetics [30].

Exchange current density (j_0) refers to the current density at the equilibrium position where the cathode and anode currents are equal. It commonly reflects the intrinsic electron transfer rate between the electrode and the electrolyte, and thus can be used to evaluate HER kinetics. Generally, a good electrocatalyst exhibits a high j_0. Currently, j_0 cannot be directly measured by electrochemical methods and can only be calculated through the Tafel equation. Extrapolate the Tafel curve to where it intersects the X-axis, and the corresponding current density is the j_0.

6.2.5 Electrochemically Active Surface Area

Electrochemically active surface area (ECSA) is a crucial parameter for evaluating electrocatalyst performance, reflecting the number of active sites on the catalyst [31]. It can be measured through different methods, including integrating redox peaks, hydrogen underpotential deposition, CO adsorption/desorption, and C_{dl}. Typically, cyclic voltammetry (CV) is employed to calculate the ECSA by measuring the C_{dl} of the catalysts [32]. The formula is as follows:

$$\mathrm{ECSA} = \frac{C_{dl}}{C_s}$$

where C_s is the specific capacitance of the sample. C_{dl} can be determined by the capacitance current in the non-Faraday region obtained from CV curves at various scan rates, with scan rate ($mV\,s^{-1}$) as the X-axis and the absolute value of the current difference (mA) at intermediate potential values as the Y-axis. This approach yields several linearly related points, and their slope after linear fitting represents the C_{dl} of the electrode. It is important to note that ECSA is only meaningful for comparison when used within the same reaction system and among similar materials.

6.2.6 Mass and Specific Activities

Mass activity (MA) is an important parameter for evaluating the performance of electrocatalysts, primarily influenced by the loading of the electrode material [33]. MA is calculated by normalizing the current to the mass loading of the catalyst. It is frequently used as a crucial metric for assessing precious metal catalysts.

Specific activity (SA) is usually normalized to the geometric area of the electrode, the Brunauer–Emmett–Teller (BET) surface area, or the ECSA of the catalyst. SA normalized to geometric area is only applicable to planar and smooth surfaces, while SA normalized to BET surface area often leads to inappropriate estimates because the surface area measured by BET is based on N_2 adsorption and desorption, which is not necessarily electrochemically active. SA normalized to ECSA is more reflective of the intrinsic activity of the catalyst and therefore has received more attention than the former ones. Unfortunately, it is somewhat difficult to accurately measure and compare the ECSA of different materials, which makes ECSA-normalized SA a less practical evaluation parameter. In addition, a benchmark overpotential should be defined to determine MA and SA for better comparison [34].

6.2.7 Turnover Frequency

The turnover frequency (TOF) represents the number of product moles generated by each active site in a unit of time, reflecting the intrinsic activity of the catalyst. The calculation formula for TOF is as follows [35]:

$$\mathrm{TOF} = \frac{j(\eta) \cdot AN_A}{S \cdot nF}$$

A represents the superficial electrode area, N_A represents Avogadro's number ($N_A = 6.022 \times 10^{23}$), j stands for the current density, η is overpotential (V), S is the number of surface sites, n is the number of electrons transferred per molecule, and F is the Faraday constant. It is important to note that the TOF value varies with overpotentials, thus it is essential to state the corresponding overpotential value when describing the TOF value [33].

6.2.8 Faraday Efficiency

Faradaic efficiency (FE) is defined as the ratio of the actual amount of hydrogen produced to the theoretically predicted amount. The theoretical hydrogen production is determined based on charge conservation and material conservation, while the actual gas production can be measured by the water displacement method or gas chromatography.

Another method involves testing using the rotating ring disk electrode (RRDE), utilizing a glassy carbon disk and a Pt ring assembly [24]. The calculation is performed using the following formula:

$$\mathrm{FE} = \frac{\frac{I_R/n_R}{N_{CL}}}{I_D/n_D} = \frac{I_R n_D}{I_D n_R N_{CL}}$$

where I_R and I_D represent the ring current and disk current, respectively. n_R and n_D denote the number of transferred electrons (For HER, both values are 2). N_{CL} represents the collection efficiency of the RRDE.

6.2.9 Stability and Durability

Stability reflects the variation in the current or potential of the catalyst during long-term operation. It is characterized by monitoring the change in current/voltage at a constant potential/current using chronoamperometry [36, 37]. Durability focuses on the resistance of catalysts to performance degradation over multiple cycles. Through the accelerated durability test (ADT), CV testing is conducted for 10 000 cycles within a specific potential range, and the LSV polarization curve after testing is compared with the one before testing to evaluate the durability of catalysts. Stability and durability have different focuses and complement each other in evaluating catalyst performance, collectively reflecting the operational status of catalysts in practical applications.

6.3 Advanced Electrocatalysts for HER

In acidic environments, noble metal catalysts exhibit intrinsic advantages in HER compared to non-noble metals. However, due to their limited availability and high costs, researchers are concentrating on improving the utilization efficiency of these metals to enhance economic viability. In the context of alkaline HER, Pt exhibits suboptimal performance owing to an additional water dissociation step. Transition metals (i.e., Fe, Co, Ni, Mo, etc.) exhibit considerable potential owing to their versatile compositions and tunable electronic structures. Furthermore, in response to actual application demands (particularly in complex settings like SWE), there is an increasing emphasis on designing corrosion-resistant catalysts that function effectively under neutral conditions. Building upon this analysis, we will explore the design strategies and recent advancements in catalyst research across acidic, alkaline, and neutral environments.

6.3.1 Noble Metal-based Electrocatalysts for Acidic HER

Among current water electrolysis technologies, acidic proton exchange membrane water electrolysis (PEMWE) has significant advantages in voltage efficiency, current density, response time, gas purity, and the development of large stacks [38]. The utilization of perfluorosulfonic polymer solid electrolytes gives rise to a highly corrosive acidic environment, prone to causing issues such as degradation and agglomeration of the electrocatalyst. To guarantee the performance and longevity of the electrolyzer, the cathodic catalyst materials are mainly noble metals and their alloys. Regarding the high cost and scarcity of noble metal-based catalysts, the current mainstream design approach is to construct acidic HER electrocatalysts with ultra-low noble metal loading to reduce costs while maintaining high activity and durability. The design of high-performance catalysts must consider two aspects: (i) increasing exposed catalytic active sites; (ii) improving intrinsic activity. To achieve these two points, some strategies have been developed to regulate the atomic-level structure of catalysts, such as alloying, doping, and regulating morphology and size. In addition, lattice strain engineering or surface and interface engineering will also produce unique electronic interactions, thereby improving catalytic performance.

6.3.1.1 Alloying

Alloying noble metals with transition metals can effectively alter the *d*-band center of the noble metal, thereby optimizing the adsorption behavior on the active sites [39]. For example, Kuang et al. optimized the adsorption energy of H* intermediates on the catalyst surface by alloying Pt with Fe to reduce the d-band center of the Pt 5d orbital, resulting in a low overpotential of 13 mV at –10 mA cm^{-2} in 0.5 M H_2SO_4 [40]. Recently, high-entropy alloys (HEAs) have garnered extensive attention due to their capacity to form novel and tunable active sites among adjacent multiple elements. Furthermore, the inherent high-entropy mixture state and slow diffusion tendency of HEAs bestow upon them outstanding thermal stability [41, 42]. Zhao et al. synthesized highly dispersed sub-2 nm PtRuMoFeCoNi HEA quantum dots (HEA-QDs) by rapid microwave radiation [43]. X-ray absorption fine structure (XAFS) measurement and density functional theory (DFT) calculations indicate that the robust electron transfer from metal atoms like Fe, Co, Ni, and Mo in HEA-QDs to Pt and Ru sites enhances the local electron density and reduces the adsorption strength of hydrogen molecules, thereby significantly boosting HER kinetics (Figure 6.2a,b). The PEMWE with HEA-QDs as the cathode exhibits a low cell voltage of 1.65 V at 1.0 A cm^{-2} with an ultra-low noble metal loading ($100\,\mu g_{Pt+Ru}\,cm^{-2}$), and only decays by $-12.2\,\mu V\,h^{-1}$ for 1000 hours. The energy consumption and hydrogen production cost of the HEA-QDs cathode cell is 3.98 kWh Nm^{-3} H_2(1.0 A cm^{-2}) and \$0.88/kg H_2, respectively, exceeding the 2030 target of the U.S. Department of Energy [44].

6.3.1.2 Heteroatom Doping

Heteroatom doping can change the electronic band structure and local charge distribution of noble metal-based catalysts, thus improving the catalytic activity [45]. As a typical example, Qiao's group reported a transition metal-doped RuIr alloy

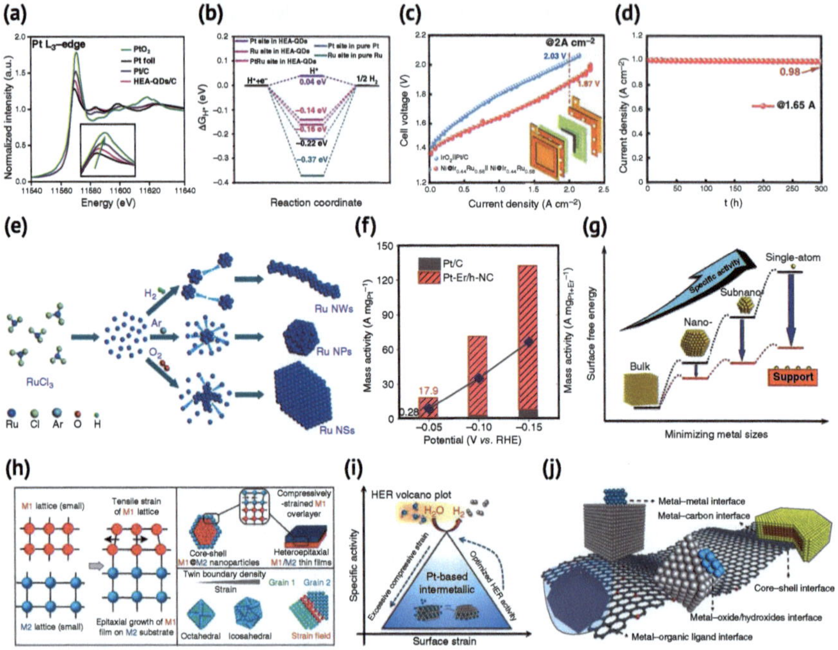

Figure 6.2 Noble metal-based hydrogen evolution electrocatalysts in acidic media. (a) X-ray absorption near edge structure (XANES) spectra of Pt L_3-edge. (b) Calculate ΔG_{H^*} profiles for Pt/Ru sites on HEA-QDs, pure Pt, and pure Ru. *Source:* Reproduced from Ref. [43], with permission of Royal Society of Chemistry. (c) PEMWE performance of Ni@$Ir_{0.44}Ru_{0.56}$ and Pt/C-IrO_2. (d) PEMWE stability test at constant voltage (1.65 V). *Source:* Reproduced from Ref. [47], with permission of John Wiley & Sons. (e) Schematic illustration of the synthetic routes of Ru NWs, Ru NSs, and Ru NPs. *Source:* Reproduced from Ref. [50], with permission of John Wiley & Sons. (f) Comparisons of the mass activity at different potentials of Pt-Er/h-NC. *Source:* Reproduced from Ref. [52], with permission of American Chemical Society. (g) Schematic illustration of the surface free energy and specific activity changes as the metal size decreases. *Source:* Reproduced from Ref. [53], with permission of American Chemical Society. (h) Different types of lattice strain. *Source:* Reproduced from Ref. [55], with permission of John Wiley & Sons. (i) Schematic illustration of the relationship between the specific activity of the Pt-based intermetallic and the surface strain. *Source:* Reproduced from Ref. [56], with permission of American Chemical Society. (j) Schematic diagram of the five interfaces of the catalyst. *Source:* Reproduced from Ref. [59], with permission of Elsevier.

electrocatalyst (M-RuIr, M = Co, Ni, Fe), owing to the special physicochemical properties induced by electronic structure regulation, the obtained M-RuIr exhibited high HER catalytic performance [46]. Among them, due to its optimized chemical state/valence state, Co-RuIr exhibited an ultra-low overpotential of 13 mV at a current density of $-10\,mA\,cm^{-2}$ in 0.1 M $HClO_4$. In another report, Ni-doped IrRu alloy electrocatalyst as the cathode delivered a low cell voltage ($1.87\,V\,@\,2\,A\,cm^{-2}$) and excellent stability (>200 h) in the practical PEMWE, showing great application prospects in water electrolysis technology (Figure 6.2c,d) [47]. This excellent performance is attributed to the fact that Ni doping effectively improves the d-band center of IrRu alloy and balances the hydrogen binding strength of the active sites.

6.3.1.3 Size and Dimension Regulation

Abundant exposed active sites and efficient electron transport can be achieved through dimension control, such as one-dimensional (1D) nanowires (NWs), two-dimensional (2D) nanosheets (NSs), and three-dimensional (3D) assemblies/nanoarrays [48, 49]. As a typical example, Chen et al. developed a gas-controlled formation method by adjusting the growth thermodynamics and kinetics to synthesize Ru NWs, NSs, and nanoparticles (NPs) in different atmospheres, respectively (Figure 6.2e) [50]. Since 1D and 2D materials have higher electron mobility than bulk materials, Ru NWs and Ru NSs exhibit outstanding HER performance.

Metal nanoclusters (MNCs) are also endowed with satisfactory HER performance due to their unique quantum size effects, surface geometry effects, and electronic properties. Nevertheless, owing to the high surface energy of MNC, it is prone to agglomerate and form NPs during the preparation and catalytic processes, thereby causing a decrease in catalytic performance. The anchoring of active sites through the interaction between MNCs and the support offers a prospective solution for solving the catalyst deactivation. For example, Wan et al. prevented the deactivation of Pt clusters by immobilizing them in the mesoporous channels of carbon carriers. The prepared catalyst exhibited an overpotential of 20.7 mV at $-10\,mA\,cm^{-2}$ in acidic conditions [51]. In addition, the introduction of a second metal as an auxiliary can also prevent the aggregation of MNCs during the reaction. For example, Chen et al. introduced Er atoms into the Pt subsurface in a dispersed manner via an elastic-space restricted strategy, yielding high atom-efficiency Pt-Er clusters. In an acidic medium, these Pt-Er clusters anchored on hollow nitrogen-doped carbon (Pt-Er/h-NC) exhibited much better MA than commercial Pt/C catalysts (Figure 6.2f) [52]. The size of active sites in heterogeneous catalysts is considered to be a key parameter that significantly affects their catalytic performance. Compared with NPs and MNCs, the metal atoms in single-atom catalysts (SACs) are uniformly dispersed on support materials in an isolated state, and the atomic utilization in a specific reaction can reach 100% (Figure 6.2g) [53]. In addition, many new properties are induced when the particle size reaches the single-atom size, such as dramatically increased surface free energy, quantum size effects, unsaturated coordination environments, and metal-support interaction. For example, Liu et al. developed an atomically dispersed Pt on onion-like carbon nanospheres (Pt_1/OLC) catalyst with only 0.27 wt% Pt loading, which afforded a ~43 times higher MA $(7.40\,A\,mg^{-1})$ than commercial Pt/C [54]. The first-principle calculations indicate that the tip enhancement effect of Pt_1/OLD catalysts at equilibrium potential optimizes the binding energy of the surface Pt sites for protons, accelerating the mass transfer on the catalyst surface.

6.3.1.4 Lattice Strain Engineering

Lattice strain is essentially the deviation of the equilibrium spacing between atoms, thus changing the electronic structure of the catalyst. For post-transition metals, when tensile strain is exerted on them, it leads to a narrower d-orbital bandwidth and a higher Fermi level. Conversely, compressive strain can weaken the adsorption capacity of late-transition metals. Therefore, lattice strain engineering, such as dislocation strain, geometric strain, and defect strain, caused by grain boundaries in epitaxially grown materials has become a cutting-edge method to control the electronic and geometric configuration of metal sites (Figure 6.2h) [55]. Recently, Zhong et al. constructed a series of Pt-based intermetallic compounds with strained Pt skins (IMC@Pt) and established a volcanic correlation between strain and acidic HER activity (Figure 6.2i) [56]. Theoretical and experimental data showed that when the surface is subjected to a compressive strain of less than 4%, the HER activity increases with the increase in the degree of compression; when the compression exceeds a certain range, this optimization decreases. In another typical example, Ir NPs with rich-Frank partial dislocations (FPDs) were synthesized using a high-temperature thermal shock method [57]. Such FPDs can produce a compressive strain of about 4%, which weakens the adsorption of Ir to hydrogen intermediates, thereby significantly improving the intrinsic activity of the catalyst.

6.3.1.5 Surface and Interface Engineering

Surface and interface engineering can induce new physicochemical properties, such as more surface-active sites, electronic effects, and strong synergistic effects [58–60]. Tuning the phase structure of noble metal catalysts can effectively change the HER activity due to their unique electronic structure and exposed surface facets. Noble metal heterostructures with unconventional heterophase demonstrate significant application potential due to phase boundaries and synergistic effects among different phases [61]. For example, Zhang's group prepared core–shell nanostructures with unconventional fcc-2H-fcc heterophases (fcc: face-centered cubic; 2H: hexagonal close-packed) by a wet chemical seeding method [62]. Thanks to the unique atomic arrangement and strain optimization of the electronic structure between the heterophase structures, the heterophase $Pd_{45}@Ir_{55}$ nanodendrites exhibited excellent electrocatalytic performance for HER in acidic conditions with a low η_{10} of 11.0 mV. There are five main interfaces involved in heterogeneous catalysis models, namely, metal–organic ligand, metal–carbon, metal oxide/hydroxides, metal–metal, and core–shell interfaces (Figure 6.2j) [63]. The exposed boundaries in the metal-support interface can participate in the reaction as new catalytic active sites, while other unexposed interfaces can indirectly affect the catalytic activity by regulating the electronic structure of the supported metal species. For example, Bao's group reported that strong electronic metal-support interactions (EMSI) between Pt SACs and 2D MoS_2 can effectively regulate the electronic structure, thereby greatly improving the HER activity [64]. Similarly, Chen et al. studied Pt SACs supported on 2D transition metal sulfur/selenides (MoS_2, WS_2, $MoSe_2$, and WSe_2) to establish the structure–activity relationship by correlating the acidic/alkaline HER activity with the average oxidation state of single-atom Pt and the Pt–H/Pt–OH interaction [65]. The core anchoring chalcogen (S, Se) and the neighboring transition metal (Mo, W) can synergistically regulate the d-orbital state of single-atom Pt through EMSI, thereby changing the adsorption energy of reactants on metal atoms and affecting HER activity.

The hydrogen evolution performance of noble metal-based electrocatalysts in acidic electrolytes has been summarized in Table 6.1. To advance the cathodic HER in acidic water electrolysis, further research is still required in the following aspects.

Table 6.1 Comparison of HER performance for reported noble metal-based catalysts in acidic electrolytes.

Catalysts	Electrolyte	η_{10} (mV)	MA @ η (A $mg_{noble\ metal}^{-1}$ @ mV)	Stability @ $-J$ (h @ mA cm^{-2})	Reference
PtRuMoFeCoNi HEA-QDs	0.5 M H_2SO_4	11	/	>1000 @ 1000	[43]
Ni@$Ir_{0.44}Ru_{0.56}$	0.5 M H_2SO_4	30	/	250 @ 200	[47]
Pt_5/HMCS	0.5 M H_2SO_4	21	12.8 @ 30	/	[51]
Pt-Er alloy	0.5 M H_2SO_4	25	17.9 @ 50	>1200 @ 1000	[52]
Pt_1/OLC	0.5 M H_2SO_4	38	7.40 @ 38	>100 @ 10	[54]
Pt_3V@Pt	0.1 M $HClO_4$	13	1.53 @ 20	/	[56]
Dr-Ir/C	0.5 M H_2SO_4	12	6.6 @ 50	200 @ 10	[57]
Pd_{45}@Ir_{55}	0.1 M $HClO_4$	11	1.16 @ 20	4 @ 10	[62]
Pt-SA_S/WS_2	0.5 M H_2SO_4	32	130.2 @ 100	/	[65]
Pt-AC/DG	0.1 M $HClO_4$	21	15.55 @ 50	/	[66]
Ru@V-RuO_2/C HMS	0.5 M H_2SO_4	47	/	/	[67]

First, to balance the cost and activity of noble metals, strategies such as exposing more active sites or increasing the activity of unit active sites are essential. On the other hand, some low-cost and efficient non-noble metal electrocatalysts should be developed, such as perovskite oxides. Second, developing advanced in situ characterization techniques to study catalyst deactivation mechanisms under operating conditions, coupled with theoretical calculations to design stable catalysts, is key to moving from lab-scale to industrial applications. Finally, it is necessary to strengthen the basic research of electrolyzers' components to promote the industrial application of acidic water electrolysis technology [68]. Optimizing key components, such as anode, PEM, porous transport layer, and bipolar plate, is the key to developing efficient and low-cost PEMWE. Future research should focus on overcoming existing limitations and exploring strategies such as modifying PEM functional groups and optimizing thickness, using protective coatings (e.g., Pt, Au) in bipolar plates, and improving coating methods to obtain high-quality membrane electrode assemblies (MEAs).

6.3.2 Transition Metal-based Electrocatalysts for Alkaline HER

Compared with acidic water electrolysis, alkaline water electrolysis has been a comprehensive development due to its cost-effectiveness such as simple electrolyzer structure and negligible reliance on noble metal-based catalysts. As early as 1888, Russian physicist Dmitry Lachinov developed the industrial method for hydrogen production through alkaline water electrolysis (Figure 6.3a) [69]. Alkaline (ALK) electrolysis is by far the most mature technology, often using high

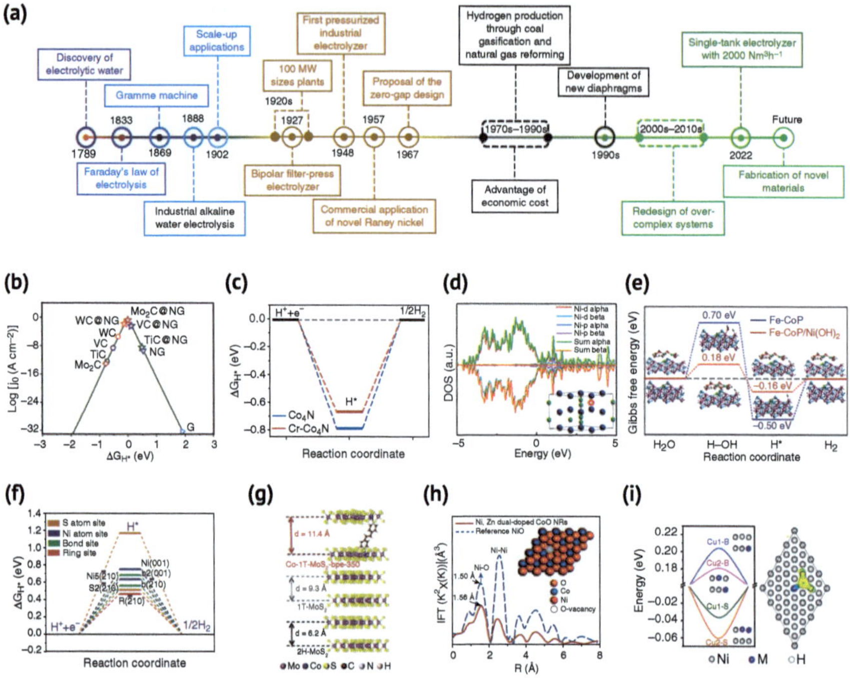

Figure 6.3 Transition-metal-based hydrogen evolution electrocatalysts in alkaline media. (a) The development course of alkaline water electrolysis. *Source:* Adapted from Ref. [69]. (b) A volcano curve of the HER activity on TMC@NG model electrocatalysts. *Source:* Reproduced from Ref. [72], with permission of John Wiley & Sons. (c) Free energies of H adsorption on Co_4N and Cr–Co_4N. *Source:* Reproduced from Ref. [75], with permission of John Wiley & Sons. (d) Total and partial electronic density of states calculated for Ni_3N_{1-x}. The Fermi level is set at 0 eV. The inset shows the atomic structure model of Ni_3N_{1-x}. *Source:* Reproduced from Ref. [76], with permission of John Wiley & Sons, CC BY 4.0. (e) Calculated Gibbs free energy diagrams for alkaline HER at bare Fe-CoP surface and Fe-CoP/$Ni(OH)_2$ hybrid surface. For insets in (e), the light pink, pale blue, bright green, light green, red, and grey balls depict Co, P, Fe, Ni, O, and H atoms, respectively, in DFT models. *Source:* Reproduced from Ref. [81], with permission of John Wiley & Sons. (f) Calculated free-energy diagram of HER over ($2\bar{1}0$) and (001) surfaces at equilibrium potential. *Source:* Reproduced from Ref. [84], with permission of American Chemical Society. (g) The schematic illustration of interlayer spacing of the prepared catalysts. *Source:* Reproduced from Ref. [85], with permission of John Wiley & Sons. (h) Fourier transformation of the EXAFS spectra. The inset in (h) shows a schematic of the Ni_x–O_v cluster. *Source:* Reproduced from Ref. [86], with permission of John Wiley & Sons. (i) The formation energy of Cu for different positions. *Source:* Reproduced from Ref. [87], with permission of John Wiley & Sons.

specific surface Ni or NiCo alloys as HER catalysts. However, most transition metals exhibit either too strong or too weak H binding ability, showing much lower performance than the precious metal catalysts [70]. Metal atoms in transition metal-based compounds form strong chemical bonds with N, C, or P atoms, which contributes to regulating the H binding energy, stabilizing the intermediates, and reducing the reaction energy barrier. Some even exhibit alkaline HER activity comparable to that of PGMs, having attracted much attention. Next, we describe the characteristics of different transition metal compounds by introducing different nonmetallic atoms as the basis for classification to guide researchers in this field.

6.3.2.1 Transition Metal Carbides

Transition metal carbides (TMCs) have noble metal-like electronic configuration and wide pH applicability. More importantly, the high electronic conductivity of TMCs highlights its inherent advantages as an electrocatalyst, in comparison with the semi-conductivity of other transition metal compounds [71]. In 1973, Levy and Boudart discovered that the carbon atoms in tungsten carbide would change the electron distribution of tungsten atoms. Further DFT calculations show that the hybridization of metal d orbitals with carbon s and p orbitals broadens the d band and induces Pt-like d-band characteristics. However, their HER performance is limited by the electron defect d orbitals, which hinders the desorption of H* and OH*. In this regard, researchers have been committed to optimizing the electronic structure through structural and interface engineering to achieve the balance of hydrogen adsorption and desorption. Yang et al. [72] proposed a general strategy that introduces nitrogen-doped graphene (NG) to weaken the interactions of TMCs with H*, balancing the adsorption and desorption of hydrogen, and thereby synergically enhancing HER kinetics (Figure 6.3b).

6.3.2.2 Transition Metal Nitrides

Transition metal nitrides (TMNs) are similar to TMCs and have electronic properties similar to Pt [73]. The difference is that the M—N bond in TMNs changes the density of the metal d-band states and shrinks the d-band. The compression of d-band typically creates vacancies or defects, which may result in more active electronic states on the metal surface [74]. However, single-phase TMNs have unsatisfactory performance due to insufficient M—H/M—OH bond strength. Given this, most researchers focus on improving the intrinsic electrocatalytic activity of TMNs, such as alloying, doping, vacancy, heterostructure, and hybridization. For example, Yao et al. regulated the water and hydrogen binding ability of Co_4N by Cr doping (Cr-Co_4N/CC) to accelerate the alkaline reaction kinetics, achieving an ultra-low overpotential of 21 mV at −10 mA cm^{-2} (Figure 6.3c) [75]. In addition, Liu et al. synthesized nitrogen vacancy-rich Ni_3N_{1-x} through plasma-enhanced nitridation of commercial Ni foam (NF). The presence of nitrogen vacancies effectively enhanced the adsorption of water molecules and improved the adsorption–desorption behavior of intermediate-adsorbed hydrogen, thereby improving the HER activity (Figure 6.3d) [76].

6.3.2.3 Transition Metal Phosphides

Phosphorus and metal sites can serve as proton acceptor sites and hydride acceptor sites, respectively, making transition metal phosphides (TMPs) unique dual-active site HER catalysts. When M combines with P, it will form rich M—M bonds, P—P bonds, and M—P bonds, which also makes TMPs have both good electrical conductivity and the characteristics of noble metals. Specifically, the introduction of the P atom shrinks the d-band of the metal atom, increasing electronic state density near the Fermi level [77]. However, the performance of TMP-based HER catalysts is still limited by poor corrosion resistance [78]. A variety of strategies have been developed to further improve the catalytic activity while enhancing stability, including component engineering, nanostructure engineering, and surface and interface engineering [79]. For example, Sun's group prepared a series of self-supporting TMP nanoarrays and showed excellent activity and stability in water electrolysis [80]. Since then, extensive research on TMPs for HER has been carried out. Besides, Yu et al. rationally adopted interface engineering, using sodium hypophosphite as the P source, and prepared Fe-CoP/$Ni(OH)_2$ catalyst through the phosphating process. The catalyst exhibits strong electronic interactions at the interface, significantly improving the interfacial reactivity of reactants and intermediates with the catalyst surface (Figure 6.3e) [81].

6.3.2.4 Transition Metal Dichalcogenides

Transition metal dichalcogenides (TMDCs) have been widely studied as electrocatalysts for HER due to their advantages such as low cost, good conductivity, and good catalytic performance [82]. In particular, 2D TMDs with high surface area and tunable electronic properties have Pt-like properties, making them a potential alternative to noble metal HER electrocatalysts. The structure of TMDCs can be divided into two parts: the base plane and the edge, in which the hydrogen evolution activity of the base plane is significantly lower than that of the edge. The base plane usually has a more stable and uniform electronic structure, while the edge usually has a higher number of unsaturated bonds or a higher electron density. However, most of the TMDCs area is the base surface, so activating the base surface activity is of great significance in improving the overall HER performance [83]. Feng et al. in situ grew $(2\bar{1}0)$ high-index faceted Ni_3S_2 nanosheet arrays on NF. Since the S sites (such as S1 and S2) on the $(2\bar{1}0)$ surface are located at the step edges and have a low coordination number, the surface has the smallest ΔG_{H^*} and exhibits outstanding HER activity (Figure 6.3f) [84]. Liu et al. modified 1T-MoS_2 through a synergistic strategy of ligand and doping, which expanded the interlayer spacing of Co-1T-MoS_2 and promoted the exposure of active sites (Figure 6.3g) [85].

6.3.2.5 Other Transition Metal-based Electrocatalysts

Other transition metal compounds, such as transition metal oxides, have poor catalytic activity for HER, so relatively few reports have been reported. Efforts have also been made to improve their intrinsic activity. For example, Qiao's group broke the limitation of the unattractive hydrogen desorption ability of pure TMOs by doping CoO with both Ni and Zn atoms, tuning the bulk electronic structure to enhance the conductivity, thereby accelerating the alkaline HER (Figure 6.3h) [86]. In addition, TM alloying is also one of the effective strategies to improve HER performance. Wang et al. demonstrated by DFT calculations that ΔG_{H^*} is determined by the interaction of H with Ni and M atoms, and found a potential NiCu electrocatalyst close to the optimal ΔG_{H^*} (Figure 6.3i) [87].

TM-based catalysts are crucial for the development of alkaline water electrolysis technology. Table 6.2 summarizes the TM-based HER electrocatalysts in alkaline

Table 6.2 Comparison of HER performance for reported transition-metal-based catalysts in 1.0 M KOH solution.

Catalysts	η_{10}**(mV)**	**Stability @ −J (h @ mA cm^{-2})**	**Reference**
Mo_2C@NG/CNT	135	20 @ 10	[72]
Cr–Co_4N/CC	21	>10 @ 100	[75]
Ni_3N_{1-x}/NF	55	50 @ 50	[76]
CoP/CC	209	/	[80]
Fe-CoP/$Ni(OH)_2$	91	/	[81]
Ni_3S_2/NF	223	>200 @ 10	[84]
Co-1T-MoS_2	118	50 @ 10	[85]
Ni, Zn-CoO NRs	53	/	[86]
O-NiCu	23	100 @ 10	[87]
Ni-GF/VC	128	20 @ 10	[88]
NiS	95	/	[89]
15% CoNiPt NFs	25	/	[90]
CN@NiCoS/NF	4.6	>1000 @ 100	[91]
C-Co_2P	30	60 @ 10	[92]

electrolytes. So far, the development of alkaline HER electrocatalysts has made remarkable progress, but there are still some problems worthy of further study. First, emerging catalyst design strategies, including machine learning, artificial intelligence, and big data analysis, offer valuable support in the development of alkaline HER catalysts, facilitating faster design and optimization processes. Second, in situ characterization techniques enable the identification of active sites, capture of reaction intermediates, and determination of reaction pathways, while also revealing changes in the surface structure of the catalysts at the atomic level. Third, considering the advantages of anion exchange membrane (AEM) electrolysis technology, attention should focus on both the development of high-performance catalysts and the design of MEAs, including the preparation of nanostructured MEAs and the optimization of gas diffusion layers (GDLs) to enhance catalyst utilization and mass transfer efficiency.

6.3.3 Neutral HER

Electrocatalytic hydrogen production typically uses acidic or alkaline electrolytes: acidic conditions provide a high concentration of protons or hydronium ions as reactants, while alkaline conditions enable the production of high-purity hydrogen. However, HER-based hydrogen production under extreme pH conditions faces issues such as corrosive environments, costly exchange membranes, and the need for acid/base-resistant electrocatalysts. In contrast, neutral electrolytes have significant advantages, with mild reactions, low catalyst requirements, and no damage to the equipment, which effectively extends the service life of the equipment. Against this background, SWE has become a hot topic of research. Seawater resources are abundant, and in the process of hydrogen production by electrolysis, seawater usually does not require complicated purification treatment, so the cost of hydrogen production is relatively low. Furthermore, it can effectively integrate with existing offshore wind and solar power facilities to enable sustainable hydrogen production, providing storable hydrogen fuel and electricity to coastal communities [93].

Despite the many advantages of SWE, the complexity of seawater electrolytes causes a series of problems [94, 95]. Figure 6.4 shows the main problems and potential solutions. For example, the anodic Cl^- oxidation kinetics are slow and poison the active sites of the catalyst, resulting in a huge overpotential. Enhancing the chemical inertness of the catalyst toward Cl^- provides a promising solution [96]. Koper's group deposited a layer of MnO_x on IrO_x to act as a diffusion barrier, preventing Cl^- reactions while allowing the transport of water, protons, and O_2 [97]. Besides, the deposition of Mg^{2+}/Ca^{2+} at the cathode is also a problem that cannot be ignored. Severe metal cation precipitation blocks the surface active sites and

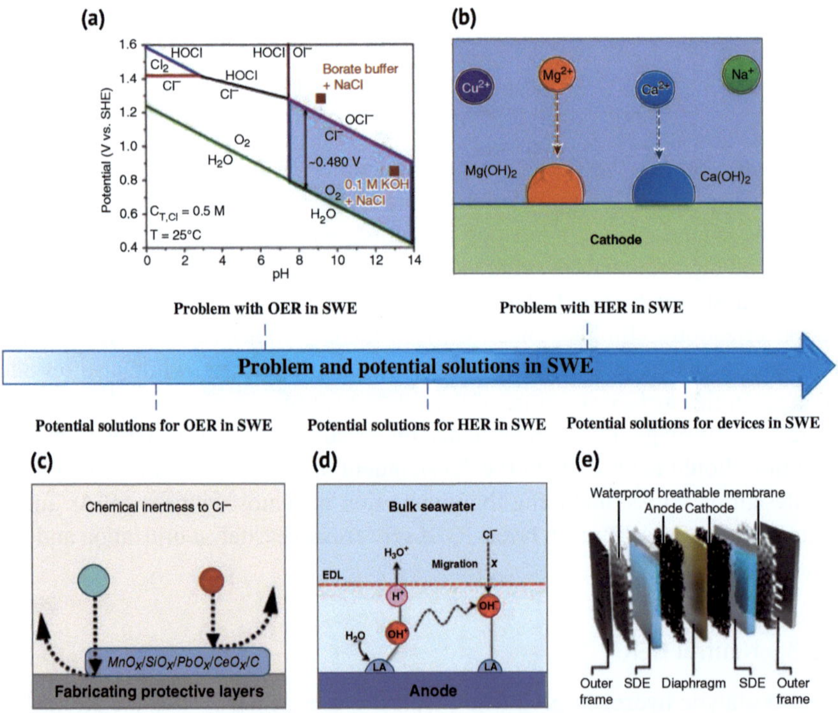

Figure 6.4 Problem and potential solutions in SWE. (a) Problem with OER in SWE. *Source:* Reproduced from Ref. [94], with permission of John Wiley & Sons. (b) Problem with HER in SWE. *Source:* Reproduced from Ref. [95], with permission of John Wiley & Sons. (c) Potential solutions for OER in SWE. *Source:* Reproduced from Ref. [96], with permission of John Wiley & Sons. (d) Potential solutions for HER in SWE. *Source:* Reproduced from Ref. [98], with permission of American Association for the Advancement of Science. (e) Potential solutions for devices in SWE. *Source:* Reproduced from Ref. [100], with permission of Springer Nature.

leads to irreversible activity loss [98]. Qiao's group introduced a Lewis acid layer to bind OH^-, reducing Mg^{2+}/Ca^{2+} precipitation and improving catalyst stability [99]. Device optimization is also an effective strategy to address these challenges. Xie's group introduced a self-permeable waterproof membrane and a self-wetting electrode, allowing seawater to undergo purification through a liquid–gas–liquid phase change before entering the electrolyzer. This process effectively removes impurities from seawater, preventing their interference with the catalyst and enhancing overall system performance [100].

In summary, catalysts for SWE have been reported one after another, and their performance has been gradually improved (Table 6.3). Even so, there is a gap compared with acidic and alkaline water electrolysis. To advance neutral SWE for hydrogen production, three development directions deserve attention: improving the activity and stability of catalysts in neutral seawater, combining material design with device optimization, and clarifying the HER mechanism under neutral conditions. Neutral SWE holds great potential, and these directions will establish a solid foundation for reliable and sustainable hydrogen production.

6.4 Summary and Perspective

This chapter initially presents a comprehensive overview of the principal evaluation metrics for HER catalysts, followed by an in-depth examination of design strategies and recent advancements in research about catalysts functioning in acidic, alkaline, and neutral environments. While notable progress has been achieved in enhancing HER catalytic performance, substantial challenges persist regarding long-term stability, cost-effectiveness, and adaptability to practical applications. Therefore, the current challenges and prospects of electrocatalytic HER are summarized to provide robust guidance for subsequent research endeavors (Figure 6.5).

Table 6.3 Comparison of neutral HER performance for the reported electrocatalysts.

Catalysts	Electrolyte	η_{10}(mV)	Stability @ $-J$ (h @ mA cm^{-2})	Reference
Cr_2O_3-CrO_x	Natural seawater	/	100 @ 500	[99]
$Co_{0.31}Mo_{1.69}C$/MXene/NC	Natural seawater	208	225 @ 500	[101]
Porous CrO_x/Cu-Ni	1.0 M PBS	48	24 @ 100	[102]
Ru/Nb_2O_5	1.0 M PBS	31	/	[103]
$Pt@CoO_x$	1.0 M PBS	82	24 @ 10	[104]
Ru-WO_{3-x}	1.0 M PBS	19	30 @ 20	[105]
$RuSe_x$-RuNC	1.0M KPi	29	40 @ 100	[106]
$CrO_x@Ni_2P$-Ni_5P_4/NF	1.0 M PBS	109	500 @ 50	[107]
Pt_{SA}-Pt_C/NDPCM	1.0 M PBS	14	60 @ 100	[108]
Am-Mo-$NiS_{0.5}Se_{0.5}$	1.0 M PBS	48	300 @ 10	[109]

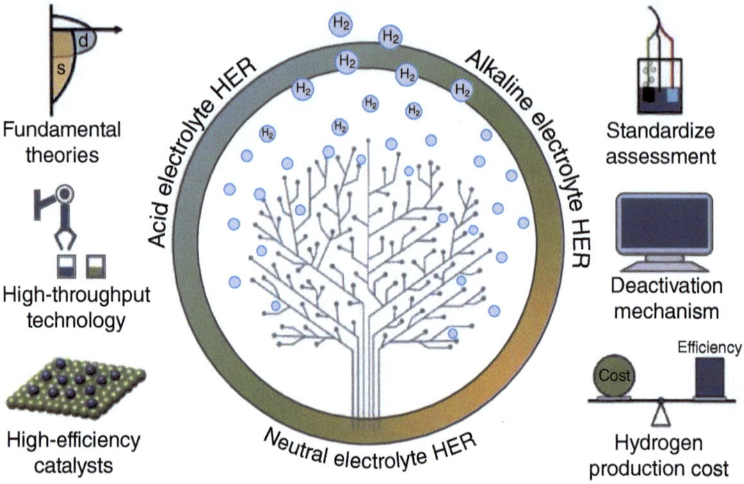

Figure 6.5 Summary and perspectives for HER.

6.4.1 Strengthen Fundamental Theories

A primary challenge in the field of HER lies in the enhancement of fundamental theories, which is essential for guiding researchers in designing catalyst structures that exhibit greater efficiency and stability. To achieve this goal, it is imperative to further advance sophisticated theoretical calculation methods alongside in situ characterization techniques. These advancements will enable a precise elucidation of the surface reaction dynamics while capturing the structural evolution of active sites and their interactions with reactants, thereby illuminating the fundamental theories related to HER.

6.4.2 Develop High-throughput Technology

The rational design and synthesis of HER electrocatalysts are dependent on prior knowledge, and this approach to catalyst development avoids a significant number of repetitive experiments. Nevertheless, as the complexity of the catalysts escalates, the scientific capacity for rationally designing materials is gradually challenged. In contrast, high-throughput screening technology explores catalysts with different elements and structures through parallelized and automated experimental operations, rapidly constructing a library of HER electrocatalysts. Through the combination of data-driven analysis, it offers effective design directions for researchers.

6.4.3 Synthesis High-efficiency Catalysts

Although high-throughput screening techniques can promptly identify and combine candidate elements, innovative synthetic approaches are still requisite for the production of catalysts that take into account both cost and efficiency. These approaches must be featured by precise control of the atomic-level structure to ensure that the catalyst has maximal atomic utilization efficiency, enhanced intrinsic activity, and outstanding structural stability. In this respect, although stability still awaits improvement, the synthesis of high-density SACs is a direction worthy of investment in the future. Furthermore, how to effectively upscale laboratory-scale synthesis methods and preserve the performance of the catalysts will be the key to achieving large-scale catalyst production. Regrettably, there are few studies in this regard.

6.4.4 Standardize Performance Assessment

The synthesis and performance evaluation of electrocatalysts are significantly influenced by various experimental parameters, including trace elements, the mass, and method of catalyst loading, standardization of reference and counter electrodes, systematic errors in testing instruments, environmental pressure, and temperature, as well as electrolyte configuration. The confused experimental procedures have resulted in ambiguous catalyst compositions and implausible performance data, which can severely mislead subsequent research efforts. Therefore, the unification of the three-electrode testing standards will guarantee that each laboratory can test the performance of catalysts under the same conditions, thereby making the results comparable. Furthermore, a notable disparity exists between laboratory tests and industrial applications. Laboratories commonly employ a three-electrode system for HER testing, whereas industrial-grade applications depend on complex apparatuses. The dissimilarities between these two testing approaches frequently lead to suboptimal outcomes when laboratory results are transformed into industrial applications. Consequently, evaluating the performance of catalysts in devices represents another significant development trend in this domain.

6.4.5 Monitor Deactivation Mechanism

Current methodologies for investigating catalyst stability decline primarily involve analyzing changes in phase and surface morphology. While this approach provides valuable insights into macroscopic differences between pre- and post-use catalysts, it frequently falls short of elucidating the true microscopic mechanisms underlying catalyst deactivation under operational conditions. Therefore, the development of in situ monitoring systems capable of tracking dynamic changes in composition and structure throughout stability testing is crucial for comprehensively understanding catalyst deactivation mechanisms, optimizing catalyst design, and ultimately advancing toward highly durable catalysts. Regrettably, research efforts focused on this area remain markedly limited.

6.4.6 Reduce Hydrogen Production Costs

The large-scale application of HER is contingent not only upon the efficiency of the catalyst, but also on cost-effectiveness. In the future, the cost of hydrogen production requires reduction through multiple approaches, including the lowering of electricity prices by employing renewable energy sources to drive the electrolysis process. Additionally, it is of paramount importance to decrease the manufacturing and maintenance costs of electrolyzers, encompassing the reduction of production costs of core components such as membranes, electrodes, and catalysts. Simultaneously, the decoupling of the anode catalytic reaction is regarded as a significant strategy for enhancing the overall hydrogen production efficiency. In conventional water electrolysis reactions, the oxygen generated at the anode has a low value and incurs substantial energy consumption. Through decoupling the anode reaction to generate high-added-value chemicals, the economics of the electrolysis process can be prominently improved. This will facilitate the concurrent production of hydrogen and high-added-value products, further propelling the commercialization of hydrogen production through electrolysis of water.

Acknowledgment

This work was supported by the National Natural Science Foundation of China (22409028).

References

1 Zhang, J.C., Ma, C.N., Jia, S.Y., et al. (2023). Electrocatalysts design guided by active intermediates of hydrogen evolution reaction. *Adv. Energy Mater.* 13 (43): 2302436.
2 Wang, Q., Hisatomi, T., Jia, Q., et al. (2016). Scalable water splitting on particulate photocatalyst sheets with a solar-to-hydrogen energy conversion efficiency exceeding 1%. *Nat. Mater.* 15 (6): 611.
3 Quan, L., Jiang, H., Mei, G., et al. (2024). Bifunctional electrocatalysts for overall and hybrid water splitting. *Chem. Rev.* 124 (7): 3694.
4 Kempler, P.A., Coridan, R.H., and Luo, L. (2024). Gas evolution in water electrolysis. *Chem. Rev.* 124 (19): 10964.
5 Sun, X., Wang, S., Hou, Y., et al. (2023). Self-supporting Metal–Organic Framework-based hydrogen and oxygen electrocatalysts. *J. Mater. Chem. A* 11 (25): 13089.
6 Jiao, Y., Zheng, Y., Jaroniec, M., et al. (2015). Design of electrocatalysts for oxygen- and hydrogen-involving energy conversion reactions. *Chem. Soc. Rev.* 44 (8): 2060.
7 Zhu, J., Hu, L., Zhao, P., et al. (2019). Recent advances in electrocatalytic hydrogen evolution using nanoparticles. *Chem. Rev.* 120 (2): 851.
8 Ledezma-Yanez, I., Wallace, W.D.Z., Sebastián-Pascual, P., et al. (2017). Interfacial water reorganization as a pH-dependent descriptor of the hydrogen evolution rate on platinum electrodes. *Nat. Energy* 2 (4): 17031.
9 Zheng, Y., Jiao, Y., Zhu, Y., et al. (2014). Hydrogen evolution by a metal-free electrocatalyst. *Nat. Commun.* 5 (1): 3783.
10 Daliran, S., Oveisi, A.R., Peng, Y., et al. (2022). Metal-organic framework (MOF)-, covalent-organic framework (COF)-, and porous-organic polymers (POP)-catalyzed selective CH bond activation and functionalization reactions. *Chem. Soc. Rev.* 51 (18): 7810.

11 Zhang, J., Chen, G., Mullen, K., et al. (2018). Carbon-rich nanomaterials: fascinating hydrogen and oxygen electrocatalysts. *Adv. Mater.* 30: 1800528.

12 Diaz, L.L. (2022). Green hydrogen economy: game-changing technologies to transform the world's energy supply. A division of the American Chemical Society. Retrieved from https://www.cas.org/resources/cas-insights/sustainability/green-hydrogen-econ.

13 Zheng, J., Nash, J., Xu, B., et al. (2018). Perspective – towards establishing apparent hydrogen binding energy as the descriptor for hydrogen oxidation/evolution reactions. *J. Electrochem. Soc.* 165 (2): H27.

14 Schmidt, T.J., Ross Jr, P.N., and Markovic, N.M. (2002). Temperature dependent surface electrochemistry on Pt single crystals in alkaline electrolytes: Part 2. The hydrogen evolution/oxidation reaction. *J. Electroanal. Chem.* 524: 252.

15 Goyal, A. and Koper, M.T.M. (2021). The interrelated effect of cations and electrolyte pH on the hydrogen evolution reaction on gold electrodes in alkaline media. *Angew. Chem. Int. Ed.* 60 (24): 13452.

16 Li, P., Jiang, Y., Hu, Y., et al. (2022). Hydrogen bond network connectivity in the electric double layer dominates the kinetic pH effect in hydrogen electrocatalysis on Pt. *Nat. Catal.* 5 (10): 900.

17 Zhang, B., Zhang, L., Tan, Q., et al. (2020). Simultaneous interfacial chemistry and inner Helmholtz plane regulation for superior alkaline hydrogen evolution. *Energy Environ. Sci.* 13 (9): 3007.

18 Rajan, A.G. and Carter, E.A. (2020). Microkinetic model for pH-and potential-dependent oxygen evolution during water splitting on Fe-doped β-NiOOH. *Energy Environ. Sci.* 13 (12): 4962.

19 Subbaraman, R., Tripkovic, D., Strmcnik, D., et al. (2011). Enhancing hydrogen evolution activity in water splitting by tailoring Li^+-$Ni(OH)_2$-Pt interfaces. *Science* 334 (6060): 1256.

20 Seh, Z.W., Kibsgaard, J., Dickens, C.F., et al. (2017). Combining theory and experiment in electrocatalysis: insights into materials design. *Science* 355 (6321): eaad4998.

21 Zou, X. and Zhang, Y. (2015). Noble metal-free hydrogen evolution catalysts for water splitting. *Chem. Soc. Rev.* 44 (15): 5148.

22 Wang, Q. and Astruc, D. (2020). State of the art and prospects in metal-organic framework (MOF)-based and MOF-derived nanocatalysis. *Chem. Rev.* 120 (2): 1438.

23 Zhang, X., Zhang, M., Deng, Y., et al. (2021). A stable low-temperature H_2-production catalyst by crowding Pt on α-MoC. *Nature* 589 (7842): 396.

24 Wu, Z.-P., Lu, X.F., Zang, S.-Q., et al. (2020). Non-noble-metal-based electrocatalysts toward the oxygen evolution reaction. *Adv. Funct. Mater.* 30 (15): 1910274.

25 Zhao, G., Rui, K., Dou, S.X., et al. (2018). Heterostructures for electrochemical hydrogen evolution reaction: a review. *Adv. Funct. Mater.* 28 (43): 1803291.

26 Kawashimam, K., Márquez, R.A., Son, Y.J., et al. (2023). Accurate potentials of Hg/HgO electrodes: practical parameters for reporting alkaline water electrolysis overpotentials. *ACS Catal.* 13 (3): 1893.

27 Yu, J., He, Q., Yang, G., et al. (2019). Recent advances and prospective in ruthenium-based materials for electrochemical water splitting. *ACS Catal.* 9 (11): 9973.

28 Lu, X.F., Zhang, S.L., Sim, W.L., et al. (2021). Phosphorized $CoNi_2S_4$ yolk-shell spheres for highly efficient hydrogen production via water and urea electrolysis. *Angew. Chem. Int. Ed.* 60 (42): 22885.

29 Wang, X., Kolen'ko, Y.V., Bao, X.Q., et al. (2015). One-step synthesis of self-supported nickel phosphide nanosheet array cathodes for efficient electrocatalytic hydrogen generation. *Angew. Chem. Int. Ed.* 54 (28): 8188.

30 Vrubel, H., Moehl, T., Gratzel, M., et al. (2013). Revealing and accelerating slow electron transport in amorphous molybdenum sulphide particles for hydrogen evolution reaction. *Chem. Commun.* 49 (79): 8985.

31 Lu, X.F., Yu, L., Zhang, J., et al. (2019). Ultrafine dual-phased carbide nanocrystals confined in porous nitrogen-doped carbon dodecahedrons for efficient hydrogen evolution reaction. *Adv. Mater.* 31 (30): 1900633.

32 McCrory, C.C., Jung, S., Ferrer, I.M., et al. (2015). Benchmarking hydrogen evolving reaction and oxygen evolving reaction electrocatalysts for solar water splitting devices. *J. Am. Chem. Soc.* 137 (13): 4347.

33 Anantharaj, S., Ede, S.R., Karthick, K., et al. (2018). Precision and correctness in the evaluation of electrocatalytic water splitting: revisiting activity parameters with a critical assessment. *Energy Environ. Sci.* 11 (4): 744.

34 Wei, C. and Xu, Z.J. (2018). The comprehensive understanding of as an evaluation parameter for electrochemical water splitting. *Small Methods* 2 (11): 1800168.

35 Benck, J.D., Hellstern, T.R., Kibsgaard, J., et al. (2014). Catalyzing the hydrogen evolution reaction (HER) with molybdenum sulfide nanomaterials. *ACS Catal.* 4 (11): 3957.

36 Dang, Q., Lin, H.P., Fan, Z.L., et al. (2021). Iridium metallene oxide for acidic oxygen evolution catalysis. *Nat. Commun.* 12 (1): 6007.

37 Zhou, Y., Guo, W., Xing, L., et al. (2024). Keys to unravel the stability/durability issues of platinum-group-metal catalysts toward oxygen evolution reaction for acidic water splitting. *ACS Cent. Sci.* 10 (11): 2006.

38 Li, J., Zeng, J., Zhao, F., et al. (2024). A review on highly efficient Ru-based electrocatalysts for acidic oxygen evolution reaction. *Energy Fuels* 38 (13): 11521.

39 Liu, H.-l., Nosheen, F., and Wang, X. (2015). Noble metal alloy complex nanostructures: controllable synthesis and their electrochemical property. *Chem. Soc. Rev.* 44 (10): 3056.

40 Kuang, P., Ni, Z., Zhu, B., et al. (2023). Modulating the d-band center enables ultrafine Pt_3Fe alloy nanoparticles for pH-universal hydrogen evolution reaction. *Adv. Mater.* 35 (41): 2303030.

41 Li, H., Lai, J., Li, Z., et al. (2021). Multi-sites electrocatalysis in high-entropy alloys. *Adv. Funct. Mater.* 31 (47): 2106715.

42 Yao, Y., Dong, Q., Brozena, A., et al. (2022). High-entropy nanoparticles: synthesis-structure-property relationships and data-driven discovery. *Science* 376 (6589): eabn3103.

43 Zhao, H., Liu, M., Wang, Q., et al. (2024). Strong transboundary electron transfer of high-entropy quantum-dots driving rapid hydrogen evolution kinetics. *Energy Environ. Sci.* 17 (18): 6594.

44 US Department of Energy (2023). *U.S. National Clean Hydrogen Strategy and Roadmap*. https://www.hydrogen.energy.gov/library/roadmaps-vision/clean-hydrogen-strategy-roadmap.

45 Zhang, A., Liang, Y., Zhang, H., et al. (2021). Doping regulation in transition metal compounds for electrocatalysis. *Chem. Soc. Rev.* 50 (17): 9817.

46 Shan, J., Ling, T., Davey, K., et al. (2019). Transition-metal-doped RuIr bifunctional nanocrystals for overall water splitting in acidic environments. *Adv. Mater.* 31 (17): 1900510.

47 Xie, Y., Feng, Y., Pan, S., et al. (2024). Electrochemical leaching of Ni dopants in IrRu alloy electrocatalyst boosts overall water splitting. *Adv. Funct. Mater.* 34 (41): 2406351.

48 Luo, Y., Zhang, Z., Chhowalla, M., et al. (2022). Recent advances in design of electrocatalysts for high-current-density water splitting. *Adv. Mater.* 34 (16): 2108133.

49 Wang, C., Yang, F., and Feng, L. (2023). Recent advances in Iridium-based catalysts with different dimensions for the acidic oxygen evolution reaction. *Nanoscale Horiz.* 8 (9): 1174.

50 Chen, L., Li, Y., and Liang, X. (2021). $Ar/H_2/O_2$-controlled growth thermodynamics and kinetics to create zero-, one-, and two-dimensional ruthenium nanocrystals towards acidic overall water splitting. *Adv. Funct. Mater.* 31 (11): 2007344.

51 Wan, X.-K., Wu, H.B., Guan, B.Y., et al. (2020). Confining sub-nanometer Pt clusters in hollow mesoporous carbon spheres for boosting hydrogen evolution activity. *Adv. Mater.* 32 (7): 1901349.

52 Chen, G., Chen, W., Lu, R., et al. (2023). Near-atomic-scale superfine alloy clusters for ultrastable acidic hydrogen electrocatalysis. *J. Am. Chem. Soc.* 145 (40): 22069.

53 Yang, X.-F., Wang, A., Qiao, B., et al. (2013). Single-atom catalysts: a new frontier in heterogeneous catalysis. *Acc. Chem. Res.* 46 (8): 1740.

54 Liu, D., Li, X., Chen, S., et al. (2019). Atomically dispersed platinum supported on curved carbon supports for efficient electrocatalytic hydrogen evolution. *Nat. Energy* 4 (6): 512.

55 Hou, Z., Cui, C., Li, Y., et al. (2023). Lattice-strain engineering for heterogenous electrocatalytic oxygen evolution reaction. *Adv. Mater.* 35 (39): 2209876.

56 Zhong, Z., Tu, Y., Zhang, L., et al. (2024). Surface strain effect on electrocatalytic hydrogen evolution reaction of Pt-based intermetallics. *ACS Catal.* 14 (5): 2917.

57 Liu, P., Zhang, X., Fei, J., et al. (2024). Frank partial dislocations in coplanar Ir/C ultrathin nanosheets boost hydrogen evolution reaction. *Adv. Mater.* 36 (11): 2310591.

58 Wu, B. and Zheng, N. (2013). Surface and interface control of noble metal nanocrystals for catalytic and electrocatalytic applications. *Nano Today* 8 (2): 168.

59 Yang, Y., Luo, M., Zhang, W., et al. (2018). Metal surface and interface energy electrocatalysis: fundamentals, performance engineering, and opportunities. *Chem* 4 (9): 2054.

60 Zhang, J., Zhang, Q., and Feng, X. (2019). Support and interface effects in water-splitting electrocatalysts. *Adv. Mater.* 31 (31): 1808167.

61 Chen, Y., Lai, Z., Zhang, X., et al. (2020). Phase engineering of nanomaterials. *Nat. Rev. Chem.* 4 (5): 243.

62 Ge, Y., Wang, X., Chen, B., et al. (2022). Preparation of fcc-2H-fcc heterophase Pd@Ir nanostructures for high-performance electrochemical hydrogen evolution. *Adv. Mater.* 34 (4): 2107399.

63 Zhu, Y.P., Guo, C., Zheng, Y., et al. (2017). Surface and interface engineering of noble-metal-free electrocatalysts for efficient energy conversion processes. *Acc. Chem. Res.* 50 (4): 915.

64 Deng, J., Li, H., Xiao, J., et al. (2015). Triggering the electrocatalytic hydrogen evolution activity of the inert two-dimensional MoS_2 surface via single-atom metal doping. *Energy Environ. Sci.* 8 (5): 1594.

65 Shi, Y., Ma, Z.-R., Xiao, Y.-Y., et al. (2021). Electronic metal–support interaction modulates single-atom platinum catalysis for hydrogen evolution reaction. *Nat. Commun.* 12 (1): 3021.

66 Cheng, Q., Hu, C., Wang, G., et al. (2020). Carbon-defect-driven electroless deposition of Pt atomic clusters for highly efficient hydrogen evolution. *J. Am. Chem. Soc.* 142 (12): 5594.

67 Li, Y., Wang, W., Cheng, M., et al. (2023). Arming Ru with oxygen-vacancy-enriched RuO_2 sub-nanometer skin activates superior bifunctionality for pH-universal overall water splitting. *Adv. Mater.* 35 (24): 2206351.

68 Du, Y., Liu, J., Chen, J., et al. (2024). Design principle and regulation strategy of noble metal-based materials for practical proton exchange membrane water electrolyzer. *Adv. Energy Mater.* 2404113.

69 Yu, Z.Y., Duan, Y., Feng, X.Y., et al. (2021). Clean and affordable hydrogen fuel from alkaline water splitting: past, recent progress, and future prospects. *Adv. Mater.* 33 (31): 2007100.

70 Greeley, J., Jaramillo, T.F., Bonde, J. et al. (2006). Computational high-throughput screening of electrocatalytic materials for hydrogen evolution. *Nat. Mater.* 5 (11): 909.

71 Gao, Q., Zhang, W., Shi, Z., et al. (2018). Structural design and electronic modulation of transition-metal-carbide electrocatalysts toward efficient hydrogen evolution. *Adv. Mater.* 31 (2): 1802880.

72 Yang, C., Shen, K., Zhao, R., et al. (2022). Balance effect: a universal strategy for transition metal carbides to enhance hydrogen evolution. *Adv. Funct. Mater.* 32 (5): 2108167.

73 Zhong, Y., Xia, X., Shi, F., et al. (2016). Transition metal carbides and nitrides in energy storage and conversion. *Adv. Sci.* 3 (5): 1500286.

74 Chen, J.G. (1996). Carbide and nitride overlayers on early transition metal surfaces: preparation, characterization, and reactivities. *Chem. Rev.* 96 (4): 1477.

75 Yao, N., Li, P., Zhou, Z., et al. (2019). Synergistically tuning water and hydrogen binding abilities over Co_4N by Cr doping for exceptional alkaline hydrogen evolution electrocatalysis. *Adv. Energy Mater.* 9 (41): 1902449.

76 Liu, B., He, B., Peng, H.Q., et al. (2018). Unconventional nickel nitride enriched with nitrogen vacancies as a high-efficiency electrocatalyst for hydrogen evolution. *Adv. Sci.* 5 (8): 1800406.

77 Shi, Y. and Zhang, B. (2016). Correction: Recent advances in transition metal phosphide nanomaterials: synthesis and applications in hydrogen evolution reaction. *Chem. Soc. Rev.* 45 (6): 1781.

78 Li, Y., Dong, Z., and Jiao, L. (2019). Multifunctional transition metal-based phosphides in energy-related electrocatalysis. *Adv. Energy Mater.* 10 (11): 1902104.

79 Liu, D., Xu, G., Yang, H., et al. (2023). Rational design of transition metal phosphide-based electrocatalysts for hydrogen evolution. *Adv. Funct. Mater.* 33 (7): 2208358.

80 Tian, J., Liu, Q., Asiri, A.M., et al. (2014). Self-supported nanoporous cobalt phosphide nanowire arrays: an efficient 3d hydrogen-evolving cathode over the wide range of pH 0–14. *J. Am. Chem. Soc.* 136 (21): 7587.

81 Yu, X., Zhao, J., and Johnsson, M. (2021). Interfacial engineering of nickel hydroxide on cobalt phosphide for alkaline water electrocatalysis. *Adv. Funct. Mater.* 31 (25): 2101578.

82 Guo, Y., Park, T., Yi, J.W., et al. (2019). Nanoarchitectonics for transition-metal-sulfide-based electrocatalysts for water splitting. *Adv. Mater.* 31 (17): 1807134.

83 Fang, P., Zhu, M., Liu, J., et al. (2023). Making ternary-metal hydroxysulfide catalyst via cathodic reconstruction with ion regulation for industrial-level hydrogen generation. *Adv. Energy Mater.* 13 (35): 2301222.

84 Feng, L.-L., Yu, G., Wu, Y., et al. (2015). High-index faceted Ni_3S_2 nanosheet arrays as highly active and ultrastable electrocatalysts for water splitting. *J. Am. Chem. Soc.* 137 (44): 14023.

85 Liu, H.J., Zhang, S., Chai, Y.M., et al. (2023). Ligand modulation of active sites to promote cobalt-doped 1T-MoS_2 electrocatalytic hydrogen evolution in alkaline media. *Angew. Chem. Int. Ed.* 62 (48): e202313845.

86 Ling, T., Zhang, T., Ge, B., et al. (2019). Well-dispersed nickel- and zinc-tailored electronic structure of a transition metal oxide for highly active alkaline hydrogen evolution reaction. *Adv. Mater.* 31 (16): 1807771.

87 Wang, J., Xin, S., Xiao, Y., et al. (2022). Manipulating the water dissociation electrocatalytic sites of bimetallic nickel-based alloys for highly efficient alkaline hydrogen evolution. *Angew. Chem. Int. Ed.* 61 (30): e202202518.

88 Yang, C., Zhao, R., Xiang, H., et al. (2020). Ni-activated transition metal carbides for efficient hydrogen evolution in acidic and alkaline solutions. *Adv. Energy Mater.* 10 (37): 2002260.

89 Ding, X., Liu, D., Zhao, P., et al. (2024). Dynamic restructuring of nickel sulfides for electrocatalytic hydrogen evolution reaction. *Nat. Commun.* 15 (1): 5336.

90 Pan, Y., Gao, J., Lv, E., et al. (2023). Integration of alloy segregation and surface Co-O hybridization in carbon-encapsulated CoNiPt alloy catalyst for superior alkaline hydrogen evolution. *Adv. Funct. Mater.* 33 (41): 2303833.

91 Li, M., Li, H., Fan, H., et al. (2024). Engineering interfacial sulfur migration in transition-metal sulfide enables low overpotential for durable hydrogen evolution in seawater. *Nat. Commun.* 15 (1): 6154.

92 Xu, W., Fan, G., Zhu, S. et al. (2021). Electronic structure modulation of nanoporous cobalt phosphide by carbon doping for alkaline hydrogen evolution reaction. *Adv. Funct. Mater.* 31 (48): 2107333.

93 Tong, W., Forster, M., Dionigi, F., et al. (2020). Electrolysis of low-grade and saline surface water. *Nat. Energy* 5 (5): 367.

94 Dionigi, F., Reier, T., Pawolek, Z., et al. (2016). Design criteria, operating conditions, and nickel-iron hydroxide catalyst materials for selective seawater electrolysis. *ChemSusChem* 9 (9): 962.

95 Yu, Z. and Liu, L. (2024). Recent advances in hybrid seawater electrolysis for hydrogen production. *Adv. Mater.* 36 (13): 2308647.

96 Liu, Y., Wang, Y., Fornasiero, P., et al. (2024). Long-term durability of seawater electrolysis for hydrogen: from catalysts to systems. *Angew. Chem. Int. Ed.* 63 (47): e202412087.

97 Vos, J.G., Wezendonk, T.A., Jeremiasse, A.W., et al. (2018). MnO_x/IrO_x as selective oxygen evolution electrocatalyst in acidic chloride solution. *J. Am. Chem. Soc.* 140 (32): 10270.

98 Jin, H., Xu, J., Liu, H., et al. (2023). Emerging materials and technologies for electrocatalytic seawater splitting. *Sci. Adv.* 9 (42): eadi7755.

99 Guo, J., Zheng, Y., Hu, Z., et al. (2023). Direct seawater electrolysis by adjusting the local reaction environment of a catalyst. *Nat. Energy* 8 (3): 264.

100 Xie, H., Zhao, Z., Liu, T., et al. (2022). A membrane-based seawater electrolyser for hydrogen generation. *Nature* 612 (7941): 673.

101 Wu, X., Zhou, S., Wang, Z., et al. (2019). Engineering multifunctional collaborative catalytic interface enabling efficient hydrogen evolution in all pH range and seawater. *Adv. Energy Mater.* 9 (34): 1901333.

102 Dinh, C.T., Jain, A., de Arquer, F.P.G., et al. (2018). Multi-site electrocatalysts for hydrogen evolution in neutral media by destabilization of water molecules. *Nat. Energy* 4 (2): 107.

103 Chen, X.H., Li, X.L., Li, T., et al. (2024). Enhancing neutral hydrogen production by disrupting the rigid hydrogen bond network on Ru nanoclusters through Nb_2O_5-mediated water reorientation. *Energy Environ. Sci.* 17 (14): 5091.

104 Zhai, L., She, X., Zhuang, L., et al. (2022). Modulating built-in electric field via variable oxygen affinity for robust hydrogen evolution reaction in neutral media. *Angew. Chem. Int. Ed.* 61 (14): e202116057.

105 Chen, J., Chen, C., Qin, M., et al. (2022). Reversible hydrogen spillover in Ru-WO_{3-x} enhances hydrogen evolution activity in neutral pH water splitting. *Nat. Commun.* 13 (1): 5382.

106 Sun, K., Wu, X., Zhuang, Z., et al. (2022). Interfacial water engineering boosts neutral water reduction. *Nat. Commun.* 13 (1): 6260.

107 Zhang, P., Man, H., Huo, Y., et al. (2024). CrO_x modified nickel phosphides electrocatalyst for stable hydrogen evolution reaction in neutral media. *Adv. Funct. Mater.* 34 (49): 2409365.

108 Yang, W., Li, M., Zhang, B., et al. (2023). Interfacial microenvironment modulation boosts efficient hydrogen evolution reaction in neutral and alkaline. *Adv. Funct. Mater.* 33 (45): 2304852.

109 Wang, Y., Li, X., Huang, Z., et al. (2023). Amorphous Mo-doped $NiS_{0.5}Se_{0.5}$ nanosheets@crystalline $NiS_{0.5}Se_{0.5}$ nanorods for high current-density electrocatalytic water splitting in neutral media. *Angew. Chem. Int. Ed.* 62 (6): e202215256.

7

Electrochemical Oxygen Evolution Reaction

Yifan Ma[1,2#], Junhua Li[1#], Ying Tang[1#], Hang An[1,2#], Xiaopeng Wang[1]*

[1]*College of Materials Science and Engineering, Sichuan University, Chengdu, China*
[2]*Department of Materials Science and Engineering, National University of Singapore, Singapore*

[#]These authors contributed equally to this work.
[*]Corresponding author: wangxiaopeng0620@163.com

7.1 Introduction

Hydrogen (H_2) energy is a zero-carbon emission renewable clean energy source with high energy density [1–3]. The chemical energy per mass of hydrogen ($39.4\,kWh\,kg^{-1}$) is three times larger than that of other chemical fuels. Moreover, it is necessary to achieve carbon neutrality in 2050 or earlier to limit global temperature increase at the end of this century to 1.5 °C compared to pre-industrial times [1–5]. Considering the historical development of energy carriers toward more hydrogen-rich fuels and the necessity to avoid the carbon dioxide emission that causes the greenhouse effect, hydrogen energy will play a crucial role in the future energy structure. It is estimated that H_2 demand could increase six-fold by 2050 [6].

Electrolysis of water is an established technology and is used today to produce high-purity hydrogen (green hydrogen: no carbon emissions in the process of producing green hydrogen) [7–16]. Moreover, electricity-driven water splitting can facilitate the storage of intermittent electrical energy (such as wind power and solar power, etc.) in the form of hydrogen gas, which will promote cross-regional and cross-seasonal optimization of heterogeneous energy allocation [17]. As shown in Figure 7.1a, the electrolysis of water for hydrogen production technology will play a crucial role in the entire application of hydrogen energy production [18].

The mechanism of electrolyzing water to produce hydrogen is simple, as shown in the following reaction: $H_2O = H_2 + 1/2O_2$. Furthermore, as shown in Figure 7.1b, this reaction includes cathodic hydrogen evolution reaction (HER) and anodic oxygen evolution reaction (OER). The minimum electrical energy required for electrolyzing water is determined by the change in free energy of the underlying reaction. The theoretical free energy change for the decomposition of water into hydrogen and oxygen is $237\,kJ\,mol^{-1}$, which requires a theoretical decomposition

Electrochemical Energy Conversion via Small Molecule Catalysis, First Edition.
Edited by Zhicheng Zhang and Yuanmiao Sun.

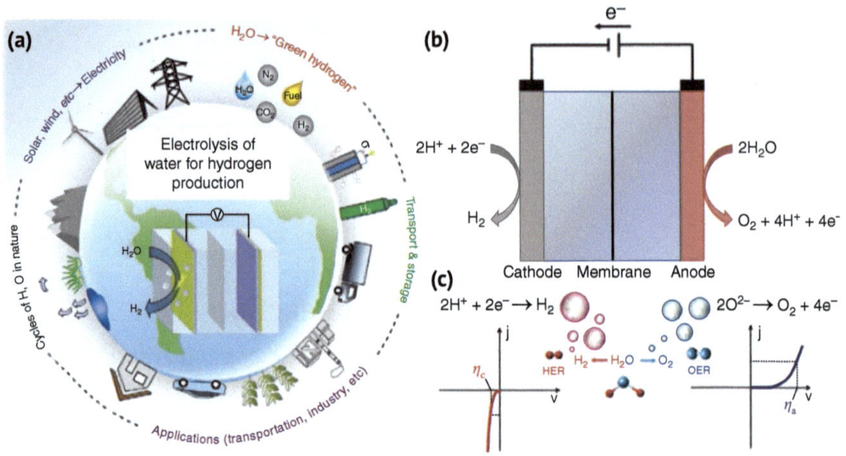

Figure 7.1 (a) A schematic showing the crucial role of the electrolysis of water for "green hydrogen" production technology coupled with renewable electricity. *Source:* Ref. [18] / John Wiley & Sons. (b) Schematic illustration of electrocatalytic water splitting. *Source:* [17] / Royal Society of Chemistry. (c) The polarization curves for HER (left) and OER (right), the η_c and η_a are the overpotentials for cathode and anode at the same current (j), respectively. *Source:* [27] / Royal Society of Chemistry.

voltage of 1.23 V (25 °C and 1 bar pressure). However, the voltage required for electrolyzing water is much higher than 1.23 V due to the slow reaction kinetics on the catalyst surface [19–26].

As a half-reaction of electricity-driven water splitting, the OER is the major bottleneck due to the sluggish kinetics of this four-electron transfer reaction ($2H_2O - 4e^- = O_2 + 4H^+$) as shown in Figure 7.1c. Taking the OER process with metal as the active center, which is the most common in alkaline solutions, as an example, the four-electron transfer process [27] is shown as follows:

$$OH^- + {}^* \rightarrow HO^* + e^- \tag{7.1}$$

$$HO^* \rightarrow O^* + e^- + H^+ \tag{7.2}$$

$$O^* + OH^- \rightarrow HOO^* + e^- \tag{7.3}$$

$$HOO^* \rightarrow {}^* + O_2(g) + e^- + H^+ \tag{7.4}$$

where * represents the adsorption site on the surface of electrocatalysts. The OER process described earlier typically involves four concerted proton–electron transfer reactions. Equation (7.1) describes that OH^- first adsorbs on the surface O vacancy site. Then, HO* species continue the deprotonation process to form O*. Following this event, the O* undergoes a subsequent reaction to form the HOO* intermediate with another OH^-. Finally, oxygen gas (O_2) evolves through the deprotonation of HOO* with the regeneration of the active site. Meanwhile, it is worth noting that a proton is injected into the electrolyte, eventually combining with a transferred electron at the cathode.

Compared with the HER process, which only involves two-electron transfer, the four-electron transfer OER needs large overpotentials, leading to the high cost and high energy consumption of electrocatalytic water splitting [24–26, 28]. Thus, the OER process governs the overall efficiency of electrochemical water splitting. Developing low-cost and robust OER catalysts is critical to solving this efficiency problem in water splitting. Currently, numerous efforts have been devoted to developing high-performance electrocatalysts for improving the efficiency of the OER process. For instance, RuO_2 and IrO_2 are two kinds of typical OER electrocatalysts and they exhibit excellent OER catalytic activity in both acidic and alkaline electrolytes [29, 30]. However, the instability and high cost (belonging to the precious metal oxides) determine that they are not practical for large-scale production. The non-noble metal oxide catalysts containing transition metals, and various metal oxide families such as perovskite, spinel, and layer structure type-family, have been extensively studied. The low-cost, easily synthesized, stable, environmentally benign, and highly efficient transition metal-based oxides are becoming good candidates for electrocatalytic OER [31–35]. Up to now, enormous effort has been devoted to finding high-performance electrocatalysts, which has also stimulated the invention of new techniques to investigate the properties of materials or the fundamental mechanism of the OER. This accumulated knowledge establishes the foundation of the mechanism of the OER. These contents will be detailed in the subsequent sections.

7.2 The State-of-the-art Characterization Techniques

A comprehensive analysis of OER electrocatalysts typically involves various characterization techniques encompassing electrochemical measurements and structural and morphological characterizations [36–39]. Electrochemical characterization begins with cyclic voltammetry (CV), which provides insights into redox behavior and active site generation. With increasing cycling, the oxidation peaks become more obvious, suggesting an increase in electrochemically active sites – a crucial factor in OER enhancement. To accurately determine the electrochemically active surface area (ECSA), CV measurements are conducted in non-Faradaic regions across various scan rates [40]. In OER research, linear sweep voltammetry (LSV) is widely used to evaluate the activity by identifying key metrics such as overpotential – the minimum extra voltage needed to drive the OER at a specific current density (at 10 mA cm^{-2}). Lower overpotential values indicate higher OER performance, making LSV a critical tool for assessing and comparing the performance of different OER catalysts. Electrochemical impedance spectroscopy (EIS) offers valuable information on charge transfer resistance and electrolyte diffusion resistance, shedding light on kinetic limitations and interfacial properties that influence OER performance [41–43]. Finally, chronopotentiometry (CP) is employed to evaluate the long-term stability of electrocatalysts, particularly post-surface reconstruction, to evaluate their durability under sustained reaction conditions [44, 45].

In addition to electrochemical characterization, we also employ the morphology and structure characterizations to analyze the variation of electrocatalysts [46–50].

Various techniques can effectively capture morphological changes during the OER process. Scanning electron microscopy (SEM) and transmission electron microscopy (TEM) are widely used to investigate particle morphology and average size [51]. In previous studies, pristine electrocatalysts displayed clear lattice fringes, while post-electrochemical activation showed the formation of an amorphous layer on the particle surface, with the bulk structure remaining unchanged. Additionally, in situ TEM enables real-time observation of surface reconstruction on the particle surface during OER. Some studies have combined TEM with fast Fourier transform (FFT) or nanobeam diffraction (NBD) to reveal detailed structural changes in electrocatalysts throughout surface reconstruction. To investigate the bulk structure information, several techniques are employed, involving X-ray diffraction (XRD), Brunauer–Emmett–Teller (BET) analysis, Raman spectroscopy, and extended X-ray absorption fine structure (EXAFS) have been utilized. For analyzing the electronic state of electrocatalysts, characterizations such as X-ray photoelectron spectroscopy (XPS), X-ray absorption near edge structure (XANES), and electron paramagnetic resonance (EPR) are commonly used [52]. In situ and operando characterization techniques have revolutionized OER studies by enabling real-time, nondestructive monitoring of the reaction process [49, 53, 54]. These methods allow researchers to detect reaction intermediates, providing crucial insights into surface reconstruction and the fundamental catalytic mechanisms.

Yu Sun et al. [55] proposed an A-site management strategy to accelerate the phase reconstruction during the OER process. In situ electrochemical Raman measurements were performed to explore the phase transformation into $La_{1-x}Ce_xNiO_3$ ($x = 0$, 0.05, 0.1, and 0.15) during the OER process, as shown in Figure 7.2a. In situ electrochemical Raman spectroscopy revealed significant phase changes as the applied potential increased. Specifically, the Raman peaks associated with $LaNiO_3$ at 211 and 426 cm^{-1} diminished at 1.45 V (vs. RHE), while new peaks at 574 and 551 cm^{-1}, corresponding to NiOOH, emerged. For the Ce-doped $LaNiO_3$ ($x = 0.1$), phase transformation occurred at a lower potential of 1.35 V (vs. RHE), demonstrating that Ce doping promoted phase reconstruction and enhanced OER performance. Additionally, BET analysis showed that $La_{0.9}Ce_{0.1}NiO_3$ had a higher specific surface area ($13.32\,m^2\,g^{-1}$) compared to $LaNiO_3$ ($10.12\,m^2g^{-1}$), suggesting more active sites and further contributing to improved OER activity.

X-ray absorption spectroscopy (XAS) has become a crucial technique for investigating the electronic and geometric structures during surface reconstruction, particularly under OER conditions. XAS consists of two regions: XANES and EXAFS. XANES reveals changes in the valence states and electronic structure of metal atoms during redox reactions, which lead to the formation of metal oxide/hydroxide intermediates and their subsequent transformation into metal oxyhydroxides. EXAFS, on the other hand, provides information on the local atomic environment, including bond distances and coordination numbers, offering insights into the structural changes occurring during the OER process. Fabbri E et al. [56] employed in situ XAS measurements showed that during the OER process, the Co *K*-edge position of the $Ba_{0.5}Sr_{0.5}Co_{0.8}Fe_{0.2}O_{3-\delta}$ synthesized by the flame spray method (BSCF-FS) catalyst moved toward the high energy direction with the increase of applied potential, indicating that the oxidation state of Co increased,

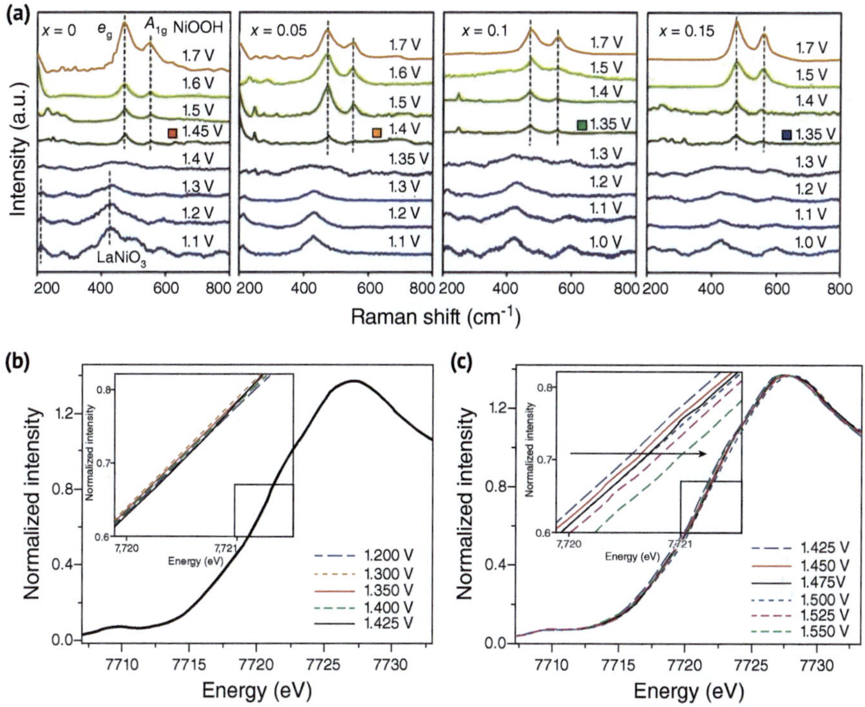

Figure 7.2 (a) In situ Raman spectra of $La_{1-x}Ce_xNiO_3$ (x = 0, 0.05, 0.1, and 0.15) during the OER process (1.0–1.7 V vs. RHE). *Source:* Ref. [55] / John Wiley & Sons. (b and c) XANES spectra recorded at the Co *K*-edge of the BSCF-FS electrode during anodic potential holding from 1.2 to 1.55 V RHE in 0.1 M KOH. *Source:* Ref. [56], Springer Nature.

as shown in Figure 7.2b,c. Combined with EXAFS analysis, a CoO(OH) layer was formed on the catalyst surface with the increase of voltage.

7.3 OER Mechanisms

Understanding the reaction mechanism plays a vital role in designing advanced OER catalysts. Typically, there are four most widely accepted mechanisms for OER: adsorbate evolution mechanism (AEM), lattice oxygen oxidation mechanism (LOM), oxide path mechanism (OPM), and coupled oxygen evolution mechanism (COM). These mechanisms differ in their molecular interactions and oxidation processes on the catalyst surface, each contributing uniquely to the overall understanding and advancement of OER catalyst development.

7.3.1 Adsorbate Evolution Mechanism

As shown in Figure 7.3a, the AEM pathway in the OER utilizes the metal as the redox center, involving a sequence of four electron–proton transfer steps occurring over metal–ion centers on the catalyst surface. Initially, an adsorbed water molecule

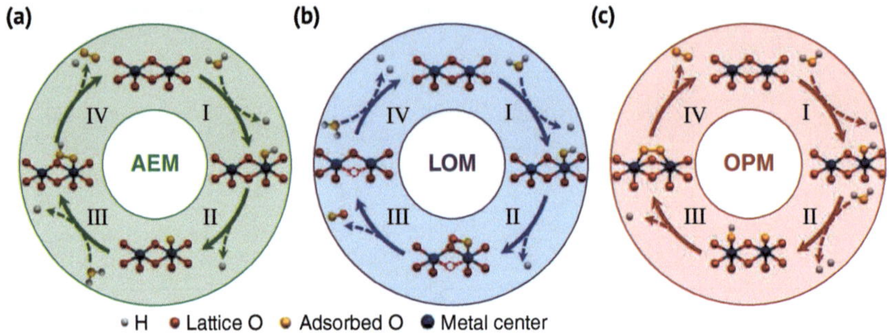

Figure 7.3 A survey of the catalytic reaction pathway, including (a) AEM, (b) LOM, and (c) OPM. *Source:* Ref. [74] / Springer Nature.

dissociates, forming a HO* intermediate. This is followed by a deprotonation step, producing an O* species. Subsequently, the O* intermediate undergoes a nucleophilic attack by another water molecule, leading to the formation of the HOO* species, which ultimately results in the release of an O_2 molecule. In theory, an optimal OER catalyst would require four reaction steps at zero potential, with identical reaction-free energies for each step [57]. However, achieving this thermochemically ideal condition is nearly impossible in practice due to the varying adsorption energies of OER intermediates.

In 2011, Koper introduced a general scaling relationship between the binding energies of HO* and HOO*, with a nearly constant difference ($\Delta G_{HOO^*} - \Delta G_{HO^*} = 3.2 \pm 0.2\,eV$) [58]. Since either ΔG_{HO*} or ΔG_{HOO^*} can serve as the rate-determining step (RDS) in most OER processes, this binding energy difference between HO* and HOO* is widely considered a key descriptor of OER activity [59]. For instance, when the formation of HO* faces a higher energy barrier, the formation of HOO* tends to have a correspondingly lower energy barrier, and vice versa. To achieve optimal OER activity, the binding strengths of reaction intermediates at the active sites must be balanced – not too strong or too weak – to maintain appropriate adsorption and desorption energies. Ideally, the binding energy (1.6 eV) should be evenly distributed between the HO* to O* and O* to HOO* steps. This would result in a minimum theoretical overpotential of 370 mV for an ideal catalyst, calculated as [$3.2 - 2 \times 1.2$]. In this context, overcoming or mitigating the limitations of this scaling relationship is considered a promising strategy to further improve the electrocatalytic OER performance of materials. This can be achieved by adjusting the electronic configuration of the surface structure to optimize the binding strengths of oxygenated intermediates. Notably, RuO_2 and IrO_2 exhibit proximity to ideal values on the plot while maintaining relative stability under acidic OER conditions. The AEM mechanism explains the variability observed in OER activity (as determined by DFT) and provides a set of descriptors, including $\Delta G_O - \Delta G_{HO*}$ [57], the strength of the OH—$M_{2+\delta}$ bond [60], e_g orbital occupancy [61], and the O *p*-band center [62].

Although the AEM pathway has been extensively studied in both homogeneous and heterogeneous systems, the multistep OER mechanism has not yet been fully validated experimentally due to the complexity of the multielectron transfer process. Several challenges remain, such as understanding the origin of the varying binding energies between oxygenated species and catalysts, as well as identifying the precise location of the active sites.

7.3.2 Lattice Oxygen Oxidation Mechanism

Oxygen isotope labeling experiments have demonstrated that lattice oxygen participates in the OER catalytic cycle of certain metal oxide electrocatalysts [63, 64]. Therefore, a second four-electron transfer mechanism, known as the LOM pathway, has been proposed, with its catalytic cycle illustrated in Figure 7.3b. When the electronic states near the Fermi level correspond to oxygen, catalysis follows the LOM mechanism, with lattice oxygen (O_L) acting as the redox center [52, 65]. In the LOM pathway, the initial two step-forming O* and HO* species resemble those in the AEM process. However, the next step involves the coupling of surface O* species with lattice oxygen from the catalyst structure, leading to direct O—O bond formation. Following this, surface vacancies are replenished by water molecules, forming HO* species, with the removal of a proton through a one-electron oxidation step.

Compared with the AEM pathway, the LOM cycle does not form OOH*, breaking the proportional relationship between OH and OOH ($\Delta G_{OOH} - \Delta G_{OH} = 3.2 \pm 0.2\,eV$) that imposes a limitation in AEM, where the lowest theoretical overpotential is 0.37 V. Additionally, the presence of oxygen within the catalyst lattice acts as an intermediary, reducing the need for extensive coverage of reaction intermediates. Consequently, the O—O coupling step becomes less demanding (typically the RDS in AEM), suggesting that LOM theoretically exhibits superior performance compared to AEM. However, it is important to acknowledge that each scenario has its pros and cons. Unlike the stable active centers observed in AEM, catalyst structures in LOM undergo continuous dynamic changes throughout the reaction process. The appearance of oxygen vacancies in LOM can lead to metal dissolution, thereby reducing the stability of the catalyst. The primary difference between the two mechanisms lies in the involvement of active lattice oxygen in the OER process within LOM. In practice, AEM and LOM often coexist during the OER process, leading to some degree of competition and even transitions between the two mechanisms [66, 67]. Regardless of the specific mechanism employed, the O—O bonding process in AEM and the deprotonation steps in LOM directly impact the kinetics of OER [68]. It is worth noting that in acidic OER conditions, due to the limited availability of OH^- ions directly involved in deprotonation, water adsorption and dissociation must first occur at active sites, introducing additional steps and kinetic barriers [69].

7.3.3 Oxide Path Mechanism

In addition to AEM and LOM pathways, the OPM mechanism has been proposed to explain why both the OER activity and stability exceed the theoretical limit [70]. In short, the OPM pathway enables direct O-O radical coupling without creating oxygen vacancy defects or involving additional intermediates like HOO*. In this mechanism (Figure 7.3c), only O* and HO* serve as OER intermediates. Crucially, well-positioned active metal sites work together to dissociate water and facilitate *O radical coupling, leading to O_2 formation without relying on lattice oxygen. As a result, the OPM pathway imposes stricter requirements on the geometric configuration of metal active sites [71, 72]. Multinuclear structures offer greater redox flexibility, which is beneficial for OER processes involving multielectron transfer. Dual-metal sites with optimal atomic spacing are especially important, as they can facilitate O—O radical coupling with a low energy barrier [73].

Compared to the AEM and LOM mechanisms, the OPM mechanism demands more precise geometric configurations of metal active sites, functioning only when the atomic spacing is optimal. However, under ideal conditions, the OPM mechanism can break the scaling relationship of reaction intermediates without compromising stability. Its reaction pathway involves only *O and *OH intermediates, avoiding the formation of *OOH, which enables direct O—O radical coupling to produce oxygen. This provides a promising strategy for enhancing OER performance through the careful design of metal active site geometry.

7.3.4 Coupled Oxygen Evolution Mechanism

The traditional AEM and LOM mechanisms involve distinct metal or oxygen species as redox centers in the electron transfer pathway, without simultaneous interactions. However, Wang et al. [68] introduced a light-triggered COM that creates a novel electron transfer pathway in NR-NiOOH, as shown in Figure 7.4. This mechanism bypasses the oxygen—oxygen bonding in the AEM and the deprotonation process in the LOM by leveraging a light-induced reversible transformation from octahedral to tetrahedral coordination, allowing for the concurrent participation of both metal and oxygen. In catalytic reactions based on the COM mechanism, the metal acts as the redox center during deprotonation, while oxygen serves as the redox center during the formation of oxygen—oxygen bonds. The in-situ synchrotron absorption spectra reveal that under light illumination, the oxidation state of nickel in NR-NiOOH gradually decreases, indicating that the metal acts as the center for oxidation–reduction. The researchers attribute this phenomenon to the light-triggered transformation from octahedral to tetrahedral coordination. Furthermore, they observed the generation of nonbonded oxygen during this phase transition, suggesting that oxygen may also function as a redox center in the catalytic reaction. To further investigate this, they conducted TMA molecular probe and ^{18}O isotope tests, which confirmed that oxygen indeed serves as a redox

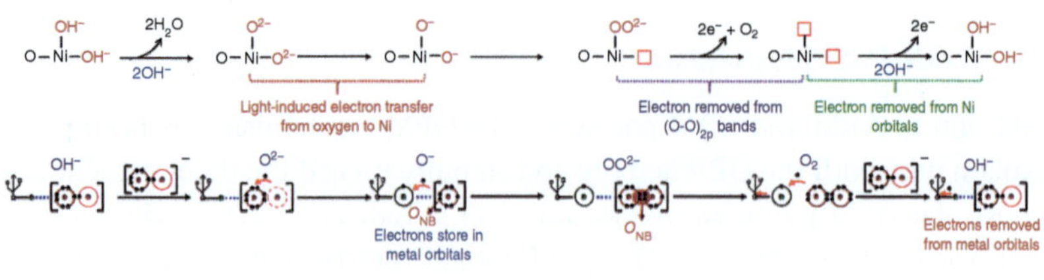

Figure 7.4 Electron transfer pathways of COM routes. *Source:* Ref. [68] / Springer Nature.

center in the catalytic process. Based on these results, they successfully demonstrated that in the COM mechanism, both metal and oxygen act simultaneously as redox centers.

They subsequently employed theoretical calculations to validate the feasibility of the COM mechanism. They argue that the essence of the COM mechanism lies in the reversible transformation between two phases, facilitated by the transfer of oxygen excited by light to the metal dz^2 orbital. This electron transfer pathway is closely linked to the distortion of the crystal structure. Specifically, when the degree of crystal distortion is low, electron transfer from the oxygen band to the metal dz^2 orbital is not possible. Conversely, when the crystal structure is significantly distorted, this electron transfer can occur.

The proposal of the COM mechanism addresses the limitations of traditional OER mechanisms and has been shown to be applicable to various catalysts, extending beyond just NR-NiOOH. This discovery offers new insights and research avenues for the study of water splitting and the development of artificial photosynthesis devices.

7.4 The Mechanism and Development Status of Some Typical Energy Conversion Reactions

Water splitting, a pivotal process for sustainable hydrogen production, comprises three primary methods: electrolysis, photocatalytic water splitting, and photoelectrochemical (PEC) water splitting. Electrolysis utilizes electrical energy to split water molecules into hydrogen and oxygen, offering a mature, highly controllable technology but with significant energy demands. Photocatalytic water splitting directly harnesses sunlight and semiconductor photocatalysts, offering sustainability but facing challenges in efficiency and catalyst stability. PEC water splitting integrates features of both, utilizing light-absorbing materials and an electrical bias to improve efficiency, though it demands sophisticated material designs. Among these, the OER is a critical step in all methods, determining overall efficiency due to its high energy barrier and slow kinetics, highlighting the need for efficient OER catalysts to advance these technologies.

7.4.1 Photocatalytic Water Splitting

Photocatalytic water splitting primarily involves the use of semiconductors to harness solar energy and split water into hydrogen and oxygen [75]. The process consists of three key steps (Figure 7.5a):

Light absorption: A semiconductor absorbs light with energy equal to or greater than its bandgap, exciting electrons from the valence band (VB) to the conduction band (CB), leaving behind holes in the VB. Charge separation and transport: The photogenerated electron–hole pairs must migrate to the semiconductor surface without recombining. Effective charge separation is critical for maintaining high efficiency. Redox reactions: At the surface, photogenerated electrons reduce protons to hydrogen, while holes oxidize water to oxygen.

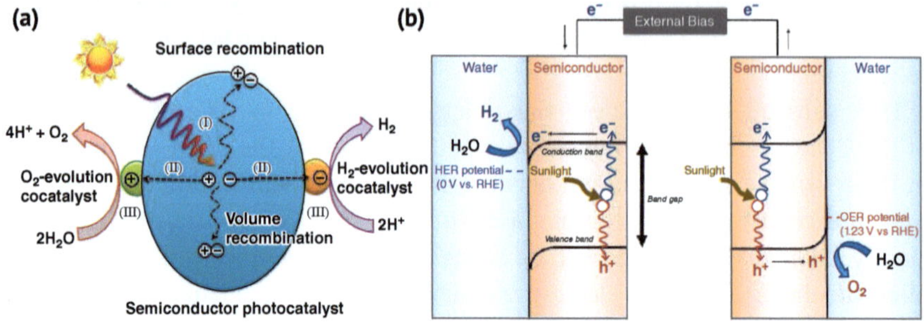

Figure 7.5 (a) Photocatalytic water-splitting process. *Source:* Ref. [87] / John Wiley & Sons. (b) Operational principle of PEC water splitting with two semiconductor electrodes. *Source:* Ref. [88] / MDPI, CC BY 4.0.

Cathode half-reaction (reduction):

$$2H^+ + 2e^- \rightarrow H_2 \tag{7.5}$$

Anode half-reaction (oxidation):

$$H_2O + 2h^+ \rightarrow 0.5O_2 + 2H^+ \tag{7.6}$$

The main challenges in photocatalytic water splitting are low quantum efficiency, limited solar utilization, and poor stability of photocatalysts [76]. Low quantum efficiency arises from the rapid recombination of photogenerated electron–hole pairs, which significantly reduces the availability of charge carriers for redox reactions. Limited solar utilization is a result of the wide bandgaps of many photocatalysts, such as TiO_2 [77], which restrict light absorption to the UV region, covering only a small fraction of the solar spectrum. Additionally, the poor stability of photocatalysts, including issues like photo-corrosion in sulfides or structural degradation under operational conditions, further hampers long-term performance and scalability [78, 79]. Addressing these issues is critical for advancing practical applications of photocatalytic water splitting.

In water electrolysis, electrical energy is used to split water into hydrogen and oxygen. At the cathode, hydrogen gas is produced by the reduction of water,

$$2H_2O + 2e^- \rightarrow H_2 + 2OH^- \tag{7.7}$$

while at the anode, oxygen gas is generated through the oxidation of water.

$$2H_2O \rightarrow O_2 + 4H^+ + 4e^- \tag{7.8}$$

This process transforms electrical energy into chemical energy, producing hydrogen and oxygen gases. The efficiency of the process depends on the applied voltage and the performance of the electrolyzer.

7.4.2 Electrolysis

Electrocatalytic water splitting primarily involves two reactions HER and OER; as mentioned earlier, the kinetics of OER is much slower than HER, so OER has always been a hot spot in the study of hydropower electrolysis. As mentioned before, the study on the mechanism of OER has been increasingly mature and refined.

7.4.3 Photoelectrochemical Water Splitting

PEC water splitting utilizes semiconductor photoelectrodes to harness sunlight, generating electron–hole pairs that drive redox reactions to produce hydrogen and oxygen. At the anode, the OER oxidizes water to oxygen, while at the cathode, the HER reduces protons to hydrogen. PEC water splitting combines the direct solar utilization of photocatalysis and the efficiency of electrolysis in an integrated system, offering higher solar-to-hydrogen efficiency and reduced energy input. In PEC water splitting, applying a bias overcomes energy barriers, enhances charge separation, and improves reaction efficiency; however, challenges such as material stability and charge carrier recombination persist. Meanwhile, OER is still particularly important as it is the rate-limiting step due to its complex four-electron transfer mechanism, making efficient catalysts crucial for lowering overpotential and enhancing kinetics [19–27, 68]. Recent advancements emphasize improving photoelectrodes through nanocomposites [80, 81], bandgap engineering [82], and in situ modifications to boost light absorption [83], charge separation [84], and catalytic activity [85]. However, challenges such as low efficiency, material stability, and high recombination rates of charge carriers remain barriers to large-scale application.

The PEC water-splitting process (Figure 7.5b) begins when the semiconductor absorbs light, generating electron–hole pairs. These pairs are then separated by an internal electric field or an external bias, with electrons moving to the conduction band and holes to the valence band. Finally, at the semiconductor–water interface, the electrons and holes participate in the HER and OER, respectively, completing the overall water-splitting process. This is consistent with the basic principle of photocatalytic water splitting, However, the photocatalyst for photo-hydrolysis of water that requires semiconductor materials to meet its conduction band level (E_{CB}) has a lower potential than the hydrogen evolutionary reaction (HER) (0 V vs. RHE), and the valence band level (E_{VB}) has a higher potential than the oxygen evolutionary reaction (OER) (1.23 V vs. RHE). The band gap (Eg) is greater than 1.23eV, and the overpotential of HER and OER usually requires 1.5–1.8 eV [86]. While PEC shares the principle of using light to generate electron–hole pairs like photocatalytic water splitting, its distinct feature lies in the separation of the two electrodes, allowing independent optimization of the light absorber and catalyst.

7.5 Catalysis Evaluation and Prediction

In the study of OER catalysts, theoretical calculations are key tools for understanding mechanisms, evaluating OER performance and optimizing material design. Density functional theory (DFT) is one of the most used quantum chemical methods [89]. Moreover, it is widely used in the study of the electronic structure and reaction mechanism of electrocatalyst surfaces. First, we can calculate the adsorption energy of reaction intermediates (OH*, O*, and OOH*) on the catalyst surface through DFT, find a balance point of catalyst adsorption energy, and design excellent catalysts. Second, we can calculate the free energy of each reaction step based on DFT, determine the rate-limiting step of the OER process, and estimate the theoretical overpotential. In addition, molecular dynamics (MD) can also be used to study the behavior of catalysts in certain environments, particularly where catalyst surface reconstruction is involved or when ions in the aqueous solution interact with the catalyst surface. We can predict the structural stability of the electrocatalyst and the dynamic changes in surface atomic arrangement under OER reaction conditions through MD simulations.

OER catalysts hold great promise, especially in advancing clean energy technologies. These catalysts are critical for enabling efficient, low-cost, and stable energy conversion processes, such as water splitting for hydrogen production and metal–air batteries. Key future research areas include developing non-precious metal catalysts to lower costs, enhancing catalyst activity and durability for real-world applications, and leveraging theoretical modeling and machine learning to accelerate material discovery and design. As technology progresses, OER catalysts will be integral to build sustainable energy systems.

7.6 Conclusion

This chapter emphasizes the pivotal role of the OER in green hydrogen production and its significance for a sustainable, carbon-neutral energy future. Despite its potential, OER is a bottleneck in water splitting because of its slow kinetics and high overpotential, highlighting the need for efficient, cost-effective catalysts. Advances in characterization techniques, mechanistic studies, and computational modeling have enhanced the understanding of OER processes and informed catalyst optimization. Furthermore, integrating OER into hydrogen production technologies, such as electrolysis, photocatalysis, and PEC water splitting, highlights its versatility and importance. Ongoing innovation in catalyst design and the adoption of multidisciplinary approaches are crucial to addressing current challenges and realizing the full potential of hydrogen energy systems.

Acknowledgment

This work was supported by the Young Scientists Fund of the National Natural Science Foundation of China (52402292).

References

1 Wei, S., Sacchi, R., Tukker, A. et al. (2024). Future environmental impacts of global hydrogen production. *Energy Environ. Sci.* 17 (6): 2157.

2 Terlouw, T., Rosa, L., Bauer, C. et al. (2024). Future hydrogen economies imply environmental trade-offs and a supply-demand mismatch. *Nat. Commun.* 15 (1): 7043.

3 Blay-Roger, R., Bach, W., Bobadilla, L.F. et al. (2024) Natural hydrogen in the energy transition: Fundamentals, promise, and enigmas. *Renew. Sust. Energ. Rev.* 189: 113888.

4 Sovacool, B.K., Del Rio, D.F., Herman, K. et al. (2024). Reconfiguring European industry for net-zero: a qualitative review of hydrogen and carbon capture utilization and storage benefits and implementation challenges. *Energy Environ. Sci.* 17: 3523.

5 Kamran, M. and Turzyński, M. (2024). Exploring hydrogen energy systems: a comprehensive review of technologies, applications, prevailing trends, and associated challenges. *J. Energy Storage*. 96: 112601.

6 IEA (2021). Net Zero by 2050, IEA, Paris https://www.iea.org/reports/net-zero-by-2050, Licence: CC BY 4.0

7 Zhang, Y., Mu, X., Liu, Z. et al. (2024). Twin-distortion modulated ultra-low coordination PtRuNi-O_x catalyst for enhanced hydrogen production from chemical wastewater. *Nat. Commun.* 15 (1): 10149.

8 Wang, X., Yang, Q., Singh, S. et al. (2024). Topological semimetals with intrinsic chirality as spin-controlling electrocatalysts for the oxygen evolution reaction. *Nat. Energy*. 10.1038/s41560-024-01674-9.

9 Bao, D., Huang, L., Gao, Y. et al. (2024). Dynamic creation of a local acid-like environment for hydrogen evolution reaction in natural seawater. *J. Am. Chem. Soc.* 146 (50), 34711–34719.

10 Sun, X., Shen, W., Liu, H. et al. (2024). Corrosion-resistant NiFe anode towards kilowatt-scale alkaline seawater electrolysis. *Nat. Commun.* 15 (1): 10351.

11 Zhang, J., Dang, J., Zhu, X. et al. (2023). Ultra-low Pt-loaded catalyst based on nickel mesh for boosting alkaline water electrolysis. *Appl. Catal. B-Environ.* 325: 122296.

12 Zhang, X., Zhong, H., Zhang, Q. et al. (2024). High-spin Co^{3+} in cobalt oxyhydroxide for efficient water oxidation. *Nat. Commun.* 15 (1): 1383.

13 Ye, P., Fang, K., Wang, H. et al. (2024). Lattice oxygen activation and local electric field enhancement by co-doping Fe and F in CoO nanoneedle arrays for industrial electrocatalytic water oxidation. *Nat. Commun.* 15 (1): 1012.

14 Wang, Z., Li, J., Zhang, Q. et al. (2024). Facilitating Formate Selectivity via Optimizing e_g^* Band Broadening in NiMn Hydroxides for Ethylene Glycol Electro Selectivity*Angew. Chem. Int. Ed.* 63 (44): e202411517.

15 Li, J., Wu, C., Wang, Z. et al. (2024). Unveiling the pivotal role of dx^2–y^2 electronic states in nickel-based hydroxide electrocatalysts for methanol oxidation. *Angew. Chem. Int. Ed.* 63 (25): e202404730.

16 Yao, R., Sun, K., Zhang, K. et al. (2024). Stable hydrogen evolution reaction at high current densities via designing the Ni single atoms and Ru nanoparticles linked by carbon bridges. *Nat. Commun.* 15 (1): 2218.

17 Song, J., Wei, C., Huang, Z.F. et al. (2020). A review on fundamentals for designing oxygen evolution electrocatalysts. *Chem. Soc. Rev.* 49 (7): 2196.

18 Luo, Y., Zhang, Z., Chhowalla, M. et al. (2022). Recent advances in design of electrocatalysts for high-current-density water splitting. *Adv. Mater.* 34 (16): 2108133.

19 Wu, C., Wang, X., Tang, Y. et al. (2023). Origin of surface reconstruction in lattice oxygen oxidation mechanism based-transition metal oxides: a spontaneous chemical process. *Angew. Chem. Int. Ed.* 62 (21): e202218599.

20 Tang, Y., Wu, C., Zhang, Q. et al. (2023). Accelerated surface reconstruction through regulating the solid-liquid interface by oxyanions in perovskite electrocatalysts for enhanced oxygen evolution. *Angew. Chem. Int. Ed.* 62 (37): e202309107.

21 Shen, X., Li, H., Ma, T. et al. (2024). Construction of heterojunction-rich metal nitrides porous nanosheets electrocatalyst for alkaline water/seawater splitting at large current density. *Small* 20 (23): 10535.

22 Rong, C., Wang, S., Shen, X. et al. (2024). Defect-balanced active and stable Co_3O_{4-x} for proton exchange membrane water electrolysis at ampere-level current density. *Energy Environ. Sci.* 17 (12): 4196.

23 Jiang, C., He, H., Guo, H. et al. (2024). Transfer learning guided discovery of efficient perovskite oxide for alkaline water oxidation. *Nat. Commun.* 15 (1): 6301.

24 Deng, P.J., Liu, Y., Liu, H. et al. (2024). Layered double hydroxides with carbonate intercalation as ultra-stable anodes for seawater splitting at ampere-level current density. *Adv. Energy Mater.* 14 (23): 2400053.

25 Zhai, P., Wang, C., Zhao, Y. et al. (2023). Regulating electronic states of nitride/hydroxide to accelerate kinetics for oxygen evolution at large current density. *Nat. Commun.* 14 (1): 1873.

26 Wu, T., Sun, Y., Ren, X. et al. (2023). Reconstruction of thiospinel to active sites and spin channels for water oxidation. *Adv. Mater.* 35 (2): e2207041.

27 Suen, N.T., Hung, S.F., Quan, Q. et al. (2017). Electrocatalysis for the oxygen evolution reaction: recent development and future perspectives. *Chem. Soc. Rev.* 46 (2): 337.

28 Wang, F., Zou, P., Zhang, Y. et al. (2023). Activating lattice oxygen in high-entropy LDH for robust and durable water oxidation. *Nat. Commun.* 14 (1): 6019.

29 Li, A., Kong, S., Adachi, K. et al. (2024). Atomically dispersed hexavalent iridium oxide from MnO_2 reduction for oxygen evolution catalysis. *Science* 384 (6696): 666.

30 Boakye, F.O., Harrath, K., Zhang, D. et al. (2024). Synergistic engineering of dopant and support of Ru oxide catalyst enables ultrahigh performance for acidic oxygen evolution. *Adv. Funct. Mater.* 34 (48): 2408714.

31 Xu, L., Yuan, B., Min, L. et al. (2022). Preparation of NiCo-LDH@NiCoV-LDH interconnected nanosheets as high-performance electrocatalysts for overall water splitting. *Int. J. Hydrog. Energy*. 47 (35): 15583.

32 Wu, Z.-Y., Chen, F.-Y., Li, B. et al. (2022). Non-iridium-based electrocatalyst for durable acidic oxygen evolution reaction in proton exchange membrane water electrolysis. *Nat. Mater.* 22 (1): 100.

33 Xie, Q., Ren, D., Bai, L. et al. (2023). Investigation of nickel iron layered double hydroxide for water oxidation in different pH electrolytes. *Chin. J. Catal.* 44: 127.

34 Peng, L., Yang, N., Yang, Y. et al. (2021). Atomic cation-vacancy engineering of nife-layered double hydroxides for improved activity and stability towards the oxygen evolution reaction. *Angew. Chem. Int. Ed.* 60 (46): 24612.

35 Kang, J., Qiu, X., Hu, Q. et al. (2021). Valence oscillation and dynamic active sites in monolayer NiCo hydroxides for water oxidation. *Nat. Catal.* 4 (1): 1050.

36 Zeng, F., Mebrahtu, C., Liao, L. et al. (2022). Stability and deactivation of OER electrocatalysts: A review. *J. Energy Chem.* 69: 301.

37 McCrory, C.C., Jung, S., Ferrer, I. M. et al. (2015). Benchmarking hydrogen evolving reaction and oxygen evolving reaction electrocatalysts for solar water splitting devices. *J. Am. Chem. Soc.* 137 (13): 4347.

38 Wei, C., Rao, R. R., Peng, J. et al. (2019). Recommended practices and benchmark activity for hydrogen and oxygen electrocatalysis in water splitting and fuel cells. *Adv. Mater.* 31 (31): 1806296.

39 Hong, W.T., Risch, M., Stoerzinger, K.A. et al. (2015). Toward the rational design of non-precious transition metal oxides for oxygen electrocatalysis. *Energy Environ. Sci.* 8 (5): 1404.

40 Wei, C., Sun, S., Mandler, D. et al. (2019). Approaches for measuring the surface areas of metal oxide electrocatalysts for determining their intrinsic electrocatalytic activity. *Chem. Soc. Rev.* 48 (9): 2518.

41 García-Osorio, D.A., Jaimes, R., Vazquez-Arenas, J. et al. (2017). The kinetic parameters of the oxygen evolution reaction (OER) calculated on inactive anodes via EIS transfer functions:• OH formation. *J. Electrochem. Soc.* 164 (11): E3321.

42 Ye, C., Wang, M.Q., Bao, S.J. et al. (2019). Micropore-boosted layered double hydroxide catalysts: EIS analysis in structure and activity for effective oxygen evolution reactions. *ACS Appl. Mater. Interfaces.* 11 (34): 30887.

43 Lyons, M.E. and Brandon, M.P. (2009). The significance of electrochemical impedance spectra recorded during active oxygen evolution for oxide covered Ni, Co and Fe electrodes in alkaline solution. *J. Electroanal. Chem.* 631 (1–2): 62.

44 Marquez, R.A., Espinosa, M., Kalokowski, E. et al. (2024). A guide to electrocatalyst stability using lab-scale alkaline water electrolyzers. *ACS Energy Lett.* 9 (2): 547.

45 Li, H., Lin, Y., Duan, J. et al. (2024). Stability of electrocatalytic OER: from principle to application. *Chem. Soc. Rev.* 21 (53): 10709.

46 Wang, X., Zhong, H., Xi, S. et al. (2022). Understanding of oxygen redox in the oxygen evolution reaction. *Adv. Mater.* 34 (50): 2107956.

47 Yoon, W.S., Haas, O., Muhammad, S. et al. (2014). In situ soft XAS study on nickel-based layered cathode material at elevated temperatures: a novel approach to study thermal stability. *Sci. Rep.* 4 (1): 6827.

48 Ketenoglu, D. (2022). A general overview and comparative interpretation on element-specific X-ray spectroscopy techniques: XPS, XAS, and XRS. *Xray Spectrom.* 51 (5–6): 422.

49 Liu, D., Shadike, Z., Lin, R. et al. (2019). Review of recent development of in situ/operando characterization techniques for lithium battery research. *Adv. Mater.* 31 (28): 1806620.

50 de Groot, F.M., de Smit, E., van Schooneveld, M.M. et al. (2010). In-situ scanning transmission X-ray microscopy of catalytic solids and related nanomaterials. *ChemPhysChem.* 11 (5): 951.

51 Grimaud, A., Demortière, A., Saubanère, M. et al. (2016). Activation of surface oxygen sites on an iridium-based model catalyst for the oxygen evolution reaction. *Nat. Energy.* 2 (1): 1.

52 Zhong, H., Zhang, Q., Yu, J. et al. (2023). Fundamental understanding of structural reconstruction behaviors in oxygen evolution reaction electrocatalysts. *Adv. Energy Mater.* 13 (31): 2301391.

53 Li, X., Yang, X., Zhang, J. et al. (2019). In situ/operando techniques for characterization of single-atom catalysts. *ACS Catal.* 9 (3): 2521.

54 Wu, J., Wang, S., Ji, R. et al. (2024). In situ characterization techniques for electrochemical nitrogen reduction reaction. *ACS Nano* 18 (32): 20934.

55 Sun, Y., Li, R., Chen, X. et al. (2021). A-site management prompts the dynamic reconstructed active phase of perovskite oxide OER catalysts. *Adv. Energy Mater.* 11 (12): 2003755.

56 Fabbri, E., Nachtegaal, M., Binninger, T. et al. (2017). Dynamic surface self-reconstruction is the key of highly active perovskite nano-electrocatalysts for water splitting. *Nat. Mater.* 16 (9): 925.

57 Man, I.C., Su, H.Y., Calle-Vallejo, F. et al. (2011). Universality in oxygen evolution electrocatalysis on oxide surfaces. *ChemCatChem.* 3 (7): 1159.

58 Koper, M.T. (2011). Thermodynamic theory of multi-electron transfer reactions: Implications for electrocatalysis. *J. Electroanal. Chem.* 660 (2): 254.

59 Montoya, J.H., Seitz, L.C., Chakthranont, P. et al. (2017). Materials for solar fuels and chemicals. *Nat. Mater.* 16 (1): 70.

60 Subbaraman, R., Tripkovic, D., Chang, K.C. et al. (2012). Trends in activity for the water electrolyser reactions on 3 d M (Ni, Co, Fe, Mn) hydr (oxy) oxide catalysts. *Nat. Mater.* 11 (6): 550.

61 Suntivich, J., May, K.J., Gasteiger, H.A. et al. (2011). A perovskite oxide optimized for oxygen evolution catalysis from molecular orbital principles. *Science* 334 (6061): 1383.

62 Grimaud, A., May, K.J., Carlton, C.E. et al. (2013). Double perovskites as a family of highly active catalysts for oxygen evolution in alkaline solution. *Nat. Commun.* 4 (1): 2439.

63 Lee, S., Banjac, K., Lingenfelder, M. et al. (2019). Oxygen isotope labeling experiments reveal different reaction sites for the oxygen evolution reaction on nickel and nickel iron oxides. *Angew. Chem. Int. Ed.* 131 (30): 10401.

64 Zhang, L., Wang, L., Wen, Y. et al. (2020). Boosting neutral water oxidation through surface oxygen modulation. *Adv. Mater.* 32 (31): 2002297.

65 Ma, Q. and Mu, S. (2023). Acidic oxygen evolution reaction: mechanism, catalyst classification, and enhancement strategies. *Interd. Mater.* 2 (1): 53.

66 Tan, X., Zhang, M., Chen, D. et al. (2023). Electrochemical etching switches electrocatalytic oxygen evolution pathway of $IrOx/Y_2O_3$ from adsorbate evolution mechanism to lattice-oxygen-mediated mechanism. *Small* 19 (44): 2303249.

67 Chen, R., Wang, Z., Chen, S. et al. (2023). Activating lattice oxygen in spinel oxides via engineering octahedral sites for oxygen evolution. *ACS Energy Lett.* 8 (8): 3504.

68 Wang, X., Xi, S., Huang, P. et al. (2022). Pivotal role of reversible NiO_6 geometric conversion in oxygen evolution. *Nature* 611 (7937): 702.

69 Rong, C., Dastafkan, K., Wang, Y. et al. (2023). Breaking the activity and stability bottlenecks of electrocatalysts for oxygen evolution reactions in acids. *Adv. Mater.* 35 (49): 2211884.

70 Hao, S., Liu, M., Pan, J. et al. (2020). Dopants fixation of Ruthenium for boosting acidic oxygen evolution stability and activity. *Nat. Commun.* 11 (1): 5368.

71 Kodera, M., Kawahara, Y., Hitomi, Y. et al. (2012). Reversible O–O bond scission of peroxodiiron (III) to high-spin oxodiiron (IV) in dioxygen activation of a diiron center with a bis-tpa dinucleating ligand as a soluble methane monooxygenase model. *J. Am. Chem. Soc.* 134 (32): 13236.

72 Okamura, M., Kondo, M., Kuga, R. et al. (2016). A pentanuclear iron catalyst designed for water oxidation. *Nature* 530 (7591): 465.

73 Lin, C., Li, J.L., Li, X. et al. (2021). In-situ reconstructed Ru atom array on α-MnO_2 with enhanced performance for acidic water oxidation. *Nat. Catal.* 4 (12): 1012.

74 Ji, Q., Tang, B., Zhang, X. et al. (2024). Operando identification of the oxide path mechanism with different dual-active sites for acidic water oxidation. *Nat. Commun.* 15 (1): 8089.

75 Nishioka, S., Osterloh, F.E., Wang, X. et al. (2023). Photocatalytic water splitting. *Nat Rev Methods Primers.* 3 (1): 42.

76 Jafari, T., Moharreri, E., Amin, A.S. et al. (2016). Photocatalytic water splitting-the untamed dream: a review of recent advances. *Molecules* 21 (7): 900.

77 Molaei, M.J. (2024). Recent advances in hydrogen production through photocatalytic water splitting: a review. *Fuel* 365: 131159.

78 Maeda, K. and Domen, K. (2010). Photocatalytic water splitting: recent progress and future challenges. *J. Phys. Chem. Lett.* 1 (18): 2655.

79 Ye, K.-H., Wang, Z., Gu, J. et al. (2017). Carbon quantum dots as a visible light sensitizer to significantly increase the solar water splitting performance of bismuth vanadate photoanodes. *Energy Environ. Sci.* 10 (3): 772.

80 Mittal, S., Khosya, M., Faraz, M. et al. (2024). 2D SnS_2-CNT nanocomposite photoelectrode for enhanced photoelectrochemical water splitting performance. *Int. J. Hydrog. Energy.* 81: 878.

81 Lai, Y.-C. and Tsai, Y.-C. (2012). An efficient 3C-silicon carbide/titania nanocomposite photoelectrode for dye-sensitized solar cell. *Chem. Commun.* 48 (53): 6696.

82 Sołtys-Mróz, M., Syrek, K., Pierzchała, J. et al. (2020). Band gap engineering of nanotubular Fe_2O_3-TiO_2 photoanodes by wet impregnation. *Appl. Surf. Sci.* 517: 146195.

83 Kim, T.W., Ping, Y., Galli, G. et al. (2015). Simultaneous enhancements in photon absorption and charge transport of bismuth vanadate photoanodes for solar water splitting. *Nat. Commun.* 6 (1): 8769.

84 Wang, S., Chen, P., Bai, Y. et al. (2018). New $BiVO_4$ dual photoanodes with enriched oxygen vacancies for efficient solar-driven water splitting. *Adv. Mater.* 30 (20): 1800486.

85 Wang, X.H., Li, J.-G., Kamiyama, H. et al. (2006). Wavelength-sensitive photocatalytic degradation of methyl orange in aqueous suspension over iron(III)-doped TiO_2 nanopowders under UV and visible light irradiation. *J. Phys. Chem. B.* 110 (13): 6804.

86 Wu, H., Zhang, L., Du, A. et al. (2022). Low-bias photoelectrochemical water splitting via mediating trap states and small polaron hopping. *Nat. Commun.* 13 (1): 6231.

87 Liu, X. and Zhuang, H. (2021). Recent progresses in photocatalytic hydrogen production: design and construction of Ni-based cocatalysts. *Int. J. Energy Res.* 45 (2): 1480.

88 Son, M.K. (2023). Key strategies on Cu_2O photocathodes toward practical photoelectrochemical water splitting. *Nanomaterials* 13 (24): 3142.

89 Kumar, S.G.H., Bozal-Ginesta, C., Wang, N. et al. (2024). From computational screening to the synthesis of a promising OER catalyst. *Chem. Sci.* 15: 10556.

8

Electrochemical Hydrogen/Liquid Fuel Oxidation Reaction

Ran Wang[1], *Huan Pang*[1,2*]

[1] *School of Chemistry and Chemical Engineering, Yangzhou University, Yangzhou, Jiangsu, China*
[2] *State Key Laboratory of Coordination Chemistry, Nanjing University, Nanjing, Jiangsu, China*

*Corresponding author: panghuan@yzu.edu.cn

8.1 Introduction

The extensive use of fossil fuels has led to the emission of vast amounts of greenhouse gases, exerting profound negative impacts on the global ecological environment. As global warming intensifies and extreme weather events occur more frequently, efforts to mitigate climate change have been significantly enhanced worldwide. Consequently, advancing global decarbonization efforts and transitioning to sustainable energy systems have become pressing priorities [1–3]. Among various technologies, fuel cells have emerged as an essential pathway to address global energy and environmental challenges due to their high efficiency, clean operation, low-temperature functionality, and zero-carbon emissions [4, 5]. Fuel cells convert the chemical energy of fuels directly into electricity through electrochemical reactions, with their fundamental working principles involving the anodic fuel oxidation reaction (FOR), the cathodic oxygen reduction reaction (ORR), and ion transport processes [6]. Hydrogen, with its high energy density and the clean byproduct of water is widely regarded as an ideal zero-carbon fuel. Proton exchange membrane fuel cells (PEMFCs), powered by hydrogen, have become a central focus in fuel cell research and have demonstrated broad application potential in renewable energy utilization, portable electronic devices, and large-scale power systems [7, 8]. However, challenges such as the storage, transportation, and refueling of hydrogen, especially for large-scale applications, pose significant technical and economic barriers. Ensuring the safe and efficient transport and storage of hydrogen remains a critical issue to be addressed.

Electrochemical Energy Conversion via Small Molecule Catalysis, First Edition.
Edited by Zhicheng Zhang and Yuanmiao Sun.

In contrast to hydrogen, liquid-phase small-molecule fuels, such as methanol, ethanol, and formic acid, offer a more efficient and convenient alternative [9]. These liquid fuels can be directly used in fuel cells to generate clean electricity, making them particularly suitable for portable electronics and large-scale power applications. Compared to hydrogen, liquid fuels exhibit higher volumetric energy densities and significant advantages in storage, transport, and refueling. As a result, direct liquid fuel cells (DLFCs) have gained increasing research attention [4, 10]. Depending on the type of fuel, DLFCs can be categorized into the following types: (1) direct alcohol fuel cells, including methanol, ethanol, isopropanol, ethylene glycol, and glycerol; (2) direct formic acid fuel cells; and (3) other types, such as direct hydrazine and dimethyl ether fuel cells. Among these, DLFCs based on methanol, ethanol, and formic acid are the most extensively studied.

The practical performance of fuel cells largely depends on the efficiency and stability of the anodic electrochemical FOR [11–13]. This reaction typically involves the oxidation of fuels (e.g., hydrogen, alcohols, or formic acid) to generate electrons, protons, and corresponding byproducts. The efficiency of this process directly determines the energy conversion efficiency, fuel utilization, and overall performance of the fuel cell [14, 15]. Therefore, investigating the catalytic mechanisms and optimizing the performance of FORs are crucial for overcoming technical bottlenecks in fuel cell applications. Such efforts also provide theoretical and experimental guidance for designing highly efficient and cost-effective catalysts.

This chapter systematically examines the reaction mechanisms and catalyst advancements for the anodic oxidation of hydrogen and several representative liquid fuels, such as methanol, ethanol, and formic acid. The discussion focuses on the application prospects and technical challenges of these fuels in fuel cells, particularly regarding enhancing catalytic efficiency, reducing catalyst costs, and improving long-term stability. Additionally, this chapter summarizes recent research progress, highlighting its significance for advancing clean energy technologies and achieving sustainable energy transitions.

8.2 Hydrogen Fuel Oxidation Reaction

8.2.1 Hydrogen Oxidation Reaction Mechanism

The hydrogen oxidation reaction (HOR) is a critical process at the anode of hydrogen fuel cells [16, 17]. Studies on HOR mechanisms typically classify them based on acidic and alkaline media, as the reaction pathways differ in these environments [18, 19]. The net stoichiometric equations for HOR in acidic and alkaline conditions are as follows:

$$H_2 \rightarrow 2H^+ + 2e^- \text{ (Acidic medium)}$$

$$H_2 + 2OH^- \rightarrow 2H_2O + 2e^- \text{ (Alkaline medium)}$$

On the microscopic level, HOR is widely recognized to involve two elementary steps: dissociative adsorption and electrochemical desorption (Volmer reaction).

These processes can be described in detail for acidic and alkaline environments as follows [20, 21]:

In acidic medium:

$$H_2 + 2^* \leftrightarrow 2H_{ad} \text{ (Tafel)}$$

$$H_2 + {}^* \leftrightarrow H_{ad} + H^+ + e^- \text{ (Heyrovsky)}$$

$$H_{ad} \leftrightarrow {}^* + H^+ + e^- \text{ (Volmer)}$$

In alkaline medium:

$$H_2 + 2^* \leftrightarrow 2H_{ad} \text{ (Tafel)}$$

$$H_2 + OH^- + {}^* \leftrightarrow H_{ad} + H_2O + e^- \text{ (Heyrovsky)}$$

$$H_{ad} + OH^- \leftrightarrow {}^* + H_2O + e^- \text{ (Volmer)}$$

Here, * represents the hydrogen adsorption site on the catalyst and H_{ad} denotes the adsorbed hydrogen. The HOR mechanism begins with the adsorption and dissociation of H_2 on the electrode surface to form H_{ad} [22, 23]. This process can occur through two pathways (Figure 8.1a): (1) Tafel step involves the dissociation of H_2 into two H_{ad} on adjacent sites without electron transfer [20, 24]. This pathway is

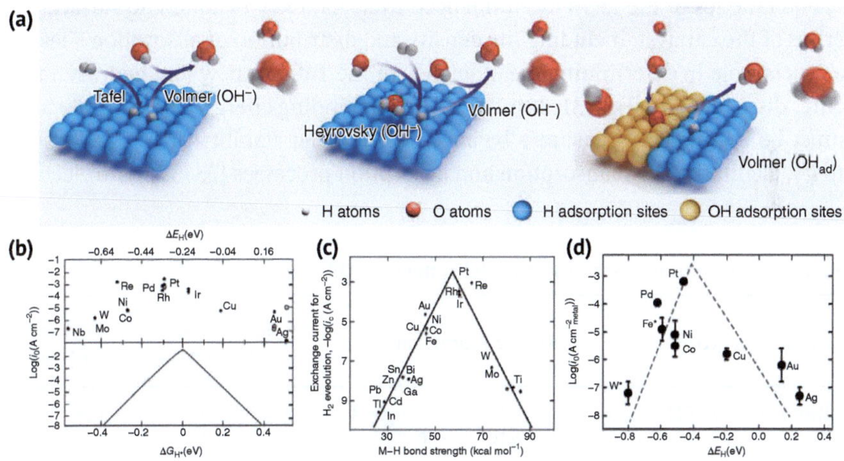

Figure 8.1 (a) HOR routes in alkaline media: Tafel–Volmer, Heyrovsky–Volmer mechanism with OH⁻ only in the electrolyte and Volmer step with OH_{ad} adsorbed on the catalyst surface. *Source:* Reproduced from [24] / with permission of American Chemical Society. (b) Exchange current density for HER in acid, plotted as a function of the calculated hydrogen chemisorption energy per atom. *Source:* Reproduced from [25] / with permission of IOP Publishing. (c) Exchange current density for HER in acid, plotted as a function of the strength of the intermediate metal–hydrogen bond. *Source:* Reproduced from [26] / with permission of Elsevier. (d) Exchange current density for HER in base, plotted as a function of the calculated hydrogen binding energy.
Source: [27] / Royal Society of Chemistry.

effective on surfaces with high site density but may be limited by geometric constraints. (2) Heyrovsky step involves dissociation of H_2 coupled with proton and electron transfer, forming H_{ad} and H^+. This pathway is effective even on catalysts with sparse sites but may face kinetic limitations due to coupled transfer processes. The H_{ad} formed is subsequently oxidized to H^+ and e^- in the Volmer step, which is critical for completing the HOR.

8.2.1.1 HOR Reaction Kinetics

Understanding the microscopic kinetics of HOR and identifying its rate-determining step (RDS) are essential for optimizing catalysts and improving reaction efficiency [28]. The Tafel slope, an important parameter derived from experimental data, provides insights into the reaction pathway and its limiting step. HOR kinetics in acidic media are relatively fast, with a commonly reported Tafel slope of 30 mV dec^{-1} for Pt-based catalysts (Table 8.1). However, this value often reflects mass transport limitations in rotating disk electrode (RDE) systems rather than the intrinsic reaction mechanism. Studies using hydrogen pump systems reveal a Tafel slope of approximately 120 mV dec^{-1}, indicating that the Volmer step is the RDS in acidic media [21]. In alkaline media, HOR kinetics are slower due to limited proton transfer efficiency. Pt catalysts exhibit a Tafel slope of 118 mV dec^{-1}, suggesting a Tafel–Volmer pathway with the Volmer step as the RDS [29]. This step involves the oxidation and desorption of H_{ad}, which is kinetically hindered by the lower availability of protons in alkaline conditions [30].

The kinetics of the HOR are influenced by several key factors. The surface properties of the catalyst, including the density and distribution of adsorption sites, play a crucial role in determining the efficiency of the Tafel step, where hydrogen molecules dissociate into H_{ad} [31, 32]. The hydrogen binding energy (HBE) of the catalyst must be optimized to ensure a balance between the stability of H_{ad} and its reactivity, facilitating both adsorption and desorption processes [33–35]. The electrolyte

Table 8.1 Kinetic expressions and Tafel slope for HOR.

Mechanism (RDS)	Kinetic expressions	Tafel slope expressions	Tafel slopes (mV dec^{-1})
Tafel–Volmer (RDS)	$i = i_0\left[e^{\frac{\alpha F}{RT}\eta} - e^{\frac{(1-\alpha)F}{RT}\eta}\right]$	$\frac{2.303RT}{(1-\alpha)F}$	118
Tafel (RDS)–Volmer	$\eta = \frac{RT}{2F}\ln\left(1 + \frac{i}{i_\tau}\right)$	$\frac{2.303RT}{2F}$	30
Heyrovsky–Volmer (RDS)	$i = i_0\left[e^{\frac{\alpha F}{RT}\eta} - e^{\frac{(2-\alpha)F}{RT}\eta}\right]$	$\frac{2.303RT}{(2-\alpha)F}$	39
Heyrovsky (RDS) –Volmer	$i = i_0\left[e^{\frac{(1+\alpha)\alpha F}{RT}\eta} - e^{\frac{(\alpha-1)F}{RT}\eta}\right]$	$\frac{2.303RT}{(1-\alpha)F}$	118
Diffusion control	$\eta_d = -\frac{RT}{2F}\ln\left(1 - \frac{i_d}{i_l}\right)$	$\frac{2.303RT}{2F}$	30

Note: the transfer coefficient α is typically assumed to be 0.5, and the temperature is considered as 298 K.
Source: Adapted from [36].

environment, whether acidic or alkaline, significantly affects proton transfer rates, which directly impact the overall reaction kinetics. Additionally, the electronic structure of the catalyst is critical for modulating charge transfer processes and determining reaction pathways. Addressing these factors is essential for designing catalysts that exhibit high efficiency across varying electrolyte conditions. For instance, tailoring the electronic structure and surface morphology of catalysts can enhance their activity during the Volmer step, a common kinetic bottleneck in HOR [34, 35]. Such insights provide a foundation for developing advanced catalytic systems optimized for specific operational environments.

8.2.1.2 HBE Theory

Understanding the fundamental reaction mechanism and RDS of the HOR highlights the importance of kinetic descriptors in catalyst design. Alkaline HOR and hydrogen evolution reaction (HER) are interrelated reverse processes that share common intermediates, with complementary strategies for performance optimization. Together, they drive advancements in hydrogen energy technologies. Studies by Nørskov, Trasatti, and Gasteiger groups have evaluated HER activity on various monometallic surfaces, revealing a volcanic relationship between HER kinetics and HBE in acidic media (Figure 8.1b, c) [25, 26]. Platinum (Pt) demonstrated the highest activity at the apex of this relationship, which is consistent with the Sabatier principle. While HBE values under vacuum differ from those in real electrochemical environments, corrections for solvent effects and surface electric fields allow HBE to serve as a reliable descriptor for HER activity on certain metal surfaces. Extending this concept to alkaline media, Sheng et al. [27] showed that HBE could also predict HOR/HER activity, producing a similar volcanic trend (Figure 8.1d). This approach enables efficient screening of highly active HOR/HER catalysts using computationally derived HBE values. HBE theory is a crucial tool for elucidating the relationship between a catalyst's hydrogen adsorption capability and its catalytic activity [37]. HBE refers to the free energy released when a hydrogen atom adsorbs onto a catalyst surface, serving as a key parameter to evaluate adsorption strength [38–40]. An ideal catalyst should exhibit an HBE value close to zero, striking a balance between effective hydrogen adsorption and prompt desorption post-reaction. Catalysts with overly low HBE values struggle to desorb hydrogen atoms, limiting reaction rates, while those with excessively high HBE values face difficulties in hydrogen molecule activation, hindering reaction initiation. Thus, achieving an optimal HBE is critical for high catalytic activity.

HBE theory guides catalyst optimization by enabling researchers to modulate the electronic structure of catalysts. Techniques such as doping, surface modification, and alloying effectively tune HBE values toward the ideal range [36, 41, 42]. For instance, noble metal catalysts like Pt exhibit near-optimal HBE values and demonstrate superior catalytic performance [6]. In contrast, non-noble metal catalysts, such as Ni or Co, often deviate from ideal HBE values and require further optimization to improve performance [43, 44]. Such strategies not only reduce costs but also expand the applicability of catalysts across diverse reaction environments. Despite its simplicity and utility, HBE theory has limitations. It does not fully account for factors such as electrolyte properties, electric field effects, or interfacial complexities, which are critical in practical applications [45, 46]. Therefore, HBE should be considered alongside complementary experimental and theoretical approaches. For example, combining HBE theory with electrochemical testing, advanced characterization techniques, and computational simulations can provide a more comprehensive understanding of catalyst behavior under real-world conditions [47]. This integrated approach enhances the rational design of HOR catalysts, ensuring improvements in both activity and stability. By leveraging HBE theory and multidimensional research, scientists can develop efficient, cost-effective HOR catalysts, thereby advancing technologies such as fuel cells and contributing to the broader adoption of clean energy solutions.

8.2.2 HOR Catalyst Research Progress

Currently, platinum (Pt) is universally recognized as the most active catalyst for the HOR in both acidic and alkaline media [48, 49]. However, due to the scarcity and high cost of platinum, its large-scale application in HOR catalysis remains limited [50]. The HOR reaction catalyzed by Pt exhibits a kinetic rate in alkaline media that is more than two orders of magnitude lower than in acidic media. This leads to the need for more platinum in alkaline conditions to achieve commercially feasible catalytic activity [20]. To address these challenges and reduce the cost of hydrogen fuel cells, developing cheap and efficient alternative catalysts has become a crucial research direction.

8.2.2.1 Catalysts in Acidic Media

In acidic media, noble metal catalysts (such as platinum and its alloys) are the core of HOR research [51]. Platinum-based catalysts exhibit outstanding catalytic performance due to their efficient hydrogen dissociation ability ($H_2 \rightarrow 2H_{ad}$) and rapid surface hydrogen desorption kinetics ($H_{ad} \rightarrow H^+ + e^-$) [52]. Significant progress has been made in electronic structure modulation and hydrogen adsorption energy optimization through alloying platinum with other metals such as Ru, Ir, and Pd [53, 54]. For example, Pt–Ru alloy catalysts not only show higher HOR activity but also exhibit stronger resistance to CO poisoning [18]. Despite these advancements, the HOR reaction in acidic media still faces kinetic limitations, especially with regard to proton transfer steps that influence the reaction rate. Furthermore, stability issues in strongly acidic environments, such as catalyst dissolution and particle agglomeration, significantly reduce catalyst lifetime [55]. The scarcity and high cost of platinum further limit its widespread use. Therefore, developing efficient, stable, and cost-effective HOR catalysts remains a key challenge in the field.

In recent years, the development of single-atom catalysts has provided a breakthrough in this area [56–58]. By dispersing individual noble metal atoms on carbon-based materials, single-atom catalysts greatly improve atomic utilization while maintaining excellent catalytic performance. Additionally, the design of bifunctional catalysts, which combine resistance to poisoning with high HOR activity, attempts to reduce the interference of impurities (such as CO) in the fuel [23, 51, 59]. However, even with these strategies, achieving high activity and long-term stability for catalysts under acidic conditions remains challenging.

8.2.2.2 Catalysts in Alkaline Media

In contrast to acidic media, the study of HOR catalysts in alkaline media presents different challenges [60, 61]. Compared to platinum, palladium has a stronger hydrogen adsorption ability in alkaline media. Alloying palladium with other metals, such as Ni, Cu, and Co, further improves its performance. For example, Yuan et al. [60] demonstrated that the Pd–Ni alloy catalyst exhibits enhanced HOR catalytic activity (Figure 8.2a, b). Through density functional theory (DFT) calculations, they explained that the strong synergy between Pd single atoms (SAs)/Pd dimer and Ni substrate can modulate the binding strength of proton (H)/hydroxyl (OH), thereby significantly reducing the activation energy barrier of a decisive reaction step (Figure 8.2c–e). In alkaline media, the HOR reaction benefits from the participation of OH^- ions, which can facilitate the desorption of reaction intermediates, accelerating the reaction rate.

There are significant differences in HOR catalysts between acidic and alkaline media. In acidic media, platinum-based catalysts remain the primary choice due to their excellent reaction kinetics, but their high cost and stability issues limit further development. In contrast, non-noble metal catalysts in alkaline media, particularly palladium-based catalysts, provide a more economical option and demonstrate excellent catalytic performance [62, 63]. However, the HOR reaction rate in alkaline media is lower, mainly due to the influence of hydrogen oxide intermediates. Future research should focus on optimizing the catalyst surface chemistry and adsorption behaviors, developing bifunctional catalysts that combine high activity

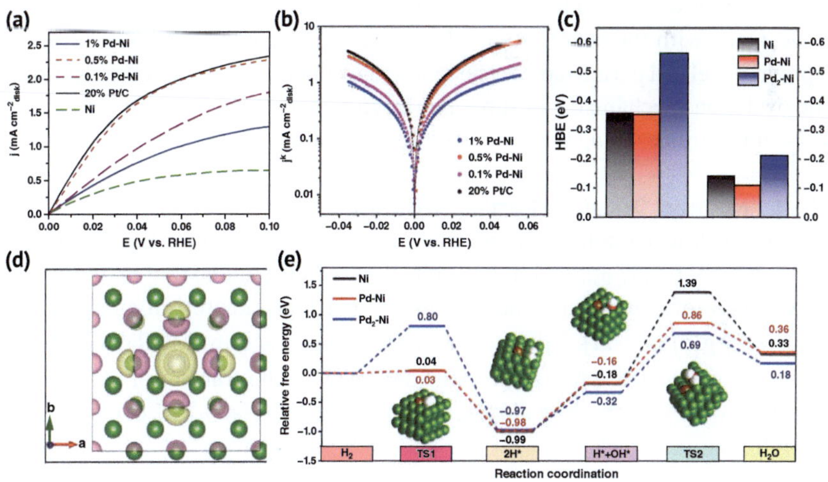

Figure 8.2 (a) HOR polarization curves of Ni, x% Pd–Ni (x = 0.1, 0.5, and 1), and 20% Pt/C. (b) Tafel plots of the kinetic current densities. (c) Calculated HBE on the surface of Ni, Pd–Ni, and Pd_2–Ni models. (d) Differential charge density diagram of Pd–Ni (top view). (e) Free energy diagrams of the key intermediates involved in HOR on Ni, Pd–Ni, and Pd_2–Ni surfaces. *Source:* [60] / John Wiley & Sons / CC BY-NC 4.0.

with anti-poisoning performance and gaining a deeper understanding of the reaction mechanisms in both acidic and alkaline media. Simultaneously, enhancing catalyst durability and overall performance through support design and interface engineering will provide broader prospects for the application of HOR catalysts in fuel cells and other clean energy technologies.

8.3 Liquid Fuel Oxidation Reaction

8.3.1 Liquid Fuel Oxidation Mechanism

Liquid fuel cells have gained significant attention in recent years due to their high energy density, ease of operation, and excellent adaptability to low-temperature environments [64, 65]. Compared to traditional hydrogen fuel cells, liquid fuel cells can directly use liquid fuels as energy carriers, which are easier to store and transport. This makes them highly suitable for applications in portable electronics, transportation, and distributed energy systems [4, 66, 67]. Among various liquid fuels, methanol, ethanol, and formic acid are considered the most promising anode fuels. These fuels are not only widely available and inexpensive but also offer high theoretical energy densities.

Despite the many advantages of liquid fuel cells, their large-scale application is still limited by the performance of the anode-side electrocatalytic oxidation reactions (EOR). At the anode, fuel molecules must undergo oxidation reactions facilitated by catalysts to release electrons. However, these oxidation reactions often involve complex multi-electron transfers, bond cleavages, and formations, generating multiple reaction intermediates and by-products [68, 69]. Therefore, a deeper understanding of the electrocatalytic oxidation mechanisms of liquid fuels at the anode, along with the development of efficient and durable catalysts, is critical for improving fuel cell performance. The following sections will discuss the electrocatalytic oxidation mechanisms of three typical fuels—methanol, ethanol, and formic acid—at the anode, along with the latest research advancements in this area.

8.3.1.1 Methanol Oxidation Reaction Mechanism

Methanol is a water-soluble liquid fuel and one of the earliest liquid fuels proposed and practically applied in fuel cells. It is widely studied due to its abundant sources, ease of storage and transportation, and high theoretical energy density (6.1 kWh kg^{-1}) [70, 71]. The methanol oxidation reaction (MOR) at the anode is a multistep electrochemical process that involves the transfer of six electrons, ultimately producing carbon dioxide (CO_2) and water (H_2O). The overall reaction is as follows:

$$CH_3OH + H_2O \rightarrow CO_2 + 6H^+ + 6e^-$$

$$E^0 = 0.016\ \text{V (vs. SHE)}$$

In acidic electrolytes, platinum (Pt) catalysts, known for their excellent electrocatalytic activity and chemical stability, are considered the core catalysts for MOR

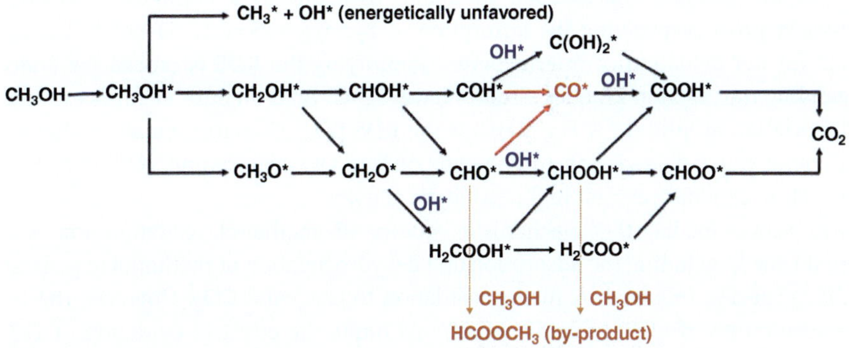

Figure 8.3 A dual-pathway mechanism for electrocatalytic MOR, where * represents the adsorption site on the catalyst surface, the red arrows indicate the CO pathway, and the others the CO-free pathway. *Source:* [72] / John Wiley & Sons.

research. The electrocatalytic oxidation of methanol on Pt surfaces can be summarized in the following major steps:

1. Adsorption and Dehydrogenation of Methanol
   Methanol molecules first adsorb onto the catalyst surface and undergo dehydrogenation, forming a series of adsorbed intermediates, including hydroxymethyl (CH_2OH^*), formaldehyde (CH_2O^*), formyl (CHO^*), and carbon monoxide (CO^*) (Figure 8.3) [72]. Among these, CO^* is the primary poisoning intermediate, which strongly adsorbs onto the Pt active sites and significantly inhibits subsequent oxidation reactions. The main reactions are as follows:

   $$Pt + CH_3OH \rightarrow Pt\text{-}CH_2OH + H^+ + e^-$$

   $$Pt + Pt\text{-}CH_2OH \rightarrow Pt\text{-}CHOH + H^+ + e^-$$

   $$Pt + Pt\text{-}CHOH \rightarrow Pt\text{-}CHO + H^+ + e^-$$

   $$Pt\text{-}CHO \rightarrow Pt\text{-}CO + H^+ + e^-$$

2. Adsorption and Dissociation of Water
   Water molecules adsorb and dissociate on the Pt catalyst surface to form oxidizing intermediate species (Pt-OH), which provide an oxygen source for further oxidation of CO:

   $$Pt + H_2O \rightarrow Pt\text{-}OH + H^+ + e^-$$

3. CO Oxidation to CO_2
   The adsorbed CO (CO_{ad}) is oxidized to CO_2 and desorbs upon reaction with Pt-OH. This process regenerates the Pt active sites and is the key step in MOR, but it is typically slow and considered the rate-limiting step:

   $$Pt\text{-}OH + Pt\text{-}CO \rightarrow 2Pt + CO_2 + H^+ + e^-$$

From the reaction mechanism, methanol oxidation is a complex six-electron transfer process involving the adsorption, dehydrogenation, C—H bond cleavage, and further oxidation of intermediates. Identifying the RDS is crucial for understanding the reaction kinetics. Studies have shown that, on pure Pt surfaces, water dissociation around 0.7 V (vs. RHE) is the RDS [73]. Therefore, catalysts that can promote water dissociation to generate surface OH species and accelerate CO_{ad} oxidation generally exhibit higher catalytic activity.

In acidic media, the mechanistic studies of methanol oxidation are well-established, including the adsorption and dehydrogenation of methanol to generate CH_xO species, followed by further oxidation to CO_{ad} and CO_2. However, the low concentration of OH^- in acidic conditions limits the effective oxidation of CO_{ad} due to the difficulty in water dissociation. In contrast, in alkaline media, OH^- ions directly participate in the reaction, significantly enhancing the dehydrogenation of methanol and producing soluble intermediates (e.g., $HCOO^-$), thus reducing surface poisoning. Moreover, the reaction rate between OH^- ions and CO_{ad} is higher in alkaline media, resulting in superior catalyst performance compared to acidic conditions. However, the study of methanol oxidation in alkaline media is still incomplete, especially in terms of the behavior of complex intermediates and their mechanisms, which remain largely unexplored.

Although the kinetics of MOR is faster in alkaline media, future research should focus on understanding the generation and transformation pathways of intermediates, the activation mechanisms of OH^- ions, and the dynamic behavior of catalyst surface active sites [74, 75]. These studies will provide new insights into the MOR mechanism in alkaline media and offer theoretical and practical guidance for designing efficient and stable catalysts, thus advancing the development and practical application of direct methanol fuel cells.

8.3.1.2 Ethanol Oxidation Reaction Mechanism

Due to the toxicity of methanol and the environmental issues arising from its miscibility with water, the further development of direct methanol fuel cells (DMFCs) has been somewhat limited [76, 77]. In this context, ethanol has emerged as an alternative fuel and has gained increasing attention. Compared to methanol, ethanol has a higher energy density (7.9 $kW\,h\,kg^{-1}$) and lower toxicity. Moreover, ethanol is easily produced through simple processes and can be derived from renewable resources such as biomass, making it widely available and environmentally compatible [78, 79]. Therefore, direct ethanol fuel cells (DEFCs) are considered a promising research direction for next-generation green energy technologies.

In acidic media, the ethanol oxidation reaction typically occurs on platinum (Pt)-based catalysts. The reaction mechanism can be summarized in two main pathways:

1. Complete oxidation pathway

   $$CH_3CH_2OH \rightarrow [CH_3CH_2OH]_{ad} \rightarrow C_{1ad}, C_{2ad} \rightarrow CO_2$$

   In the complete oxidation pathway, ethanol is fully oxidized to produce CO_2 and H_2O. This pathway involves several dehydrogenation steps and the cleavage of the C—C bond.

2. Partial oxidation pathway

$$CH_3CH_2OH \rightarrow [CH_3CH_2OH]_{ad} \rightarrow CH_3CHO \rightarrow CH_3COOH$$

In this pathway, ethanol first adsorbs onto the catalyst surface and undergoes dehydrogenation to form acetaldehyde (CH_3CHO), which is then further oxidized to acetic acid (CH_3COOH).

In practical applications, the ethanol oxidation process is highly complex, involving C_1 and C_2 intermediates (such as CO and CH_3COOH), which contribute differently to the formation of CO_2. Although extensive research has been conducted on the mechanism of ethanol oxidation, some key scientific issues remain unresolved. For example, there is still debate over whether acetic acid (CH_3COOH) is further oxidized from acetaldehyde (CH_3CHO) or formed directly in a single step. Additionally, the structure of the reaction intermediates, particularly whether they contain two carbon atoms (C_2) or one carbon atom (C_1), is still inconclusive. From a thermodynamic perspective, the electrochemical oxidation onset potential of ethanol is similar to that of methanol, being 0.084 and 0.046 V (vs. SHE), respectively [80]. However, from a kinetic standpoint, the complete oxidation of ethanol to CO_2 involves a 12-electron transfer, while methanol only requires 6 electrons. Furthermore, the complete oxidation of ethanol requires C—C bond cleavage, which has a significantly higher activation energy compared to the cleavage of C—H bonds, making the reaction more challenging.

Although significant research has been conducted on ethanol oxidation in acidic media, challenges such as catalyst poisoning and low efficiency of the complete oxidation pathway remain. In contrast, alkaline media, with their higher OH^- concentration, can promote ethanol dehydrogenation and accelerate the conversion of intermediates. As a result, alkaline media are considered an important direction for DEFC research. However, the reaction mechanism under alkaline conditions is not yet fully understood, particularly with regard to C—C bond cleavage and the dynamic behavior of intermediates, which requires further investigation.

8.3.1.3 Formic Acid Oxidation Reaction Mechanism

Although liquid methanol has a relatively high energy density, its high permeability through Nafion membranes and the toxicity of methanol vapor limit the application of methanol fuel cells. Formic acid, as a potential alternative fuel, has garnered increasing attention. As a C_1 small molecule, formic acid plays an important role as an intermediate in the MOR, making it an ideal model for studying the electrochemical oxidation mechanisms of organic small molecules. Compared to methanol, formic acid is a liquid at room temperature and has lower permeability through Nafion membranes, which reduces the risk of cross-contamination [81]. Additionally, formic acid is used as a food additive, has low toxicity, and can be easily produced from renewable resources such as biomass. Although formic acid has a lower energy density ($1.7\ \text{kWh kg}^{-1}$) than methanol, its lower permeability and environmental friendliness make it a promising fuel, especially in low-temperature fuel cell applications.

The electrochemical oxidation reaction of formic acid has been widely studied and is generally considered to follow a bifurcated reaction mechanism. The main reaction pathways include the following two routes:

1. Direct Pathway
   In the direct pathway, formic acid first loses a proton (H^+) and an electron (e^-), forming a weakly adsorbed COOH* intermediate. The COOH* intermediate is then further oxidized to CO_2. The reaction is as follows:

   $$HCOOH + M \rightarrow COOH_{ad} + 2H^+ + 2e^-$$

   $$COOH_{ad} + M \rightarrow CO_2 + 2H^+ + 2e^-$$

   In this process, formic acid undergoes dehydrogenation to form the COOH* intermediate, which is subsequently oxidized to CO_2. The advantage of this pathway is that it does not generate CO as an intermediate, thereby effectively avoiding catalyst poisoning and exhibiting better catalytic performance.
2. Indirect Pathway
   In the indirect pathway, formic acid first undergoes a dehydration reaction to generate a CO* intermediate. The CO* intermediate then reacts with oxygen-containing species (such as OH*) to form CO_2, similar to the MOR mechanism. The reaction process is as follows:

   $$HCOOH + M \rightarrow M\text{-}CO_{ad} + 2H_2O$$

   $$M + H_2O \rightarrow M\text{-}OH_{ad} + H^+ + e^-$$

   $$M\text{-}CO_{ad} + M\text{-}OH_{ad} \rightarrow 2M + CO_2 + H^+ + e^-$$

   This pathway is similar to the MOR mechanism, where the CO* intermediate is formed and subsequently reacts with oxygenated species (such as OH*) to produce CO_2. However, CO* is strongly adsorbed and occupies the active sites on the catalyst surface, hindering further reactions. Therefore, the oxidation of CO* can only proceed at higher potentials, where the generated OH* species facilitate the oxidation of CO* to CO_2. The key difference between the direct and indirect pathways lies in the formation of the CO intermediate [82]. The direct pathway avoids CO formation, preventing catalyst poisoning and thus showing higher catalytic activity. In contrast, the CO* intermediate in the indirect pathway strongly adsorbs onto the catalyst surface, causing the oxidation of formic acid at low potentials to be hindered [83]. Therefore, in the indirect pathway, improving the catalyst's efficiency for oxidizing CO* is critical [84]. By designing catalysts that promote formic acid dehydrogenation while preventing the accumulation of CO intermediates, the efficiency and stability of formic acid fuel cells can be greatly enhanced.

8.3.2 Liquid Fuel Oxidation Catalyst Research Progress

8.3.2.1 Pt-based and Pd-based Electrocatalysts

Electrocatalysts play a critical role in liquid FORs, and their optimization has always been a central research focus. In the early stages, nano-sized Pt black catalysts were widely used in fuel cell anode and cathode reactions. However, Pt black catalysts are prone to agglomeration during the catalytic process, leading to a significant decrease in catalytic efficiency. This not only increases the demand for Pt loading but also raises the cost. Therefore, to improve the utilization efficiency of precious metals, researchers have started to load Pt nanoparticles onto various carbon supports [85, 86]. Common carbon supports include carbon nanotubes, carbon nanofibers, graphene, and carbon black, which provide a larger specific surface area and more electrochemical active sites, thereby enhancing the catalytic activity and stability of Pt nanoparticles [87, 88]. In recent years, metal oxide materials have emerged as a promising new class of catalyst supports due to their strong metal-support interactions and excellent corrosion resistance, and they are expected to replace traditional carbon-based supports [89–91]. Despite the high catalytic performance of Pt catalysts in liquid FORs, their high cost and sensitivity to CO poisoning in certain reactions (e.g., methanol and ethanol oxidation) still limit their widespread application. To address CO poisoning issues in the oxidation of different small liquid molecules, researchers have alloyed Pt with other elements (such as Ru, Co, Zn, Ni, etc.) to create alloy catalysts [92–96]. During the catalytic reaction, alloy elements not only facilitate water dissociation but also adsorb OH* at lower potentials, helping to oxidize toxic intermediates such as CO*. Furthermore, the incorporation of additional elements into binary Pt-based catalysts leads to the formation of multicomponent catalysts. Li et al. [97] prepared concave PtCo nanocrosses (PtCo CNCs) through a one-pot hydrothermal method with iminodiacetic acid (IDA) as a structure-directing agent (Figure 8.4a). The PtCo CNCs, featuring six 3D concave arms, showed superior MOR activity, stability, and CO_{ad} tolerance compared to commercial Pt black and nanoparticles (Figure 8.4b). This further adjusts the electronic structure of Pt's outer layer, thereby reducing its affinity for toxic intermediates during the reaction and enhancing the electrocatalytic activity and stability of Pd-based catalysts.

Pd-based catalysts exhibit higher electrocatalytic activity than Pt-based catalysts, especially in small molecule FORs, in alkaline media. Pd-based catalysts show extremely high catalytic activity in the electrocatalytic oxidation of formic acid and maintain stability over a broader pH range [99, 100]. From a practical cost perspective, Pd reserves on Earth are nearly twice as abundant as Pt, and Pd-based catalysts also exhibit better CO tolerance during reactions. Due to these advantages, researchers have invested significant efforts in the development of Pd-based catalysts [101–103]. Similar to the development of Pt-based catalysts, the focus in the development of Pd-based catalysts has been on improving the utilization of Pd, enhancing electrocatalytic activity, and ensuring long-term stability. Liang et al. [98] synthesized 3D Au@Pt-Pd hemispherical structures (H-Ss) with anisotropic morphology and element distribution through a BO_2^--assisted method (Figure 8.4c–f). This approach enhances the coordination environment of Pt-Pd, improving electrocatalytic activity and stability for various oxidation reactions. The work offers a promising strategy for designing advanced anisotropic nanostructures as efficient electrocatalysts.

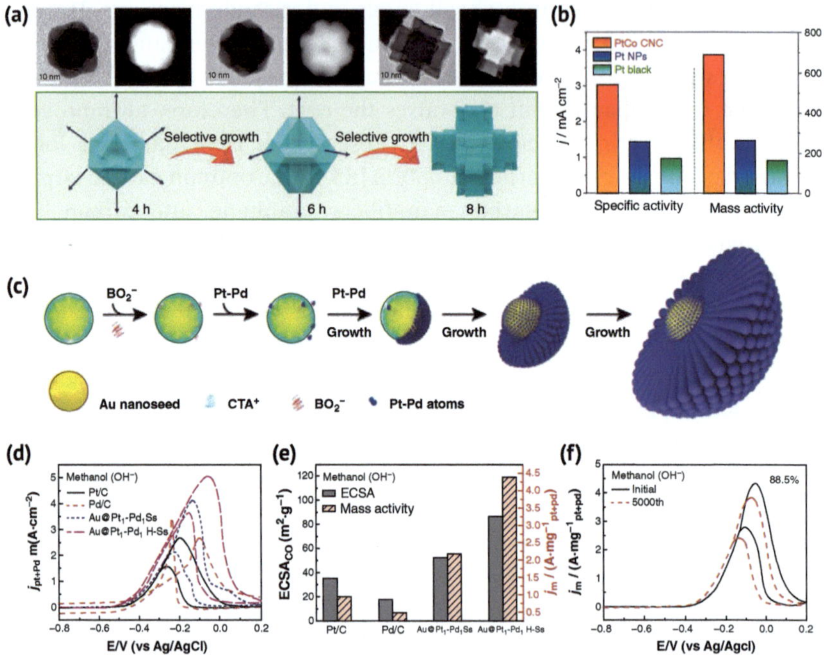

Figure 8.4 (a) TEM and schematic illustration of PtCo nanocrystals at different reaction time. (b) Contrast histograms of mass activity and specific activity. *Source:* [97] / with permission of Elsevier. (c) Schematic diagram of the growth mechanism of Au@Pt-Pd H-Ss. (d) CV curves. (e) Comparison of $ECSA_{CO}$ and MOR mass activity of the commercial Pt/C catalyst, commercial Pd/C catalyst, Au@Pt_1Pd_1 Ss, and Au@Pt_1Pd_1 H-Ss. (f) CV curves before and after 5000 cycles for MOR. *Source:* [98] / John Wiley & Sons.

8.3.2.2 Transition Metal-based Catalysts

In recent years, transition metal-based catalysts have gradually become a research hotspot in DLFCs. Although precious metal catalysts (such as Pt and Pd) exhibit excellent catalytic performance in the oxidation reactions of organic small molecules like methanol, ethanol, and formic acid, their high cost and catalyst poisoning issues still restrict their large-scale application. As a result, researchers have begun exploring low-cost non-precious metal electrode catalysts and striving to apply them in DLFCs. Common non-precious metal catalysts include transition metals, transition metal sulfides, metal complexes, and carbon-based catalysts doped with nonmetals [104].

Among transition metal catalysts, nickel is considered one of the most promising electrode catalysts, widely used in HER and alcohol oxidation reactions [105]. Nickel catalysts exhibit high catalytic activity, good conductivity, low cost, and good chemical stability. Moreover, nickel compounds (such as NiO, $Ni(OH)_2$, etc.) also demonstrate good catalytic performance [106, 107]. Therefore, nickel-based alloy catalysts and various modified nickel-based catalysts have been widely applied. Cobalt and cobalt oxides (Co_3O_4 and CoO) have been used as emerging catalysts owing to their exceptional intrinsic characteristics such as excellent structural flexibility, tunable chemical properties, and higher activity [108, 109]. To improve the catalyst's activity and stability, researchers have begun exploring the combination of transition metals such as nickel and cobalt with other metals (e.g., Ru, Fe, and Cu) to form binary or multicomponent catalysts [110]. Through the synergistic effect between metals, these multimetallic catalysts can significantly optimize catalytic performance. For example, Ni-Co binary catalysts exhibit higher catalytic activity than single-metal catalysts due to the electronic transfer between Ni and Co [111]. Moreover, the composition and structure of binary catalysts can be adjusted, making them adaptable to different reaction conditions and offering broader applicability and longer lifespan [112]. The development of multimetallic catalysts provides a powerful solution to replace traditional precious metal catalysts, driving the research of low-cost and highly efficient catalysts. Xu et al. [113] explored Co-Pt-Ru for methanol oxidation, achieving fourfold higher activity than PtRu/C and higher DMFC power density. $Co_{88}Pt_6Ru_6$/NC, after acid-induced transformation to a porous structure, retained superior performance (Figure 8.5a–c). Li et al. [114] reported that MnO_x/CP demonstrates exceptional stability for electrocatalytic hydrogen production when coupled with glycerol oxidation reaction (GOR) (Figure 8.5d). The addition of glycerol enhances the stability of MnO_2 by inhibiting both the

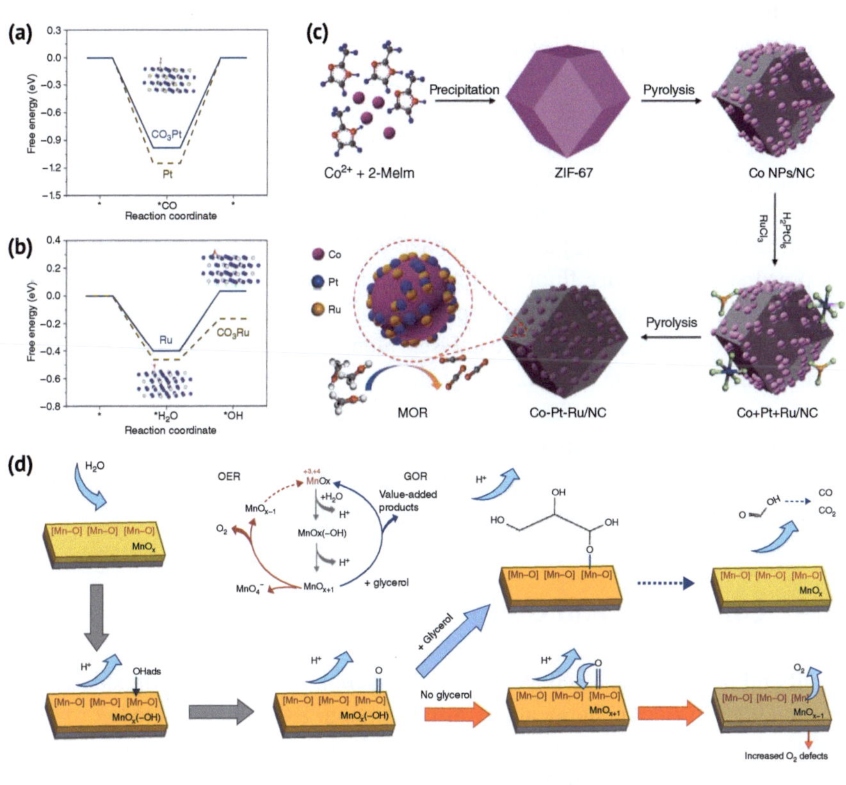

Figure 8.5 (a and b) Free energy diagrams of Pt and Co_3Pt for CO adsorption and Ru and Co_3Ru for H_2O and OH adsorptions. (c) Schematic of the Co–Pt–Ru/NC synthesis. *Source:* Reproduced from [113] / with permission of American Chemical Society. (d) Schematics of the proposed mechanism for GOR and OER on MnO_x/CP catalysts in acid media. *Source:* [114] / John Wiley & Sons.

lattice oxygen process and MnO_2 dissolution. Future work should enhance stability, size uniformity, and acid resistance, offering new strategies for MOR catalysts.

8.3.2.3 Nonmetal Doped Carbon-based Catalysts

Nonmetal doped carbon-based catalysts, as emerging non-precious metal catalysts, have shown significant potential in the catalytic application of DLFCs [115–117]. Different dopants alter the electroneutrality of the carbon structure, as each dopant exhibits different electronegativity values. For example, nitrogen (N) has an electronegativity of 3.04, oxygen (O) has 3.44, boron (B) has 2.04, fluorine (F) has 3.98, phosphorus (P) has 2.19, sulfur (S) has 2.58, and carbon (C) has 2.55 [118]. Therefore, compared to C, dopants with more positive (B) or negative (N) electronegativity can lead to uneven charge distribution in the carbon structure. Nitrogen-doped carbon-based catalysts are one of the most common doping methods. Nitrogen atoms, due to their atomic radius being similar to that of carbon atoms, can easily replace carbon atoms in the lattice of carbon-based catalytic materials [119]. Furthermore, the lone pair electrons of nitrogen atoms conjugate with the π bonds in the carbon material lattice, making nitrogen-doped carbon catalysts exhibit excellent electrocatalytic performance. Matter et al. [120] successfully prepared nitrogen-doped nonmetallic carbon catalysts, opening a new direction in the research of fuel cell electrode catalysts. Gong's team [121] prepared nitrogen-doped carbon nanotube arrays by pyrolyzing phthalic acid iron in ammonia gas. The synthesized vertically aligned nitrogen-containing carbon nanotubes (VA-NCNTs) exhibited high ORR activity and strong durability, with insensitivity to CO poisoning. In addition, the highly electronegative fluorine element has also attracted attention from researchers and has been applied in fuel cell nonmetallic catalysts. For instance, Xu et al. [122] used inexpensive carbon black and ammonium fluoride as precursors to prepare fluorine-doped carbon material catalysts by pyrolysis at 1000 °C. This catalyst exhibited superior catalytic activity in alkaline conditions compared to commercial Pt/C catalysts, along with excellent tolerance to poisoning and stability, making it one of the most promising nonmetal catalysts for fuel cells. These studies show that nonmetal-doped carbon-based catalysts, especially nitrogen and fluorine-doped carbon materials, are gradually becoming an important direction in the field of liquid FOR catalysts. They offer great potential in catalytic performance, cost-effectiveness, and durability.

8.4 Conclusion and Outlook

HOR and liquid FORs, such as methanol, ethanol, and formic acid oxidation, play pivotal roles in electrochemical energy conversion. HOR is central to hydrogen fuel cells, while liquid FORs are critical for liquid fuel cells. These reactions not only determine the energy conversion efficiency of the cells but also have significant implications for environmental protection and sustainable development. Despite the superior catalytic performance of Pt-based catalysts, which have long dominated these reactions, their high cost, scarcity, and susceptibility to poisoning by intermediates limit their widespread application. Thus, the search for low-cost, efficient, and stable alternative catalysts has become a key focus in fuel cell research, yielding significant progress. Currently, transition metal-based catalysts and carbon-based catalysts exhibit excellent electrocatalytic performance and are emerging as promising substitutes for platinum catalysts. The development of these alternatives addresses both the economic and environmental challenges posed by precious metal catalysts.

With the continuous advancement of in situ characterization techniques, researchers are now able to precisely probe the structural changes and reaction dynamics of catalysts during the reaction process. Techniques such as in situ infrared spectroscopy, electrochemical scanning probe microscopy, and synchrotron radiation have become essential tools for elucidating the catalytic mechanisms. These methods not only allow for a deeper understanding of the interactions between catalysts and reactants but also enable the tracking of surface structural changes during the reaction, providing critical experimental data for catalyst optimization. Meanwhile, DFT and other computational methods offer robust theoretical support for understanding key mechanisms, such as electron transfer, adsorption, and desorption, during catalysis. By integrating experimental and theoretical studies, researchers can efficiently design catalysts with improved activity and stability. These technologies enable the rapid identification of potential catalyst candidates from vast datasets, significantly reducing experimental time and costs. When coupled with an in-depth understanding of reaction mechanisms, these tools allow for the precise selection of catalysts with high activity and stability, accelerating the development and application of advanced catalysts.

In conclusion, the synergistic advancements in catalyst design, in situ characterization techniques, and computational methods have greatly accelerated the development of low-cost, efficient, and stable catalysts. These breakthroughs not only promote the practical application of electrochemical energy technologies but also contribute significantly to the energy transition and sustainable development goals.

Acknowledgment

This work was supported by the National Natural Science Foundation of China (52371240, U1904215), Changjiang scholars program of the Ministry of Education (Q2018270).

References

1 Lu, H., Tournet, J., Dastafkan, K. et al. (2021). Noble-metal-free multicomponent nanointegration for sustainable energy conversion. *Chem. Rev.* 121 (17): 10271.

2 Radpour, S., Gemechu, E., Ahiduzzaman, M. et al. (2021). Developing a framework to assess the long-term adoption of renewable energy technologies in the electric power sector: the effects of carbon price and economic incentives. *Renew. Sustain. Energy Rev.* 152: 111663.

3 Zhang, G., Li, X., Liu, Y. et al. (2024). Metal-organic framework derived micro-/nano-materials: precise synthesis and clean energy applications. *Inorg. Chem. Front.* 11 (19): 6275.

4 Shaari, N., Kamarudin, S.K., Bahru, R. et al. (2021). Progress and challenges: review for direct liquid fuel cell. *Int. J. Energy Res.* 45 (5): 6644.

5 Alias, M.S., Kamarudin, S.K., Zainoodin, A.M. et al. (2020). Active direct methanol fuel cell: an overview. *Int. J. Hydrog. Energy* 45 (38): 19620.

6 Su, L., Wu, H., Zhang, S. et al. (2024). Insight into intermediate behaviors and design strategies of platinum group metal-based alkaline hydrogen oxidation catalysts. *Adv. Mater.* 37 (4): 2414628.

7 Pan, M., Pan, C., Li, C. et al. (2021). A review of membranes in proton exchange membrane fuel cells: transport phenomena, performance and durability. *Renew. Sustain. Energy Rev.* 141: 110771.

8 Xiao, F., Wang, Y.-C., Wu, Z.-P. et al. (2021). Recent advances in electrocatalysts for proton exchange membrane fuel cells and alkaline membrane fuel cells. *Adv. Mater.* 33 (50): 2006292.

9 An, Y., Lei, T., Jiang, W. et al. (2024). Research progress on photocatalytic, electrocatalytic and photoelectrocatalytic selective oxidation of 5-hydroxymethylfurfural. *Green Chem.* 26 (21): 10739.

10 Chang, J., Wang, G., Zhang, W. et al. (2022). Atomically dispersed catalysts for small molecule electrooxidation in direct liquid fuel cells. *J. Energy Chem.* 68: 439.

11 Zhou, Y., Xu, W., Wei, Z. et al. (2024). Molecular iridium catalyzed electrochemical formic acid oxidation: mechanistic insights. *Angew. Chem. Int. Ed.* 137 (1): e202412901.

12 Li, S., Ajmal, S., Zhou, X. et al. (2024). Mixed-dimensional partial dealloyed PtCuBi/C as high-performance electrocatalysts for methanol oxidation with enhanced CO tolerance. *Small* 20 (22): 2309226.

13 Shi, L., Liu, D., Lin, X. et al. (2024). Stable and high-performance flow H_2-O_2 fuel cells with coupled acidic oxygen reduction and alkaline hydrogen oxidation reactions. *Adv. Mater.* 36 (23): 2314077.

14 Zhang, Y., Chen, T., Alia, S. et al. (2016). Single-molecule nanocatalysis shows in situ deactivation of Pt/C electrocatalysts during the hydrogen-oxidation reaction. *Angew. Chem. Int. Ed.* 128 (9): 3138.

15 Nose, M., Kinumoto, T., Choo, H.-S. et al. (2009). Electrochemical oxidation of highly oriented pyrolytic graphite in sulphuric acid solution under potential pulse condition. *Fuel Cells* 9 (3): 284.

16 Miao, R., Chen, L., Shao, L. et al. (2019). Electron transfer to decorated graphene oxide particles. *Angew. Chem. Int. Ed.* 58 (36): 12549.

17 Pulidindi, I.N., et al. (2023). Recent advances in electrocatalysts for hydrogen oxidation reaction in alkaline electrolytes. In: *Electrocatalysis for Membrane Fuel Cells: Methods, Modeling and Applications*, p. 205–226. Wiley.

18 Wang, S., Fu, L., Huang, H. et al. (2023). Local oxidation induced amorphization of 1.5-nm-thick Pt-Ru nanowires enables superactive and CO-tolerant hydrogen oxidation in alkaline media. *Adv. Funct. Mater.* 33 (43): 2304125.

19 Su, L., Gong, D., Yao, N. et al. (2021). Modification of the intermediate binding energies on Ni/Ni_3N heterostructure for enhanced alkaline hydrogen oxidation reaction. *Adv. Funct. Mater.* 31 (49): 2106156.

20 Sheng, W., Gasteiger, H.A., and Shao-Horn, Y. (2010). Hydrogen oxidation and evolution reaction kinetics on platinum: acid vs alkaline electrolytes. *J. Electrochem. Soc.* 157 (11): B1529.

21 Durst, J., Siebel, A., Simon, C. et al. (2014). New insights into the electrochemical hydrogen oxidation and evolution reaction mechanism. *Energy Environ. Sci.* 7 (7): 2255.

22 Yang, K., Ma, H., Ren, R. et al. (2024). Multidimensional electrochemistry decodes the operando mechanism of hydrogen oxidation. *Angew. Chem. Int. Ed.* 63 (24): e202318389.

23 Huang, Z., Lu, R., Zhang, Y. et al. (2023). A highly efficient pH-universal HOR catalyst with engineered electronic structures of single Pt sites by isolated Co atoms. *Adv. Funct. Mater.* 33 (47): 2306333.

24 Yao, Z.-C., Tang, T., Jiang, Z. et al. (2022). Electrocatalytic hydrogen oxidation in alkaline media: from mechanistic insights to catalyst design. *ACS Nano* 16 (4): 5153.

25 Nørskov, J.K., Bligaard, T., Logadottir, A. et al. (2005). Trends in the exchange current for hydrogen evolution. *J. Electrochem. Soc.* 152 (3): J23.

26 Trasatti, S. (1972). Work function, electronegativity, and electrochemical behaviour of metals: III. Electrolytic hydrogen evolution in acid solutions. *J. Electroanal. Chem. Interfacial Electrochem.* 39 (1): 163.

27 Sheng, W., Myint, M.G., Chen, J. et al. (2013). Correlating the hydrogen evolution reaction activity in alkaline electrolytes with the hydrogen binding energy on monometallic surfaces. *Energy Environ. Sci.* 6 (5): 1509.

28 Jin, Y., Gong, D., Luo, J. et al. (2023). Regulation of rate-determining step on $(Rh_xNi_{1-x})_2P$ for enhanced alkaline hydrogen oxidation reaction. *ChemCatChem* 15 (15): e202300219.

29 Shen, L., Lu, B., Li, Y. et al. (2020). Interfacial structure of water as a new descriptor of the hydrogen evolution reaction. *Angew. Chem. Int. Ed.* 59 (50): 22397.

30 St. John, S., Atkinson, R.W.I., Unocic, R.R. et al. (2015). Ruthenium-alloy electrocatalysts with tunable hydrogen oxidation kinetics in alkaline electrolyte. *J. Phys. Chem. C* 119 (24): 13481.

31 Guo, Y., He, W., Tan, X. et al. (2024). Promoting the Volmer step of alkaline hydrogen oxidation reaction by incorporating Zn single atoms into Ru lattice and carbon frameworks. *Adv. Funct. Mater.* 35 (1): 2412142.

32 Mu, X.Q., Liu, S.L., Zhang, M.Y. et al. (2024). Symmetry-broken Ru nanoparticles with parasitic Ru-Co dual-single atoms overcome the Volmer step of alkaline hydrogen oxidation. *Angew. Chem. Int. Ed.* 63 (12): e202319618.

33 Zhang, L., Li, J., Yang, Z. et al. (2021). Phosphorus-doping-induced optimization of atomic hydrogen binding energy in $MoSe_2$ under high coverage for efficient electrocatalytic hydrogen evolution reactions. *ChemistrySelect* 6 (6): 1305.

34 Liu, C., Zhang, G., Yu, L. et al. (2018). Oxygen doping to optimize atomic hydrogen binding energy on NiCoP for highly efficient hydrogen evolution. *Small* 14 (22): 1800421.

35 Li, J., Ghoshal, S., Bates, M.K. et al. (2017). Experimental proof of the bifunctional mechanism for the hydrogen oxidation in alkaline media. *Angew. Chem. Int. Ed.* 129 (49): 15800.

36 Tian, X., Zhao, P., and Sheng, W. (2019). Hydrogen evolution and oxidation: mechanistic studies and material advances. *Adv. Mater.* 31 (31): 1808066.

37 Bligaard, T., Nørskov, J.K., Dahl, S. et al. (2004). The Brønsted-Evans-Polanyi relation and the volcano curve in heterogeneous catalysis. *J. Catal.* 224 (1): 206.

38 Mu, X., Liu, S., Chen, L. et al. (2023). Alkaline hydrogen oxidation reaction catalysts: insight into catalytic mechanisms, classification, activity regulation and challenges. *Small Struct.* 4 (4): 2200281.

39 Zeng, Y., Wang, X., Hu, Y. et al. (2024). Research progress of electrocatalysts for hydrogen oxidation reaction in alkaline media. *Carbon Neutraliz.* 3 (4): 710.

40 Li, X.-Y., Zhang, Z.-M., Zhuang, X.-X. et al. (2024). Electrolyte effects in electrocatalytic kinetics. *Chin. J. Chem.* 42 (24): 3533.

41 Zhao, G., Chen, J., Sun, W. et al. (2021). Non-platinum group metal electrocatalysts toward efficient hydrogen oxidation reaction. *Adv. Funct. Mater.* 31 (20): 2010633.

42 Dong, Y., Sun, Q., Zhan, C. et al. (2023). Lattice and surface engineering of ruthenium nanostructures for enhanced hydrogen oxidation catalysis. *Adv. Funct. Mater.* 33 (5): 2210328.

43 Duan, Y., Zhang, X.-L., Gao, F.-Y. et al. (2023). Interfacial engineering of Ni/V_2O_3 heterostructure catalyst for boosting hydrogen oxidation reaction in alkaline electrolytes. *Angew. Chem. Int. Ed.* 62 (10): e202217275.

44 Ni, W., Wang, T., Schouwink, P.A. et al. (2020). Efficient hydrogen oxidation catalyzed by strain-engineered nickel nanoparticles. *Angew. Chem. Int. Ed.* 59 (27): 10797.

45 Ruck, S., Hutzler, A., Thiele, S. et al. (2024). Highly active NiRu/C cathode catalyst synthesized by displacement reaction for anion exchange membrane water electrolysis. *Small Methods* 2024: 2401179.

46 Cui, Z., Ren, Z., Ma, C. et al. (2024). Dilute RuCo alloy synergizing single Ru and Co atoms as efficient and CO-resistant anode catalyst for anion exchange membrane fuel cells. *Angew. Chem. Int. Ed.* 136 (28): e202404761.

47 Zhang, J., Zhu, W., Huang, T. et al. (2021). Recent insights on catalyst layers for anion exchange membrane fuel cells. *Adv. Sci.* 8 (15): 2100284.

48 Antolini, E. (2007). Platinum-based ternary catalysts for low temperature fuel cells: Part I. Preparation methods and structural characteristics. *Appl. Catal. B Environ.* 74 (3): 324.

49 Mahoney, E.G., Sheng, W., Yan, Y. et al. (2014). Platinum-modified gold electrocatalysts for the hydrogen oxidation reaction in alkaline electrolytes. *ChemElectroChem* 1 (12): 2058.

50 Kelly, T.G., Lee, K.X., and Chen, J.G. (2014). Pt-modified molybdenum carbide for the hydrogen evolution reaction: from model surfaces to powder electrocatalysts. *J. Power Sources* 271: 76.

51 Yan, L., Wang, D., Li, M. et al. (2024). Hexa-atom Pt catalyst fabricated by a ligand engineering strategy for efficient hydrogen oxidation reaction. *Angew. Chem. Int. Ed.* 136 (43): e202410832.

52 Ramaker, D.E. and Roth, C. (2015). Nature of the intermediate binding sites in hydrogen oxidation/evolution over Pt in alkaline and acidic media. *ChemElectroChem* 2 (10): 1582.

53 Lv, Q., Liu, D., Zhu, W. et al. (2024). Iridium-based alkaline hydrogen oxidation reaction electrocatalysts. *Chem.- Eur. J.* 30 (37): e202400838.

54 Fraser, K., Dosaev, K., Savinova, E. et al. (2024). Assessing electrocatalyst-ionomer interactions: HOR/HER activity of Pt/C catalyst layers incorporating poly(imidazolium)s. *ChemCatChem* 16 (2): e202301304.

55 Park, J., Kim, H., Kim, S. et al. (2024). Boosting alkaline hydrogen oxidation activity of Ru single-atom through promoting hydroxyl adsorption on Ru/WC1-interfaces. *Adv. Mater.* 36 (4): 2308899.

56 Zhang, H., Wu, F., Huang, R. et al. (2024). Symmetry evolution induced 2D Pt single atom catalyst with high density for alkaline hydrogen oxidation. *Adv. Mater.* 36 (31): 2404672.

57 Gao, Y., Liu, B., and Wang, D. (2023). Microenvironment engineering of single/dual-atom catalysts for electrocatalytic application. *Adv. Mater.* 35 (31): 2209654.

58 Li, Y., Wei, Z., Sun, Z. et al. (2024). Sulfur modified carbon-based single-atom catalysts for electrocatalytic reactions. *Small* 20 (38): 2401900.

59 Mu, X.Q., Liu, S.L., Zhang, M.Y. et al. (2024). Symmetry-broken Ru nanoparticles with parasitic Ru-Co dual-single atoms overcome the Volmer step of alkaline hydrogen oxidation. *Angew. Chem. Int. Ed.* 136 (12): e202319618.

60 Yuan, Y., Wu, X.-Q., Yin, X. et al. (2024). Dilute Pd-Ni alloy through low-temperature pyrolysis for enhanced electrocatalytic hydrogen oxidation. *Angew. Chem. Int. Ed.* 136 (52): e202412680.

61 Jin, Y., Su, L., Jia, H. et al. (2023). Nickel-based electrocatalysts for hydrogen oxidation reaction under alkaline electrolytes. *ChemCatChem* 15 (9): e202300056.

62 Yang, D., Li, B., Zhang, H. et al. (2013). High performance by applying IrCo/C nanoparticles as an anode catalyst for PEMFC. *Fuel Cells* 13 (2): 309.

63 Zhang, S., Ren, R., Cao, J. et al. (2024). Ru-MnO heterostructure clusters toward efficient and CO-tolerant alkaline hydrogen oxidation reaction. *Adv. Energy Mater.* 2024: 2404266.

64 Sun, Q., Xu, H., and Du, Y. (2020). Recent achievements in noble metal catalysts with unique nanostructures for liquid fuel cells. *ChemSusChem* 13 (10): 2540.

65 Yu, X. and Manthiram, A. (2018). Scalable membraneless direct liquid fuel cells based on a catalyst-selective strategy. *ENERGY Environ. Mater.* 1 (1): 13.

66 Fang, C., Zhao, J., Ding, Q. et al. (2018). Ultrathin-branched Pt grown on quasi-sphere Pd with enhanced electrocatalytic performances. *ChemistrySelect* 3 (5): 1531.

67 Sheng, Y., Yang, R., Xie, J. et al. (2024). Energy-saving ambient electrosynthesis of nylon-6 precursor coupled with electrocatalytic upcycling of polyethylene terephthalate. *Small* 20 (47): 2404477.

68 Zhang, J., Lu, S., Xiang, Y. et al. (2020). Intrinsic effect of carbon supports on the activity and stability of precious metal based catalysts for electrocatalytic alcohol oxidation in fuel cells: a review. *ChemSusChem* 13 (10): 2484.

69 Neurock, M. (2010). First-principles modeling for the electro-oxidation of small molecules. In: *Handbook of Fuel Cells*. Wiley.

70 Shen, L., Ying, J., Tian, G. et al. (2021). Ultralong PtPd alloyed nanowires anchored on graphene for efficient methanol oxidation reaction. *Chem. Asian J.* 16 (9): 1130.

71 Kelly, C.H.W., Benedetti, T.M., Alinezhad, A. et al. (2019). Controlling metallic nanoparticle redox properties for improved methanol oxidation reaction electrocatalysis. *ChemCatChem* 11 (24): 5989.

72 Wang, J., Zhang, B., Guo, W. et al. (2023). Toward electrocatalytic methanol oxidation reaction: longstanding debates and emerging catalysts. *Adv. Mater.* 35 (26): 2211099.

73 Kim, K.S., Hong, Y., Kim, H.C. et al. (2019). Ultrathin-polyaniline-coated Pt-Ni alloy nanooctahedra for the electrochemical methanol oxidation reaction. *Chem.-Eur. J.* 25 (29): 7185.

74 Chen, D.-J. and Tong, Y.J. (2015). Irrelevance of carbon monoxide poisoning in the methanol oxidation reaction on a PtRu electrocatalyst. *Angew. Chem. Int. Ed.* 54 (32): 9394.

75 Zhang, T., Sun, Y., Li, X. et al. (2020). PtPdAg hollow nanodendrites: template-free synthesis and high electrocatalytic activity for methanol oxidation reaction. *Small Methods* 4 (1): 1900709.

76 Bai, S., Xu, Y., Cao, K. et al. (2021). Selective ethanol oxidation reaction at the Rh-SnO_2 interface. *Adv. Mater.* 33 (5): 2005767.

77 Xu, J. and Wang, Z. (2023). Mechanistic investigations of electrochemical ethanol oxidation reaction by in situ Raman spectroscopy. *ChemElectroChem* 10 (19): e202300370.

78 Zhou, M., Jin, B., Kong, W. et al. (2024). Dual polarization of Ni sites at VO_x-Ni_3N interface boosts ethanol oxidation reaction. *Adv. Sci.* 11 (40): 2407473.

79 Su, Y., Yao, C., Zhang, Q. et al. (2019). Palladium nanoparticles supported on B-doped carbon nanocage as electrocatalyst toward ethanol oxidation reaction. *ChemElectroChem* 6 (20): 5211.

80 Bueno, O.V.M., San-Miguel, M.A., and da Silva, E.Z. (2023). Ethanol oxidation reaction mechanism on gold nanowires from density functional theory. *ChemPhysChem* 24 (6): e202200723.

81 Lee, C.-W., Jung, S.Y., Ryu, J.H. et al. (2024). Empowering tri-functional palladium's catalytic activity and durability in electrocatalytic formic acid oxidation reaction via innovative self-caging and alloying strategies. *Adv. Sci.* 11 (46): 2405725.

82 Chen, Q., Ren, Y., Jin, H. et al. (2023). Icosahedral Pd nanocrystal catalysts with dispersed Bi atoms via photochemical route for enhanced formic acid oxidation reaction. *ChemNanoMat* 9 (8): e202300127.

83 Ge, T., Cui, W., Gao, B. et al. (2023). Supercritical CO_2 mediated multi-scale structural engineering in PdCu/C for boosting electrocatalytic formic acid oxidation. *ChemCatChem* 15 (21): e202300936.

84 Chen, M.-X., Liu, Y., Song, T.-W. et al. (2022). Intermetallic PdCd core promoting CO tolerance of Pd shell for electrocatalytic formic acid oxidation. *Chin. J. Chem.* 40 (18): 2161.

85 Dai, L., Chang, D.W., Baek, J.-B. et al. (2012). Carbon nanomaterials for advanced energy conversion and storage. *Small* 8 (8): 1130.

86 Antolini, E. (2009). Carbon supports for low-temperature fuel cell catalysts. *Appl. Catal. B Environ.* 88 (1): 1.

87 Zhao, S., Li, Y., Yin, H. et al. (2015). Three-dimensional graphene/Pt nanoparticle composites as freestanding anode for enhancing performance of microbial fuel cells. *Sci. Adv.* 1 (10): e1500372.
88 Yang, W., Wang, Y., Li, J. et al. (2010). Polymer wrapping technique: an effective route to prepare Pt nanoflower/carbon nanotube hybrids and application in oxygen reduction. *Energy Environ. Sci.* 3 (1): 144.
89 Huang, T., Mao, S., Zhou, G. et al. (2015). A high-performance catalyst support for methanol oxidation with graphene and vanadium carbonitride. *Nanoscale* 7 (4): 1301.
90 Li, D., Li, Y., Liu, X. et al. (2019). $NiAl_2O_4$ spinel supported Pt catalyst: high performance and origin in aqueous-phase reforming of methanol. *ACS Catal.* 9 (10): 9671.
91 Ho, V.T.T., Pan, C.-J., Rick, J. et al. (2011). Nanostructured $Ti_{0.7}Mo_{0.3}O_2$ support enhances electron transfer to Pt: high-performance catalyst for oxygen reduction reaction. *J. Am. Chem. Soc.* 133 (30): 11716.
92 Sheng, G., Chen, J., Ye, H. et al. (2018). Hollow PdCo alloy nanospheres with mesoporous shells as high-performance catalysts for methanol oxidation. *J. Colloid Interface Sci.* 522: 264.
93 Zhang, F.-Y., Sheng, T., Tian, N. et al. (2017). Cu overlayers on tetrahexahedral Pd nanocrystals with high-index facets for CO_2 electroreduction to alcohols. *Chem. Commun.* 53 (57): 8085.
94 Jiang, K., Wang, P., Guo, S. et al. (2016). Ordered PdCu-based nanoparticles as bifunctional oxygen-reduction and ethanol-oxidation electrocatalysts. *Angew. Chem. Int. Ed.* 128 (31): 9176.
95 Xu, H., Yan, B., Zhang, K. et al. (2017). Facile fabrication of novel PdRu nanoflowers as highly active catalysts for the electrooxidation of methanol. *J. Colloid Interface Sci.* 505: 1.
96 Lv, H., Wang, Y., Lopes, A. et al. (2019). Ultrathin PdAg single-crystalline nanowires enhance ethanol oxidation electrocatalysis. *Appl. Catal. B Environ.* 249: 116.
97 Li, Z., Jiang, X., Wang, X. et al. (2020). Concave PtCo nanocrosses for methanol oxidation reaction. *Appl. Catal. B Environ.* 277: 119135.
98 Liang, W., Wang, Y., Zhao, L. et al. (2021). 3D anisotropic Au@Pt-Pd hemispherical nanostructures as efficient electrocatalysts for methanol, ethanol, and formic acid oxidation reaction. *Adv. Mater.* 33 (30): 2100713.
99 Xu, C., Shen, P.K., and Liu, Y. (2007). Ethanol electrooxidation on Pt/C and Pd/C catalysts promoted with oxide. *J. Power Sources* 164 (2): 527.
100 Wang, Y., Zou, S., and Cai, W.-B. (2015). Recent advances on electro-oxidation of ethanol on Pt- and Pd-based catalysts: from reaction mechanisms to catalytic materials. *Catalysts* 5 (3): 1507.
101 Xi, Z., Erdosy, D.P., Mendoza-Garcia, A. et al. (2017). Pd nanoparticles coupled to $WO_{2.72}$ nanorods for enhanced electrochemical oxidation of formic acid. *Nano Lett.* 17 (4): 2727.
102 Wu, H., Li, H., Zhai, Y. et al. (2012). Facile synthesis of free-standing Pd-based nanomembranes with enhanced catalytic performance for methanol/ethanol oxidation. *Adv. Mater.* 24 (12): 1594.

103 Zhang, X., Zhu, J., Tiwary, C.S. et al. (2016). Palladium nanoparticles supported on nitrogen and sulfur dual-doped graphene as highly active electrocatalysts for formic acid and methanol oxidation. *ACS Appl. Mater. Interfaces* 8 (17): 10858.

104 Guo, X., Feng, S., Peng, Y. et al. (2025). Emerging insights into the application of metal-organic framework (MOF)-based materials for electrochemical heavy metal ion detection. *Food Chem.* 463: 141387.

105 Zhuang, X., Zhang, S., Tang, Y. et al. (2023). Recent progress of MOF/MXene-based composites: synthesis, functionality and application. *Coord. Chem. Rev.* 490: 215208.

106 Pieta, I.S., Rathi, A., Pieta, P. et al. (2019). Electrocatalytic methanol oxidation over Cu, Ni and bimetallic Cu-Ni nanoparticles supported on graphitic carbon nitride. *Appl. Catal. B Environ.* 244: 272.

107 Wang, G., Ling, Y., Lu, X. et al. (2012). Solar driven hydrogen releasing from urea and human urine. *Energy Environ. Sci.* 5 (8): 8215.

108 Zhu, R., Jiang, Y., Sun, B. et al. (2024). MOF-derived Co-Mo bimetallic heterostructures for the selective trapping and conversion of polysulfides in lithium-sulfur batteries. *Inorg. Chem. Front.* 11 (23): 8290.

109 Hu, Y., Zhang, S., Xu, K. et al. (2025). Nano-metal-organic frameworks and nano-covalent-organic frameworks: controllable synthesis and applications. *Chem.- Asian J.* 20 (1): e202400896.

110 Zhu, R., Xie, L., Zhang, Y. et al. (2024). Hf-doped CoP hollow nanocubes as high-performance electrocatalyst for oxygen evolution reaction. *Inorg. Chem.* 63 (28): 13093.

111 Zhu, H., Zhang, S., Huang, Y.-X. et al. (2013). Monodisperse $M_xFe_{3-x}O_4$ (M = Fe, Cu, Co, Mn) nanoparticles and their electrocatalysis for oxygen reduction reaction. *Nano Lett.* 13 (6): 2947.

112 Umeshbabu, E., Rajeshkhanna, G., Justin, P. et al. (2015). Magnetic, optical and electrocatalytic properties of urchin and sheaf-like $NiCo_2O_4$ nanostructures. *Mater. Chem. Phys.* 165: 235.

113 Xu, H., Zheng, D.J., Iriawan, H. et al. (2024). A cobalt-platinum-ruthenium system for acidic methanol oxidation. *Chem. Mater.* 36 (14): 6938.

114 Li, Y., Wei, X., Han, S. et al. (2021). MnO_2 electrocatalysts coordinating alcohol oxidation for ultra-durable hydrogen and chemical productions in acidic solutions. *Angew. Chem. Int. Ed.* 133 (39): 21634.

115 Zhu, J., Xiao, M., Song, P. et al. (2018). Highly polarized carbon nano-architecture as robust metal-free catalyst for oxygen reduction in polymer electrolyte membrane fuel cells. *Nano Energy* 49: 23.

116 Xue, L., Li, Y., Liu, X. et al. (2018). Zigzag carbon as efficient and stable oxygen reduction electrocatalyst for proton exchange membrane fuel cells. *Nat. Commun.* 9 (1): 3819.

117 Ding, H., Liu, Z., Xie, J. et al. (2024). Ion exchange-mediated 3D cross-linked ZIF-L superstructure for flexible electrochemical energy storage. *Angew. Chem. Int. Ed.* 136 (36): e202410255.

118 Zhou, N., Wang, N., Wu, Z. et al. (2018). Probing active sites on metal-free, nitrogen-doped carbons for oxygen electroreduction: a review. *Catalysts* 8 (11): 509.

119 Wang, S., Yu, D., Dai, L. et al. (2011). Polyelectrolyte-functionalized graphene as metal-free electrocatalysts for oxygen reduction. *ACS Nano* 5 (8): 6202.
120 Matter, P.H. and Ozkan, U.S. (2006). Non-metal catalysts for dioxygen reduction in an acidic electrolyte. *Catal. Lett.* 109 (3): 115.
121 Gong, K., Du, F., Xia, Z. et al. (2009). Nitrogen-doped carbon nanotube arrays with high electrocatalytic activity for oxygen reduction. *Science* 323 (5915): 760.
122 Sun, X., Zhang, Y., Song, P. et al. (2013). Fluorine-doped carbon blacks: highly efficient metal-free electrocatalysts for oxygen reduction reaction. *ACS Catal.* 3 (8): 1726.

9

Catalysts for Electrocatalytic Oxygen Reduction Reaction

*An-Liang Wang**

School of Chemistry and Chemical Engineering, Shandong University, Jinan, China

*Corresponding author: alwang@sdu.edu.cn

9.1 Introduction

The ORR is at the heart of many energy conversion devices, including proton exchange membrane fuel cells and metal-air batteries [1]. However, the difficulties in O_2 activation, O—O bond cleavage, and the removal of oxide result in sluggish kinetics and the occurrence of side reaction (i.e., oxygen reduction to hydrogen peroxide), which significantly hinders the total power density output of these energy conversion devices [2]. Consequently, there is a pressing need for electrocatalysts that can expedite the ORR kinetics, improve the selectivity and stability. To date, Pt has been widely recognized as the benchmark for ORR catalysis. However, the scarcity of Pt on Earth and the high cost of Pt-based catalysts pose significant obstacles to their wide application. Therefore, the development of cost-effective ORR electrocatalysts with high activity, selectivity, and stability is crucial for the practical implementation of these advanced energy conversion devices.

In this chapter, the ORR mechanism is first introduced. Subsequently, two primary approaches for the evaluation of ORR catalysts are briefly outlined. Following this, developed ORR catalysts are summarized and classified into Pt-based catalysts, Pd-based catalysts, non-platinum group metal catalysts, non-metal catalysts, and SACs. Finally, a concise conclusion and outlook are presented for advanced ORR catalysts.

9.2 The ORR Mechanism

Based on the number of electrons transferred, the ORR pathways can be classified into two types: one is the $4e^-$ transfer pathway, and the other is the two-electron ($2e^-$) transfer pathway [3]. In the $4e^-$ transfer pathway, O_2 is finally reduced to

Electrochemical Energy Conversion via Small Molecule Catalysis, First Edition.
Edited by Zhicheng Zhang and Yuanmiao Sun.

H_2O or (OH^-) in an acidic (or alkaline) electrolyte. In contrast, in the 2e⁻ transfer pathway, O_2 is reduced to H_2O_2 (or HO_2^-) in an acidic (or alkaline) electrolyte. The 4e⁻ transfer pathway is recognized as the favorable pathway since it avoids voltage loss and degradation of the proton-conducting polymer electrolyte associated with the 2e⁻ transfer pathway. Therefore, the 4e⁻ transfer pathway is primarily discussed in the following sections.

In the 4e⁻ transfer pathway, both associative and dissociative mechanisms are considered, depending on whether the absorbed O_2 reactant dissociates before or after reduction. The associative mechanism involves four elementary steps with three distinct intermediates (i.e., *OOH, *O, and *OH). In contrast, only intermediates of *O and *OH are involved in the dissociative pathway (Figure 9.1). Studies have indicated that oxygen coverage or O_2 adsorption strength plays a pivotal role in determining the ORR mechanism. Specifically, high oxygen coverage favors the formation of *OOH through the combination of absorbed O_2 with H (associative mechanism), whereas low oxygen coverage facilitates the O—O cleavage to form *OH (dissociation mechanism) [4, 5].

For both dissociative and associative pathways, the key oxygen-containing intermediates (i.e., *O and *OH) are involved. Since they attach to the surface through the O atom in a similar fashion, their adsorption energies exhibit a linear relationship, referred to as the scaling relationship [7, 8]. Hence, it is generally accepted that the oxygen adsorption energy (ΔE_O) is widely recognized as the descriptor for ORR activity. A correlation between ΔE_O and ORR activity, known as "volcano plots," has been established [9]. An ideal catalyst should bind oxygen with moderate strength [4, 10]. Specifically, if the interaction between catalyst surface and oxygen is too strong, the removal of *O and *OH will be impeded, while a weak interaction will limit the adsorption of O_2 and subsequent dissociation to *O. It is further acknowledged that the Pt surface binds oxygen too tightly compared with the optimal value, which leaves room for enhancement. Additionally, the type of catalysts, the electrolyte, and applied potentials significantly affect the ORR mechanism.

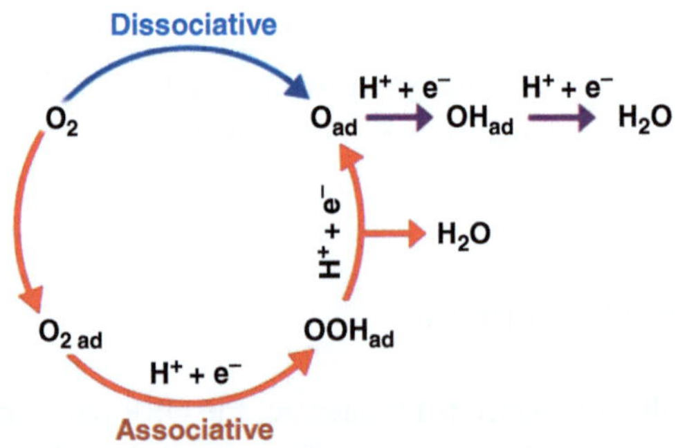

Figure 9.1 Schematic illustration for 4e⁻ transfer ORR pathways. The blue and red arrows represent the dissociative and associative pathways, respectively. The purple arrows represent the reactions that take place in both pathways. *Source:* Reproduced from [6] / with permission of American Chemical Society.

9.3 ORR Catalyst Characterization Techniques

Two primary approaches have been used to characterize the ORR performance: rotating disk electrode (RDE) and membrane electrode assembly (MEA) tests (Figure 9.2). Due to the easy operability and inexpensiveness, RDE methods have been extensively employed for the evaluation of the intrinsic performance of catalysts, and are more suitable for theoretical studies. During RDE tests, to minimize the impact of mass transfer on ORR activity measurements, glassy carbon electrodes with catalysts are rotated to increase the mass transfer rates of O_2 at the electrode surface. The intrinsic activity (kinetic current without mass transfer effect) of the catalysts can be derived according to the Koutecky–Levich equation:

$$\frac{1}{j}=\frac{1}{j_\mathrm{k}}+\frac{1}{j_\mathrm{lc}}=\frac{1}{j_\mathrm{k}}+\frac{1}{0.62n\mathrm{F}AC_0^*D_0^{2/3}\nu^{-1/6}\omega^{1/2}} \tag{9.1}$$

where j, j_k, and j_lc are the measured current density, kinetic current density, and diffusion-limited current density, respectively. The diffusion-limited current is determined by the number of electrons transferred (n), the Faraday constant (F), the electrode's geometric area (A), the concentration of dissolved O_2 in solution (C_0^*), the diffusion coefficient of O_2 (D_0), the kinetic viscosity of the solution (ν), and the rotation speed of the electrode (ω). j_k is generally extrapolated from the Koutecky–Levich plot (j^{-1} vs $\omega^{1/2}$) at different speeds. Based on polarization curves under different rotating speeds, j_k can be extracted from the Koutecky–Levich equation and further converted into specific activity and mass activity. Additionally, the half-wave potential ($E_{1/2}$), which is the potential to achieve a 1/2 current density of j_{lc}, is also widely used to evaluate the ORR activity. A larger $E_{1/2}$ indicates a high catalytic activity and vice versa.

To assess the practical utility of ORR catalysts in fuel cell applications, constructing a MEA and a single MEA fuel cell is crucial for the evaluation of advanced ORR catalysts. The MEA comprises the following components: conductive gas diffusion layer (GDL), which allows gases to permeate through, catalyst layer (CL), and membrane that functions as electrolyte. The performance of the catalyst layer primarily

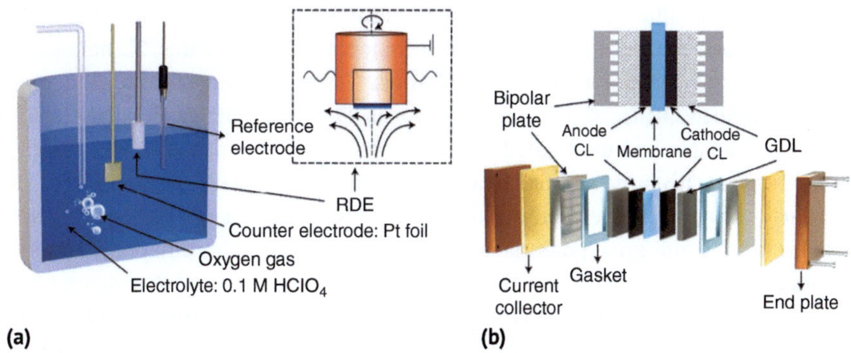

Figure 9.2 (a) A diagrammatic representation of RDE configuration. (b) Single-fuel-cell configuration. The MEA consists of gas diffusion layer (GDL), catalyst layer (CL), and membrane. *Source:* [11] / Springer Nature.

determines the power density of the fuel cell. Typically, the anode catalyst is the specially designed catalyst, while the cathodic catalyst is commercial Pt/C. The MEA tests can be conducted using either pure oxygen/H_2 or air/H_2 flow [10]. Evaluation under the pure oxygen conditions offers a more accurate assessment of the intrinsic activity, as they are less affected by mass transfer issues. Conversely, the MEA tests using air as the oxygen source directly reflect the practical performance of the fuel cell under a working environment. Generally, the test conditions for MEAs align with those specified in the Department of Energy (DOE) protocol. The efficiency of fuel cell, including mass activity and rated power, can be achieved from polarization plots, while accelerated durability test (ADT) is employed to evaluate the stability. The DOE targets are commonly used as benchmarks for assessing catalysts in fuel cell devices.

9.4 The Category of ORR Catalyst

The ORR catalysts are crucial for the practical implementation of fuel cells and metal-air battery. Over the past decades, numerous advanced catalysts have been designed and synthesized for ORR. These catalysts can be categorized into several groups: Pt-based catalysts, Pd-based catalysts, non-platinum group metal catalysts, metal-free catalysts, and SACs.

9.4.1 Pt-based Metal Catalysts

The current state-of-the-art catalyst for ORR is Pt nanoparticles supported on a carbon black substrate. However, it suffers from low abundance, high cost, low catalytic activity, and susceptibility to CO poisoning. Consequently, tremendous efforts have been dedicated to developing cost-effective ORR electrocatalysts with enhanced activity and improved stability, aiming to facilitate the widespread application of these energy conversion technologies.

9.4.1.1 Pt Monometallic, Bimetallic, and Trimetallic Catalysts

For Pt monometallic catalysts, researchers have been devoted to the synthesis of various nanostructures, including one-dimensional (1D) nanowires, 2D nanosheet/nanoplate, and 3D polyhedrons/nanoframes. These Pt nanostructures demonstrate enhanced ORR activity due to improved Pt utilization, accelerated electron and mass transfer, and modulated electronic structure. The ultralong and ultrathin Pt nanowires with large surface area can exhibit excellent ORR performance [12]. Similarly, the controllable synthesis of concave or convex Pt nanocrystals exposed to specific facets holds great potential to form the unsaturated coordination centers on distinct facets for ORR catalysis. For instance, Yu et al. synthesized Pt concave nanocubes using Br^- as a capping agent, the concave nanocubes enclosed by {720}, {510}, and {830} facets, exhibited enhanced ORR activity [13]. Unfortunately, despite their improved specific activity, the mass activity of Pt concave nanocubes is relatively poor due to their large size (>15 nm). Except for the crystal facet, the particle size of Pt nanoparticle significantly influences the ORR activity. Shao and coworkers investigated the size-dependent specific and mass activity on the Pt particles in the range of 1–5 nm. They found that the maximal mass activity was achieved on Pt particles with 2.2 nm in diameter [14].

Tremendous work has shown that alloying Pt with other metals can boost the ORR activity and reduce the usage of Pt. The introduction of a second or third metal significantly reduces the usage of Pt and affects the electronic structure of Pt alloy through the strain, ligand, and ensemble effects, thereby modifying the Pt-O intermediate adsorption. In 1993, Mukerjee et al. demonstrated that enhanced ORR kinetics could be achieved by alloying Pt with transition metal (i.e., Ni, Cr, and Co) [15]. Stamenkovic and his coworkers further explored the fundamental relationship between the surface electronic structure (d-band center) and ORR activity using a series of Pt_3M (M=Ni, Co, Fe, Ti, V) alloy surfaces. Their findings revealed a volcano-type relationship between the specific activity of Pt_3M alloys and the d-band center, which suggests that too strong and too weak adsorption of oxygen intermediates would limit the ORR kinetics (Figure 9.3a, b). Notably, Pt_3Co, Pt_3Ni, and Pt_3Fe were located at the top position of volcano, indicating the higher ORR activity

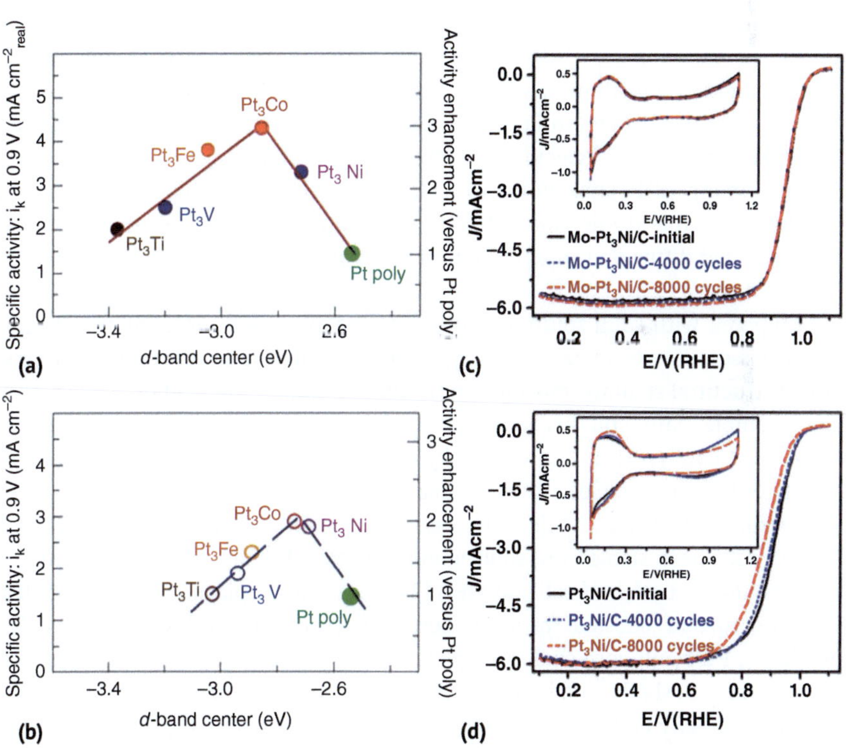

Figure 9.3 (a) Relationships between experimentally measured specific activity for the ORR on Pt_3M surfaces in 0.1 M $HClO_4$ at 333 K vs. the d-band center position for the Pt-skin (a) and Pt-skeleton (b) surfaces. *Source:* [16] / Springer Nature. ORR polarization curves and (inset) corresponding cyclic voltammograms of (c) the octahedral $MoPt_3Ni/C$ catalyst and (d) the octahedral Pt_3Ni/C catalyst before, after 4000, and after 8000 potential cycles. *Source:* Reproduced from [22] / with permission of American Association for the Advancement of Science.

than other alloys and Pt catalyst [16]. Inspired by these findings, PtM alloys have garnered significant attention, the size, shape, and component of PtM alloys were modulated to enhance the ORR performance [17]. Jiang et al. successfully synthesized a class of sub-nanometer Pt and Pt alloy (PtCo, PtNi, and PtNiCo) nanowires with a diameter of only 0.8 nm and the length of the nanowires can be controlled from 9 to 35 nm. Due to their atomic-level characteristics and alloy effect, PtNiCo nanowires exhibited high ORR activity and stability, with the mass and specific activities of 4.20 A mg^{-1} and 5.11 mA cm^{-2}, respectively, which are 32.3 and 26.9 times higher than those of commercial Pt/C catalyst [18]. Additionally, Pt-based alloys with hollow nanocage and nanoframes were reported to improve the ORR activity. For instance, hollow PtPd nanocages were prepared by depositing a few atomic layers of Pt on Pd nanocrystals and then etching away the Pd templates, the etching process involved the formation of vacancies. By using this method, PtPd cubic and octahedral nanocages enclosed by {100} and {111} facets, respectively, were prepared and exhibited a significantly improved ORR activity [19]. Similarly, Pt_3Ni rhombohedral nanoframes achieved 36-fold enhancement in mass activity and a factor of 22-fold enhancement in specific activity [20].

The incorporation of an additional element into PtM alloy has been found to further enhance the ORR performance [21]. Huang et al. explored the ORR performance of transition metal (M=V, Cr, Mn, Fe, Co, Mo, W)-doped Pt_3Ni octahedra, and found that Mo-Pt_3Ni exhibited the best ORR performance. Density functional theory (DFT) calculations revealed that Mo-doping increased the oxygen binding energy of Pt sites on the (111) facet, thereby enhancing the ORR activity. Furthermore, the Mo dopant stabilized both Ni and Pt atoms against dissolution and inhibited diffusion through the formation of strong Mo—Pt and Mo—Ni bonds (Figure 9.3c, d) [22].

Compared with solid solution alloys, intermetallics with order atomic arrangement possess low formation entropy, strong interatomic interactions, and exceptional structural stability, raising significant attention in the field of ORR catalysis. For example, Sun and his team successfully synthesized $L1_0$-FePt as an efficient ORR catalyst in 0.1 M $HClO_4$ by annealing the MgO-coated dumbbell-like Fe_3O_4_FePt nanoparticles at 700 °C. In acidic solutions, only Fe atoms were etched out, resulting in the formation of core/shell $L1_0$-FePt@Pt after further annealing at 400 °C. These stable core-shell nanoparticles exhibited much enhanced ORR performance in both half-cell and MEA [23]. Moreover, replacing Fe in $L1_0$-FePt@Pt with Co further boosted the ORR performance can be further promoted [24]. Similarly, PtPb@Pt core-shell nanoplate with the intermetallic core and uniform four layers of Pt shell could induce large biaxial strains, leading to high ORR specific and mass activity with 7.8 mA cm^{-2} and 4.3 A mg^{-1} at 0.9 V [25]. Recently, a series of intermetallics, including Pt_3Ga [26], Pt_2CuGa [27], and $Pt_{0.2}Pd_{1.8}Ge$ [28], have been reported to accelerate the ORR kinetics, showcasing the versatility and potential of intermetallics in ORR catalysis.

9.4.1.2 Pt-based High-entropy Alloy Catalysts

High-entropy alloy (HEA) electrocatalysts have garnered considerable attention in the field of electrocatalysis due to their unique properties, such as high entropy effect, lattice distortion effect, sluggish diffusion, and "cocktail" effect [29–31]. The modulations of compositions and shapes are two main strategies to enhance the performance of HEA catalysts. Tao et al. proposed a low-temperature wet-chemical method for the synthesis of HEA ultrathin ribbons, allowing for precise control of components and concentrations in HEA. The meticulously designed PtPdIrRuAg HEA exhibited significant improvement in catalyzing the ORR, while PtPdIrRuAuAg HEA showed promising performance as a cathode in Li—O_2 batteries. Theoretical calculations suggested that the concentration of highly dynamic reduction elements (Pd, Pt, Ag, Au) was critical to realize superior electroactivity for HEAs, while the relatively inert oxidative elements (Ir, Ru, Rh, Os) promoted the site-to-site electron transfer efficiency but potentially led to local aggregations [32]. Additional HEA catalysts, including PtFeCoNiMn [33] and FeNiCuCoPt [34], have been reported to exhibit promoted ORR activity. Recently, structurally ordered HEA nanostructures have been shown to accelerate the ORR kinetics. Feng et al. first utilized ordered HEA intermetallic (HEI) in ORR catalysis. They found that structurally ordered PtIrFeCoCu HEI (PIFCC-HEI) could achieve a mass activity of 7.14 A ${mg_{noble\ metals}}^{-1}$ at 0.85 V and demonstrated extraordinary stability over 60 000 cycles. Furthermore, the fuel cell assembled by this HEI could deliver a peak density of 1.73 W cm^{-2} and almost no working voltage decay after 80 hours operation (Figure 9.4) [35]. Lately, they also demonstrated an ultra-small ordered PtFeCoNiCuZn HEA with excellent performance for both ORR and fuel cell [36]. Meanwhile, Zhang et al. developed an efficient machine learning strategy to rationally design a $Pt(FeCoNiCu)_3$ HEA intermetallic for ORR, which exhibited excellent ORR performance [37].

9.4.1.3 Pt-based Heterostructures

Utilizing the strong metal-support interaction effect has emerged as a promising way to enhance the ORR performance. The Li group reported a dealloying process to fabricate the interface between dealloyed PtNi alloy and amorphous NiB membrane. They discovered that the coating NiB membrane worked as an electron acceptor to tune the d band center of PtNi and resulted in a 27-fold enhancement in mass activity compared to commercial Pt/C catalyst [38]. Furthermore, searching for suitable support for Pt or Pt-based alloys has garnered significant attention. Supports such as tungsten carbide (WC) [39], $Co_6Mo_6C_2$ [40], and titanium nickel nitride (TiNiN) [41], were found to enhance the ORR performance of Pt due to the highly dispersed Pt atoms and the down-shift of d-band center induced by the interface. Furthermore, constructing an interface can also improve the stability of ORR catalysts. Zhao et al. designed a graphene-nanopocket-encaged PtCo nanocatalyst (PtCo@Gnp). The presence of a non-contacting graphene nanopocket allowed for electrochemical accessibility, restricted the coalescence, and prevented the oxidative dissolution, diffusion, and Ostwald ripening. As a result, the PtCo@Gnp enabled highly robust PEMFCs at an ultralow Pt loading of 0.07 mg_{Pt} cm^{-2} (cathode loading of 0.060 mg_{Pt} cm^{-2}), delivering a mass activity of 1.21 A ${mg_{Pt}}^{-1}$, a rated power of 13.2 W ${mg_{Pt}}^{-1}$, and a mass activity retention of 73%, all exceeding the DOE target 2020 (Figure 9.5). Moreover, they successfully assembled a 90 kW fuel cell device using only 6.8 g Pt [42].

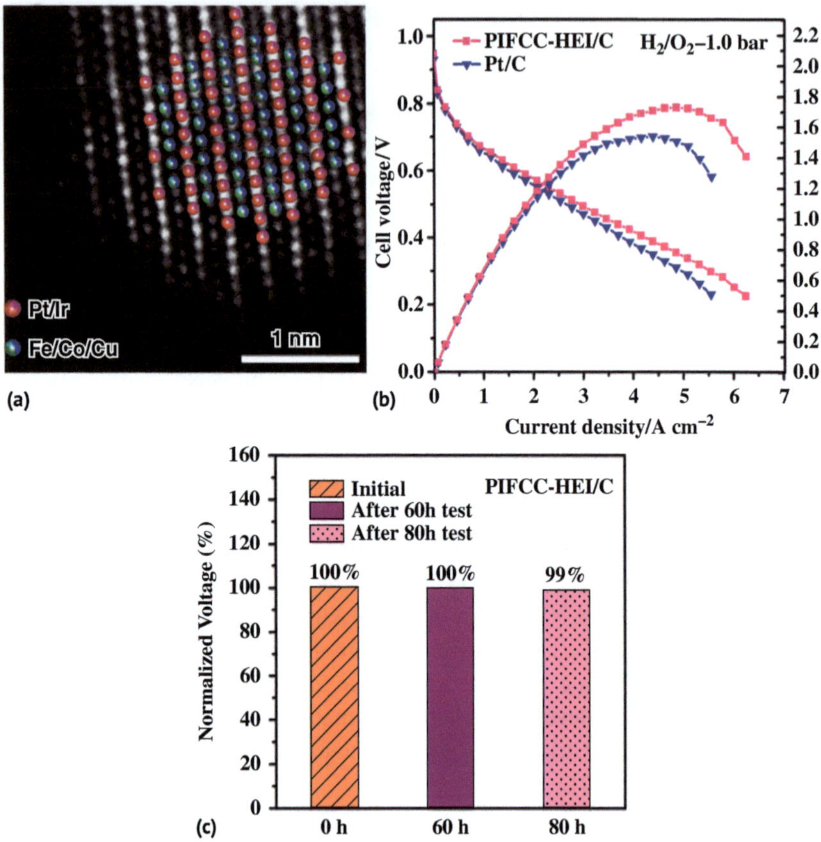

Figure 9.4 (a) Atomically resolved HAADF-STEM image of PIFCC-HEI nanoparticle. (b) Polarization and power density curves of the H_2/O_2 fuel cell with PIFCC-HEI/C and commercial Pt/C as the cathode ($0.10_{\mathrm{mgnoble\ metal}}\,\mathrm{cm}^{-2}$) under O_2 pressure of 1.0 bar. (c) Quantitative comparisons of the stability at different times. *Source:* Reproduced from [35] / with permission of American Chemical Society.

9.5 Pd-Based Metal Catalysts

Pd-based materials have attracted extensive attention as potential substitutes for Pt-based catalysts because of their similar chemical properties and low cost compared with Pt [43, 44]. For a monometallic Pd catalyst, it is difficult to achieve the ORR catalytic performance close to Pt catalyst, as the binding strength of O on the surface of Pd is too strong, hindering the reduction of oxygen [45]. Consequently, research has predominantly focused on alloying Pd with other transition metals and fine-tuning its shape to enhance ORR activity while further reducing catalyst costs. The enhanced ORR performance can be ascribed to an increased number of active sites and modified electronic structure resulting from strain effects, ligand effects with suitable adsorption energy of O intermediate [46]. For instance, Luo et al. successfully synthesized PdMo bimetallene with a thin-sheet structure, which enabled

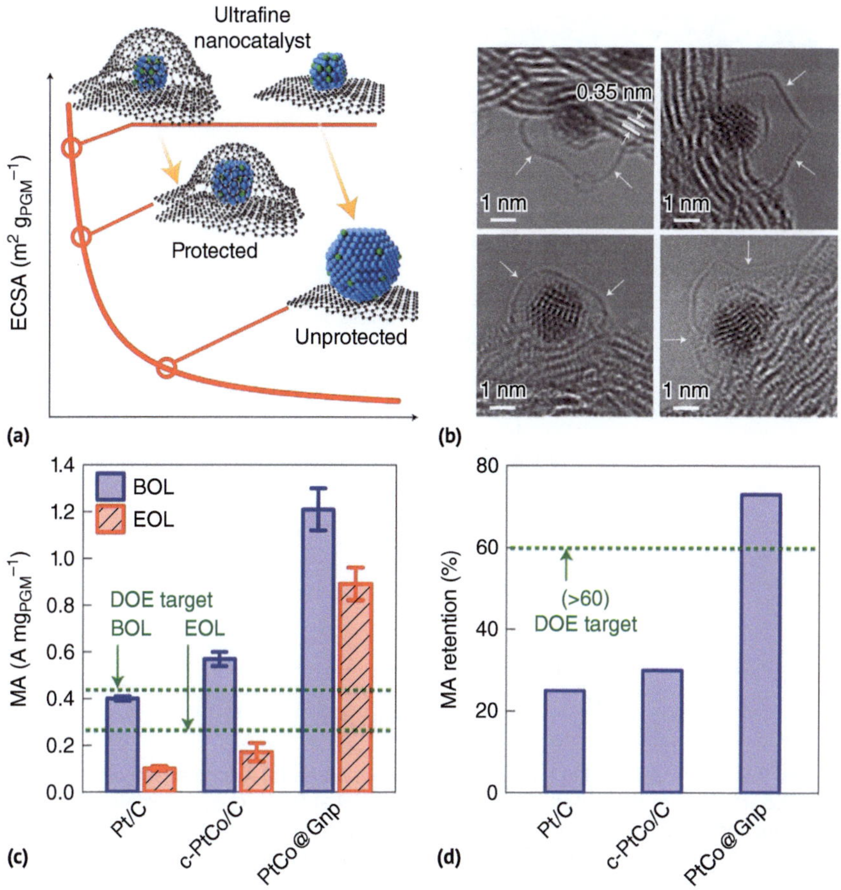

Figure 9.5 (a) Schematic illustration of ultrafine nanocatalysts encaged in graphene pockets and their impact on ECSA retention after an accelerated durability test. (b) Comparison of mass activity (MA) (b) and MA retention (c). BOL and EOL represent beginning-of-life and end-of-life, respectively, and error bars represent the standard deviation. *Source:* [42] / Springer Nature.

a large electrochemically active surface area and high atomic utilization. PdMo bimetallene exhibited a mass activity of 16.37 A mg_{Pd}^{-1} at 0.9 V in alkaline electrolytes, which is 78 and 327 times higher than those of commercial Pt/C and Pd/C, respectively. Theoretical calculations revealed that the alloying effect, strain effect due to the curved geometry, and quantum size effect caused by the ultrathin structure tuned the electronic structure for optimized oxygen binding (Figure 9.6) [47]. Other Pd-based alloy nanostructures, including PdCo sponge [48], PdNi nanoframe [49], PdFe nanoleaves [50], PdTe hexagonal nanoplates [51], and PdCu nanosheets [52], have also been explored. Furthermore, defect engineering and interface constructing have been employed to enhance the ORR performance [53, 54]. Zhang's work demonstrated that the Pd_4Sn nanowires with rich surface defects could accelerate the ORR kinetics [54]. Additionally, the ordered atomic arrangements of Pd and other transition metals (i.e., Pb [55], Sb [56], Cu [57]) in Pd-based intermetallics

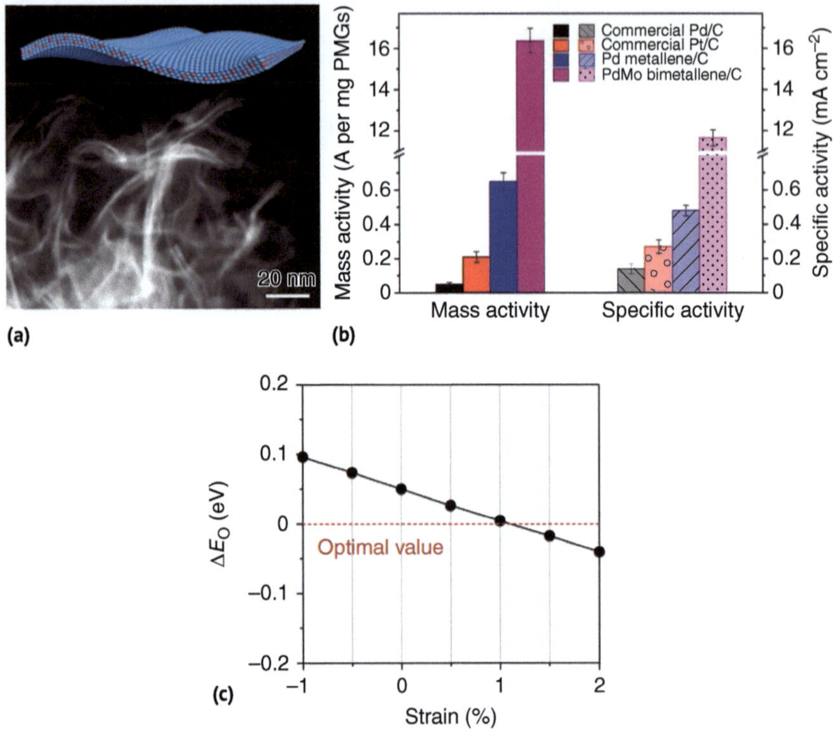

Figure 9.6 (a) HAADF-STEM image of PdMo bimetallene. (b) A comparison of the mass and specific activities of the stated catalysts in 0.1 M KOH at 0.9 V. (c) ΔE_O of PdMo bimetallene as a function of compressive (negative) and tensile (positive) strains. The horizontal red line indicates the optimal ΔE_O value. *Source:* [47] / Springer Nature.

can further enhance the ORR performance, due to the favored geometrical structure and interatomic interaction.

9.6 Non-platinum Group Metal Catalysts

Replacement of noble metal catalysts is a key research objective in the field of ORR catalysis. Many metal-nonmetal compounds, including oxides, chalcogenides, nitrides, oxynitrides, and carbides, have been reported to exhibit excellent ORR catalytic activity and stability. Among metal-oxide-based ORR catalysts, spinel (AB_2O_4) and perovskite (ABO_3) structures have been considerably considered [58–61]. Liang et al. demonstrated a hybrid material consisting of Co_3O_4 nanocrystals grown on reduced graphene oxide as excellent ORR catalyst in alkaline electrolyte due to the synergistic chemical coupling effect between Co_3O_4 and the support [62]. Suntivich et al. further demonstrated that the ORR activity for perovskite oxides correlated with σ*-orbital (e_g) occupation and the extent of B-site transition-metal–oxygen covalency, which serves as a secondary activity descriptor (Figure 9.7) [60]. By adjusting the element at the A and B sites in spinel/perovskite-based catalysts, the performance can be tailored accordingly [63].

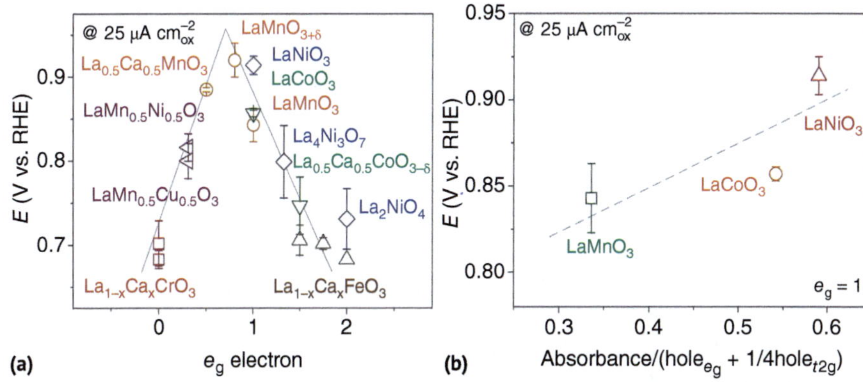

Figure 9.7 (a) Potentials at 25 μA cm_{ox}^{-2} as a function of e_g orbital in perovskite-based oxides. Data symbols vary with type of B ions (Cr, red; Mn, orange; Fe, gray; Co, green; Ni, blue; mixed compounds, purple), where $x = 0$ and 0.5 for Cr, and 0, 0.25 and 0.5 for Fe. Error bars represent standard deviations. (b) Potentials at 25 μA $cm^{-2}{}_{ox}$ as a function of the B–O covalency, which is estimated by the normalized absorbance at e_g-filling = 1. Error bars represent standard deviations of at least three measurements. *Source:* [60] / Springer Nature.

Transition metal chalcogenides have been the subject of considerable research. Among them, Co-based chalcogenides have attracted huge attention, including Co-based sulfides [64], Co-based selenide [65], and Co-based antimonide [66]. However, the challenges posed by their low conductivity and activity decay pose significant limitations on their practical applications. To address these issues, carbon materials are often employed as supports for Co-based chalcogenides, which can tune the particle size, prevent the particle agglomeration, and improve the charge transport [67]. For instance, the Dai group reported a hybrid ORR catalyst, comprising $Co_{1-x}S$ nanoparticles nucleated and grown on the surface of reduced graphene oxides (RGO). This catalyst demonstrated the highest ORR performance among all Co-based chalcogenide ORR catalysts at the time of its publication [68].

Apart from abovementioned Co-based chalcogenide ORR catalysts, Ru-based chalcogenides [69] and the hybrids combining oxides with chalcogenides [70] have also been proposed as promising ORR catalysts. For instance, uniform $RuSe_{2+\delta}$ nanotubes reported by Camargo et al. exhibited excellent ORR activity and methanol tolerant performance [71]. Additionally, a micro-box heterostructure comprising Co_3O_4 and CoS demonstrated a comparable ORR activity to commercial Pt/C, but superior stability [72].

Transition metal carbides have been explored as alternative ORR catalysts due to their high electric conductivity, corrosion resistance and "platinum" like behavior for hydrogen and oxygen adsorption. WC [73], VC [74], Fe_3C [75], and FeMoC [76] supported on carbon materials (i.e., N-doped graphene, high-defective graphitic layers) have shown excellent ORR activity. Notably, hollow spheres comprising Fe_3C nanoparticles encased by graphitic layers exhibited excellent ORR performance in both acidic and alkaline electrolytes. The outer graphitic layer could protect the carbide nanoparticles from losing catalytic activity [75].

Transition metal nitrides have been extensively investigated as ORR catalysts. Cao et al. explored the monometallic (δ-MoN, Mo_5N_6, and Mo_2N) and bimetallic molybdenum nitrides ($Co_{0.6}Mo_{1.4}N_2$) as ORR catalysts. Their findings revealed the crystal-phase-dependent ORR performance in molybdenum nitrides, with hexagonal molybdenum nitrides exhibiting superior ORR activity compared to rock salt counterparts. Further improvement could be achieved for bimetallic molybdenum nitrides ($Co_{0.6}Mo_{1.4}N_2$) [77]. Luo and coworkers also found that doped VN (Cu, Ni, Co, Fe, Mn, Cr, Ti doped VN) showed greatly enhanced ORR activity, and Co-doped VN exhibited the best performance, which was almost comparable to that of commercial Pt/C. Mechanistic study revealed that Co doping significantly increased the proportion of V with low-valence state and facilitated the formation of new O_2 adsorption sites with high reactivity [78]. Transition metal oxynitride catalysts have attracted huge attention for ORR catalysis due to their high electrical conductivity and corrosion resistance. Monometallic oxynitrides (i.e., MnON [79], TiON [80], HfON [81]) and bimetallic oxynitrides (i.e., NbFeON [82], CoMoON [83]) were extensively studied. The ORR performance of these catalysts is significantly influenced by the nitrogen content and the ratio of the two metals. Miura group investigated the ORR activities of manganese oxynitrides with different nitrogen content and found that the increase in the nitrogen content increased the valence of Mn and resulted in improved ORR activity [79]. Cao et al. synthesized carbon-supported $Co_xMo_{1-x}O_yN_z$ catalysts by a solution impregnation method combined with ammonolysis, and investigated the ratio of Co/Mo on ORR performance. They discovered that $Co_{0.50}Mo_{0.50}O_yN_z$/C demonstrated the best ORR activity in both acidic and alkaline conditions [83].

9.7 Metal-free Catalysts

In the quest for alternatives to Pt, a diverse array of heteroatom-doped carbon-based materials have emerged as promising candidates for ORR catalysis, and featured with the abundance of resources, low cost, high catalytic activity, remarkable stability and tolerance to methanol crossover [84]. Various forms of carbon, including graphene, graphdiyne, carbon nanotubes, carbon quantum dots, carbon nanoribbons, carbon nanocages, and rod-like mesoporous carbon, have been extensively investigated [85]. The incorporation of heteroatoms, such as N, B, and S or B–N and N–S co-doping, could induce charge transfer between the doping sites and adjacent carbon atoms, and create the active sites for ORR catalysis. Among these heteroatom-doped carbon catalysts, N-doped carbon materials stand out as the most promising candidates as alternatives to commercial Pt catalysts [86]. Since Gong and coworkers reported the first instance of N-doped carbon nanotubes exhibiting outstanding ORR catalysis and provided detailed understanding of catalytic mechanism based on N-doping induced charge transfer [87], heteroatom doped carbon materials have garnered significant interest. Efforts have been focused on regulating the types of carbon, morphologies, and dopant species to enhance ORR activity. For instance, Xia et al. reported hollow frameworks of N-doped carbon nanotubes (NCNTs) derived from ZIF-67 as an active and stable electrocatalyst for ORR. They attributed the exceptional activity to the chemical compositions

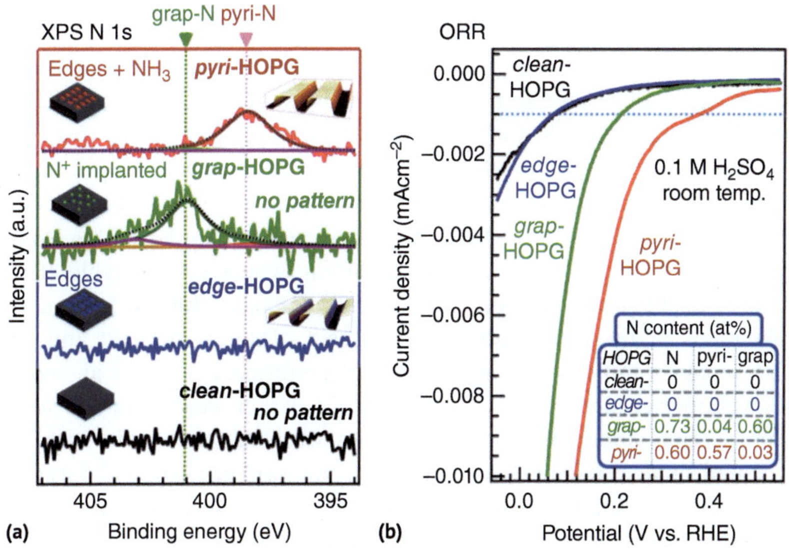

Figure 9.8 (a) N 1s XPS spectra of model catalysts. (b) ORR results for model catalysts corresponding to (a). Nitrogen contents of the model catalysts are shown as the inset in (b). *Source:* Reproduced from [86] / with permission of American Association for the Advancement of Science.

and structures of the NCNTs, as well as their robust framework structure [88]. However, the residual trace amount of Co in NCNTs may be responsible for the enhanced performance, and the active sites of NCNTs are unclear. In the same year, Guo and coworkers designed and synthesized four distinct types of model catalysts with well-defined structures based on highly oriented pyrolytic graphite (HOPG): pyridinic N-dominated HOPG, graphitic N-dominated HOPG, edges patterned on the surface without N, and clean HOPG. After rigorous testing, they demonstrated that pyridinic N created the active sites for ORR in acidic media and the carbon atoms with Lewis basicity next to pyridinic N were the ORR active sites (Figure 9.8) [89].

9.8 Single-atom Catalysts

SACs have recently gained prominence as a cutting-edge research area for ORR catalysis. With the maximum atomic utilization efficiency and modulated coordination environment of metallic site, SACs hold great potential for reducing the usage of PGM and promoting the ORR activity. The intrinsic catalytic activity and selectivity of SACs depend on several aspects, including the type of metal atomic centers, the surrounding environment of these center metal atoms, and the geometric configuration of the support. Depending on the metal atomic centers, SACs utilized for ORR are broadly categorized into Pt SACs, Fe SACs, Co SACs and others. Additionally, dual atom catalysts (DACs) with modified electronic structure and enhanced ORR catalytic activity are also demonstrated.

9.8.1 Pt-based Single-atom Catalysts

Atomically dispersed Pt atoms, featuring maximum atomic utilization efficiency, have attracted huge attention for ORR catalysis. A notable example includes pyridinic-N anchored Pt SACs (P-N1-Pt1), which were reported as a cost-effective, efficient and durable ORR catalyst (Figure 9.9). When integrated into a single cell, a high-power density of 680 mW cm^{-2} was achieved at 80 °C with a Pt loading of 0.09 mg$_{Pt}$ cm^{-2} [90]. A similar finding was demonstrated by depositing atomically dispersed Pt atoms on N-doped porous carbon, where the favorable electronic structure of Pt-N4 sites contributed to exceptional ORR performance. [91] Modulations of the coordination environment of single-atom Pt and support structure can significantly impact the ORR performance. Liu et al. reported a carbon-supported defect-anchored Pt SAC, in which single atom Pt anchored by four carbon atoms in carbon divacancies served as main active site. This Pt SAC-based single cell delivered an ultrahigh platinum utilization of 0.09 g$_{Pt}$ kW^{-1} at 80°C [92]. Additionally, atomic P-modified Pt-N_3 sites were also sought as excellent ORR catalysts [93]. Beyond the carbon-based support, metal and metal nitrides can be used as support of Pt single atoms. Zhang et al. demonstrated the fabrication of Pt single atoms on Pd octahedral nanocrystals through atomic layer deposition to form Pt/Pd single atom alloy (SAAs), which exhibited significantly enhanced ORR activity compared with commercial Pt/C. DFT calculations indicated that the relatively low binding energy of OH on Pt/Pd SAAs contributed to their superior ORR performance [94]. Yang and coworkers reported Pt single-atom

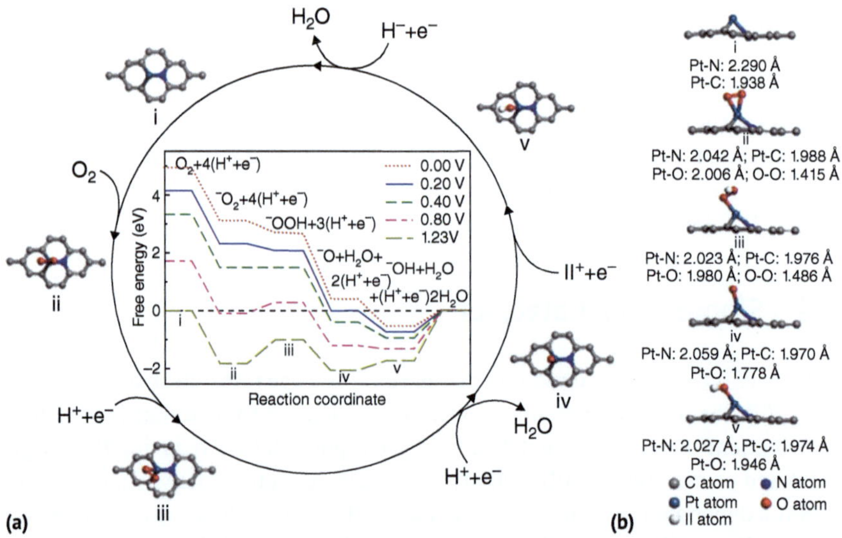

Figure 9.9 The proposed reaction pathways for complete oxygen reduction reaction on the P-N1-Pt1 catalyst. (a) Top view, (b) side view, and bond lengths. The inset in cycle (a) shows the free energy diagram for ORR on the P-N1-Pt1 catalyst in acidic media. *Source:* [90] / Springer Nature / CC BY 4.0.

supported on titanium nitride nanoparticles. However, this type of Pt SAC predominantly produced hydrogen peroxide during the ORR process. They demonstrated that the absence of Pt ensemble sites changed the reaction pathway into a two-electron-transfer process, which is in contradiction with the conclusions from other reports [95].

9.8.2 Fe Single-atom Catalysts

The development of a noble-metal-free ORR catalyst comparable to or exceeding the current Pt-based catalysts is crucial for development of fuel cell. Among the promising alternatives, non-precious SACs composed of Fe-N-C moieties have ranked as one of the most promising alternatives to manifest encouraging activity due to their cost-minimization and ultrahigh metal utilization (ca. 100%) [96–98]. Wan et al. reported a concave-shaped Fe-N-C SACs (TPI@Z8(SiO_2)-650-C) in high-performance proton exchange membrane fuel cell (PEMFC). Endowed with high intrinsic activity and dense Fe-N_4 moieties, Fe-N-C SACs assembled PEMFC achieved current densities of $0.022\,A\,cm^{-2}$ at 0.9 V (without iR drop) and $0.047\,A\,cm^{-2}$ at 0.88 V, surpassing the DOE 2018 target (Figure 9.10) [99]. The report by Yang and coworkers demonstrated that single-atom Fe catalysts supported on nitrogen-doped graphitic carbons (SA-Fe/NG) possessed outstanding ORR catalytic activity and stability in both acidic and alkaline media. Moreover, the SA-Fe/NG-assembled PEMFC exhibited a high power-density of $823\,mW\,cm^{-2}$ [100]. Despite these impressive achievements, Fe-N-C SACs face a significant durability challenge in PEMFCs, posing a hurdle for practical applications. Further studies into the poor stability of Fe-N-C SACs has revealed that the degradation of the organic ionomers and the PEM induced by Fe-involved Fenton reactions, leading to a deterioration of catalytic performance [101].

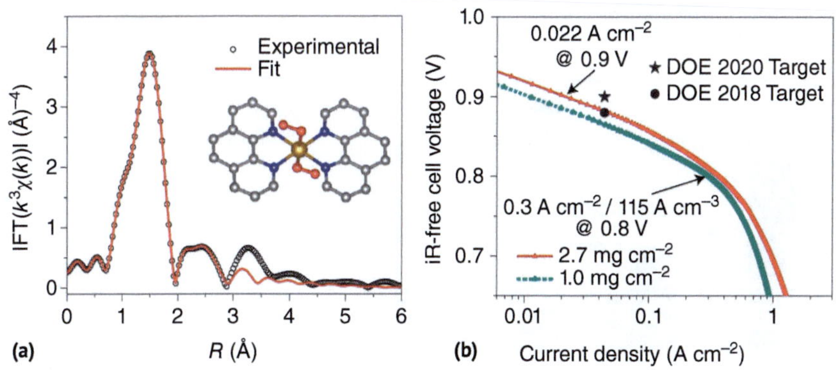

Figure 9.10 (a) A comparison between a Fourier-transformed experimental extended X-ray fine structure (EXAFS) spectrum and a fitting spectrum that is simulated based on the predicted FeN_4C_8 structure shown in inset. (b) Tafel plot for determination of TPI@Z8(SiO_2)-650-C activity at $0.9\,V_{iR-free}$ and volumetric activity at $0.8\,V_{iR-free}$ measured under 1 bar H_2—O_2. *Source:* [99] / Springer Nature.

9.8.3 Co Single-atom Catalysts

Co SACs have emerged as promising ORR catalysts as an alternative to Fe SACs due to their reduced sensitivity to Fenton effect [102]. Wang et al. reported atomically dispersed Co-N_4 sites with comparable ORR performance to Fe-N-C SAC in acidic media, and the fuel cell assembled by this atomic Co SAC delivered high power density and durability [103]. The electronic structure of single-atom Co active sites can be tailored by modulating the types of coordinated atoms, which contributes to enhanced ORR performance. Yin et al. developed atomically dispersed Co SACs on N-doped carbon (Co SACs/N-C) by the pyrolysis of predesigned bimetallic Zn/Co MOF. During this process, Co was reduced by carbonization of the organic linker and Zn was evaporated at high temperature. The obtained Co-N_2C_2 SACs exhibited superior ORR performance to commercial Pt/C and most reported non-precious metal catalysts (Figure 9.11) [104]. More recently, the same group designed a MOFs-derived Co SAC, which comprises Co_1-N_3PS moiety supported on hollow carbon polyhedron (Co_1-N_3PS/HC). Due to the asymmetric planar structure of Co_1-N_3PS and unique electronic density of single-atom Co active sites, Co_1-N_3PS/HC exhibited

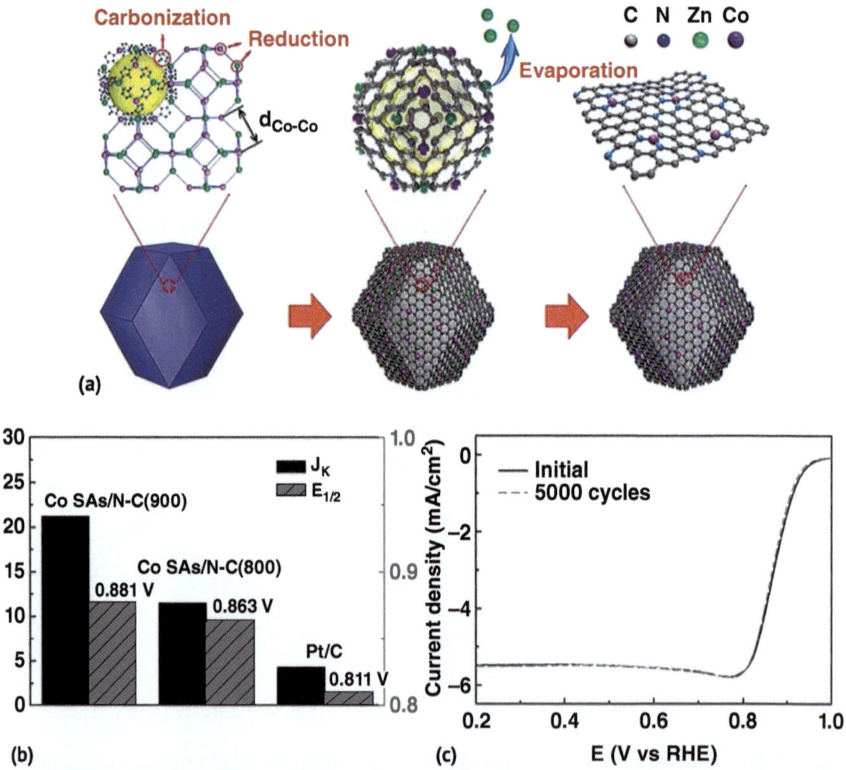

Figure 9.11 (a) The schematic illustration for the synthesis of Co SACs/N-C. (b) RDE polarization curves of Co SAs/N-C, Co NPs-N/C, and Pt/C in O_2-saturated 0.1 m KOH with a sweep rate of 10 mV S^{-1} and 1600 rpm. (c) LSV curves of Co SAs before and after 5000 potential cycles in O_2-saturated 0.1 m KOH. *Source:* [104] / John Wiley & Sons.

exceptional ORR performance in alkaline solution, with a positive half-wave potential of 0.92 V and an ultra-high kinetic current density of 28.0 mA cm^{-2}, exceeding those of commercial Pt/C and almost all non-precious ORR electrocatalysts. Impressively, Co_1-N_3PS/HC also displayed outstanding performance in acidic media and Zn-air battery [105].

9.8.4 Other Single-Atom Catalysts

Except for the aforementioned types of SACs, various other SACs have been proposed, including Zn SACs [106], Ir SACs [107], Y SACs [108], Cu SACs [109], and others. Li et al. reported an atomically dispersed Zn-N-C catalyst with an ultrahigh Zn loading of 9.33 wt%, which exhibited comparable ORR activity to that of Fe-N-C catalysts in both acidic and alkaline media [110]. Yang and coworkers fabricated Cu-N-C SACs with a uniform and well-defined Cu-N_4 structure, which displayed comparable activity and superior stability compared to commercial Pt/C. Mechanistic studies revealed the dynamic evolution of Cu-N_4 to Cu-N_3, and finally to HO-Cu-N_2 and valence of Cu changed from +2 to +1 under ORR conditions. Other research groups also employed Cu SACs for enhanced ORR catalysis [109].

A class of hybrid catalysts consisting of atomically dispersed sites and metal nanoparticles or nanoclusters, have been proposed to boost the ORR performance [111–113]. For instance, hybrid catalysts consisting of Fe single atoms and Pd nanoclusters were reported to boost the ORR and PEMFC performance (Figure 9.12a and b). Mechanistic study revealed that the strong electronic interaction between Fe single atom and Pd nanoclusters induced the transition of Fe^{2+} spin-state from low to intermediate, leading to the occurrence of a dissociative pathway and accelerating the removal of *OH [113].

In recent years, dual atom catalysts (DACs) with homometal or heterometal sites have attracted extensive attention [114]. By constructing dual-metal atom catalytic centers, the electronic structure, reaction mechanism, and catalytic activity can be fine-tuned. Wang et al. designed a highly-efficient ORR catalyst of Fe-N_4/Pt-N_4@NC comprising Fe-N_4 and Pt-N_4 dual atomic sites (Figure 9.12c). Theoretical calculations revealed that the Pt-N_4 acted as a modulator to tune the electronic structure of the Fe-N_4 active site and enhance both O_2 activity and the adsorption of ORR intermediates. Moreover, they also explored the modulation effect on other dual atomic catalysts, such as Co-N_4/Pt-N_4 and Mn-N_4/Pt-N_4, and demonstrated that Pt-N_4 showed weak and negative modulation effects on Co-N_4 and Mn-N_4 sites, respectively [115]. Another example of a Pt-Co DAC was demonstrated by Zhang et al., they synthesized a locally distributed atomic Pt-Co nitrogen-doped carbon catalyst (A-CoPt-NC). This catalyst exhibited high ORR activity (267 times higher than commercial Pt/C) and durability, significantly outperforming A-Co-NC and A-Pt-NC. Theoretical calculations suggested that the charge distribution and the d orbital shift resulting from synergistic effects of atomic Pt-Co metal centers were responsible for the enhanced ORR activity [116]. Apart from noble metals, dual-metal atomic sites comprising of transition metals have also been explored. Liu et al. developed a "pre-constrained metal twins" strategy to fabricate Fe-Co

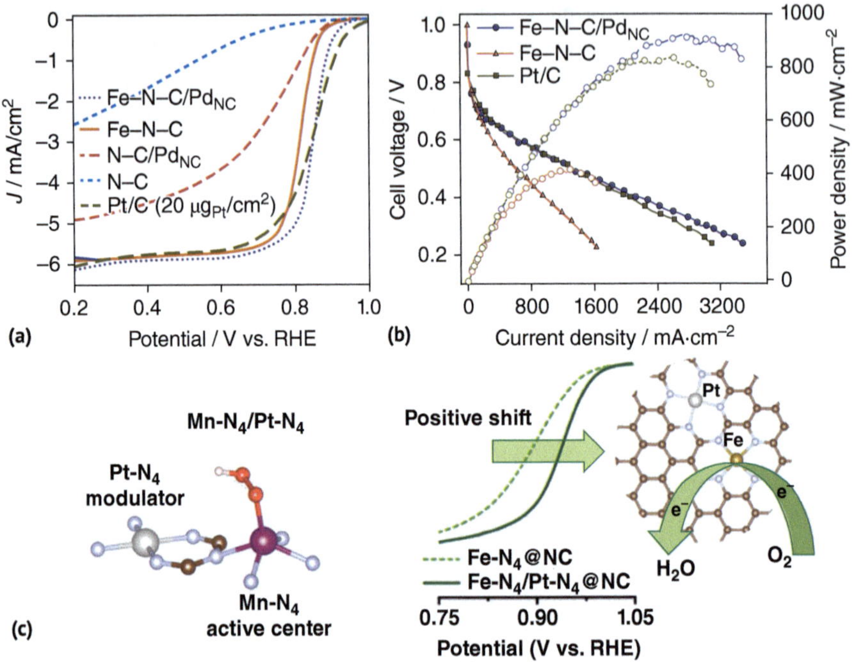

Figure 9.12 (a) ORR polarization curves measured on RDE in O_2-saturated 0.1 M $HClO_4$ electrolyte for different catalysts. (b) H_2—O_2 fuel cell performances of Fe–N–C/Pd$_{NC}$, Fe–N–C, and Pt/C catalyst at 80 °C. *Source:* Reproduced with permission of Wei et al. [113], Cell Press. (c) The schematic illustration of DACs (Mn-N4/Pt-N4) and enhanced ORR performance compared with Fe-N4@NC. *Source:* [115] / John Wiley & Sons.

dual atom catalyst embedded in N-doped graphitic carbon (FeCo-DACs/NC). They encapsulated FeCo binuclear phthalocyanines dimers into ZIF-8, subsequently annealed to obtain the FeCo-DACs/NC. This DAC exhibited excellent performance in ORR and Zn-air batteries. DFT calculations indicated that the electronic interaction between Fe and Co optimized the d-band center and free energy of intermediate states [117].

9.9 Conclusion and Outlook

In this chapter, the ORR mechanism, primarily focusing on 4e$^-$ transfer pathways, is first introduced, which necessitates the usage of advanced ORR catalysts. Subsequently, the performance evaluation methods for ORR catalysts, including RDE and MEA tests, are demonstrated. At last, the category of ORR catalysts and the influencing factors are described.

Despite the remarkable progress made thus far, several crucial challenges persist and warrant attention:

1. Currently, the assessment of most ORR catalysts predominantly relies on RDE tests. However, substantial discrepancies in measurement environments and

conditions between RDE and MEA tests create a significant gap in translating electrochemical performance from RDE to MEA. Therefore, concerted efforts should be committed to the design of efficient ORR catalysts for the MEA tests. Additionally, for practical applications, it is imperative to produce advanced catalysts with well-defined and uniform structures on a large scale.

2. Uncovering the structure-function relationship necessitates the precise characterization of catalyst's structure. The current methods used to characterize the structure are mainly based on ex situ techniques. More in-situ or operando characterization technologies, including in situ XAS, Raman, FTIR, should be performed to probe the dynamic changes of catalysts under working potential conditions. Furthermore, the integration of experimental and theoretical calculations, augmented by powerful tools like machine learning, can effectively provide the guidance for the rational design of more advanced ORR catalysts.
3. Regarding precious-metal catalysts, Pt-based catalysts are still dominant. However, their high cost, resource scarcity, susceptibility to poisoning, and limited stability due to the surface reconstruction and chemical leaching are posing significant challenges for their applications. Furthermore, the high activity of Pt-based catalysts in laboratory conditions is often difficult to replicate in practical MEA. Therefore, significant efforts should be taken to explore alternatives catalysts to overcome these limitations.
4. For nonprecious-metal catalysts, SACs appear to be attractive. However, their low density of active sites necessitates increasing loading to increase the number of active sites, which can lead to an excessively thick ORR catalytic layer and a substantial increase in mass-transfer resistance. Furthermore, most SACs are derived from pyrolyzed carbonaceous materials, where the high-temperature treatment inevitably limits the number of active sites and results in inhomogeneous distribution, which decreases the ORR activity. Consequently, achieving mass production of SACs with abundant and uniformly dispersed single active centers using non-pyrolysis synthesis method is indispensable.

Acknowledgment

This work was supported by the National Natural Science Foundation of China (22005176).

References

1 Kulkarni, A., Siahrostami, S., Patel, A., et al. (2018). Understanding catalytic activity trends in the oxygen reduction reaction. *Chem. Rev.* 118 (5): 2302–2312.

2 Nie, Y., Li, L., and Wei, Z. (2015). Recent advancements in Pt and Pt-free catalysts for oxygen reduction reaction. *Chem. Soc. Rev.* 44 (8): 2168–2201.

3 Wang, X., Li, Z., Qu, Y., et al. (2019). Review of metal catalysts for oxygen reduction reaction: from nanoscale engineering to atomic design. *Chem* 5 (6): 1486–1511.

4 Nørskov, J. K., Rossmeisl, J., Logadottir, A., et al. (2004). Origin of the overpotential for oxygen reduction at a fuel-cell cathode. *J. Phys. Chem. B* 108 (46): 17886–17892.

5 Guo, S., Zhang, S., and Sun, S. (2013). Tuning nanoparticle catalysis for the oxygen reduction reaction. *Angew. Chem. Int. Ed.* 52 (33): 8526–8544.

6 Liu, Z., Zhao, Z., Peng, B. et al. (2020). Beyond extended surfaces: understanding the oxygen reduction reaction on nanocatalysts. *J. Am. Chem. Soc.* 142 (42): 17812–17827.

7 Abild-Pedersen, F., Greeley, J., Studt, F., et al. (2007). Scaling properties of adsorption energies for hydrogen-containing molecules on transition-metal surfaces. *Phys. Rev. Lett.* 99 (1): 016105.

8 Calle-Vallejo, F., Martínez, J.I., García-Lastra, J.M., et al. (2012). Physical and chemical nature of the scaling relations between adsorption energies of atoms on metal surfaces. *Phys. Rev. Lett.* 108 (11): 116103.

9 Bligaard, T., Nørskov, J.K., Dahl, S., et al. (2004). The Brønsted–Evans–Polanyi relation and the volcano curve in heterogeneous catalysis. *J. Catal.* 224 (1): 206–217.

10 Yu, W., Porosoff, M.D., and Chen, J.G. (2012). Review of pt-based bimetallic catalysis: from model surfaces to supported catalysts. *Chem. Rev.* 112 (11): 5780–5817.

11 Fan, J., Chen, M., Zhao, Z., et al. (2021). Bridging the gap between highly active oxygen reduction reaction catalysts and effective catalyst layers for proton exchange membrane fuel cells. *Nat. Energy* 6 (5): 475–486.

12 Xia, B.Y., Wu, H.B., Yan, Y., et al. (2013). Ultrathin and ultralong single-crystal platinum nanowire assemblies with highly stable electrocatalytic activity. *J. Am. Chem. Soc.* 135 (25): 9480–9485.

13 Yu, T., Kim, D.Y., Zhang, H., et al. (2011). Platinum concave nanocubes with high-index facets and their enhanced activity for oxygen reduction reaction. *Angew. Chem. Int. Ed.* 50 (12): 2773–2777.

14 Shao, M., Peles, A., and Shoemaker, K. (2011). Electrocatalysis on platinum nanoparticles: particle size effect on oxygen reduction reaction activity. *Nano Lett.* 11 (9): 3714–3719.

15 Mukerjee, S. and Srinivasan, S. (1993). Enhanced electrocatalysis of oxygen reduction on platinum alloys in proton exchange membrane fuel cells. *J. Electroanal. Chem.* 357 (1): 201–224.

16 Stamenkovic, V.R., Mun, B.S., Arenz, M., et al. (2007). Trends in electrocatalysis on extended and nanoscale Pt-bimetallic alloy surfaces. *Nat. Mater.* 6 (3): 241–247.

17 Li, M., Zhao, Z., Cheng, T., et al. (2016). Ultrafine jagged platinum nanowires enable ultrahigh mass activity for the oxygen reduction reaction. *Science* 354 (6318): 1414–1419.

18 Jiang, K., Zhao, D., Guo, S., et al. (2017). Efficient oxygen reduction catalysis by subnanometer Pt alloy nanowires. *Sci. Adv.* 3 (2): e1601705.

19 Zhang, L., Roling, L.T., Wang, X., et al. (2015). Platinum-based nanocages with subnanometer-thick walls and well-defined, controllable facets. *Science* 349 (6246): 412–416.

20 Chen, C., Kang, Y., Huo, Z., et al. (2014). Highly crystalline multimetallic nanoframes with three-dimensional electrocatalytic surfaces. *Science* 343 (6177): 1339–1343.

21 Li, X., Duan, X., Zhang, S., et al. (2024). Strategies for achieving ultra-long ORR durability—Rh activates interatomic interactions in alloys. *Angew. Chem. Int. Ed.* 63 (24): e202400549.

22 Huang, X., Zhao, Z., Cao, L., et al. (2015). High-performance transition metal–doped Pt_3Ni octahedra for oxygen reduction reaction. *Science* 348 (6240): 1230–1234.

23 Li, J., Xi, Z., Pan, Y.-T., et al. (2018). Fe stabilization by intermetallic $L1_0$-FePt and Pt catalysis enhancement in $L1_0$-FePt/Pt nanoparticles for efficient oxygen reduction reaction in fuel cells. *J. Am. Chem. Soc.* 140 (8): 2926–2932.

24 Li, J., Sharma, S., Liu, X., et al. (2019). Hard-magnet $L1_0$-CoPt nanoparticles advance fuel cell catalysis. *Joule* 3 (1): 124–135.

25 Bu, L., Zhang, N., Guo, S., et al. (2016). Biaxially strained PtPb/Pt core/shell nanoplate boosts oxygen reduction catalysis. *Science* 354 (6318): 1410–1414.

26 Gui, R., Cheng, H., Wang, M., et al. (2024). Symmetry-induced regulation of Pt strain derived from Pt_3Ga intermetallic for boosting oxygen reduction reaction. *Adv. Mater.* 36 (11): 2307661.

27 Liu, X., Zhao, Z., Liang, J., et al. (2023). Inducing covalent atomic interaction in intermetallic Pt alloy nanocatalysts for high-performance fuel cells. *Angew. Chem.* 62 (23): e202302134.

28 Mondal, S., Bagchi, D., Riyaz, M., et al. (2022). In situ mechanistic insights for the oxygen reduction reaction in chemically modulated ordered intermetallic catalyst promoting complete electron transfer. *J. Am. Chem. Soc.* 144 (26): 11859–11869.

29 Yao, Y., Huang, Z., Xie, P., et al. (2018). Carbothermal shock synthesis of high-entropy-alloy nanoparticles. *Science* 359 (6383): 1489–1494.

30 Ren, J.-T., Chen, L., Wang, H.-Y., et al. (2023). High-entropy alloys in electrocatalysis: from fundamentals to applications. *Chem. Soc. Rev.* 52 (23): 8319–8373.

31 Löffler, T., Ludwig, A., Rossmeisl, J., et al. (2021). What makes high-entropy alloys exceptional electrocatalysts? *Angew. Chem. Int. Ed.* 60 (52): 26894–26903.

32 Tao, L., Sun, M., Zhou, Y., et al. (2022). A general synthetic method for high-entropy alloy subnanometer ribbons. *J. Am. Chem. Soc.* 144 (23): 10582–10590.

33 Wang, K., Chen, R., Yang, H., et al. (2024). The elements selection of high entropy alloy guided by thermodynamics and the enhanced electrocatalytic mechanism for oxygen reduction reaction. *Adv. Funct. Mater.* 34 (7): 2310683.

34 Pan, Y., Shan, X., Cai, F., et al. (2024). Accelerating the discovery of oxygen reduction electrocatalysts: high-throughput screening of element combinations in Pt-based high-entropy alloys. *Angew. Chem. Int. Ed.* 63 (37): e202407116.

35 Feng, G., Ning, F., Pan, Y., et al. (2023). Engineering structurally ordered high-entropy intermetallic nanoparticles with high-activity facets for oxygen reduction in practical fuel cells. *J. Am. Chem. Soc.* 145 (20): 11140–11150.

36 Chen, T., Qiu, C., Zhang, X., et al. (2024). An ultrasmall ordered high-entropy intermetallic with multiple active sites for the oxygen reduction reaction. *J. Am. Chem. Soc.* 146 (1): 1174–1184.

37 Zhang, L., Zhang, X., Chen, C., et al. (2024). Machine learning-aided discovery of low-Pt high entropy intermetallic compounds for electrochemical oxygen reduction reaction. *Angew. Chem. Int. Ed.* 163 (51): e202411123.

38 He, D., Zhang, L., He, D., et al. (2016). Amorphous nickel boride membrane on a platinum–nickel alloy surface for enhanced oxygen reduction reaction. *Nat. Commun.* 7: 12362.

39 Hsu, I.J., Hansgen, D.A., McCandless, B.E., et al. (2011). Atomic layer deposition of Pt on tungsten monocarbide (WC) for the oxygen reduction reaction. *J. Phys. Chem. C* 115 (9): 3709–3715.

40 Ma, X., Meng, H., Cai, M., et al. (2012). Bimetallic carbide nanocomposite enhanced pt catalyst with high activity and stability for the oxygen reduction reaction. *J. Am. Chem. Soc.* 134 (4): 1954–1957.

41 Tian, X., Luo, J., Nan, H., et al. (2016). Transition metal nitride coated with atomic layers of Pt as a low-cost, highly stable electrocatalyst for the oxygen reduction reaction. *J. Am. Chem. Soc.* 138 (5): 1575–1583.

42 Zhao, Z., Liu, Z., Zhang, A., et al. (2022). Graphene-nanopocket-encaged PtCo nanocatalysts for highly durable fuel cell operation under demanding ultralow-Pt-loading conditions. *Nat. Nanotechnol.* 17 (9): 968–975.

43 Wang, T., Chutia, A., Brett, D.J.L., et al. (2021). Palladium alloys used as electrocatalysts for the oxygen reduction reaction. *Energy Environ. Sci.* 14 (5): 2639–2669.

44 Xu, B., Zhang, Y., Li, L., et al. (2022). Recent progress in low-dimensional palladium-based nanostructures for electrocatalysis and beyond. *Coord. Chem. Rev.* 459: 214388.

45 Damjanovic, A. and Brusić, V. (1967). Oxygen reduction at Pt-Au and Pd-Au alloy electrodes in acid solution. *Electrochim. Acta* 12 (9): 1171–1184.

46 Suo, Y., Zhuang, L., and Lu, J. (2007). First-principles considerations in the design of Pd-alloy catalysts for oxygen reduction. *Angew. Chem. Int. Ed.* 46 (16): 2862–2864.

47 Luo, M., Zhao, Z., Zhang, Y., et al. (2019). PdMo bimetallene for oxygen reduction catalysis. *Nature* 574 (7776): 81–85.

48 Tominaka, S., Hayashi, T., Nakamura, Y., et al. (2010). Mesoporous PdCo sponge-like nanostructure synthesized by electrodeposition and dealloying for oxygen reduction reaction. *J. Mater. Chem.* 20 (34): 7175–7182.

49 Yang, Y. and Soroka, I.L. (2022). PdNi nanoframework and nanochain catalysts with enhanced oxygen reduction reaction performance. *ChemCatChem* 14 (24): e202200941.

50 Zhang, Z., More, K.L., Sun, K., et al. (2011). Preparation and characterization of PdFe nanoleaves as electrocatalysts for oxygen reduction reaction. *Chem. Mater.* 23 (6): 1570–1577.

51 Zhang, Y., Huang, B., Luo, G., et al. (2020). Atomically deviated Pd-Te nanoplates boost methanol-tolerant fuel cells. *Sci. Adv.* 6 (31): eaba9731.

52 Liu, Z., Xue, J., and Li, Y. (2023). Ultrathin PdCu nanosheet as bifunctional electrocatalysts for formate oxidation reaction and oxygen reduction reaction. *Small Methods* 7 (7): 2300021.

53 Lyu, Z., Zhang, X.-G., Wang, Y., et al. (2021). Amplified interfacial effect in an atomically dispersed RuOx-on-Pd 2D inverse nanocatalyst for high-performance oxygen reduction. *Angew. Chem. Int. Ed.* 60 (29): 16093–16100.

54 Zhang, Y., Huang, B., Shao, Q., et al. (2019). Defect engineering of palladium–tin nanowires enables efficient electrocatalysts for fuel cell reactions. *Nano Lett.* 19 (10): 6894–6903.

55 Guo, J., Gao, L., Tan, X., et al. (2021). Template-directed rapid synthesis of pd-based ultrathin porous intermetallic nanosheets for efficient oxygen reduction. *Angew. Chem. Int. Ed.* 60 (19): 10942–10949.

56 Zhang, Y., Liu, X., Liu, T., et al. (2022). Rhombohedral Pd–Sb nanoplates with Pd-terminated surface: An efficient bifunctional fuel-cell catalyst. *Adv. Mater.* 34 (31): 2202333.

57 Jiang, K., Wang, P., Guo, S., et al. (2016). Ordered PdCu-based nanoparticles as bifunctional oxygen-reduction and ethanol-oxidation electrocatalysts. *Angew. Chem. Int. Ed.* 55 (31): 9030–9035.

58 Kim, J., Ko, W., Yoo, J.M., et al. (2022). Structural insights into multi-metal spinel oxide nanoparticles for boosting oxygen reduction electrocatalysis. *Adv. Mater.* 34 (8): 2107868.

59 Wang, H., Liu, R., Li, Y., et al. (2018). Durable and efficient hollow porous oxide spinel microspheres for oxygen reduction. *Joule* 2 (2): 337–348.

60 Suntivich, J., Gasteiger, H.A., Yabuuchi, N., et al. (2011). Design principles for oxygen-reduction activity on perovskite oxide catalysts for fuel cells and metal–air batteries. *Nat. Chem.* 3 (7): 546–550.

61 Ji, Q., Bi, L., Zhang, J., et al. (2020). The role of oxygen vacancies of ABO_3 perovskite oxides in the oxygen reduction reaction. *Energy Environ. Sci.* 13 (5): 1408–1428.

62 Liang, Y., Li, Y., Wang, H., et al. (2011). Co_3O_4 nanocrystals on graphene as a synergistic catalyst for oxygen reduction reaction. *Nat. Mater.* 10 (10): 780–786.

63 Zhou, Y., Sun, S., Xi, S., et al. (2018). Superexchange effects on oxygen reduction activity of edge-sharing $[Co_xMn_{1-x}O_6]$ octahedra in spinel oxide. *Adv. Mater.* 30 (11): 1705407.

64 Zhao, W.-W., Bothra, P., Lu, Z., et al. (2019). Improved oxygen reduction reaction activity of nanostructured cos_2 through electrochemical tuning. *ACS Appl. Energy Mater.* 2 (12): 8605–8614.

65 Xu, X., Wang, X., Huo, S., et al. (2024). Modulation of phase transition in cobalt selenide with simultaneous construction of heterojunctions for highly-efficient oxygen electrocatalysis in zinc–air battery. *Adv. Mater.* 36 (8): 2306844.

66 Wu, G., Cui, G., Li, D., et al. (2009). Carbon-supported $Co_{1.67}Te_2$ nanoparticles as electrocatalysts for oxygen reduction reaction in alkaline electrolyte. *J. Mater. Chem.* 19 (36): 6581–6589.

67 Xiao, J., Kuang, Q., Yang, S., et al. (2013). Surface structure dependent electrocatalytic activity of Co_3O_4 anchored on graphene sheets toward oxygen reduction reaction. *Sci. Rep.* 3 (1): 2300.

68 Wang, H., Liang, Y., Li, Y., et al. (2011). $Co_{1-x}S$—Graphene hybrid: a high-performance metal chalcogenide electrocatalyst for oxygen reduction. *Angew. Chem. Int. Ed.* 50 (46): 10969–10972.

69 Kim, M.-G., Kim, S.-H., Jang, J.-H., et al. (2023). Nanotubular geometry for stabilizing metastable 1T-phase Ru dichalcogenides. *Adv. Energy Mater.* 13 (3): 2203133.

70 Yan, Y., Li, K., Chen, X., et al. (2017). Heterojunction-assisted $Co_3S_4@Co_3O_4$ core–shell octahedrons for supercapacitors and both oxygen and carbon dioxide reduction reactions. *Small* 13 (47): 1701724.

71 Camargo, P.H.C., Peng, Z., Lu, X., et al. (2009). Synthesis and application of $RuSe_{2+\delta}$ nanotube as a methanol tolerant electrocatalyst for the oxygen reduction reaction. *J. Mater. Chem.* 19 (7), 1024–1030.

72 Min, K., Kim, S., Lee, E., et al. (2021). A hierarchical Co_3O_4/CoS microbox heterostructure as a highly efficient bifunctional electrocatalyst for rechargeable Zn–air batteries. *J. Mater. Chem. A* 9 (32): 17344–17352.

73 Guo, J., Mao, Z., Yan, X., et al. (2016). Ultrasmall tungsten carbide catalysts stabilized in graphitic layers for high-performance oxygen reduction reaction. *Nano Energy* 28: 261–268.

74 Huang, T., Mao, S., Pu, H., et al. (2013). Nitrogen-doped graphene–vanadium carbide hybrids as a high-performance oxygen reduction reaction electrocatalyst support in alkaline media. *J. Mater. Chem. A* 1 (43): 13404–13410.

75 Hu, Y., Jensen, J.O., Zhang, W., et al. (2014). Hollow spheres of iron carbide nanoparticles encased in graphitic layers as oxygen reduction catalysts. *Angew. Chem. Int. Ed.* 53 (14): 3675–3679.

76 Chen, M., Liu, J., Zhou, W., et al. (2015). Nitrogen-doped graphene-supported transition-metals carbide electrocatalysts for oxygen reduction reaction. *Sci. Rep.* 5 (1): 10389.

77 Cao, B., Neuefeind, J.C., Adzic, R.R., et al. (2015). Molybdenum nitrides as oxygen reduction reaction catalysts: structural and electrochemical studies. *Inorg. Chem.* 54 (5): 2128–2136.

78 Luo, J., Tian, X., Zeng, J., et al. (2016). Limitations and improvement strategies for early-transition-metal nitrides as competitive catalysts toward the oxygen reduction reaction. *ACS Catal.* 6 (9): 6165–6174.

79 Miura, A., Rosero-Navarro, C., Masubuchi, Y., et al. (2016). Nitrogen-rich manganese oxynitrides with enhanced catalytic activity in the oxygen reduction reaction. *Angew. Chem. Int. Ed.* 55 (28): 7963–7967.

80 Kinumoto, T., Sou, Y., Ono, K., et al. (2015). Preparation of fibrous titania oxynitride – carbon catalyst and oxygen reduction reaction analysis in both acidic and alkaline media. *J. Power Sources* 273: 136–141.

81 Chisaka, M., Suzuki, Y., Iijima, T., et al. (2011). Effect of synthesis route on oxygen reduction reaction activity of carbon-supported hafnium oxynitride in acid media. *J. Phys. Chem. C* 115 (42): 20610–20617.

82 Luo, J., Lu, Z., Zhang, Y., et al. (2024). Bimetallic niobium-iron oxynitride as a highly active catalyst towards the oxygen reduction reaction in acidic media. *Mater. Today Nano* 25: 100448.

83 Cao, B., Veith, G.M., Diaz, R.E., et al. (2013). Cobalt molybdenum oxynitrides: synthesis, structural characterization, and catalytic activity for the oxygen reduction reaction. *Angew. Chem. Int. Ed.* 52 (41): 10753–10757.

84 Cui, P., Zhao, L., Long, Y., et al. (2023). Carbon-based electrocatalysts for acidic oxygen reduction reaction. *Angew. Chem. Int. Ed.* 62 (14): e202218269.

85 Yang, L., Shui, J., Du, L., et al. (2019). Carbon-based metal-free ORR electrocatalysts for fuel cells: past, present, and future. *Adv. Mater.* 31 (13): 1804799.

86 Singh, S.K., Takeyasu, K., and Nakamura, J. (2019). Active sites and mechanism of oxygen reduction reaction electrocatalysis on nitrogen-doped carbon materials. *Adv. Mater.* 31 (13): 1804297.

87 Gong, K., Du, F., Xia, Z., et al. (2009). Nitrogen-doped carbon nanotube arrays with high electrocatalytic activity for oxygen reduction. *Science* 323 (5915): 760–764.

88 Xia, B.Y., Yan, Y., Li, N., et al. (2016). A metal–organic framework-derived bifunctional oxygen electrocatalyst. *Nat. Energy* 1 (1): 15006.

89 Guo, D., Shibuya, R., Akiba, C., et al. (2016). Active sites of nitrogen-doped carbon materials for oxygen reduction reaction clarified using model catalysts. *Science* 351 (6271): 361–365.

90 Liu, J., Jiao, M., Lu, L., et al. (2017). High performance platinum single atom electrocatalyst for oxygen reduction reaction. *Nat. Commun.* 8 (1): 15938.

91 Li, T., Liu, J., Song, Y., et al. (2018). Photochemical solid-phase synthesis of platinum single atoms on nitrogen-doped carbon with high loading as bifunctional catalysts for hydrogen evolution and oxygen reduction reactions. *ACS Catal.* 8 (9): 8450–8458.

92 Liu, J., Jiao, M., Mei, B. et al. (2019). Carbon-supported divacancy-anchored platinum single-atom electrocatalysts with superhigh Pt utilization for the oxygen reduction reaction. *Angew. Chem. Int. Ed.* 58 (4), 1163–1167.

93 Zhu, X., Tan, X., Wu, K.-H., et al. (2021). Intrinsic ORR activity enhancement of Pt atomic sites by engineering the d-band center via local coordination tuning. *Angew. Chem. Int. Ed.* 60 (40): 21911–21917.

94 Zhang, L., Liu, H., Liu, S., et al. (2019). Pt/Pd single-atom alloys as highly active electrochemical catalysts and the origin of enhanced activity. *ACS Catal.* 9 (10): 9350–9358.

95 Yang, S., Kim, J., Tak, Y.J., et al. (2016). Single-atom catalyst of platinum supported on titanium nitride for selective electrochemical reactions. *Angew. Chem. Int. Ed.* 55 (6): 2058–2062.

96 Martinez, U., Komini Babu, S., Holby, E.F., et al. (2019). Progress in the development of Fe-based PGM-free electrocatalysts for the oxygen reduction reaction. *Adv. Mater.* 31 (31): 1806545.

97 Zhang, H., Chung, H.T., Cullen, D.A., et al. (2019). High-performance fuel cell cathodes exclusively containing atomically dispersed iron active sites. *Energy Environ. Sci.* 12 (8): 2548–2558.

98 Jiang, R., Li, L., Sheng, T., et al. (2018). Edge-site engineering of atomically dispersed Fe–N_4 by selective C–N bond cleavage for enhanced oxygen reduction reaction activities. *J. Am. Chem. Soc.* 140 (37): 11594–11598.

99 Wan, X., Liu, X., Li, Y., et al. (2019). Fe–N–C electrocatalyst with dense active sites and efficient mass transport for high-performance proton exchange membrane fuel cells. *Nat. Catal.* 2 (3): 259–268.

100 Yang, L., Cheng, D., Xu, H., et al. (2018). Unveiling the high-activity origin of single-atom iron catalysts for oxygen reduction reaction. *Proc. Natl. Acad. Sci. U.S.A.* 115 (26): 6626–6631.

101 Liu, K., Qiao, Z., Hwang, S., et al. (2019). Mn- and N- doped carbon as promising catalysts for oxygen reduction reaction: Theoretical prediction and experimental validation. *Appl. Catal. B Environ. Energy* 243: 195–203.

102 He, Y., Hwang, S., Cullen, D.A., et al. (2019). Highly active atomically dispersed CoN_4 fuel cell cathode catalysts derived from surfactant-assisted MOFs: carbon-shell confinement strategy. *Energy Environ. Sci.* 12 (1): 250–260.

103 Wang, X.X., Cullen, D.A., Pan, Y.-T., et al. (2018). Nitrogen-coordinated single cobalt atom catalysts for oxygen reduction in proton exchange membrane fuel cells. *Adv. Mater.* 30 (11): 1706758.

104 Yin, P., Yao, T., Wu, Y., et al. (2016). Single cobalt atoms with precise N-coordination as superior oxygen reduction reaction catalysts. *Angew. Chem. Int. Ed.* 55 (36): 10800–10805.

105 Chen, Y., Gao, R., Ji, S., et al. (2021). Atomic-level modulation of electronic density at cobalt single-atom sites derived from metal–organic frameworks: enhanced oxygen reduction performance. *Angew. Chem. Int. Ed.* 60 (6): 3212–3221.

106 Song, P., Luo, M., Liu, X., et al. (2017). Zn single atom catalyst for highly efficient oxygen reduction reaction. *Adv. Funct. Mater.* 27 (28): 1700802.

107 Xiao, M., Zhu, J., Li, G., et al. (2019). A single-atom iridium heterogeneous catalyst in oxygen reduction reaction. *Angew. Chem. Int. Ed.* 58 (28): 9640–9645.

108 Ji, B., Gou, J., Zheng, Y., et al. (2023). Coordination chemistry of large-sized yttrium single-atom catalysts for oxygen reduction reaction. *Adv. Mater.* 35 (24): 2300381.

109 Yang, J., Liu, W., Xu, M., et al. (2021). Dynamic behavior of single-atom catalysts in electrocatalysis: identification of Cu-N_3 as an active site for the oxygen reduction reaction. *J. Am. Chem. Soc.* 143 (36): 14530–14539.

110 Li, J., Chen, S., Yang, N., et al. (2019). Ultrahigh-loading zinc single-atom catalyst for highly efficient oxygen reduction in both acidic and alkaline media. *Angew. Chem. Int. Ed.* 58 (21): 7035–7039.

111 Huang, H., Yu, D., Hu, F., et al. (2022). Clusters induced electron redistribution to tune oxygen reduction activity of transition metal single-atom for metal–air batteries. *Angew. Chem. Int. Ed.* 61 (12): e202116068.

112 Ao, X., Zhang, W., Zhao, B., et al. (2020). Atomically dispersed Fe–N–C decorated with Pt-alloy core–shell nanoparticles for improved activity and durability towards oxygen reduction. *Energy Environ. Sci.* 13 (9): 3032–3040.

113 Wei, X., Song, S., Cai, W., et al. (2023). Tuning the spin state of Fe single atoms by Pd nanoclusters enables robust oxygen reduction with dissociative pathway. *Chem* 9 (1): 181–197.

114 Sun, Q., Yue, X., Yu, L., et al. (2024). Well-defined Co_2 dual-atom catalyst breaks scaling relations of oxygen reduction reaction. *J. Am. Chem. Soc.* 146 (51): 35295–35304.

115 Han, A., Wang, X., Tang, K., et al. (2021). An adjacent atomic platinum site enables single-atom iron with high oxygen reduction reaction performance. *Angew. Chem. Int. Ed.* 60 (35): 19262–19271.

116 Zhang, L., Fischer, J.M.T.A., Jia, Y., et al. (2018). Coordination of atomic Co–Pt coupling species at carbon defects as active sites for oxygen reduction reaction. *J. Am. Chem. Soc.* 140 (34): 10757–10763.

117 Liu, M., Li, N., Cao, S., et al. (2022). A "Pre-Constrained Metal Twins" strategy to prepare efficient dual-metal-atom catalysts for cooperative oxygen electrocatalysis. *Adv. Mater.* 34 (7): 2107421.

10

Electrochemical Conversion of Biomass Derivatives

Tao Zhang[1]*, Jing Peng*[1,2,3*]

[1] *Institute of Technology for Carbon Neutrality, Shenzhen Institutes of Advanced Technology, Chinese Academy of Sciences, Shenzhen, China*
[2] *Faculty of Materials Science and Energy Engineering, Shenzhen University of Advanced Technology, Shenzhen, China*
[3] *Key Laboratory of Precision and Intelligent Chemistry, University of Science & Technology of China, Hefei, China*

*Corresponding author: jing.peng@siat.ac.cn

10.1 Introduction

The extensive use of fossil fuels not only accelerates the depletion of resources but also exacerbates the emission of greenhouse gases and environmental pollution issues. Therefore, the development and utilization of renewable energy resources, especially biomass resources, are of great significance for reducing dependence on fossil fuels and alleviating environmental pressure [1–3].

Biomass-derived products refer to various products obtained from biomass through physical, chemical, or biotechnological processes, including but not limited to carbohydrates, alcohols, lipids, and other organic compounds [4]. These derivatives provide a rich raw material basis for electrocatalytic oxidation. Biomass sources include agricultural waste, wood, algae, and others, all of which are widely distributed across various environmental conditions and possess good regenerative capabilities [5]. However, the effective conversion and utilization of biomass face technical challenges, especially how to efficiently convert its main components, including cellulose, hemicellulose, lignin, and other derived organic molecules, into high-value chemicals [6, 7]. Traditional thermochemical conversion methods have many limitations, including high temperature and pressure conditions, high energy consumption, low conversion efficiency, and the potential for by-products and pollutants [8, 9].

Electrochemical conversion technology has become a research hotspot in the field of biomass conversion because it efficiently converts biomass-derived substances at room temperature and pressure, with advantages such as simple operation, low energy consumption, high selectivity, and easy scalability [10–12].

Electrochemical Energy Conversion via Small Molecule Catalysis, First Edition.
Edited by Zhicheng Zhang and Yuanmiao Sun.

Secondly, the use of catalysts only requires non-precious metals, such as Co-, Ni-, and Cu-based catalysts, to achieve efficient conversion [13–15]. In particular, electrochemical oxidation can achieve the transformation of furan compounds (biomass derivatives), such as 5-hydroxymethylfurfural (HMF), into high-value chemicals like 2,5-furandicarboxylic acid (FDCA) under mild conditions, providing a new source of raw materials for the production of bio-based plastics, such as PEF [16–18]. Furthermore, the coupling systems of organic small-molecule electrochemical conversion technology with other cathodic reduction reactions provides new ideas for the energy utilization of biomass [19]. The electrooxidation of biomass derivatives, as an alternative to the oxygen evolution reaction (OER) in aqueous electrocatalytic systems, it achieves high selectivity and energy efficiency at lower potentials, reducing energy consumption and by-product generation, promoting the sustainable use of biomass resources [20–22]. For example, by coupling biomass derivatives electrooxidation with water splitting reactions, not only can the efficient conversion of biomass derivatives be achieved, but H_2 can also be produced [23]. In addition, on the cathodic half-reaction side, there can also be other reduction reactions, such as the carbon dioxide reduction reaction (CO_2RR) [24] and the nitrogen reduction reaction (NRR) [25, 26]. This coupling strategy improves the energy conversion efficiency of biomass resources and provides possibilities for diversified energy supply and clean production.

This chapter will delve into the mechanism of electrooxidation of biomass platform molecules, explore the design principles of electrocatalyst active sites, and discuss how to regulate the selectivity and conversion rates of the reaction through electrochemical parameters. At the same time, we will discuss the advantages and challenges of electrochemical coupling systems and how to achieve the efficient and clean utilization of biomass resources through system optimization. Finally, we will summarize the challenges and future research directions faced by the electrochemical conversion of biomass derivatives, including the innovation of catalysts, an in-depth understanding of reaction mechanisms and optimization of system integration.

10.2 Fundamentals of Electrooxidation of Biomass Derivatives

10.2.1 Principle and Mechanism

The electrochemical oxidation process of biomass-derived compounds is a complex electrochemical reaction that involves several key steps, specifically including the dissolution of biomass, occurrence of electrooxidation reactions, electron transfer, and redox reactions [27]. Prior to the electrooxidation reaction, it is crucial for biomass-derived compounds to be effectively dissolved in the electrolyte solution, enabling them to fully dissociate into ions, and thereby improving the conductivity of the electrolyte. Upon the application of an external voltage, the biomass-derived compounds undergo oxidation reactions at the anode. During this process, the biomass loses electrons and is converted into corresponding oxidation products, protons, and free electrons [28]. According to Faraday's law, the applied current is proportional to the extent of conversion of the reactants, allowing researchers to assess the progress and efficiency of the reaction by monitoring changes in the current. Real-time monitoring of this reaction is vital for optimizing the electrooxidation process and improving reaction yields. During the anodic oxidation process, the released electrons are transferred to the cathode via an external circuit, thereby completing the current loop [29, 30]. Efficient electron transfer can accelerate reaction rates and affect the types and quantities of oxidation products generated. At the cathode, electrons may combine with protons to facilitate reduction reactions, resulting in the formation of reduced products, such as hydrogen gas.

The electrochemical oxidation mechanism of biomass-derived compounds is relatively complex, involving multiple reaction steps and the formation of intermediates. The following are some common reaction mechanisms:

***Direct Oxidation Mechanism*:** In this mechanism, biomass-derived compounds undergo direct oxidation at the electrode surface, resulting in the formation of corresponding oxidation products [31, 32]. This mechanism typically occurs under conditions of high catalytic activity of the electrode material, such as when utilizing noble metal catalysts like Pt or Au [33, 34].

***Indirect Oxidation Mechanism*:** In certain scenarios, biomass-derived compounds may be oxidized through intermediates or oxidants, such as hydroxide ions (OH^-) or other reactive species [35–37]. This mechanism can enhance the selectivity of the reaction, leading to the formation of various hydroxylated compounds or aldehydes.

***Multiple Oxidation Pathways*:** Biomass-derived compounds may undergo multiple simultaneous oxidation pathways, which may interact with one another, resulting in the generation of different products. In such cases, the selectivity of the products is contingent upon reaction conditions, such as pH, temperature, electrode materials, and current density, among other factors [38, 39].

10.2.2 Descriptors of Performance

The descriptors of the electrooxidation performance of biomass-derived compounds are important parameters for assessing their performance in electrochemical applications. These descriptors encompass not only the physical and chemical characteristics of the materials, such as specific surface area, pore structure, and conductivity, but also the kinetic and thermodynamic features exhibited during the electrooxidation process. By systematically analyzing these descriptors, researchers can better understand the potential applications of biomass-derived compounds in fields such as renewable energy conversion and efficient catalysis. These primarily include the following aspects:

***Electrochemical Reaction Kinetic Parameters*:** The reaction kinetics of the electrooxidation process are often described by the relationship between current density and potential voltage. It is essential to focus on determining kinetic parameters such as exchange current density (j_0), activation energy of the electrode reaction, and the Tafel slope (*b*), as these indicators are crucial for understanding the electrochemical activity of materials [40]. Consequently, the mechanisms through which biomass-derived compounds promote or inhibit electrooxidation reactions under specific conditions can be elucidated.

***Specific Surface Area and Pore Structure*:** These are also regarded as important descriptors for evaluating electrooxidation performance. A larger specific surface area provides more reaction sites, thereby increasing the reaction rate, while an appropriate pore structure facilitates the rapid diffusion of reactants [27, 41, 42]. Techniques such as nitrogen adsorption–desorption testing (BET) can be used to accurately determine the specific surface area and pore characteristics of biomass-derived compounds, guiding their application in electrooxidation processes.

***Types and Distribution of Functional Groups*:** Biomass-derived compounds often contain various functional groups, such as hydroxyl, carboxyl, and phenolic groups, whose types and distribution directly influence the electrooxidation reactions. These groups can not only participate in the reaction but may also affect the conductivity and stability of the materials [13, 43]. Therefore, techniques such as X-ray photoelectron spectroscopy (XPS) and Fourier-transform infrared spectroscopy (FTIR) provide detailed information on the functional groups present in biomass-derived compounds, aiding in the analysis of their electrooxidation performance.

***Influence of Environmental Conditions*:** The performance of the electrooxidation reaction is closely related to environmental conditions, including temperature, pH, and reactant concentration, all of which affect the electrochemical behavior at the electrode interface [35]. By varying experimental conditions, researchers can obtain data on the electrooxidation performance of biomass-derived compounds in different environments, leading to a more comprehensive performance evaluation.

10.2.3 Catalogs of Biomass Derivatives

The biomass-derived platform molecules used for electrocatalytic oxidation mainly include HMF, glycerol, glucose, various alcohols, furan compounds, nitrogen-containing molecules, and methanol. These molecules play a significant role in the conversion to high-value chemicals. HMF is a key biomass platform molecule that can be electrocatalytically oxidized to produce the high-value compound FDCA with relatively fast reaction kinetics [44]. Glycerol can also be converted into valuable chemicals and is frequently used as a model compound for studying the performance and mechanisms of electrocatalytic oxidation reactions [45, 46]. Glucose can be transformed into formate, with aldehydes as its reaction intermediates [47, 48]. Various alcohols (such as methanol, ethanol, glycerol, and glucose) not only generate high-value chemicals during the electrocatalytic oxidation process but can also improve the efficiency of water electrolysis for hydrogen production by coupling with hydrogen evolution reactions [49, 50]. Moreover, furan compounds (such as furfural and HMF) can generate multiple derivatives upon electrochemical oxidation and are widely used in the chemical industry as platform molecules and biofuels. The electrocatalytic oxidation of these biomass-derived platform molecules provides an efficient pathway for the production of high-value chemicals. In this section, the electrocatalytic oxidation studies of various biomass-derived platform molecules will be introduced, along with a discussion of the reaction pathways involved in electrocatalytic oxidation.

10.2.3.1 Monomer/Polyol

The electrooxidation of alcohols and their biomass derivatives, including methanol, ethanol, enol, cyclohexanol, ethylene glycol, and glycerol, offers several advantages (Figure 10.1). First, the C—H and O—H bonds in alcohol molecules are relatively weak, requiring less energy for cleavage, which results in a theoretical oxidation potential lower than that of the OER for water, thereby reducing the necessary external voltage. Second, the electronic structure of alcohol molecules enables rapid electron transfer during the electrooxidation process, promoting reaction kinetics. Additionally, electrooxidation of alcohols typically involves fewer electron transfer steps, resulting in higher Faradaic efficiency and reduced formation of byproducts, ultimately increasing both the yield and purity of hydrogen. These characteristics make alcohol electrooxidation an ideal choice of anode reaction for efficient hydrogen production.

For example, a novel electrocatalytic method for the synthesis of adipic acid is combined with hydrogen production and enhanced reaction efficiency through a ligand modification strategy. As shown in Figure 10.2a, Li et al. developed an electrocatalytic approach for the oxidation of cyclohexanone to replace the traditional synthesis method that relies on corrosive nitric acid, thereby reducing emissions of the potent greenhouse gas nitrous oxide (N_2O) [51]. By employing a modified nickel hydroxide [$Ni(OH)_2$] catalyst and using sodium dodecylbenzene sulfonate (SDS) as a ligand, the affinity of the catalyst for cyclohexanone was increased, significantly improving the production efficiency and Faradaic efficiency

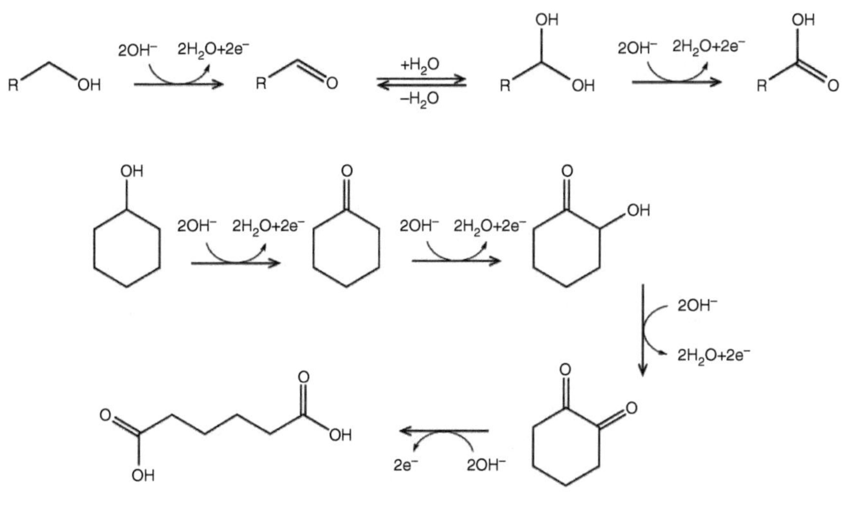

Figure 10.1 Electrochemical reaction pathways of alcohol compounds.

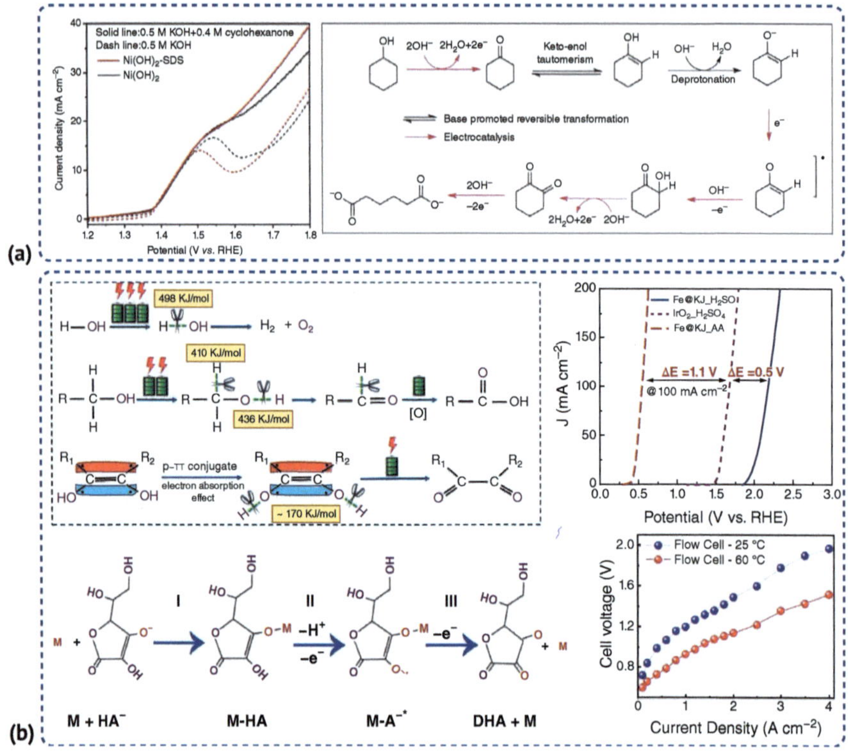

Figure 10.2 (a) Electrooxidation performance and reaction mechanism of cyclohexanol. *Source:* [51] / Springer Nature / CC BY 4.0. (b) Electrooxidation of ascorbic acid with an enol structure produces dehydroascorbic acid and couples with hydrogen generation at the cathode. *Source:* [52] / Springer Nature / CC BY 4.0.

of adipic acid while addressing the issue of cyclohexanone's low solubility in the aqueous phase (Figure 10.2a). Experimental results demonstrated that the modified catalyst achieved a 3.6-fold increase in the adipic acid production rate and a 93% Faradaic efficiency.

On the other hand, developing novel platform molecules for the electrooxidation of biomass derivatives to produce high-value-added products and hydrogen is also of significant interest. Chen et al. developed a novel hydrogen production system that combines the electrooxidation of ascorbic acid (vitamin C) with hydrogen cathode reactions [52]. This system significantly reduces the operating voltage to 1.1 V and achieves a current density of up to 2 A cm^{-2} at 60 °C, with energy consumption being only half that of traditional water electrolysis, as shown in Figure 10.2b. By using an iron single-atom catalyst, the system can produce hydrogen efficiently at a very low overpotential of just 12 mV, and the Faradaic efficiency is close to 100%. Furthermore, the byproduct of ascorbic acid, dehydroascorbic acid (DHA), has wide applications in the pharmaceutical and cosmetic industries, enhancing its economic value (Figure 10.2b) [52]. This research not only improves the efficiency of hydrogen production but also provides a new solution for the sustainable utilization of biomass, demonstrating significant practical application potential and theoretical value.

10.2.3.2 Furan Compounds

Furan compounds are considered important platform molecules derived from biomass due to their unique chemical and electronic structures. This category includes HMF, furfuryl alcohol (FFA), 2-methylfuran (2-MF), 2-furoic acid (2-FA), and cyclopentanone (CPO). Notably, HMF has been recognized as one of the most significant C_6 platform molecules since 2004 [10]. The main challenges facing research on HMF electrocatalytic oxidation include the selection and optimization of catalysts, particularly the development of efficient and selective non-precious metal catalysts; the impact of reaction conditions on product selectivity, such as how to control pH and voltage to increase the yield of FDCA; the complexity of the reaction mechanism, where the formation and conversion pathways of intermediates are not yet fully understood; and the understanding of electrocatalytic reaction kinetics, involving proton and electron transfer processes.

Wang et al. introduced a novel bipolar hydrogen production system that combines low-potential aldehyde oxidation reactions with hydrogen evolution reactions (HER), achieving hydrogen production at both the anode and cathode simultaneously under a low voltage (approximately 0.1 V) [13]. Unlike traditional high-potential aldehyde oxidation, this system allows for the release of hydrogen atoms from the aldehyde group as hydrogen gas rather than oxidizing them to water (Figure 10.3a). By utilizing a Cu catalyst, the system achieved a Faradaic efficiency (FE) of up to 100%, and the energy consumption of this system was only about 1/14 of that of conventional water electrolysis (approximately 0.35 kWh m^{-3} H_2), demonstrating its superior energy conversion efficiency.

Recent investigations reveal that benzoic acid hybridized Ni-Co hydroxide nanowire catalysts significantly enhances its electrocatalytic performance in the oxidation of HMF. Wang et al. show that the electron-withdrawing ability of benzoic acid can activate lattice hydroxyl groups, promoting the formation of high-valence Ni/Co-active sites, resulting in a current density of 111.20 mA cm^{-2} at 1.4 V, which is much higher than that of the unhybridized catalyst (Figure 10.3b) [18]. Additionally, the conversion rate of HMF, FDCA yield, and Faradaic efficiency also demonstrate excellent performance. This research showcases the potential of ligand hybridization in electrocatalysis.

10.2.3.3 Glucose

Glucose ($C_6H_{12}O_6$) is a six-carbon aldohexose sugar with a molecular structure that contains multiple hydroxyl groups and one aldehyde group. This unique molecular feature endows glucose with good oxidizing properties in electrochemical reactions. Through the process of electrochemical oxidation, glucose can be selectively converted into gluconic acid, glucaric acid, lactic acid, and formic acid (Figure 10.4) [53, 54]. Wang et al. developed a glucose-assisted Cu(I)/Cu(II) redox-coupled hydrogen production system, utilizing glucose to reduce Cu(II) to Cu(I), thereby achieving a redox cycle. In the electrochemical process, Cu(I) is oxidized to Cu(II), while glucose reacts with $Cu(OH)_2$ on the electrode surface, rapidly reducing it back to

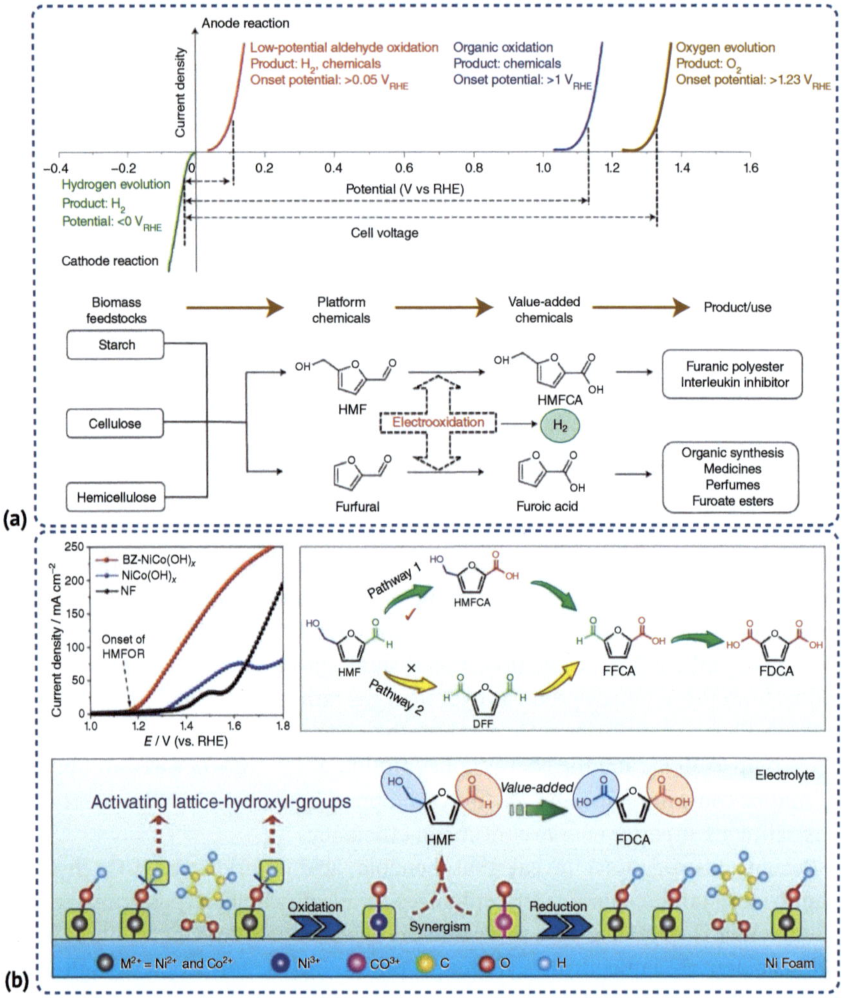

Figure 10.3 (a) Low-potential oxidation of aldehydes coupled with cathodic reactions for hydrogen production. *Source:* [13] / Springer Nature. (b) Benzoic acid ligand hybrid Ni(Co) hydroxide enhances the selectivity and efficiency of HMF electrooxidation. *Source:* [18] / John Wiley & Sons.

Cu(I) [55]. Experiments show that this system requires only 0.92 V to achieve a current density of 100 mA cm^{-2}, with an electricity consumption of 2.2 kWh per cubic meter of hydrogen, significantly lower than the 4.5 kWh required for conventional water electrolysis. This demonstrates superior energy efficiency and nearly 100% Faradaic efficiency.

From the perspective of regulating the coordination environment, a solid solution of $Ni_{0.67}Cr_{0.33}O$ with asymmetric Ni—O—Cr sites was synthesized to enhance the protonation rate during alcohol electrooxidation, thereby improving the efficiency of electrochemical devices. The alloying of NiO with chromium optimized the coordination environment of the catalytic sites, significantly increasing both

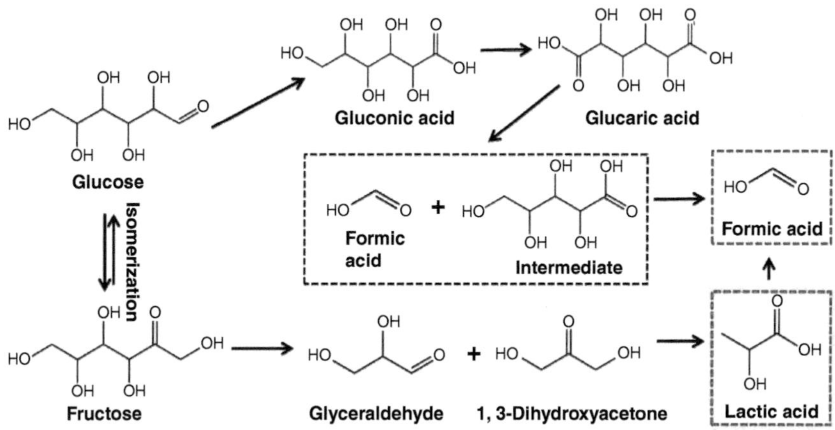

Figure 10.4 Schematic diagram of glucose electrooxidation: Generation of glucose acid, glucaric acid, lactic acid, and formic acid from glucose. *Source:* [27] / Springer Nature.

catalytic activity and stability while suppressing metal dissolution [56]. Studies indicate that the rapid protonation process during alcohol oxidation and the enhanced electron transfer mechanism are key factors in improving catalytic performance, providing new insights for enhancing the performance of electrocatalysts.

10.2.3.4 Lignin

Lignin is a complex polymer primarily found in plant cell walls, where it, along with cellulose and hemicellulose, provides structural support to plants [57]. It has a highly irregular structure and is biodegradable, although its degradation process is relatively slow. Its abundance and renewability make it an important resource for research in sustainable chemistry and energy.

Through electrocatalytic methods, the primary alcohol groups in lignin derivatives can be effectively oxidized to carboxylates. However, due to the high dissociation energy associated with the electrocatalytic oxidation cleavage of C(OH)–C, careful consideration must be given to the electrooxidation functional groups of the lignin derivatives and the design of the catalysts. For example, Duan's group developed a manganese-doped cobalt oxyhydroxide (MnCoOOH) catalyst that selectively converts lignin derivatives into carboxylates through electrochemical oxidation reactions, achieving yields as high as 80–99% [36]. The catalyst demonstrated good stability over 200 hours of operation and significantly increased current density during the oxidation reactions. They revealed how alkaline electrolytes promote the deprotonation of secondary alcohols and ketones, forming reactive nucleophiles that are oxidized by electrophilic oxygen species on the catalyst surface, thereby providing a sustainable pathway for the valorization of lignin. Furthermore, to address the issues of poor selectivity and low yields in electrocatalytic C—C bond cleavage, they introduced a novel electrocatalyst, atomically dispersed Pt-N_3C_1 sites, which can efficiently and selectively cleave C—C bonds in lignin model compounds under mild conditions [34]. The results show that the Pt_1/N-CNTs catalyst achieved a substrate conversion rate of 99% and a benzaldehyde yield of 81% in a

β-O-4 model compound, significantly outperforming traditional electrocatalysts. The article also discusses the synthesis methods, structural characteristics, and reaction mechanisms of the catalyst, highlighting that its exceptional performance is due to the Pt-N_3C_1 sites facilitating the formation of key C_β radical intermediates, thus enabling the selective cleavage of the C_α—C_β bond. This study provides new opportunities for the green and sustainable electrochemical conversion of lignin.

10.3 Cathodic Reaction in Biomass Electrooxidation System

Biomass-derived electrooxidation reactions (ECO) are a green and efficient electrocatalytic technology that utilizes biomass derivatives (such as HMF, furfural, methanol, etc.) as anodic reactants to produce high-value-added chemicals, while coupling with cathodic reactions (such as HER) to achieve efficient energy utilization [8, 9, 27]. Biomass electrooxidation reactions are typically coupled with a variety of cathodic reactions to achieve synergistic effects and efficient resource utilization. Among them, HER is the most common cathodic reaction. By coupling with the anodic oxidation of HMF to produce FDCA (2,5-furandicarboxylic acid), HER can efficiently generate hydrogen while significantly reducing the energy consumption of electrolysis [17]. Additionally, the coupling of CO_2RR with biomass oxidation can convert CO_2 into high-value-added chemicals such as formic acid and carbon monoxide, achieving carbon resource recycling [24, 58]. NRR and nitrate reduction reactions (NO_3-RR) have also gradually attracted attention, as their coupling with biomass oxidation reactions can produce ammonia or other nitrogen-containing compounds, offering new pathways for nitrogen cycling [25, 59]. These coupled reactions demonstrate the broad application prospects of biomass electrooxidation technology in energy and chemical production.

In recent years, nickel-based catalysts have achieved significant progress in biomass electrooxidation coupled reactions. Studies have shown that nickel-based catalysts [such as Ni_2P, $Ni(OH)_2$, etc.] exhibit high selectivity and faradaic efficiency (FE close to 100%) in reactions such as HMF oxidation, furfural oxidation, and methanol oxidation [37, 60, 61]. Through strategies such as heterojunction design, doping modifications, and defect engineering, the activity and stability of catalysts have been significantly enhanced. However, industrial applications still face many challenges, including the long-term stability of catalysts, the selective oxidation of complex biomass molecules, and the integrated optimization of reaction systems. Future research should focus on the design and large-scale preparation of catalysts while exploring more efficient synergistic mechanisms between anodic and cathodic reactions to drive the practical application of biomass electrooxidation coupling technology.

10.3.1 Coupled Cathodic HER

Biomass electrooxidation coupled with HER at the cathode is a green and efficient electrocatalytic technology. It involves the selective oxidation of biomass derivatives at the anode while simultaneously producing hydrogen at the cathode, achieving efficient energy utilization and the production of high-value-added chemicals [40, 62]. This technology is considered a sustainable alternative to traditional water electrolysis, as it significantly reduces energy consumption while increasing the yield of both chemicals and hydrogen. For example, over 7 billion tons of plastic have been produced globally, with 79% discarded, posing serious environmental and biological threats. Traditional plastic recycling methods, such as mechanical recycling, are inefficient, achieving a recycling rate of less than 10%, and often result in poor-quality recycled materials. Chemical recycling is considered a more promising alternative, capable of transforming waste into high-value chemicals and fuels.

Duan and his colleagues explored the electrocatalytic upgrading of polyethylene terephthalate (PET) plastics, aiming to convert them into valuable commercial chemicals, such as potassium formate and terephthalic acid, as well as hydrogen fuel, as shown in Figure 10.5 [14]. Through an electrocatalytic upstream recycling

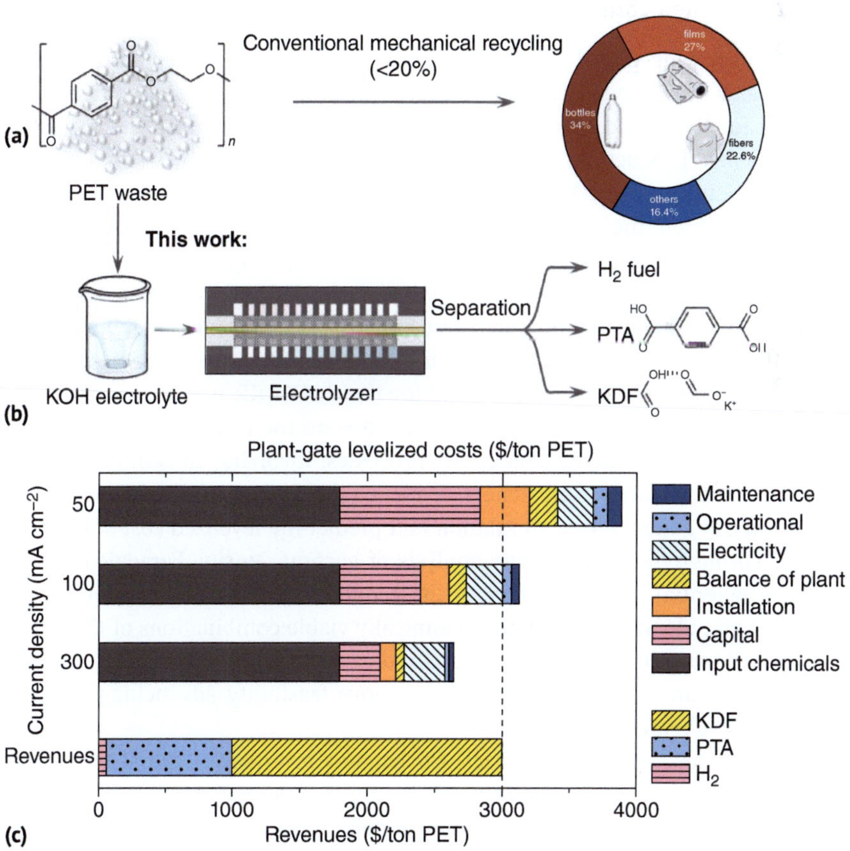

Figure 10.5 Conceptual design. (a) Traditional approach to PET recycling. (b) Electrocatalytic upcycling of PET into commodity chemicals and hydrogen fuel (Route I). (c) Technoeconomic analysis (TEA) of Route I across varying current densities [14]. *Source:* Ref. [14], Springer Nature, CC BY 4.0.

method for PET, the researchers developed a nickel-modified cobalt phosphide ($CoNi_{0.25}P$) electrocatalyst. This catalyst can selectively oxidize ethylene glycol (EG) from PET into formate at high current densities (>100 mA cm^{-2}) and with high Faradaic efficiency (>80%) in a membrane electrode assembly reactor. A preliminary techno-economic analysis was also conducted, indicating that the process is commercially viable. The study concluded that this electrocatalytic upstream recycling method for PET provides a sustainable approach to converting waste PET into value-added products, with promising economic potential under high current densities.

Current research primarily focuses on developing efficient and stable anode catalysts (such as nickel-based or ruthenium-based catalysts) and optimizing reaction conditions. Through the precise coupling of anodic and cathodic reactions, it is possible to achieve both the resource utilization of waste and the production of clean energy, thereby advancing the development of green chemistry and energy technologies.

10.3.2 Coupled Cathodic CO_2RR

With the growing global focus on reducing greenhouse gas emissions, carbon capture and utilization (CCU) technologies are gaining increasing attention, particularly for high-concentration carbon dioxide emission sources, such as power plants, cement production, and petrochemical facilities. The conversion of CO_2 into valuable chemicals can establish a carbon-neutral cycle [63].

In this research, the coupling of electrochemical CO_2RR with organic oxidation reactions (OOR) is regarded as a promising technology (Figure 10.6). This coupling not only enhances the economic efficiency of electrochemical processes but also increases the value of the products by using organic substances (such as biomass derivatives) as raw materials for cathodic reactions [19]. The oxidation of biomass derivatives, such as HMF, can be combined with CO_2RR to produce high-value-added chemicals like FDCA. This report focuses on the techno-economic analysis (TEA) of 295 electrochemical coupling processes, involving 16 cathodic reactions and 18 anodic reactions. Researchers developed a fully automated process synthesis framework to guide process simulation and predict the levelized cost of chemicals (LCC). Through global sensitivity analysis of current density, Faradaic efficiency, and overpotential, the study reveals the economic impact of different coupling reactions and identifies potential economically viable combinations of CO_2RR and organic oxidation reactions. The results indicate that coupling biomass oxidation with CO_2RR can significantly improve economic feasibility, advancing the production of clean chemicals.

Gu et al. developed a defect-rich Cu catalyst to enhance the selectivity and efficiency of the electrochemical CO_2RR toward C_{2+} alcohol products [64]. The catalyst was synthesized in a CO-rich environment, forming a structure with stepped sites that enabled high surface coverage of adsorbed *CO intermediates, thereby promoting alcohol production. Using this defect-rich Cu catalyst, the CO_2RR achieved a Faradaic efficiency of approximately 70%, with partial current densities exceeding

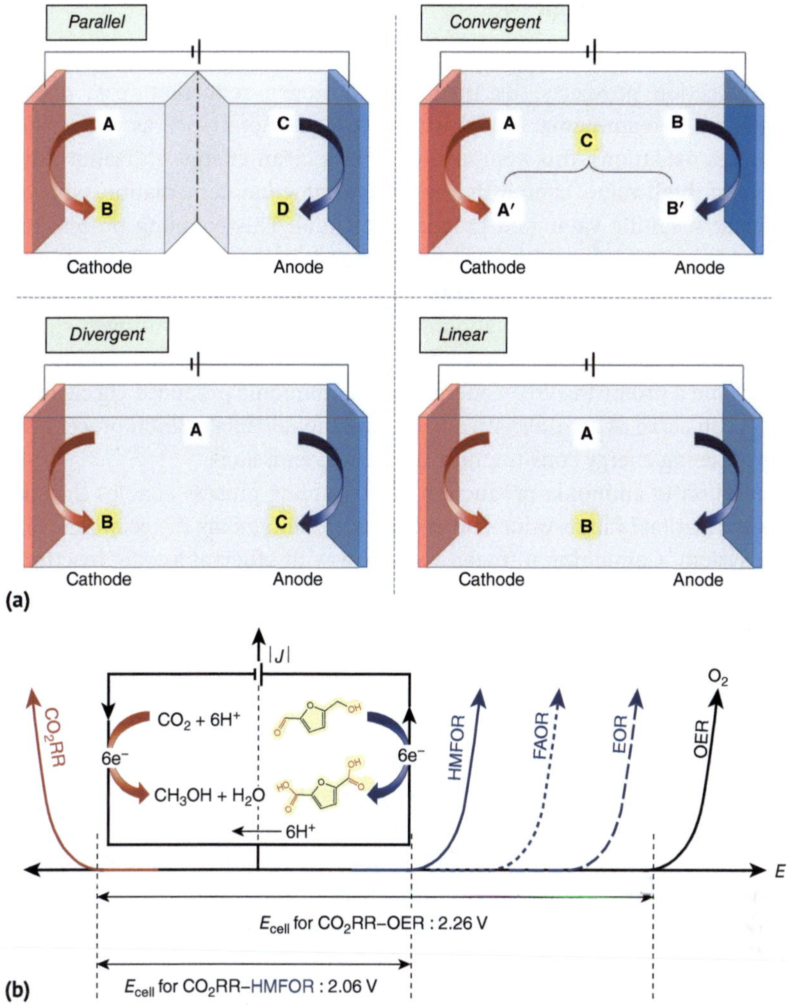

Figure 10.6 Schematic representation of the electrochemical coproduction system. (a) Four paired electrolysis configurations: parallel, convergent, divergent, and linear. (b) I–V curves and the corresponding potentials required for the cathode and anode during electrolysis. *Source:* Ref. [19], Royal Society of Chemistry, CC BY 4.0.

100 mA cm^{-2} in both flow cells and membrane electrode assembly (MEA) electrolyzers. Moreover, during continuous operation for 30 hours, a stable alcohol Faradaic efficiency of around 60% was maintained, with a C_{2+} alcohol productivity of approximately 500 mg cm^{-2} of catalyst. Through density functional theory (DFT) calculations and in situ characterization techniques, the study further revealed how the presence of defect sites enhances the local concentration of *CO intermediates, thereby improving the selectivity for alcohol production from CO_2. This research provides new insights for catalyst design in electrochemical CO_2 reduction, with significant potential for practical applications.

10.3.3 Coupled Cathodic NRR

Coupling cathodic nitrogen reduction is an electrochemical technology with extensive application prospects. By integrating nitrogen reduction (e.g., converting nitrogen gas into ammonia, N_2RR) with anodic reactions (such as water oxidation or biomass oxidation), this approach combines clean energy utilization, the production of high-value chemicals, and environmental remediation, showcasing significant scientific value and practical potential. This coupling process offers a sustainable pathway for clean energy and chemical production. Through cathodic nitrogen reduction, electrical energy can be efficiently converted into chemical energy, enabling the transformation of nitrogen gas or waste nitrogen-containing compounds (e.g., nitrates) into ammonia [65]. As an important industrial raw material and a promising green energy carrier, ammonia produced via clean energy systems can serve as an alternative to the traditional Haber-Bosch process, significantly reducing energy consumption and carbon emissions.

In addition to ammonia production, this coupling process enables the simultaneous generation of high-value-added chemicals, enhancing the economic viability of the system. Combining nitrogen reduction with efficient anodic reactions also lowers the energy input required for the overall electrolysis process. The slow kinetics of nitrogen reduction typically demand high operating potentials, but coupling the reaction with a highly efficient anodic process can reduce the system's voltage requirements, minimizing energy losses and improving overall energy efficiency. Furthermore, this technology contributes to sustainable nitrogen cycling and environmental management. By reducing nitrogen gas into ammonia or transforming nitrogen-containing pollutants such as nitrates into valuable products, it enables the recycling of waste resources. This approach helps mitigate industrial pollution, protects the environment, and enhances the sustainability of the nitrogen cycle, making it a promising solution for future energy and environmental challenges.

Qin et al. developed a novel nitrogen and boron co-doped porous carbon catalyst and applied it to electrochemical nitrogen reduction and biomass-derived compound oxidation [25]. The catalyst was synthesized via high-temperature pyrolysis and demonstrated a Faradaic efficiency of 15.2% and an ammonia production rate of 21.3 $\mu g\ h^{-1}\ mg^{-1}$ in the electrochemical nitrogen reduction reaction. Additionally, for the first time, this catalyst achieved the oxidation of 5-hydroxymethylfurfural without the use of metal-based catalysts, with a conversion efficiency of 71% and a product yield of 57%. The study revealed that nitrogen and boron co-doping not only enhanced the catalyst's specific surface area and porous structure but also formed unsaturated Lewis pairs and point defects, which improved the adsorption and activation of substrate molecules. These features endowed the catalyst with excellent performance in electrocatalytic reactions, providing new insights and directions for the design of future metal-free catalysts.

As shown in Figure 10.7, a cobalt phosphide (CoP) electrocatalyst supported on a carbon nanosheet array was developed to achieve an efficient process for electrochemical nitrate reduction to ammonia [66]. Cobalt-based double hydroxide (CoCo-LDH) was synthesized on carbon nanosheets via electrodeposition, followed

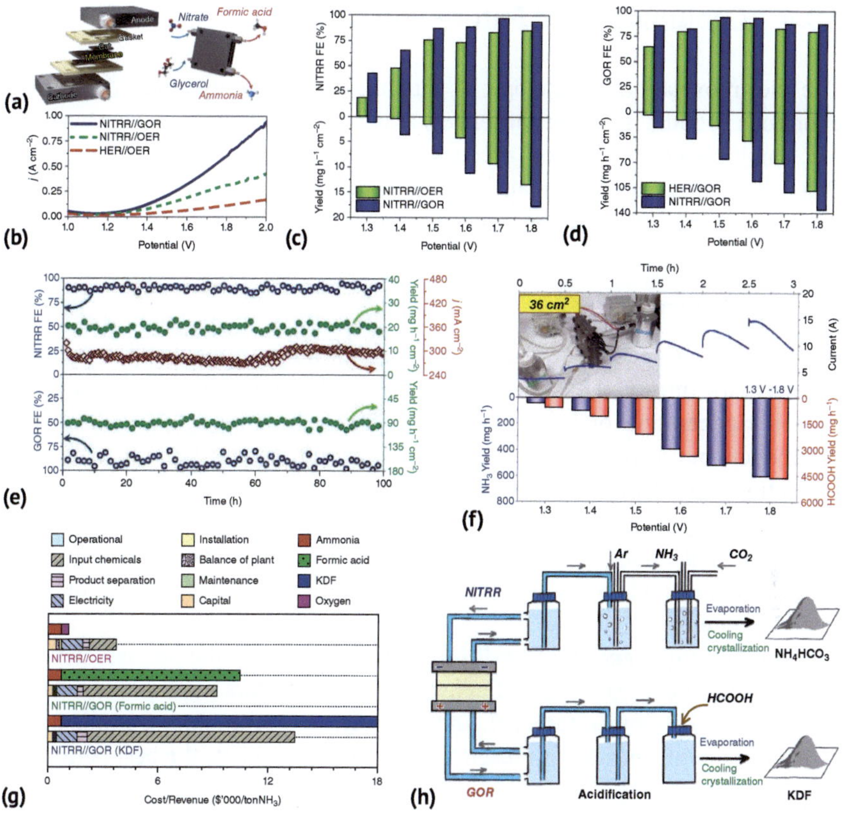

Figure 10.7 Two-electrode NITRR//GOR system and techno-economic evaluation. (a) Diagram of the MEA flow reactor used for NITRR//GOR. (b) Polarization curves comparing NITRR//GOR, NITRR//OER, and HER//OER within the MEA flow reactor. (c) Faradaic efficiency of NITRR and ammonia production rate as functions of potential for both NITRR//GOR and NITRR//OER systems. (d) Faradaic efficiency of GOR and formic acid production rate as functions of potential for NITRR//GOR and HER//GOR systems. (e) Stability testing of the NITRR//GOR system at 1.6 V. (f) Current–time (i–t) curves for the scaled-up reaction system operating from 1.3 to 1.8 V, along with ammonia and formic acid production rates at different potentials (inset: image of the scaled-up reactor used in the experiments). (g) Techno-economic comparison between NITRR//GOR and NITRR//OER systems. (h) Illustration of the production chain for generating high-value-added products. *Source:* Reproduced with permission from Ref. [66], Elsevier.

by phosphidation to form CoP. The catalyst demonstrated an outstanding Faradaic efficiency (95.1%) and ammonia yield (43.9 mg h^{-1} cm^{-2}), as well as excellent stability at −1.0 V vs. RHE. In addition, the study proposed an electrocatalytic system coupling nitrate reduction with glycerol oxidation, which significantly enhanced ammonia production (15.2 mg h^{-1} cm^{-2}) at a cell voltage of 1.8 V while simultaneously producing high-value-added formate (111.4 mg h^{-1} cm^{-2}). A preliminary techno-economic analysis indicated that this coupled system is economically viable, providing a sustainable new strategy for ammonia synthesis and utilization.

10.4 Challenges and Future Perspectives

The electrooxidation technology of biomass-derived compounds demonstrates enormous potential in renewable energy and chemical production. However, it still faces multiple challenges and avenues for development in practical applications. Below is a summary of the main challenges in the electrooxidation process, along with proposed future directions for research.

10.4.1 Challenges

***Selectivity and Activity of Catalysts*:** Current catalysts often exhibit insufficient selectivity during the biomass derivatives electrooxidation process, leading to the generation of by-products. This not only reduces the yield of target products but also increases the difficulty in subsequent separation and purification. For example, the selectivity and efficiency of HMF towards FDCA. Additionally, many catalysts may lose their activity during reactions, further diminishing catalytic efficiency. Therefore, the development of catalysts with high stability and high activity remains a significant challenge for technological advancement.

***Optimization of Reaction Conditions*:** The efficiency of the biomass derivatives electrooxidation reaction is influenced by various factors such as electrode materials, reaction temperature, pH, and electrolyte concentration. Optimizing these parameters under different reaction conditions to achieve the best performance is a complex task.

***Understanding Reaction Mechanisms*:** The current understanding of the mechanisms of electrooxidation reactions of biomass-derived compounds is still inadequate, particularly regarding the reaction pathways and formation of intermediates under different catalysts and reaction conditions. This is mainly related to multiple electron transfers and the oxidation order of different functional groups. The adoption of more in situ characterization techniques (such as in situ infrared spectroscopy and in situ Raman spectroscopy) will help to monitor the reaction processes in real time and reveal the reaction mechanisms.

***Barriers to Industrial Applications*:** While there has been progress in laboratory research, the broader application of biomass derivatives electrooxidation technology in industrial settings still encounters numerous technical and economic barriers, including the design of reactors, product separation, energy efficiency and cost-effectiveness. These factors significantly influence the large-scale promotion of biomass derivatives electrooxidation technology.

10.4.2 Future Perspectives

***Development of Novel Biomass Derivative Electrooxidation Catalysts*:** Future research should focus on developing new catalysts with high selectivity and high activity, particularly non-noble metal catalysts and multi-metal catalysts, to reduce costs and improve catalytic efficiency. Advances in nanotechnology can facilitate the design of catalysts with special structures (such as porous structures, nanoparticles, etc.) to enhance their surface area and reactivity.

Fine-tuning of Reaction Conditions: Researchers should strive to finely control reaction conditions, including optimizing electrolyte types, concentrations, pH, and current density, to achieve optimal reaction performance. For example, the selectivity of HMF at different pH, or the side reaction of OER at high current densities. By integrating computational simulations with experimental results, a deeper understanding of how various reaction conditions affect reaction pathways and product selectivity can be achieved.

In-depth Mechanistic Studies: A variety of characterization techniques should be employed to delve into the mechanisms of biomass derivatives electrooxidation reactions, particularly the formation and transformation of intermediates. This will provide robust theoretical support for catalyst design and reveal the microscopic mechanisms of reactions through combined theoretical calculations (such as DFT) and experimental findings.

Facilitating Industrial Applications: Future research should prioritize the industrial application of electrooxidation technology by developing reactors and processes suitable for large-scale production. This includes the electrooxidation/electroreduction of biomass derivatives to produce high-value chemicals, or coupling reaction systems for cathodes/anodes. Collaborating with industry will help promote the practical application of electrooxidation technology in biomass conversion and chemical production.

In summary, although the research on electrooxidation of biomass-derived compounds encounters various challenges, ongoing technological innovations, detailed mechanistic investigations, and economic evaluations hold the promise of achieving more efficient and cost-effective electrooxidation processes, providing new solutions for the utilization of renewable resources.

References

1 Akhade, S.A., Singh, N., Gutiérrez, O.Y. et al. (2020). Electrocatalytic hydrogenation of biomass-derived organics: a review. *Chem. Rev.* 120: 11370–11419.

2 Tian, C., Dorakhan, R., Wicks, J. et al. (2024). Progress and roadmap for electro-privileged transformations of bio-derived molecules. *Nat. Catal.* 7: 350–360.

3 Ragauskas, A.J., Beckham, G.T., Biddy, M.J. et al. (2014). Lignin valorization: improving lignin processing in the biorefinery. *Science* 344: 1246843.

4 Li, G., Wang, R., Pang, J. et al. (2024). Production of renewable hydrocarbon biofuels with lignocellulose and its derivatives over heterogeneous catalysts. *Chem. Rev.* 124: 2889–2954.

5 Tuck, C.O., Pérez, E., Horváth, I.T. et al. (2012). Valorization of biomass: deriving more value from waste. *Science* 337: 695–699.

6 Tuck, C.O., Pérez, E., Horváth, I.T. et al. (2015). Reductive lignocellulose fractionation into soluble lignin-derived phenolic monomers and dimers and processable carbohydrate pulps. *Energy Environ. Sci.* 8: 1748–1763.

7 Liao, Y., Koelewijn, S.F., Van den Bossche, G. et al. (2020). A sustainable wood biorefinery for low-carbon footprint chemicals production. *Science* 367: 1385–1390.

8 Fan, Y., Zou, Y., and Wang, S. (2023). Electrocatalytic oxidation of biomass derived molecules. *Sci. Bull.* 68: 2695–2699.

9 Sun, H., Xu, X., Fei, L. et al. (2024). Electrochemical oxidation of small molecules for energy-saving hydrogen production. *Adv. Energy Mater.* 14: 2401242.

10 Yang, Y. and Mu, T. (2021). Electrochemical oxidation of biomass derived 5-hydroxymethylfurfural (HMF): pathway, mechanism, catalysts and coupling reactions. *Green Chem.* 23: 4228–4254.

11 Liu, J., Li, Z., Lv, C. et al. (2024). Electrocatalytic upgrading of nitrogenous wastes into value-added chemicals: a review. *Mater. Today* 73: 208–259.

12 Luo, H., Barrio, J., Sunny, N. et al. (2021). Progress and perspectives in photo- and electrochemical-oxidation of biomass for sustainable chemicals and hydrogen production. *Adv. Energy Mater.* 11: 202101180.

13 Wang, T., Tao, L., Zhu, X. et al. (2021). Combined anodic and cathodic hydrogen production from aldehyde oxidation and hydrogen evolution reaction. *Nat. Catal.* 5: 66–73.

14 Zhou, H., Ren, Y., Li, Z. et al. (2021). Electrocatalytic upcycling of polyethylene terephthalate to commodity chemicals and H_2 fuel. *Nat. Commun.* 12: 4679.

15 Xiang, S., Dong, L., Wang, Z.Q. et al. (2022). A unique Co@CoO catalyst for hydrogenolysis of biomass-derived 5-hydroxymethylfurfural to 2,5-dimethylfuran. *Nat. Commun.* 13: 3657.

16 Cha, H.G. and Choi, K.S. (2015). Combined biomass valorization and hydrogen production in a photoelectrochemical cell. *Nat. Chem.* 7: 328–333.

17 Lei, C., Chen, Z., Jiang, T. et al. (2024). Ultra-dense supported ruthenium oxide clusters via directed ion exchange for efficient valorization of 5-hydroxymethylfurfural. *Angew. Chem. Int. Edition* 63: 202319642.

18 Liu, X., Wang, X., Mao, C. et al. (2024). Ligand-hybridization activates lattice-hydroxyl-groups of $NiCo(OH)_x$ nanowires for efficient electrosynthesis. *Angew. Chem. Int. Ed.* 63 (41): e202408109.

19 Na, J., Seo, B., Kim, J. et al. (2019). General technoeconomic analysis for electrochemical coproduction coupling carbon dioxide reduction with organic oxidation. *Nat. Commun.* 10: 5193.

20 You, B., Liu, X., Jiang, N. et al. (2016). A general strategy for decoupled hydrogen production from water splitting by integrating oxidative biomass valorization. *J. Am. Chem. Soc.* 138: 13639–13646.

21 Wang, F., Gu, Y., Tian, B. et al. (2023). Spinel-derived formation and amorphization of bimetallic oxyhydroxides for efficient electrocatalytic biomass oxidation. *J. Phys. Chem. Lett.* 14: 2674–2683.

22 Li, W., Jiang, N., Hu, B. et al. (2018). Electrolyzer design for flexible decoupled water splitting and organic upgrading with electron reservoirs. *Chem.* 4: 637–649.

23 Zhang, P., Sheng, X., Chen, X. et al. (2019). Paired electrocatalytic oxygenation and hydrogenation of organic substrates with water as the oxygen and hydrogen source. *Angew. Chem. Int. Ed.* 58: 9155–9159.

24 Bi, J., Zhu, Q., Guo, W. et al. (2022). Simultaneous CO_2 reduction and 5-hydroxymethylfurfural oxidation to value-added products by electrocatalysis. *ACS Sustain. Chem. Eng.* 10: 8043–8050.

25 Qin, Q., Heil, T., Schmidt, J. et al. (2019). Electrochemical fixation of nitrogen and its coupling with biomass valorization with a strongly adsorbing and defect optimized boron-carbon-nitrogen catalyst. *ACS Appl. Energy Mater.* 2: 8359–8365.

26 Wang, Y., He, H., Lv, H. et al. (2024). Two-dimensional single-crystalline mesoporous high-entropy oxide nanoplates for efficient electrochemical biomass upgrading. *Nat. Commun.* 15: 6761.

27 Ge, R., Li, J., and Duan, H. (2022). Recent advances in non-noble electrocatalysts for oxidative valorization of biomass derivatives. *Sci. China Mater.* 65: 3273–3301.

28 Liu, J., Fu, G., Liao, Y. et al. (2024). Electrochemical conversion of small organic molecules to value-added chemicals and hydrogen/electricity without CO_2 emission: electrocatalysts, devices and mechanisms. *eScience* 5 (1): 100267.

29 Carneiro, J. and Nikolla, E. (2019). Electrochemical conversion of biomass-based oxygenated compounds. *Ann. Rev. Chem. Biomol. Eng.* 10: 85–104.

30 Fu, Q., Xie, W., Yang, L. et al. (2024). Recent advances in electrocatalytic upgrading of biomass-derived furfural. *Chem. Eng. J.* 485: 150083.

31 Pei, A., Wang, P., Zhang, S. et al. (2024). Enhanced electrocatalytic biomass oxidation at low voltage by Ni^{2+}-O-Pd interfaces. *Nat. Commun.* 15: 5899.

32 Li, J., Ji, K., Li, B. et al. (2023). Rechargeable biomass battery for electricity storage/generation and concurrent valuable chemicals production. *Angew. Chem. Int. Ed.* 62: 202304852.

33 Li, Z., Yan, Y., Xu, S.M. et al. (2022). Alcohols electrooxidation coupled with H_2 production at high current densities promoted by a cooperative catalyst. *Nat. Commun.* 13: 147.

34 Cui, T., Ma, L., Wang, S. et al. (2021). Atomically dispersed Pt-N_3C_1 sites enabling efficient and selective electrocatalytic C–C bond cleavage in lignin models under ambient conditions. *J. Am. Chem. Soc.* 143: 9429–9439.

35 Ge, R., Wang, Y., Li, Z. et al. (2022). Selective electrooxidation of biomass-derived alcohols to aldehydes in a neutral medium: promoted water dissociation over a nickel-oxide-supported ruthenium single-atom catalyst. *Angew. Chem. Int. Ed.* 61: 202200211.

36 Zhou, H., Li, Z., Xu, S.M. et al. (2021). Selectively upgrading lignin derivatives to carboxylates through electrochemical oxidative C(OH)–C bond cleavage by a Mn-doped cobalt oxyhydroxide catalyst. *Angew. Chem. Int. Ed.* 60: 8976–8982.

37 Zhou, H., Li, Z., Xu, S.M. et al. (2019). Integrating hydrogen production with aqueous selective semi-dehydrogenation of tetrahydroisoquinolines over a Ni_2P bifunctional electrode. *Angew. Chem. Int. Ed.* 58: 12014–12017.

38 Prajapati, A., Govindarajan, N., Sun, W. et al. (2024). Mechanistic insights into the electrochemical oxidation of 5-hydroxymethylfurfural on a thin-film Ni anode. *ACS Catalysis* 14: 10122–10131.

39 Govindarajan, N., Xu, A., and Chan, K. (2022). How pH affects electrochemical processes. *Science* 375: 379–380.

40 Wang, W., Zhu, Y.B., Wen, Q. et al. (2019). Modulation of molecular spatial distribution and chemisorption with perforated nanosheets for ethanol electro-oxidation. *Adv. Mater.* 31: 1900528.

41 Barwe, S., Weidner, J., Cychy, S. et al. (2018). Electrocatalytic oxidation of 5-(hydroxymethyl)furfural using high-surface-area nickel boride. *Angew. Chem. Int. Ed.* 57: 11460–11464.

42 Fang, R., Dhakshinamoorthy, A., Li, Y. et al. (2020). Metal-organic frameworks for biomass conversion. *Chem. Soc. Rev.* 49: 3638–3687.

43 Bondue, C.J. and Koper, M.T.M. (2019). Electrochemical reduction of the carbonyl functional group: the importance of adsorption geometry, molecular structure, and electrode surface structure. *J. Am. Chem. Soc.* 141: 12071–12078.

44 Cao, X., Ding, Y., Chen, D. et al. (2024). Cluster-level heterostructure of PMo_{12}/Cu for efficient and selective electrocatalytic hydrogenation of high-concentration 5-hydroxymethylfurfural. *J. Am. Chem. Soc.* 146: 25125–25136.

45 Wu, L., Wu, Q., Han, Y. et al. (2024). Strengthening the synergy between oxygen vacancies in electrocatalysts for efficient glycerol electrooxidation. *Adv. Mater.* 36: 2401857.

46 Yan, Y., Zhou, H., Xu, S.M. et al. (2023). Electrocatalytic upcycling of biomass and plastic wastes to biodegradable polymer monomers and hydrogen fuel at high current densities. *J. Am. Chem. Soc.* 145: 6144–6155.

47 Liu, W.J., Xu, Z., Zhao, D. et al. (2020). Efficient electrochemical production of glucaric acid and H_2 via glucose electrolysis. *Adv. Mater.* 11: 265.

48 Rafaïdeen, T., Baranton, S., and Coutanceau, C. (2019). Pd-shaped nanoparticles modified by gold ad-atoms: effects on surface structure and activity toward glucose electrooxidation. *Front. Chem.* 7: 00453.

49 Zhao, B., Liu, J.W., Yin, Y.R. et al. (2019). Carbon nanofibers@NiSe core/sheath nanostructures as efficient electrocatalysts for integrating highly selective methanol conversion and less-energy intensive hydrogen production. *J. Mater. Chem. A* 7: 25878–25886.

50 Li, Y., Wei, X., Han, S. et al. (2021). MnO_2 electrocatalysts coordinating alcohol oxidation for ultra-durable hydrogen and chemical productions in acidic solutions. *Angew. Chem. Int. Ed.* 60: 21464–21472.

51 Li, Z., Li, X., Zhou, H. et al. (2022). Electrocatalytic synthesis of adipic acid coupled with H_2 production enhanced by a ligand modification strategy. *Nat. Commun.* 13: 5009.

52 Chen, Z.J., Dong, J., Wu, J. et al. (2023). Acidic enol electrooxidation-coupled hydrogen production with ampere-level current density. *Nat. Commun.* 14: 4210.

53 Ostervold, L., Bakovic, S.I.P., Hestekin, J. et al. (2021). Electrochemical biomass upgrading: degradation of glucose to lactic acid on a copper(II) electrode. *RSC Adv.* 11: 31208–31218.

54 Li, D., Huang, Y., Li, Z. et al. (2022). Deep eutectic solvents derived carbon-based efficient electrocatalyst for boosting H_2 production coupled with glucose oxidation. *Chem. Eng. J.* 430: 132783.

55 Zhang, Y., Zhou, B., Wei, Z. et al. (2021). Coupling glucose-assisted Cu(I)/Cu(II) redox with electrochemical hydrogen production. *Adv. Mater.* 33: 2104791.

56 Feng, Y., Wang, X., Ma, J. et al. (2024). A solid-solution with asymmetric Ni-O-Cr sites for boosting protonation toward anodic oxidation. *Adv. Energy Mater.* 14: 2401501.

57 Xiao, X., Sun, S., Qi, Y. et al. (2023). Selective electrocatalytic oxidation of C(OH)–C bond in lignin with Pt@CeO_2 toward the synthesis of benzoic acid. *Electrochim. Acta* 470: 143377.

58 Verma, S., Lu, S., and Kenis, P.J.A. (2019). Co-electrolysis of CO_2 and glycerol as a pathway to carbon chemicals with improved technoeconomics due to low electricity consumption. *Nat. Energy* 4: 466–474.

59 Li, C., Li, H., Zhang, B. et al. (2024). Efficient electrocatalytic oxidation of glycerol to formate coupled with nitrate reduction over Cu-doped NiCo alloy supported on nickel foam. *Angew. Chem. Int. Ed.* 63: e202411542.

60 Chen, W., Xie, C., Wang, Y. et al. (2020). Activity origins and design principles of nickel-based catalysts for nucleophile electrooxidation. *Chem* 6: 2974–2993.

61 Zhang, N., Zou, Y., Tao, L. et al. (2019). Electrochemical oxidation of 5-hydroxymethylfurfural on nickel nitride/carbon nanosheets: reaction pathway determined by in situ sum frequency generation vibrational spectroscopy. *Angew. Chem. Int. Ed.* 58: 15895–15903.

62 Xiang, K., Wu, D., Deng, X. et al. (2020). Boosting H_2 generation coupled with selective oxidation of methanol into value-added chemical over cobalt hydroxide@hydroxysulfide nanosheets electrocatalysts. *Adv. Funct. Mater.* 30: 201909610.

63 Ross, M.B., De Luna, P., Li, Y. et al. (2019). Designing materials for electrochemical carbon dioxide recycling. *Nat. Catal.* 2: 648–658.

64 Gu, Z., Shen, H., Chen, Z. et al. (2021). Efficient electrocatalytic CO_2 reduction to C_{2+} alcohols at defect-site-rich cu surface. *Joule* 5: 429–440.

65 Qin, Q., Zhao, Y., Schmallegger, M. et al. (2019). Enhanced electrocatalytic N_2 reduction via partial anion substitution in titanium oxide-carbon composites. *Angew. Chem. Int. Ed.* 58: 13101–13106.

66 Li, J., Li, H., Fan, K. et al. (2023). Electrocatalytic nitrate reduction to ammonia coupled with organic oxidation. *Chem. Catal.* 3: 100638.

11

Electrochemical CO_2 Reduction and Conversion

Jiapeng Ji[1,2], *Xiaolong Zhang*[1,2,*]

[1] *Faculty of Materials Science and Energy Engineering, Shenzhen University of Advanced Technology, Shenzhen, China*
[2] *Shenzhen Key Laboratory of Energy Materials for Carbon Neutrality, Shenzhen Institute of Advanced Technology, Chinese Academy of Sciences, Shenzhen, China*

*Corresponding author: xl.zhang@siat.ac.cn

11.1 Introduction

11.1.1 Importance of CO_2 Reduction in Climate Change and Sustainable Energy Production

Carbon dioxide (CO_2) reduction is a cornerstone of addressing climate change and advancing sustainable energy production. The escalating levels of CO_2 in the atmosphere, largely resulting from the combustion of fossil fuels, are a primary driver of global warming and its associated environmental crises, such as accelerating ice melt, rising sea levels, intensified extreme weather events, ocean acidification, and widespread biodiversity loss [1, 2]. To counteract these devastating impacts, the reduction of CO_2 emissions is imperative [3]. Sustainable energy production, encompassing renewable energy sources, such as solar, wind, geothermal, and hydropower, offers a promising pathway to reduce dependence on fossil fuels and thereby mitigate CO_2 emissions [4]. These renewable sources produce energy with significantly lower carbon footprints compared to conventional fossil fuel power generation [5]. Furthermore, advancements in carbon capture, utilization, and storage (CCUS) technologies provide additional means to capture and sequester CO_2 emissions from industrial processes and power plants, further enhancing efforts to curb CO_2 levels [6]. Through the combination of sustainable energy generation and CCUS technologies, society can make significant progress toward a more robust, sustainable, and low-carbon future (Figure 11.1) [7, 8].

Electrochemical Energy Conversion via Small Molecule Catalysis, First Edition.
Edited by Zhicheng Zhang and Yuanmiao Sun.

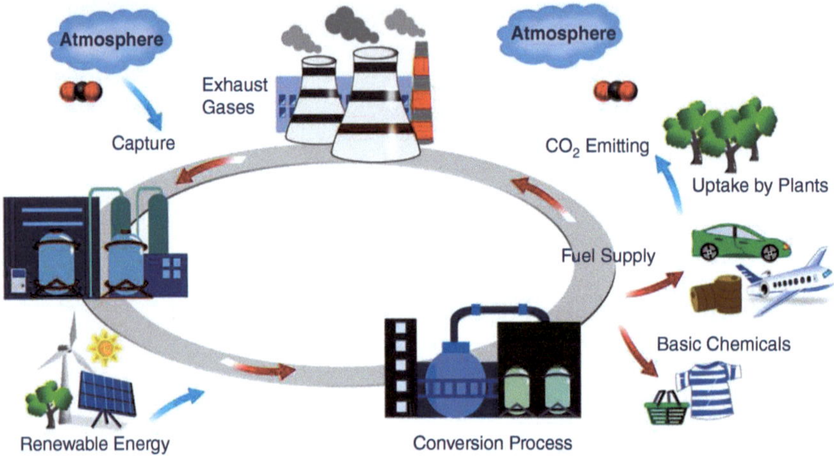

Figure 11.1 CO_2 recycling circular economy. *Source:* [8] / John Wiley & Sons.

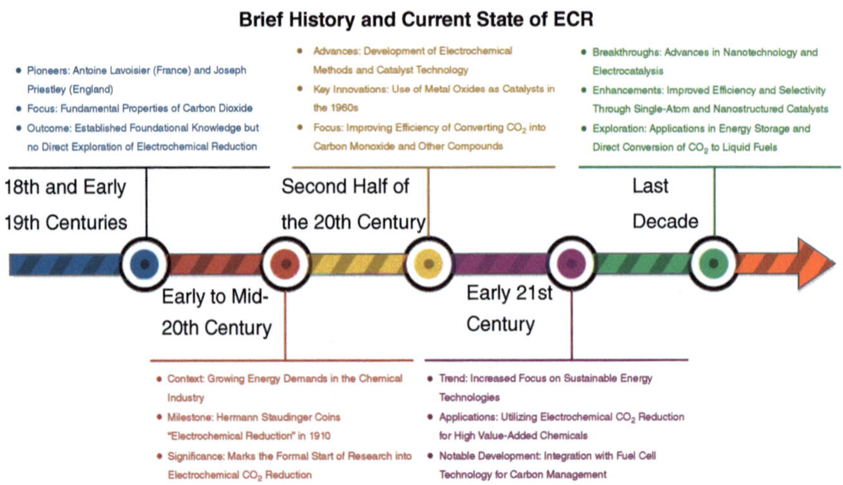

Figure 11.2 Brief history and current state of electrochemical CO_2 reduction research.

11.1.2 Brief History and Current State of Electrochemical CO_2 Reduction Research

The development of electrochemical carbon dioxide reduction (ECR) technology can be traced back to several key research stages that mark important advances from basic research to industrial applications [9, 10]. The following are the main historical nodes in the development of ECR technology (Figure 11.2).

In the 18th century, scientists began to investigate the properties of carbon dioxide, including the work of Lavoisier in France and Joseph Priestley in England, among others. Their research revealed the fundamental composition and properties of carbon dioxide, but at the time did not directly explore its application for electrochemical reduction.

The growing demand for renewable energy, driven by chemical industry expansion, has significantly increased research interest in electrochemical CO_2 reduction. For example, in 1910, the German chemist Hermann Staudinger first coined the term "electrochemical reduction," which marked the official beginning of research in the field.

Advances in electrochemical methods and catalyst design have propelled ECR research into a new phase of development. For example, in the 1960s, scientists began experimenting with using metal oxides as catalysts to improve the efficiency of converting carbon dioxide into carbon monoxide or other compounds.

As sustainable energy technologies progress, ECR has gained further attention as a promising strategy for atmospheric CO_2 mitigation. Especially with the development of fuel cell technology, ECR as a method to prepare high value-added chemicals (such as carbon monoxide and methane) has received widespread attention.

In recent years, with the development of nanotechnology and electrocatalysis, the efficiency and selectivity of ECR have been significantly improved [11]. For example, the use of single-atom catalysts (SACs) or nanostructured catalysts can significantly reduce the overpotential of the reaction and improve the reaction efficiency. In addition, the researchers explored the application of ECR in energy storage and conversion, such as the direct conversion of carbon dioxide into liquid fuels.

In short, the development of ECR technology is a process of continuous progress and deepening, involving cross-cooperation in multiple disciplinary fields. With the deepening of research and the advancement of technology, this field is expected to provide important support for solving the global climate change problem in the future.

11.2 Fundamentals of ECR

11.2.1 Thermodynamics and Kinetics of the CO_2 Reduction Reaction

The thermodynamics and kinetics of the CO_2 reduction reaction play a crucial role in understanding its feasibility and efficiency [12]. Thermodynamically, the reduction of CO_2 is challenging due to its high stability and the large amount of energy required to break the carbon—oxygen bonds. The reduction process can occur through various pathways, producing a range of products such as carbon monoxide (CO), methane (CH_4), formaldehyde (HCHO), methanol (CH_3OH), and ethylene (C_2H_4), each with its own specific thermodynamic potential and energy requirements [13]. The reduction potential for each product is influenced by factors such as the electrode material, the electrolyte used, and the reaction conditions [14]. Kinetically, the CO_2 reduction reaction is complex and involves multiple electron transfer steps, leading to the formation of various intermediates and side products. The reaction rate is influenced by the activity and stability of the catalyst, the mass transport of reactants and products, and the electrode potential [15]. Therefore, optimizing the catalyst design, electrode material, and reaction conditions is essential to enhance the kinetics of the CO_2 reduction reaction and improve the yield and selectivity of the desired products. Understanding the thermodynamics and kinetics of the CO_2 reduction reaction is critical for the development of efficient and selective catalysts and reactor designs, which are essential for the practical application of CO_2 reduction technologies.

11.2.2 Overview of Different Electrochemical Setups and Electrolytes Used for CO_2 Reduction

Electrochemical setups and electrolytes play a pivotal role in the reduction of CO_2, influencing the efficiency, selectivity, and stability of the reaction (Figure 11.3) [16, 17]. Various electrochemical setups have been developed to facilitate the reduction of CO_2, each with its unique advantages and challenges.

One common electrochemical setup for CO_2 reduction is the gas-diffusion electrode (GDE) system [18, 19]. In this setup, a porous electrode is immersed in an electrolyte solution, and CO_2 gas is supplied to the electrode through a gas-diffusion layer. The GDE system allows for efficient mass transport of CO_2 to the electrode surface, enhancing the reaction rate and selectivity. Additionally, the porous electrode structure provides a large surface area for catalyst deposition, which can

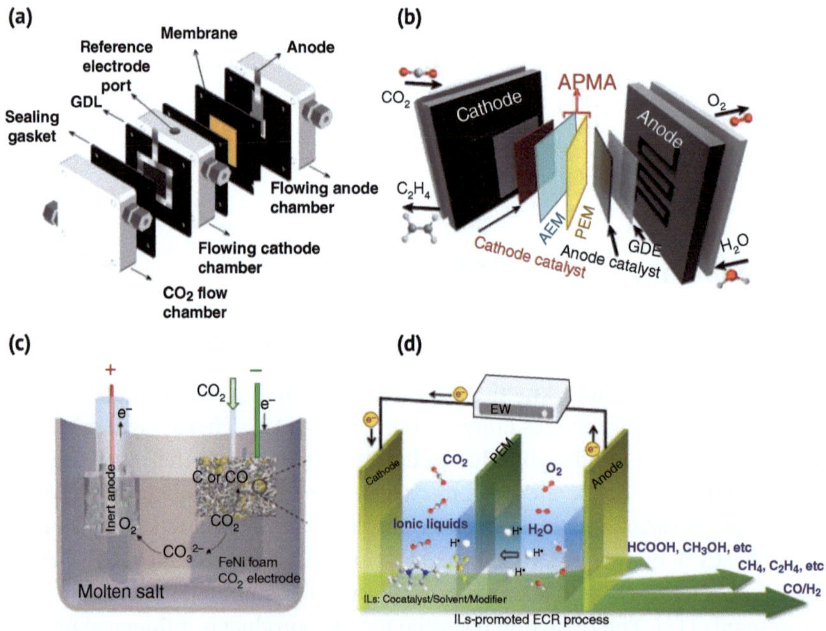

Figure 11.3 (a) Isometric view of a flow cell showing individual components. *Source:* [19] / American Chemical Society / CC-BY-NC-ND. (b) Schematic of a MEA system for ECR. *Source:* [21] / Springer Nature / CC BY 4.0. (c) The schematic diagram of CO_2 direct reduction with the FeNi foam electrode in molten salt. *Source:* Reproduced from [24] / with permission of Elsevier. (d) Illustration of ILs-promoted ECR. *Source:* [26] / American Chemical Society.

further improve the catalytic performance. Another electrochemical setup for CO_2 reduction is the membrane electrode assembly (MEA) system [20, 21]. In this setup, a catalyst layer is coated onto a gas-diffusion layer, which is then assembled with an ion-exchange membrane to form an MEA. The MEA system allows for the separation of the anode and cathode reactions, reducing the crossover of reactants and products. This can lead to improved reaction selectivity and efficiency.

In addition to the electrochemical setup, the choice of electrolyte is also crucial for the reduction of CO_2. Various electrolytes, including aqueous [22], molten salt [23, 24], and ionic liquid [25, 26] electrolytes, have been explored for this purpose. Aqueous electrolytes are the most commonly used due to their low cost, ease of handling, and high conductivity. However, they can suffer from issues such as corrosion of electrode materials and limited solubility of CO_2. Molten salt electrolytes offer high conductivity and stability at high temperatures, making them suitable for high-temperature CO_2 reduction reactions. However, they require high temperatures for operation, which can be energy-intensive and challenging to control. Ionic liquid electrolytes have emerged as a promising alternative due to their high CO_2 solubility, low volatility, and thermal stability. They can also be tailored to have specific properties, such as viscosity and ion conductivity, which can further optimize the CO_2 reduction reaction.

Overall, the choice of electrochemical setup and electrolyte is crucial for the successful reduction of CO_2. Each setup and electrolyte have its unique advantages and challenges, and the optimal choice depends on the specific reaction conditions, catalyst properties, and desired product selectivity. Future research in this area will likely focus on the development of novel electrochemical setups and electrolytes that can further enhance the efficiency, selectivity, and stability of CO_2 reduction reactions.

11.2.3 Introduction to the Concept of Overpotential and Its Significance in Electrochemical Reactions

In the realm of ECR, several key concepts are crucial for understanding the efficiency and performance of the process.

11.2.3.1 Faradaic Efficiency

Faradaic efficiency (FE) refers to the percentage of electrons that are used in a desired chemical reaction compared to the total number of electrons that flow through the electrochemical cell [27]. In the context of CO_2 reduction, this means measuring how many electrons are effectively converted into valuable products, such as carbon monoxide, methane, or ethylene, rather than being wasted in side reactions or producing unwanted byproducts like hydrogen gas. A high FE is indicative of a well-optimized electrochemical system where the majority of the input electrons are directed towards the formation of the desired products.

11.2.3.2 Overpotential

Overpotential represents the additional voltage that must be applied to drive an electrochemical reaction at a given rate beyond the thermodynamic potential required for the reaction to occur [28]. In CO_2 reduction, the overpotential is a measure of how much extra energy is needed to overcome the activation barriers and drive the conversion of CO_2 into valuable chemicals. Lower overpotentials are desirable because they indicate that less energy is wasted in the process, leading to more energy-efficient and cost-effective production of the desired products.

11.2.3.3 Current Density

Current density is also a critical parameter in ECR. It refers to the amount of electric current flowing through a unit area of the electrode surface [29]. High current densities are generally favorable because they allow for the production of larger quantities of the desired products in a shorter amount of time. However, increasing the current density can also lead to increased overpotentials and decreased Faradaic efficiencies if the system is not properly optimized. Therefore, finding the optimal current density that balances high productivity with low energy consumption and high FE is crucial for the economic viability of ECR technologies.

In summary, FE, overpotential, and current density are all essential metrics for evaluating the performance of ECR systems. By carefully monitoring and optimizing these parameters, researchers and engineers can develop more efficient and cost-effective technologies for converting CO_2 into valuable chemicals and fuels.

11.3 Catalysts for ECR

Catalysts for ECR play a pivotal role in enhancing the efficiency and selectivity of the process by lowering the activation barriers and accelerating the reaction rates of specific chemical transformations [30, 31]. These catalysts can be composed of a wide range of materials, including metals [32], metal oxides [33], and carbon-based materials [34–36], each offering unique properties and potential benefits for different reaction pathways. By carefully designing and engineering these catalysts, researchers aim to maximize the FE, minimize the overpotential, and achieve high current densities, ultimately leading to more sustainable and economically viable production of valuable chemicals and fuels from CO_2.

11.3.1 Metal-based Catalysts: Discussing the Role of Metals Such as Copper, Silver, and Gold in CO_2 Reduction

In ECR, metal catalysts play a crucial role. They effectively reduce the activation barrier for CO_2 molecule activation through their surface electronic structure and adsorption energy, thereby increasing reaction rates and optimizing selectivity. Different metals exhibit varying selectivity due to their unique electronic properties and catalytic activities (Figure 11.4) [37]. For example, some metals, such as copper, possess good C—C coupling capabilities, making them ideal for producing multicarbon products (C_{2+}), while others, like silver and zinc, tend to generate single-carbon products such as CO [38, 39]. Therefore, by judiciously selecting and optimizing metal catalysts, the final product distribution can be controlled, enabling precise synthesis of different fuels or chemicals. With advancements in

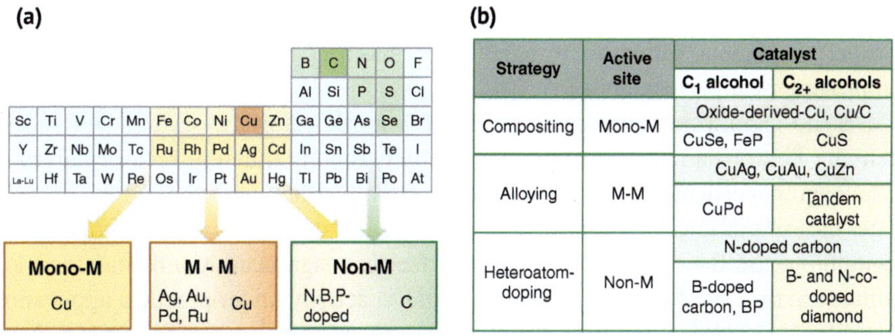

Figure 11.4 A product classification of metal catalysts for the electroreduction of CO_2. (a) The mainly reported element as component of electrocatalysts for ECR-to-alcohols; (b) component regulation strategy on electrocatalysts for ECR-to-alcohols (M denotes metal; the 'Catalyst' columns in panel (b) denote catalysts for ECR to different alcohols, where those in pale red area indicate the common catalysts active for both C_1 and C_{2+} alcohols, while catalysts in pale blue and yellow areas are active for C_1 and C_{2+} alcohols, respectively). *Source:* [37] / John Wiley & Sons.

nanotechnology and surface science, researchers can further fine-tune the surface structure, morphology, and crystal face exposure of metal catalysts, achieving significant enhancements in the catalytic performance of ECR [40, 41].

Recent research on ECR has made significant advancements in the development, synthesis, and performance optimization of various metal catalysts, including monometallic, alloy, and bimetallic catalysts. Copper (Cu) is one of the most valuable metals for research due to its high selectivity for C_{2+} products. But the reaction pathways of Cu catalysts are complex, leading to unstable product distribution and high sensitivity to reaction conditions, which creates difficulties in enhancing the selectivity and stability of Cu catalysts [42, 43]. By adjusting the morphology of Cu (such as nanoparticles, nanowires, and nanosheets) and crystal face exposure, researchers have found that certain specific morphologies and crystal planes of Cu exhibit higher catalytic activity and distinct selectivity [32, 44]. Besides Cu, other metals also exhibit unique catalytic performances. Catalysts like zinc (Zn) and silver (Ag) primarily tend to produce carbon monoxide and demonstrate high FE and selectivity. The CO generation capability of Zn shows remarkable stability, especially at moderate potentials, making it promising for industrial applications [45]. Non-precious metal catalysts such as bismuth (Bi) and indium (In) are more suitable for formic acid production, displaying excellent selectivity for formic acid generation, particularly at low potentials [46, 47]. Additionally, tin (Sn) also favors formic acid production and has advantages in water solubility and environmental friendliness, making it widely used in ECR research under neutral or near-neutral conditions [48]. In recent years, alloy catalysts have been a research hotspot. By combining two or more metals, alloy catalysts can leverage the synergistic effects between different metals to regulate catalytic performance. For example, Cu-Ag alloys can significantly improve CO generation efficiency, while Cu-Sn and Cu-Zn alloys show promising potential for selectively producing C_{2+} products [49, 50]. Moreover, alloy catalysts demonstrate significant advantages in enhancing catalyst stability and poison resistance. Thus, tuning the electronic structure and surface properties of metal catalysts through alloying has become an important strategy for improving ECR performance [51, 52].

Despite significant advancements in ECR metal catalysts, several bottlenecks remain. First, balancing selectivity and stability among different metal catalysts is challenging, particularly in the generation of multicarbon products, which necessitates the exploration of appropriate catalyst design strategies [53, 54]. Second, reaction conditions and the state of the catalyst surface can significantly influence reaction efficiency, making it crucial to maintain catalyst activity under practical application conditions [55, 56]. To address these challenges, one approach is to enhance mass transfer efficiency of reactants and products by constructing catalysts with hierarchical porous structures [57, 58]. Additionally, further improvements in the activity, selectivity, and stability of metal catalysts can be achieved through atomic-level regulation, loading of nanocatalytic materials, and optimization of electrolyte solutions [59, 60]. Furthermore, integrating advanced characterization techniques and theoretical simulations is expected to provide deep insights into the catalytic surface reaction mechanisms, guiding the rational design of catalysts [61, 62].

11.3.2 Nonmetal Catalysts: Exploring the Use of Carbon-based and Other Nonmetallic Materials

Nonmetal catalysts have unique applications in ECR. Compared to traditional metal catalysts, these catalysts are resource-rich, cost-effective, environmentally friendly, and free from precious metal components, which significantly reduces the economic cost of ECR and enhances its sustainability (Figure 11.5) [63–65]. Carbon-based catalysts, such as nitrogen-doped graphene (N-doped graphene) and nitrogen-doped carbon nanotubes (N-doped CNTs), have garnered widespread attention due to their high conductivity and abundant active sites [34–36]. Nitrogen-doped carbon materials effectively promote the activation and selective reduction of CO_2

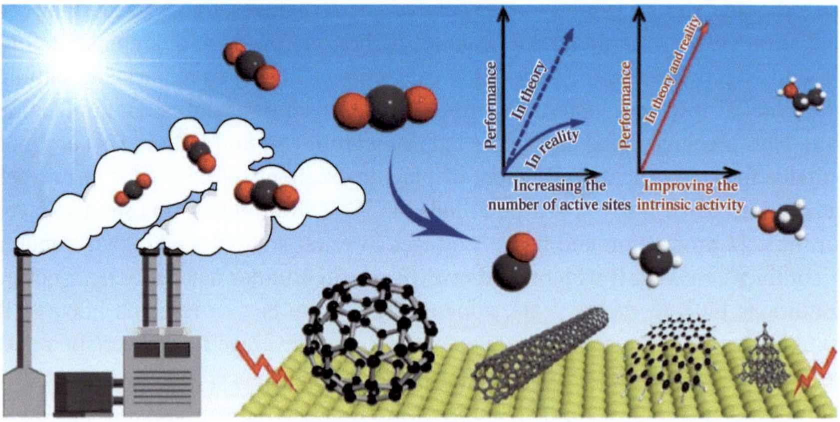

Figure 11.5 Rational design of carbon-based metal-free catalysts for electrochemical carbon dioxide reduction. *Source:* Reproduced from [65] / with permission of Elsevier.

by incorporating various nitrogen atom sites, such as pyridinic and pyrrolic nitrogen, into their structure [66]. These carbon-based materials typically produce CO at lower overpotentials and demonstrate high Faradaic efficiencies. Meanwhile, sulfur- and phosphorus-doped carbon materials show advantages in generating formic acid, while heteroatom doping can further optimize selectivity by tuning the electronic structure of carbon materials [67].

In addition to carbon-based catalysts, nonmetal catalysts such as molybdenum disulfide (MoS_2), carbon nitride (g-C_3N_4), and black phosphorus also exhibit unique catalytic properties. For example, molybdenum disulfide, due to its layered structure and abundant edge sites, shows good selectivity for CO generation and a high CO production rate at moderate potentials [68]. Carbon nitride, with its bandgap structure and high specific surface area, serves as a promising catalyst substrate for generating various reduction products [69]. Another class of promising nonmetal catalysts is black phosphorus, whose unique layered structure and high conductivity provide a favorable platform for CO_2 reduction. By constructing layered or defect structures, the active site density and reaction selectivity of black phosphorus can be significantly enhanced [70]. Additionally, phosphide materials such as nickel phosphide (Ni_2P) and cobalt phosphide (CoP) demonstrate excellent catalytic performance in formic acid production [71]. Therefore, the diversity of carbon-based and nonmetal catalysts offers broad application prospects in terms of selectivity and stability for ECR.

Despite the promising performance of nonmetal catalysts in ECR, several bottlenecks remain. First, the active sites of carbon-based and nonmetal catalysts are often limited, showing a trend of reduced activity at high current densities. Second, the selective control of different products poses significant challenges, as current catalyst designs have not fully achieved effective selectivity for multicarbon products. Additionally, the stability and durability of catalysts remain major limitations for further applications, as carbon-based and nonmetal materials may undergo structural reorganization or deactivation during prolonged reaction processes, adversely affecting their catalytic performance. To address these challenges, various strategies can be employed: on one hand, optimizing heteroatom doping and surface defect control can increase the density of active sites and enhance catalyst stability [72]; on the other hand, designing hierarchical porous structures and atomic-level modifications can promote mass transfer efficiency and selective product generation [73]. Furthermore, integrating theoretical calculations with experimental methods to study the reaction mechanisms of carbon-based and nonmetal catalysts may reveal their surface reaction pathways, providing theoretical guidance for rational design and industrial applications.

11.3.3 Hybrid and Nanocomposite Catalysts: Combining Different Materials to Enhance Catalytic Activity

Composite catalysts hold significant application potential in the field of ECR. By combining multiple materials or active components, these catalysts can not only enhance the adsorption and activation of CO_2 molecules but also optimize reaction pathways to obtain desired products [74]. The synergistic effects of the multiple components and interfacial interactions in composite catalysts can effectively reduce the overpotentials required for reactions, thereby improving selectivity, activity, and stability, leading to more efficient CO_2 reduction performance [75].

Currently, research on composite catalysts primarily focuses on the combination of different material systems and their synergistic mechanisms. For example, metal-organic frameworks (MOFs) as composite catalysts exhibit a porous structure and high specific surface area, enabling selective adsorption of CO_2 at the molecular level and facilitating electron transfer [76]. Metal–nonmetal composite catalysts leverage the electronic effects between metals and nonmetals to enhance the activity and selectivity of the catalytic processes [77]. These different types of composite catalysts each have unique characteristics, making them suitable for various product pathways and application needs. Consequently, composite catalysts have become one of the key research directions for enhancing CO_2 reduction performance, showing great potential for the efficient conversion of CO_2 into oxygenated liquid products [78].

Although composite catalysts demonstrate promising potential in CO_2 electroreduction, their practical application still faces numerous challenges. One major issue is the stability of active species during prolonged reactions, particularly as interfacial structures can degrade due to oxidation or reconstruction, leading to decreased activity. Additionally, the complex components of composite catalysts increase the difficulty of catalyst synthesis and reproducibility, which limits their scalability for industrial applications. To address these issues, strategies can be implemented to optimize the microstructure and crystal plane exposure of the catalysts, enhancing their oxidative resistance and structural stability [79]. Furthermore, developing more efficient and controllable preparation methods along with performance prediction models will ensure the consistency of the catalyst's structure and performance. These strategies will provide effective pathways for the further development of composite catalysts in the field of ECR.

11.4 Mechanisms and Pathways of ECR

The overview of mechanisms and pathways of ECR encompasses a complex interplay of various reactions and intermediates that occur at the cathode of an electrochemical cell, where CO_2 molecules are converted into a range of chemical products through the application of an electrical current in an aqueous electrolyte [31, 80]. This process begins with the adsorption of CO_2 molecules onto the surface of a catalyst, typically a metal or metal-based material, which facilitates the reduction reaction. The initial step involves the transfer of one or more electrons to the CO_2 molecule, forming intermediates such as $CO_2\bullet^-$ radicals. These intermediates can then undergo further reduction steps, either directly or through the formation of additional intermediates, to produce a variety of products including carbon monoxide (CO) [31], formate ($HCOO^-$) [81], methane (CH_4) [30], ethylene (C_2H_4) [82], ethanol (C_2H_5OH) [83], and other hydrocarbons or alcohols (Figure 11.6) [84]. The specific pathways and products formed are influenced by a range of factors, including the nature of the catalyst, the composition of the electrolyte, the

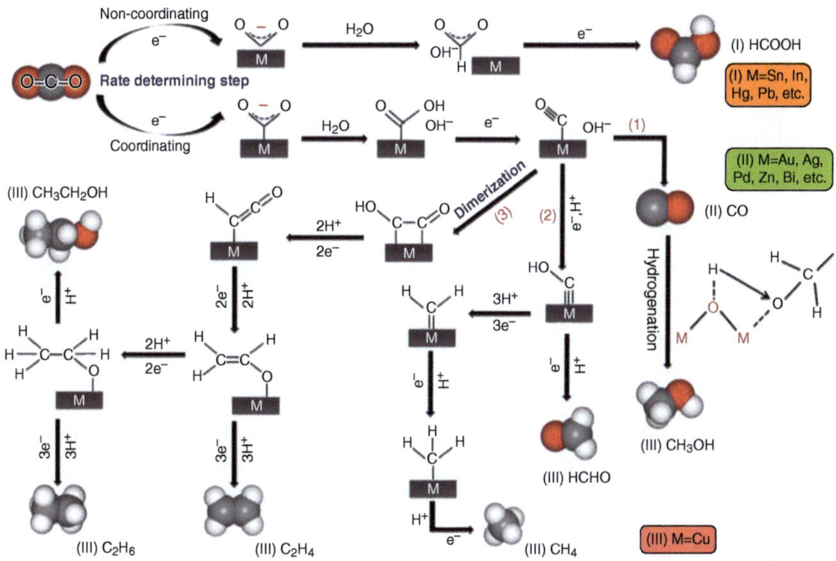

Figure 11.6 Schematic mechanism of different metal electrocatalysts for CO_2 reduction reaction in aqueous solution. *Source:* [84] / John Wiley & Sons / CC BY 4.0.

applied potential, and the reaction temperature and pressure. Understanding and controlling these factors is crucial for optimizing the energy efficiency, selectivity, and stability of the ECR process.

11.4.1 Detailed Description of the Reduction Process, Including Intermediate Steps and Possible Products

The electrocatalytic reduction of carbon dioxide to produce these products typically involves multistep electron and proton transfer processes. The formation of CO is one of the simpler processes in ECR, usually involving the transfer of two electrons and two protons. The CO_2 molecule first adsorbs onto the catalyst surface, where it undergoes single-electron reduction to generate a CO_2 intermediate. This intermediate is then further reduced to form *COOH (a carboxylic intermediate), which ultimately loses a hydroxyl group to produce CO [31]. The generation pathway of CO typically requires highly active metal catalysts (such as Au, Ag, and Zn) to facilitate the formation of $^{*}OC_2$ and stabilize the *COOH intermediate [85]. The reaction steps are as follows:

CO_2 adsorption and activation

$$CO_2 + * \rightarrow *CO_2$$

First proton-coupled electron transfer (PCET) forming *COOH

$$*CO_2 + H^+ + e^- \rightarrow *COOH$$

Second PCET step, cleaving C–OH bond to release H_2O

$$*COOH + H^+ + e^- \rightarrow *CO + H_2O$$

CO desorption (active site regeneration)

$$*CO \rightarrow CO + *$$

Overall reaction (2-electron process)

$$CO_2 + 2H^+ + 2e^- \rightarrow CO + H_2O$$

Key Notes: *denotes an active adsorption site on the catalyst surface (e.g., Au, Ag, or Zn surfaces favor CO production). $H^+ + e^-$ represents PCET (typically supplied by water or electrolyte).

The formation of formic acid involves the transfer of two electrons and protons, similar to CO production, but follows a different pathway, where CO_2 is directly reduced to HCOOH without decomposition [81]. The generation pathway of formic acid often exhibits high selectivity on catalysts such as Sn, Bi, and Pb. Stable metal catalysts can prevent further decomposition of intermediates, facilitating the production of formic acid [48]. The reaction steps are as follows:

CO_2 adsorption and activation

$$CO_2 + * \rightarrow CO_2*$$

PCET to form COOH*

$$CO_2* + H^+ + e^- \rightarrow COOH*$$

Second PCET step to form HCOOH*

$$COOH* + H^+ + e^- \rightarrow HCOOH*$$

Formic acid desorption

$$HCOOH* \rightarrow HCOOH + *$$

Overall reaction (2-electron process)

$$CO_2 + 2H^+ + 2e^- \rightarrow HCOOH$$

Key Notes: *denotes an active adsorption site on the catalyst surface (e.g., Sn, Bi, or Pb).

The generation of methanol is a complex process that involves the transfer of six electrons and protons, requiring a gradual progression at lower potentials. During methanol production, multiple intermediates, such as CO, CHO, and CH_2O, are formed [86]. The generation of methanol typically requires highly active Cu-based catalysts, as Cu can effectively adsorb and stabilize the reaction intermediates [83]. The reaction steps are as follows:

CO_2 adsorption and activation

$$CO_2 + * \rightarrow *CO_2$$

First PCET forming *COOH

$$*CO_2 + H^+ + e^- \rightarrow *COOH$$

Second PCET step to form *CO (key branching point)

$$*COOH + H^+ + e^- \rightarrow *CO + H_2O$$

*CO hydrogenation to form *CHO

$$*CO + H^+ + e^- \rightarrow *CHO$$

*CHO hydrogenation to form $*CH_2O$ (formaldehyde intermediate)

$$*CHO + H^+ + e^- \rightarrow *CH_2O$$

Final hydrogenation to form CH_3OH & desorption

$$*CH_2O + 2H^+ + 2e^- \rightarrow CH_3OH + *$$

Overall reaction (6-electron process)

$$CO_2 + 6H^+ + 6e^- \rightarrow CH_3OH + H_2O$$

Key Notes: *denotes an active adsorption site on the catalyst surface (e.g., Cu, Cu_2O, or alloy surfaces surfaces).

The generation of methane also involves the transfer of 8 electrons and 8 protons, requiring a multistep reduction of CO_2. The pathway for methane production generally proceeds through the CO intermediate, which is then gradually hydrogenated to form CH_4. The methane formation pathway is also catalyst-dependent, with Cu being particularly effective due to its ability to regulate hydrogen atom supply and facilitate stepwise hydrogenation toward methane production [30]. The reaction steps are as follows:

CO_2 adsorption and activation

$$CO_2 + * \rightarrow *CO_2$$

First PCET forming *COOH

$$*CO_2 + H^+ + e^- \rightarrow *COOH$$

Second PCET step to form *CO

$$*COOH + H^+ + e^- \rightarrow *CO + H_2O$$

*CO hydrogenation to form *CHO

$$*CO + H^+ + e^- \rightarrow *CHO$$

*CHO hydrogenation to form $*CH_3O$

$$*CHO + 2H^+ + 2e^- \rightarrow *CH_3O$$

Final hydrogenation to form CH_4 (desorption)

$$*CH_3O + 3H^+ + 3e^- \rightarrow CH_4 + H_2O + *$$

Overall reaction (8-electron process)

$$CO_2 + 8H^+ + 8e^- \rightarrow CH_4 + 2H_2O$$

Key Notes: *denotes an active adsorption site on the catalyst surface (e.g., Cu-based surfaces).

The formation of ethylene is a complex reaction in CO_2 reduction, involving the transfer of 12 electrons and 12 protons, and it primarily occurs on Cu catalysts. The process of CO_2 reduction to ethylene typically involves the polymerization of CO intermediates to form C—C bonds, which are then reduced to generate ethylene [82]. The reaction steps are as follows:

CO_2 adsorption and activation

$$CO_2 + * \rightarrow *CO_2$$

First PCET forming *COOH

$$*CO_2 + H^+ + e^- \rightarrow *COOH$$

Second PCET step to form *CO

$$*COOH + H^+ + e^- \rightarrow *CO + H_2O$$

CO dimerization (C–C coupling)

$$2{*}CO \rightarrow *COCO$$

Hydrogenation to form $*C_2H_4$

$$*COCO + 8H^+ + 8e^- \rightarrow *C_2H_4 + 2H_2O$$

Ethylene desorption

$$*C_2H_4 \rightarrow C_2H_4 + *$$

Overall reaction (12-electron process)

$$2CO_2 + 12H^+ + 12e^- \rightarrow *C_2H_4 + 4H_2O$$

Key Notes: *denotes an active adsorption site on the catalyst surface (e.g., Cu-based surfaces).

The production of ethanol, on the other hand, involves the transfer of 12 electrons and 12 protons and is a multistep reaction process, usually initiated by the further reaction of CO intermediates generated from the reduction of carbon dioxide. This pathway is complex and typically requires bifunctional catalyst like Cu, which can simultaneously promote both CO_2-to-CO conversion and subsequent CO transformation into long-chain carbon products [83]. The main intermediates for ethanol production include *CO, *CHO, and $*CH_2OH$, and this reaction process often requires strong adsorption activity. The reaction steps are as follows:

CO_2 adsorption and activation

$$CO_2 + * \rightarrow *CO_2$$

First PCET forming *COOH

$$*CO_2 + H^+ + e^- \rightarrow *COOH$$

Second PCET step to form *CO

$$*COOH + H^+ + e^- \rightarrow *CO + H_2O$$

CO dimerization (C–C coupling)

$$2{*}CO \rightarrow *COCO$$

Hydrogenation to form *COCHO

$$*COCO + H^+ + e^- \rightarrow *COCHO$$

Further hydrogenation and isomerization to form $*OCH_2CH_2OH$

$$*COCHO + 4H^+ + 4e^- \rightarrow *OCH_2CH_2OH$$

C–O bond cleavage to form CH_3CH_2OH and site regeneration

$$*OCH_2CH_2OH + H^+ + e^- \rightarrow *O + CH_3CH_2OH$$

$$*O + 2H^+ + 2e^- \rightarrow H_2O + *$$

Overall reaction (12-electron process)

$$2CO_2 + 12H^+ + 12e^- \rightarrow CH_3CH_2OH + 3H_2O$$

Key Notes: *denotes an active adsorption site on the catalyst surface (e.g., Cu-based surfaces).

The generation of other oxygenated liquid products, such as acetone and propanol, typically involves complex multistep reaction processes characterized by high selectivity and longer carbon chain structures [87]. This generation requires catalysts with well-defined interfacial structures and electronic effects to support the stability of intermediates and control reaction pathways [88]. In practical applications, the selectivity of CO_2 electroreduction products is highly dependent on the active sites of the catalysts, reaction potentials, and operating conditions. The formation pathways and required intermediates for each product differ, making it essential to select suitable catalyst materials and optimize reaction conditions to achieve efficient generation of target products.

11.4.2 Factors Influencing the Selectivity of the Reaction Towards Different Products (e.g., Methane, Ethylene, Ethanol)

The selectivity of ECR towards different products is influenced by a range of factors, which can be broadly categorized into three main groups: catalyst properties, reaction conditions, and electrolyte composition.

11.4.2.1 Catalyst Properties

The nature and properties of the catalyst play a crucial role in determining the selectivity of ECR [34]. Different catalysts have different abilities to stabilize the intermediates formed during the reduction process, which can affect the activation energies of the individual steps and thus the overall pathway. For example, some catalysts may favor the formation of CO as the primary product, while others may promote the formation of formate or methane. The selectivity towards a particular product can be influenced by factors such as the catalyst's electronic structure [89], surface morphology [90], and the presence of defects or dopants [22].

The choice of catalyst material is also important. Noble metals, such as gold and silver, have been shown to exhibit high selectivity for the formation of CO, while transition metals, such as copper, nickel, and cobalt, can promote the formation of hydrocarbons and alcohols. Carbon-based materials, such as graphene and CNTs, have also been explored as potential catalysts for ECR, with some showing promising selectivity towards specific products.

11.4.2.2 Reaction Conditions

The reaction conditions, including the applied potential, temperature, and pressure, can also affect the selectivity of ECR [91]. The applied potential determines the driving force for the reduction reaction and can influence the energetics of the various reduction steps. Higher potentials can favor the formation of more reduced products, such as methane or ethylene, while lower potentials may promote the formation of CO or formate [28].

Temperature [92] and pressure [93] can also play a role in determining the selectivity of the reaction. Increasing the temperature can enhance the kinetics of the reduction reaction, potentially leading to increased selectivity towards more reduced products. Similarly, increasing the pressure can increase the concentration of CO_2 at the catalyst surface, which can affect the adsorption and reduction of CO_2 molecules.

11.4.2.3 Electrolyte Composition

The composition of the electrolyte can also influence the selectivity of ECR [17]. The electrolyte provides the ions necessary for the reduction reaction to occur and can affect the stability and reactivity of the catalyst surface. Different electrolytes can support the formation of different products, depending on their ability to stabilize the intermediates formed during the reduction process.

For example, aqueous electrolytes containing bicarbonate ions (HCO_3^-) have been shown to promote the formation of formate, while electrolytes containing chloride ions (Cl^-) can enhance the formation of CO [94]. The pH of the electrolyte can also affect the selectivity of the reaction, with acidic electrolytes favoring the formation of CO and formate [95], and alkaline electrolytes promoting the formation of methane and other hydrocarbons [96].

In summary, the selectivity of ECR towards different products is influenced by a range of factors, including catalyst properties, reaction conditions, and electrolyte composition. Understanding and controlling these factors is crucial for optimizing the selectivity, energy efficiency, and stability of ECR technologies.

11.5 Challenges and Opportunities

11.5.1 Challenges and Opportunities for the Future Development of ECR

The electrochemical reduction of CO_2 offers a promising pathway for converting this abundant greenhouse gas into valuable chemicals and fuels. However, the process faces several challenges that need to be addressed to achieve commercial viability and scalability. At the same time, the field presents numerous opportunities for innovation and interdisciplinary collaboration, offering potential solutions to some of the most pressing global challenges.

11.5.2 Challenges in ECR

11.5.2.1 Catalyst Development and Selectivity

One of the most significant challenges in ECR is the development of catalysts that can efficiently and selectively convert CO_2 into desired products [11]. Currently, most catalysts are either too expensive, too inefficient, or lack the selectivity necessary to produce specific target molecules [97]. Achieving high selectivity is particularly challenging because CO_2 reduction can proceed through multiple reaction pathways, leading to a variety of products such as methane, ethylene, ethanol, and others [34]. Therefore, researchers must focus on designing catalysts that can favor specific reaction pathways and products while minimizing side reactions.

Moreover, the stability and durability of catalysts under electrochemical conditions are also critical issues. Many catalysts degrade over time due to corrosion, poisoning, or loss of active sites, leading to decreased efficiency and product selectivity. Developing catalysts with high stability and durability will be essential for the long-term operation and commercialization of ECR systems [79].

11.5.2.2 Reaction Conditions and Energy Efficiency

The reaction conditions, including temperature [92], pressure [93], and electrolyte composition [17], also play a crucial role in determining the efficiency and selectivity of ECR. For example, higher temperatures can increase reaction rates but may also lead to catalyst degradation and increased energy consumption. Similarly, higher pressures can increase the solubility of CO_2 in the electrolyte, but they also require more energy to maintain.

Achieving high energy efficiency is another significant challenge. The electrochemical reduction of CO_2 requires the input of electrical energy, which can be costly and environmentally burdensome if derived from fossil fuels [98]. Therefore, integrating ECR systems with renewable energy sources such as solar, wind, and geothermal power could provide a more sustainable solution. However, this integration requires overcoming technical and economic challenges, such as matching the supply and demand of renewable energy and optimizing the system for maximum energy efficiency.

11.5.2.3 Scale-up and Commercialization

Scaling up ECR systems to industrial levels presents another set of challenges. Many current systems are still at the laboratory or pilot scale, and scaling them up to produce significant quantities of products requires overcoming technical, economic, and regulatory hurdles [99]. For example, developing large-scale electrolyzers with high efficiency and durability, as well as optimizing the process for continuous operation and product recovery, are critical steps toward commercialization [100].

Moreover, the cost of producing ECR products must be competitive with conventional methods of production, such as fossil-fuel-based synthesis. This requires reducing the costs of catalysts, electrolytes, and other materials, as well as optimizing the process for maximum productivity and yield. Additionally, developing markets for these products and overcoming regulatory barriers will be essential for their widespread adoption [101].

11.5.3 Opportunities for Future Development

Despite these challenges, the field of ECR presents numerous opportunities for innovation and interdisciplinary collaboration. Here are some of the key opportunities:

11.5.3.1 Advanced Catalyst Design and Materials Science

Advances in catalyst design and materials science offer promising solutions to the challenges of selectivity, stability, and durability [98, 102, 103]. Researchers are exploring a wide range of catalyst materials, including metals, metal oxides, metal-organic frameworks (MOFs), and others, to identify those with optimal activity and selectivity for specific products. Additionally, surface engineering techniques such as doping, alloying, and nanostructuring can be used to tailor the catalyst properties and enhance its performance. Moreover, the integration of computational modeling and machine learning can accelerate the discovery and optimization of new catalyst materials [104–106]. By simulating the reaction pathways and interactions between catalysts and reactants, researchers can identify potential catalysts with high selectivity and activity. Machine learning algorithms can then be used to screen large libraries of potential catalysts and optimize their properties through iterative cycles of design, synthesis, and testing.

11.5.3.2 Electrolyte Engineering and Reaction Engineering

Electrolyte engineering and reaction engineering also offer opportunities to improve the efficiency and selectivity of ECR. Researchers are exploring a wide range of electrolyte compositions, including aqueous [22], nonaqueous [25, 26], and molten salt [23, 24] electrolytes, to identify those with optimal properties for specific reactions. For example, some electrolytes can stabilize intermediate species and facilitate electron transfer, while others can promote the formation of specific products. Reaction engineering techniques such as flow cells, membrane reactors, and microreactors can also be used to enhance the performance of ECR systems [107]. These techniques can provide better mixing, heat transfer, and mass transfer, leading to higher reaction rates and product selectivity. Additionally, the integration of ECR with value-added anode processes with other processes, such as (i) chlor-alkali and chlorate production, conversion of benzene to benzoquinone, and the epoxidation of propylene to propylene oxide; (ii) oxidation of industrial by-products; (iii) large-scale production of basic chemicals (e.g., H_2O_2); and (iv) synthesis of fine chemicals, can create more efficient and sustainable systems (Figure 11.7) [108, 109].

11.5.3.3 Interdisciplinary Collaboration and Innovation

The field of ECR benefits from interdisciplinary collaboration across various scientific and engineering disciplines. Researchers from chemistry, materials science, engineering, and other fields can work together to address the challenges of catalyst development, reaction conditions, and scale-up. By combining their expertise

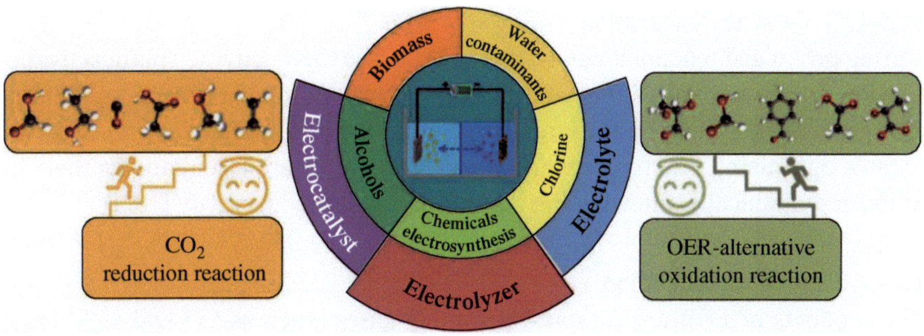

Figure 11.7 Electrochemical CO_2 reduction coupled with alternative oxidation reactions: electrocatalysts, electrolytes, and electrolyzers. *Source:* Ref. [108], with permission of Elsevier.

and perspectives, they can develop innovative solutions that overcome technical and economic barriers and pave the way for commercialization [99, 110]. Moreover, the field presents opportunities for collaboration with industries and policymakers. Industry partners can provide insights into market needs, production processes, and regulatory requirements, while policymakers can support research and development through funding, tax incentives, and regulatory frameworks. Such collaborations can help accelerate the development and deployment of ECR technologies and contribute to a more sustainable future [111].

11.5.3.4 Sustainability and Environmental Benefits

The field of ECR offers significant sustainability and environmental benefits. Through converting CO_2 into useful chemicals and fuels, this technology can contribute to carbon recycling and reduce the demand for fossil fuels. This can lead to decreased greenhouse gas emissions, improved energy security, and economic benefits through the production of high-value products [112]. Moreover, ECR can be integrated with renewable energy sources to create more sustainable systems [5]. By using solar, wind, or geothermal power to provide the electrical energy needed for the reaction, researchers can reduce the environmental footprint of the process and contribute to the transition to a low-carbon economy [4].

In conclusion, the field of ECR presents numerous challenges as well as opportunities for future progress. By addressing these challenges through advanced catalyst design, electrolyte engineering, reaction engineering, and interdisciplinary collaboration, researchers can develop more efficient and selective systems that contribute to carbon recycling and sustainability. With continued innovation and support from industry and policymakers, the future of ECR looks promising, offering potential solutions to some of the most pressing global challenges.

11.5.4 Technical Challenges: Addressing Issues Such as Catalyst Deactivation, Product Separation, and Energy Efficiency

Electrocatalytic carbon dioxide reduction faces various technical challenges, primarily including catalyst deactivation, product separation, and energy efficiency.

11.5.4.1 Catalyst Deactivation

During prolonged reactions, catalysts gradually lose their activity. This deactivation is often due to fouling of the catalyst surface, oxidation, or physical or chemical changes resulting from the accumulation of products and byproducts [80]. Solutions include optimizing the structure and composition of catalysts, enhancing their stability, or dynamically renewing the active sites to maintain activity [102].

11.5.4.2 Product Separation

The products of ECR are complex, including gases (such as CO and H_2), liquids (such as formic acid and ethanol), and solid deposits. In such multiproduct systems, efficiently separating the target product from the mixture is a significant challenge. Designing selective catalysts and adjusting reaction conditions can help reduce byproduct formation and improve the purity of target products. Additionally, employing novel separation technologies, such as membrane separation and solvent extraction, can enhance separation efficiency [100].

11.5.4.3 Energy Efficiency

The ECR process requires substantial electrical energy to drive the reactions, making energy consumption a critical factor, especially in large-scale industrial applications. To reduce energy usage, it is necessary to develop catalysts with low overpotentials, high activity, and high FE. This can be achieved by introducing bimetallic or multifunctional composite catalysts, optimizing catalyst surface sites, or employing reaction coupling methods to enhance EE [98].

Through ongoing research and innovation in these areas, there is great potential to significantly improve the practical feasibility and economic viability of electrocatalytic carbon dioxide reduction.

11.5.5 Economic Considerations: Cost–Benefit Analysis of ECR Compared to Traditional Chemical Processes

In the cost-effectiveness comparison between ECR and traditional chemical processes, the economic considerations can be addressed from several aspects (Figure 11.8) [110, 113, 114]:

11.5.5.1 Raw Material Costs

The process of ECR employs CO_2 and H_2O as primary inputs, both of which are widely available and cost-effective. Moreover, this approach generally functions under more moderate reaction parameters, sidestepping extreme conditions like elevated temperatures and pressures, thereby minimizing resource usage and lowering expenses related to machinery. In contrast, conventional techniques like thermal and photocatalytic methods frequently demand environments with intense heat or pressure, which in turn requires more robust equipment and leads to increased initial investments. Additionally, certain methods rely on the use of highly purified hydrogen and other specialized materials, which incur significant procurement and maintenance expenses [115].

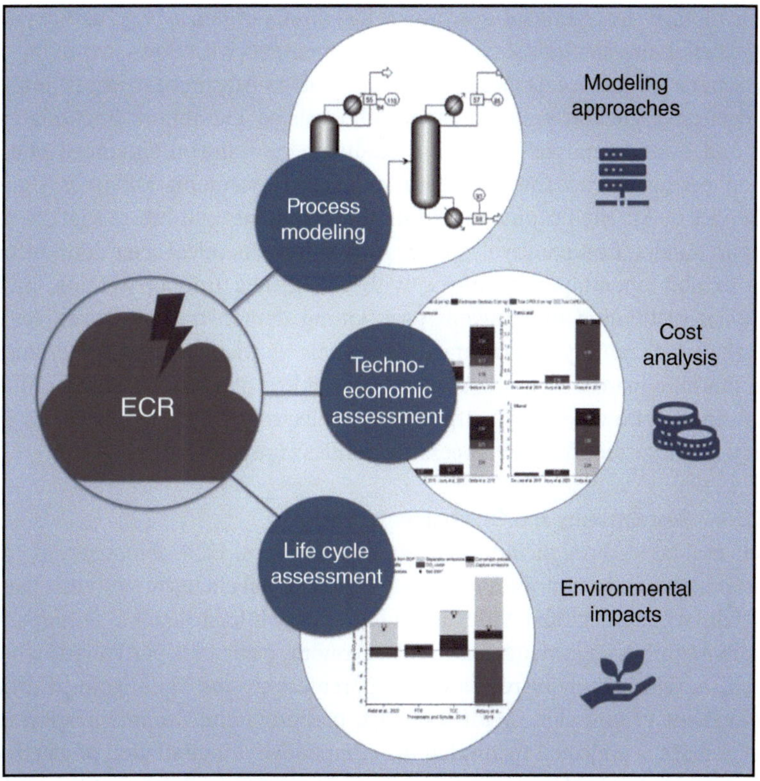

Figure 11.8 Process modeling, techno-economic assessment, and life cycle assessment of the electrochemical reduction of CO_2. *Source:* Ref. [114], Elsevier, CC BY 4.0.

11.5.5.2 Energy Consumption

ECR primarily depends on electrical energy inputs, where advancements in electrode materials and catalytic systems can effectively minimize energy consumption. Current technological limitations manifest as suboptimal conversion rates in these electrochemical systems, leading to elevated power demands per output unit. Implementing sustainable power sources for this process demonstrates dual advantages in decarbonization and operational economics. Additionally, the creation of catalytic interfaces with superior FE and reduced activation barriers presents opportunities for substantial energy optimization across production cycles. Conventional industrial approaches predominantly utilize hydrocarbon-based energy carriers, introducing both price volatility in energy expenditures and considerable greenhouse gas outputs. Although thermal and pressure intensification strategies can enhance productivity in traditional systems, their environmental impact metrics and energy-related operational expenses remain disproportionately elevated compared to electrochemical alternatives [97].

11.5.5.3 Product Separation and Purity

ECR often generates multiphase output streams containing co-existing gaseous and liquid species, which imposes significant technical challenges in post-reaction purification. For oxygenated derivatives like C_1–C_2 organics (e.g., carboxylic acids and short-chain alcohols), subsequent refinement workflows involving multi-step separation protocols become indispensable. Advanced process integration approaches, particularly electrochemically assisted membrane fractionation systems and hybrid catalytic-separation modules, have shown enhanced phase separation performance. However, their practical implementation faces significant challenges due to high capital investment requirements and substantial operational cost parameters. Comparatively, conventional petrochemical separation infrastructures exhibit operational maturity, with established unit operations, including fractional distillation and solvent partitioning demonstrating proven scalability and cost predictability at manufacturing levels. Particularly for liquid-phase outputs, thermodynamic separation techniques achieve good mass transfer efficiencies in continuous flow systems, outperforming electrochemical counterparts in both energy intensity metrics and process economics [116].

11.5.5.4 Sustainability and Environmental Costs

When evaluated through industrial ecology analyses, ECR demonstrates superior environmental impact profiles relative to conventional chemical synthesis pathways, especially when integrated with sustainable energy infrastructures. This operational paradigm exhibits alignment with circular economy principles, particularly when contextualized within evolving regulatory support mechanisms. The ongoing institutional development of emission credit exchange platforms and decarbonization-focused fiscal policies is projected to amplify the competitive differentiation of electrochemical systems in terms of ecological performance metrics. In sharp contrast, traditional architectures dependent on petrochemical feedstocks generate quantifiable carbon emissions. Through system analyses incorporating emission intensity and environmental expenditures, conventional methodologies reveal diminishing economic competitiveness under progressive carbon pricing regimes [117].

11.5.5.5 Equipment Investment and Maintenance

The infrastructure for electrochemical carbon dioxide conversion necessitates capital expenditure on specialized apparatuses, including reaction vessels and catalytic interface components. Notably, these systems exhibit operational advantages through ambient temperature/pressure working parameters that substantially mitigate equipment degradation rates. From an engineering perspective, modular design characteristics endow such electrochemical platforms with remarkable scalability, particularly advantageous for decentralized production scenarios. This contrasts sharply with conventional thermochemical processes that demand energy-intensive reaction environments characterized by elevated thermal gradients and compression requirements, inevitably escalating capital expenditure and operational maintenance overheads. Under extreme operating conditions involving thermal stress and corrosive media, conventional reactors exhibit accelerated material fatigue patterns. Enhanced safety and frequent part replacement are required to maintain process integrity [110].

To conclude, ECR demonstrates significant economic potential, particularly through the integration of sustainable power sources, enhanced catalytic performance, and progressive purification methodologies, collectively contributing to reduced production expenditures. Conventional chemical conversion pathways encounter escalating ecological and regulatory burdens stemming from greenhouse gas emissions. Consequently, with continuous technological innovations and systematic energy framework enhancements, the economic viability of ECR systems is projected to experience progressive improvement.

11.5.6 Environmental Sustainability Evaluation of Electrochemical Conversion Systems

The environmental assessment of ECR systems requires comprehensive consideration of sustainability metrics and ecological impacts. This section examines critical evaluation parameters and optimization strategies through four key dimensions.

11.5.6.1 Energy Source Optimization and Conversion Efficiency

The ecological impact of electrochemical systems is predominantly determined by their energy source characteristics. Renewable energy integration (photovoltaic and wind energy systems) can dramatically decrease process-related greenhouse gas outputs. Conversely, fossil fuel–derived electricity may result in net positive carbon emissions, potentially offsetting the system's decarbonization benefits. Enhancement of conversion efficiency through advanced electrode architectures and optimized reaction parameters can substantially improve energy utilization rates, thereby boosting system sustainability [101].

11.5.6.2 Catalytic System Selection and Durability Considerations

Catalyst production processes, particularly for noble metal–based systems, carry substantial environmental burdens. Earth-abundant transition metal catalysts (copper and zinc-based systems) offer improved sustainability profiles. Prolonging catalyst service life through material innovation and operational optimization can effectively reduce manufacturing frequency and associated environmental impacts.

11.5.6.3 Resource Utilization and Output Management

Electrochemical conversion systems typically require water resources as proton donors, necessitating careful water management strategies. The treatment and utilization of electrochemical conversion byproducts (including methane and formate derivatives) require specific consideration to minimize potential ecological impacts [118].

11.5.6.4 Comprehensive Life Cycle Analysis

Life cycle assessment (LCA) methodologies provide systematic evaluation frameworks for electrochemical conversion systems, encompassing material extraction, manufacturing processes, operational phases, and end-of-life management. These analyses enable accurate quantification of greenhouse gas emissions throughout the product lifecycle [119]. Through detailed energy flow analysis and process optimization, significant reductions in carbon emission can be achieved.

Future development pathways involve system integration with renewable energy infrastructures and establishment of circular economy models through industrial symbiosis. Process intensification, sustainable material utilization, and energy efficiency improvements collectively contribute to substantial carbon footprint reduction, supporting global carbon neutrality objectives.

11.6 Case Studies and Applications

11.6.1 Examples of ECR Systems in Laboratory and Pilot Scales

In terms of laboratory and pilot-scale examples for ECR systems, there are some successful cases and potential industrial applications.

11.6.1.1 Laboratory Scale Example

The conversion of CO_2 into fuels presents a viable approach to addressing various energy and environmental challenges. Nonetheless, due to the chemical stability of CO_2, many electrochemical and photochemical conversion processes exhibit low efficiency. In 2016, a research team led by Larry A. Curtiss and Amin Salehi-Khojin at the University of Illinois published a study titled "Nanostructured transition metal dichalcogenide electrocatalysts for CO_2 reduction in ionic liquid." They reported a nanostructured transition metal disulfide used to catalyze the electrochemical conversion of CO_2 to CO in ionic liquids. The study found that tungsten diselenide nanosheets exhibited a current density of 18.95 mA/cm^2 at a low overpotential of 54 mV, with a FE for CO of 24% and a turnover frequency for CO generation of 0.28 s^{-1}. Additionally, this catalyst was applied to a light-harvesting artificial leaf platform, capable of simultaneously oxidizing water without an external voltage. This research was published in *Science*, providing new catalytic materials and methods for the efficient conversion of CO_2 [25].

Electrochemical reduction of CO_2 to CO is the first step in synthesizing more complex carbon-based fuels using renewable electricity. However, the reaction kinetics are slow, primarily due to the low local concentration of CO_2 around the catalyst. In 2016, Edward H. Sargent's research team at the University of Toronto published "Enhanced electrocatalytic CO_2 reduction via field-induced reagent concentration." The study demonstrated that nanostructured electrodes can generate a local high electric field at low applied overpotentials, concentrating electrolyte cations and significantly increasing CO_2 concentration. Simulations and experiments confirmed that gold nanoneedles achieved a geometric current density of 22 mA/cm^2 at −0.35 V, far exceeding the performance of existing best catalysts. Additionally, palladium nanoneedle catalysts produced formic acid with over 90% FE at −0.2 V. This work, published in *Nature*, showcased the broad applicability of the field-induced reagent concentration concept [120].

The selectivity of ethylene production on copper catalysts faces challenges. In 2019, Edward H. Sargent's team published "Molecular tuning of CO_2-to-ethylene conversion." They proposed a molecular tuning strategy that functionalizes the catalyst surface with organic molecules to stabilize intermediates and enhance selectivity. Experiments showed that the modified catalysts achieved a FE of 72% in neutral media, with a current density reaching 230 mA/cm^2, and sustained stable ethylene synthesis in a MEA system for 190 hours. This work, published in *Nature*, provides new insights for optimizing heterogeneous catalysts [121].

In the same month of that year, Hailiang Wang from Yale University and Yongye Liang from Southern University of Science and Technology published "Domino electroreduction of CO_2 to methanol on a molecular catalyst." The study indicated that when cobalt phthalocyanine is immobilized on CNTs, it can efficiently catalyze the six-electron reduction of carbon dioxide to methanol, achieving over 40% FE and a current density exceeding 10 mA/cm^2. This process generates carbon monoxide as an intermediate through a series of reactions. Although catalytic activity declines over time, this decay can be suppressed by attaching electron-donating groups to the phthalocyanine ring. The catalyst exhibits stable performance in near-neutral electrolytes. This work was published in *Nature* and provides valuable insights for subsequent research on ECR to transform carbon emissions into fuels and value-added chemicals [122].

Practical electrochemical conversion of carbon dioxide requires efficient and highly selective catalysts to produce a single product at high current densities. Solid-state electrocatalysts can achieve ECR at current densities of $\geq$ 150 mA/cm^2, but maintaining high selectivity and efficiency at these high current densities remains a challenge. In 2019, Marc Robert from Paris University and colleagues published an article titled "Molecular electrocatalysts can mediate fast, selective CO_2 reduction in a flow cell." The study demonstrated that molecular ECR catalyst design can achieve high selectivity and low overpotential, but its current density is not suitable for commercial operations. Using the widely available molecular catalyst cobalt phthalocyanine, the research achieved CO selectivity exceeding 95% at 150 mA/cm^2 in a zero-gap membrane flow reactor. This study, published in *Science*, indicates the efficient performance of molecular catalysts under these operational conditions and provides new ideas for optimizing ECR catalysts and electrolyzers [123].

In 2024, Xia and others from Huazhong University of Science and Technology published an article titled "Durable CO_2 conversion in the proton-exchange membrane system." This study designed a proton-exchange membrane system that electrochemically converts carbon dioxide into formic acid, utilizing a catalyst sourced from discarded lead-acid batteries, enhancing reaction efficiency through a lattice carbon activation mechanism. In acidic electrolytes, the formation of carbonates was avoided, improving the system's stability. The system achieved over 93% FE and nearly 91% single-pass conversion efficiency at a current density of 600 mA/cm^2, operating continuously for over 5200 hours. The research indicates that the combination of this high-performance catalyst and durable membrane will promote the development of carbon neutrality technologies. This study was also published in *Nature*, highlighting how the integration of high-performance catalysts and durable membranes can advance carbon neutrality technologies [124].

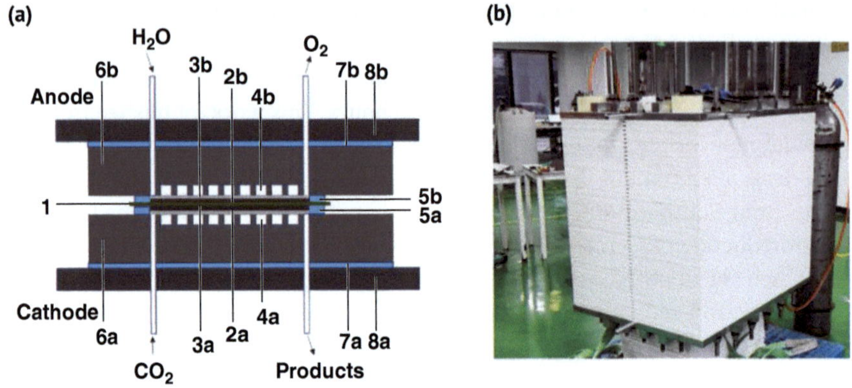

Figure 11.9 (a) Schematic illustration (cross view) of CO_2 electrolyzer: 1, membrane; 2, electrocatalyst layers (2a, cathodic; 2b, anodic); 3, current collectors (3a, cathodic, 3b, anodic); 4, flow channels (4a, cathodic, 4b, anodic); 5, seal gasket (5a, cathodic, 5b, anodic); 6, plastic frame (6a, cathodic, 6b, anodic); 7, insulation gaskets (7a, cathodic, 7b, anodic); 8, Clamp (8a, bolts, 8b nuts). *Source:* Ref. [125], American Chemical Society. (b) Pilot scale CO_2 electrolyzer. *Source:* [126] / American Chemical Society.

11.6.1.2 Pilot and Industrial Scale Examples

The electrolyzer is the core component of the CO_2 electrocatalytic reduction device. Multiple electrolyzers can be connected in series to form a stack (Figure 11.9) [125]. The design goal of the electrolyzer is to reduce resistance, lower energy consumption, and maintain long-term operation. A single electrolyzer mainly consists of components such as the MEA, gas diffusion layer, plastic frame, and bipolar plates. The plastic frame is lightweight, and the gas diffusion layer facilitates the transport of CO_2. Similarly, the anode gas diffusion layer can vent the generated O_2. Furthermore, an appropriate porosity in the diffusion layer supports mass transfer of reactants and products, while accommodating size variations during mechanical compression and operation.

The performance of ECR to produce syngas using a series of connected small electrolytic cells has been demonstrated in a pilot test at the Yitai Chemical coal-to-liquid plant in Ordos, Inner Mongolia. The single electrolyzer cell operated at a voltage of only 2.8 V, with an annual CO_2 processing capacity of 30 t/a and stability lasting up to 2000 hours [126]. Currently, several companies and teams around the world, such as OPUS-12, 3M, Dioxide Materials, DNV, CO_2 CERT, and Siemens, are dedicated to CO_2 electrolysis under industrial conditions [127–130]. However, scaling up and engineering the electrolyzer remains a significant challenge. Most current research is based on laboratory-scale fundamental studies, where the evaluation conditions for CO_2 electrolysis are highly idealized. In terms of catalysts, literature reports are mainly based on evaluations conducted under laboratory conditions involving high-purity gases, short durations, and negligible material costs. Regarding reactors, the current sizes are small, with low throughput per unit and overall high energy consumption, making it impossible to conduct joint validation with the corresponding scaling-up processes. Therefore, in most cases, data obtained from laboratory studies cannot be extrapolated to large-scale electrolyzers. In large electrolyzers, key factors such as heat management, power, flow rate, and phase control need to be investigated, which cannot be accomplished with laboratory electrolyzers. Additionally, CO_2 electrolysis technology needs to be positioned for application scenarios to distinguish it from other CCUS technologies. Overall, CO_2 electrolysis technology is a highly promising emerging technology and an important component of the CCUS field, warranting vigorous industrial research.

11.6.2 Discussion on the Potential Industrial Applications and Integration with Existing Chemical Processes

11.6.2.1 Potential Industrial Applications

11.6.2.1.1 Synthetic Fuels

Reducing CO_2 to syngas (H_2 and CO) can lead to the production of liquid fuels through chemical processes like Fischer–Tropsch synthesis, providing a renewable energy solution.

11.6.2.1.2 Chemical Feedstocks

Electrochemical reduction can convert CO_2 into chemical feedstocks such as methanol and acetic acid, which can be used to produce plastics and other chemical products.

11.6.2.2 Integration with Existing Chemical Processes

11.6.2.2.1 Combining with Hydrogen Production

Electrochemical reduction can be integrated with renewable hydrogen production processes (such as water electrolysis) to form a closed-loop carbon management system. This integration can reduce fossil fuel use and improve economic viability.

11.6.2.2.2 Co-Utilization with Traditional Chemical Processes

By combining electrochemical reduction with existing chemical processes (such as the syngas production), higher resource utilization rates can be achieved while reducing CO_2 emissions.

These examples and potential applications demonstrate the vast prospects of ECR systems and how technological innovations can drive the production of sustainable chemicals.

11.7 Future Perspectives

11.7.1 Exploring New Catalyst Materials and Reaction Conditions to Improve Efficiency and Selectivity

The prospects of ECR technology are promising, primarily reflected in innovations in catalyst materials and optimization of reaction conditions. Novel catalyst materials, such as two-dimensional materials (e.g., graphene and MXenes), SACs, and alloy and composite catalysts, can significantly enhance reaction efficiency and selectivity [131]. Additionally, research on bio-based catalysts offers new possibilities for improving catalytic efficiency [132]. In terms of optimizing reaction conditions, improvements in electrolytes, the design of efficient electrodes, and adjustments to the reaction environment can further enhance the overall performance of CO_2 reduction reactions [98].

11.7.2 Prospects for Scaling Up ECR Technology

To expand the application of ECR technology, the development of integrated systems is crucial, combining ECR with renewable energy to create efficient energy-chemical conversion systems [99]. Moreover, the development of modular electrolyzers will facilitate the flexible application of this technology in small to medium-sized plants or distributed energy systems. Policy support and economic incentives will also drive the commercialization of this technology, promoting its widespread use in industry. Interdisciplinary collaboration will further advance technological innovation, providing support for more sustainable production of chemicals and energy.

11.7.3 Policy Frameworks and Economic Incentives for Advancing Industrial ECR Sectors

The implementation of ECR systems in industrial settings is significantly influenced by policy frameworks and economic incentives. Emission control strategies, particularly carbon pricing instruments, including emission taxation schemes and cap-and-trade systems, can elevate the financial implications of greenhouse gas outputs, thereby stimulating industrial adoption of decarbonization technologies [133, 134]. Concurrently, legislative mandates for clean energy utilization can drive the incorporation of electrochemical conversion systems, particularly through the utilization of sustainable power sources for electrolytic processes. Complementary financial mechanisms, including innovation grants, fiscal benefits, and operational subsidies, can mitigate capital expenditure barriers and running costs, enhancing the attractiveness of electrochemical conversion systems for industrial operators.

In the future, the development of intersectoral cooperative networks and knowledge exchange infrastructures will significantly accelerate technology deployment. Such collaborative frameworks enable technological synergy and mutual benefits across industries, while digital knowledge platforms foster enhanced interaction between industrial stakeholders, academic researchers, and regulatory bodies, facilitating the dissemination of technical innovations and market intelligence. Moreover, international partnerships and public engagement initiatives will establish a supportive ecosystem for technology diffusion, expediting market penetration and contributing to the realization of climate mitigation objectives.

11.8 Conclusion

11.8.1 Summarizing the Key Points Discussed in this Chapter

Electrocatalytic carbon dioxide reduction technology has the potential to promote sustainable energy development by converting CO_2 into valuable chemicals (such as ethanol and methanol) and fuels. This technology effectively utilizes renewable energy and provides solutions for carbon capture and utilization (CCU). Catalysts are crucial in ECR, as they determine the selectivity and efficiency of the reactions. Currently, copper-based catalysts are widely studied due to their ability to selectively reduce CO_2 to various carbon-based compounds. Other metal catalysts, such as silver, platinum, and nickel, as well as alloy catalysts, have also been developed to improve reaction performance. Furthermore, the design of the catalyst's microstructure and morphology, such as porous structures and nanoparticles, plays an important role in enhancing catalytic activity. The reaction mechanism of ECR is complex, involving multiple steps, including the adsorption of CO_2, reduction reactions, and product desorption. Research indicates that electron transfer, proton coupling, and the adsorption states of the catalyst surface significantly influence reaction selectivity and rate. A deeper understanding of the reaction mechanism can aid in optimizing catalyst design and improving conversion efficiency.

Despite the potential of ECR technology, it still faces challenges, including insufficient reaction selectivity, poor catalyst stability, and high energy consumption. Additionally, commercial applications need to address economic viability and scalability. Therefore, researchers must continuously explore new materials, optimize reaction conditions, and enhance system integration efficiency. With the global emphasis on emission reduction targets and the rapid development of renewable energy, the prospects for ECR technology are promising. In the future, policy support, technological innovation, and increasing market demand will drive the commercialization of ECR technology. Moreover, the integration of ECR with other carbon management technologies, such as CCUS, is expected to form comprehensive solutions, providing effective pathways to achieve carbon neutrality goals.

11.8.2 Highlighting the Importance of ECR in Achieving Sustainability Goals

Electrochemical carbon dioxide reduction technology plays a significant role in achieving Sustainable Development Goals (SDGs). First, ECR technology effectively reduces greenhouse gas emissions by converting carbon dioxide into valuable chemicals and fuels, thereby mitigating the negative impacts of climate change. This aligns closely with SDGs aimed at promoting global greenhouse gas reduction efforts. Additionally, ECR technology supports the utilization of renewable energy, facilitating the realization of renewable energy targets. By employing renewable electricity for CO_2 reduction, ECR not only enhances energy efficiency but also aids in the low-carbon transition of the economy. Furthermore, this technology can create new economic opportunities, foster the development of a green economy, and promote sustainable industrial and infrastructure development.

Therefore, ECR technology is not only an effective means of addressing climate change but also a vital component for achieving comprehensive sustainable development. Increased research and application support for ECR technology can accelerate the global transition to sustainable development models, paving the way for future environmental protection and economic growth.

11.8.3 Outlook on the Future of ECR Technology

The future prospects of ECR technology are promising, mainly reflected in the following aspects.

11.8.3.1 Technological Innovation

With in-depth research on catalysts and reaction mechanisms, the development of new, efficient catalysts will significantly enhance the selectivity and conversion efficiency of CO_2 reduction. The use of nanomaterials, alloy catalysts, and porous structures will optimize reactions, reduce energy consumption, and improve product yields. Additionally, advancements in electrochemical technology will continue to improve reactor design and system integration, facilitating larger production scales.

11.8.3.2 Commercial Applications

ECR technology is expected to find commercial applications across multiple industries, including chemicals, energy, and transportation. As the market demand for low-carbon and renewable products grows, companies will increasingly focus on adopting ECR technology to produce chemicals and fuels, creating new business opportunities. Furthermore, the implementation of policy support and carbon pricing mechanisms will further incentivize companies to invest in ECR technology.

11.8.3.3 Integration with Other Technologies

In the future, ECR technology will be closely integrated with other carbon management technologies, such as carbon capture and storage, to form comprehensive carbon reduction solutions. This integration can enhance overall economic and environmental benefits, promoting the development of a circular carbon economy. Additionally, the combination of ECR with hydrogen production and renewable energy systems will further drive the energy transition and sustainable development.

11.8.3.4 Global Cooperation and Policy Support

International cooperation will play a crucial role in advancing ECR technology. Governments, research institutions, and businesses worldwide should work together to share technologies and experiences, promoting technology transfer and investment. At the same time, supportive policies and incentive measures will foster a conducive environment for the advancement of ECR technology, helping to achieve sustainable development goals.

In summary, the future outlook for ECR technology is optimistic. Through technological innovation, commercial applications, and international cooperation, it is expected to have a major impact on tackling climate change and advancing sustainable development.

References

1 Friedlingstein, P., O'Sullivan, M., Jones, M.W. et al. (2023). Global carbon budget 2023. *Earth Syst. Sci. Data*. 15: 5301–5369.

2 Zhang, Z., Wang, T., Blunt, M.J. et al. (2020). Advances in carbon capture, utilization and storage. *Appl. Energy*. 278: 115627.

3 Gössling, S. and Higham, J. (2021). The low-carbon imperative: Destination management under urgent climate change. *J. Travel Res*. 60: 1167–1179.

4 Rahman, M.M., Alam, K., and Velayutham, E. (2022). Reduction of CO_2 emissions: The role of renewable energy, technological innovation and export quality. *Energy Rep.* 8: 2793–2805.

5 Perry, S., Klemeš, J., and Bulatov, I. (2008). Integrating waste and renewable energy to reduce the carbon footprint of locally integrated energy sectors. *Energy.* 33: 1489–1497.

6 Psarras, P.C., Comello, S., Bains, P. et al. (2017). Carbon capture and utilization in the industrial sector. *Environ. Sci. Technol.* 51: 11440–11449.

7 Lin, Q., Zhang, X., Wang, T., Zheng, C. et al. (2022). Technical perspective of carbon capture, utilization, and storage. *Engineering.* 14: 27–32.

8 Goksu, A., Li, H., Liu, J., and Duyar, M.S. (2023). Nanoreactor engineering can unlock new possibilities for CO_2 tandem catalytic conversion to C–C coupled products. *Glob. Chall.* 7: 2300004.

9 Robert, M., Daasbjerg, K., and Constentin, C. (2020). *Carbon Dioxide Electrochemistry.* Royal Society of Chemistry.

10 Hudson, M.R. (2005). Electrochemical reduction of carbon dioxide. *Department of Chemistry, State University of New York at Potsdam, Potsdam New York.* 13676: 1–15.

11 Chang, B., Pang, H., Raziq, F. et al. (2023). Electrochemical reduction of carbon dioxide to multicarbon (C_{2+}) products: Challenges and perspectives. *Energy Environ. Sci.* 16: 4714–4758.

12 Yu, F., Deng, K., Du, M. et al. (2023). Electrochemical CO_2 reduction: From catalysts to reactive thermodynamics and kinetics. *Carbon Capture Sci. Technol.* 6: 100081.

13 Luo, W., Nie, X., Janik, M.J. et al. (2016). Facet dependence of CO_2 reduction paths on Cu electrodes. *ACS Catal.* 6: 219–229.

14 Cao, H., Zhang, Z., Chen, J. et al. (2022). Potential-dependent free energy relationship in interpreting the electrochemical performance of CO_2 reduction on single atom catalysts. *ACS Catal.* 12: 6606–6617.

15 Lyu, Z., Zhu, S., Xu, L. et al. (2020). Kinetically controlled synthesis of Pd–Cu Janus nanocrystals with enriched surface structures and enhanced catalytic activities toward CO_2 reduction. *J. Am. Chem. Soc.* 143: 149–162.

16 Liang, S., Altaf, N., Huang, L. et al. (2020). Electrolytic cell design for electrochemical CO_2 reduction. *J. CO_2 Util.* 35: 90–105.

17 Moura De Salles Pupo, M. and Kortlever, R. (2019). Electrolyte effects on the electrochemical reduction of CO_2. *ChemPhysChem.* 20: 2926–2935.

18 Nguyen, T.N. and Dinh, C. (2020). Gas diffusion electrode design for electrochemical carbon dioxide reduction. *Chem. Soc. Rev.* 49: 7488–7504.

19 Liu, K., Smith, W.A., and Burdyny, T. (2019). Introductory guide to assembling and operating gas diffusion electrodes for electrochemical CO_2 reduction. *ACS Energy Lett.* 4: 639–643.

20 Ge, L., Rabiee, H., Li, M. et al. (2022). Electrochemical CO_2 reduction in membrane-electrode assemblies. *Chem.* 8: 663–692.

21 She, X., Zhai, L., Wang, Y. et al. (2024). Pure-water-fed, electrocatalytic CO_2 reduction to ethylene beyond 1 000 h stability at 10 A. *Nat. Energy.* 9: 81–91.

22 Wang, Y., Han, P., Lv, X. et al. (2018). Defect and interface engineering for aqueous electrocatalytic CO_2 reduction. *Joule.* 2: 2551–2582.

23 Yi, L., Chen, J., Shao, P. et al. (2020). Molten-salt-assisted synthesis of bismuth nanosheets for long-term continuous electrocatalytic conversion of CO_2 to formate. *Angew. Chem. Int. Ed.* 59: 20112–20119.

24 Shi, H., Cai, M., Li, W. et al. (2023). Direct molten-salt electro-reduction of CO_2 in porous electrodes. *Chem. Eng. J.* 462: 142240.

25 Asadi, M., Kim, K., Liu, C. et al. (2016). Nanostructured transition metal dichalcogenide electrocatalysts for CO_2 reduction in ionic liquid. *Science.* 353: 467–470.

26 Cui, Y., He, B., Liu, X. et al. (2020). Ionic liquids-promoted electrocatalytic reduction of carbon dioxide. *Ind. Eng. Chem. Res.* 59: 20235–20252.

27 Dutta, N., Bagchi, D., Chawla, G. et al. (2024). A guideline to determine faradaic efficiency in electrochemical CO_2 reduction. *ACS Energy Lett.* 9: 323–328.

28 Chen, Y., Li, C.W., and Kanan, M.W. (2012). Aqueous CO_2 reduction at very low overpotential on oxide-derived Au nanoparticles. *J. Am. Chem. Soc.* 134: 19969–19972.

29 Gu, Y., Wei, J., Wu, X. et al. (2021). A study on improving the current density performances of CO_2 electrolysers. *Sci. Rep.* 11: 11136.

30 Jalama, K. (2017). Carbon dioxide hydrogenation over nickel-, ruthenium-, and copper-based catalysts: Review of kinetics and mechanism. *Catal. Rev.* 59: 95–164.

31 Francke, R., Schille, B., and Roemelt, M. (2018). Homogeneously catalyzed electroreduction of carbon dioxide—methods, mechanisms, and catalysts. *Chem. Rev.* 118: 4631–4701.

32 Yu, J., Wang, J., Ma, Y. et al. (2021). Recent progresses in electrochemical carbon dioxide reduction on copper-based catalysts toward multicarbon products. *Adv. Funct. Mater.* 31: 2102151.

33 Ohya, S., Kaneco, S., Katsumata, H., Suzuki, T. et al. (2009). Electrochemical reduction of CO_2 in methanol with aid of CuO and Cu_2O. *Catal. Today.* 148: 329–334.

34 Wu, J., Sharifi, T., Gao, Y., et al. (2019). Emerging carbon-based heterogeneous catalysts for electrochemical reduction of carbon dioxide into value-added chemicals. *Adv. Mater.* 31: 1804257.

35 Vasileff, A., Zheng, Y., and Qiao, S.Z. (2017). Carbon solving carbon's problems: Recent progress of nanostructured carbon-based catalysts for the electrochemical reduction of CO_2. *Adv. Energy Mater.* 7: 1700759.

36 Hu, H., Ou, J., Xu, X. et al. (2021). Graphene-assisted construction of electrocatalysts for carbon dioxide reduction. *Chem. Eng. J.* 425: 130587.

37 Zhou, L. and Lv, R. (2022). Rational catalyst design and interface engineering for electrochemical CO_2 reduction to high-valued alcohols. *J. Energy Chem.* 70: 310–331.

38 Zheng, Y., Vasileff, A., Zhou, X. et al. (2019). Understanding the roadmap for electrochemical reduction of CO_2 to multi-carbon oxygenates and hydrocarbons on copper-based catalysts. *J. Am. Chem. Soc.* 141: 7646–7659.

39 Obasanjo, C. A., Gao, J., Crane, J. et al. (2023). High-rate and selective conversion of CO_2 from aqueous solutions to hydrocarbons. *Nat. Commun.* 14: 3176.

40 Zheng, T., Jiang, K. and Wang, H. (2018). Recent advances in electrochemical CO_2-to-CO conversion on heterogeneous catalysts. *Adv. Mater.* 30: 1802066.

41 Zhai, Y., Han, P., Yun, Q. et al. (2022). Phase engineering of metal nanocatalysts for electrochemical CO_2 reduction. *EScience*. 2: 467–485.

42 Weng, Z., Wu, Y., Wang, M. et al. (2018). Active sites of copper-complex catalytic materials for electrochemical carbon dioxide reduction. *Nat. Commun.* 9: 415.

43 Kortlever, R., Shen, J., Schouten, K.J.P. et al. (2015). Catalysts and reaction pathways for the electrochemical reduction of carbon dioxide. *J. Phys. Chem. Lett.* 6: 4073–4082.

44 Lyu, Z., Zhu, S., Xie, M. et al. (2021). Controlling the surface oxidation of Cu nanowires improves their catalytic selectivity and stability toward C_{2+} products in CO_2 reduction. *Angew. Chem. Int. Ed.* 60: 1909–1915.

45 Wang, J., Zhu, Z., Lin, Y. et al. (2024). Nano-engineering in zinc-based catalysts for CO_2 electroreduction: Advances and challenges. *Carbon Neutralization*. 3: 423–440

46 Yang, S., An, H., Arnouts, S. et al. (2023). Halide-guided active site exposure in bismuth electrocatalysts for selective CO_2 conversion into formic acid. *Nat. Catal.* 6: 796–806.

47 Yang, Z., Jin, Y., Feng, Z. et al. (2024). Rational strategies for preparing highly efficient tin-, bismuth-or indium based electrocatalysts for electrochemical CO_2 reduction to formic acid. *ChemSusChem*. 18: e202401181.

48 An, X., Li, S., Hao, X. et al. (2021). Common strategies for improving the performances of tin and bismuth-based catalysts in the electrocatalytic reduction of CO_2 to formic acid/formate. *Renew. Sustain. Energy Rev.* 143: 110952.

49 Frisch, M. L., Wu, L., Atlan, C. et al. (2023). Unraveling the synergistic effects of Cu-Ag tandem catalysts during electrochemical CO_2 reduction using nanofocused X-ray probes. *Nat. Commun.* 14: 7833.

50 Ye, W., Guo, X., and Ma, T. (2021). A review on electrochemical synthesized copper-based catalysts for electrochemical reduction of CO_2 to C_{2+} products. *Chem. Eng. J.* 414: 128825.

51 Okatenko, V., Loiudice, A., Newton, M.A. et al. (2023). Alloying as a strategy to boost the stability of copper nanocatalysts during the electrochemical CO_2 reduction reaction. *J. Am. Chem. Soc.* 145: 5370–5383.

52 Liu, L., Akhoundzadeh, H., Li, M. et al. (2023). Alloy catalysts for electrocatalytic CO_2 reduction. *Small Methods*. 7: 2300482.

53 Diercks, C.S., Liu, Y., Cordova, K.E. et al. (2018). The role of reticular chemistry in the design of CO_2 reduction catalysts. *Nat. Mater.* 17: 301–307.

54 Kibria, M.G., Edwards, J.P., Gabardo, C.M. et al. (2019). Electrochemical CO_2 reduction into chemical feedstocks: From mechanistic electrocatalysis models to system design. *Adv. Mater.* 31: 1807166.

55 Wagner, A., Sahm, C.D. and Reisner, E. (2020). Towards molecular understanding of local chemical environment effects in electro- and photocatalytic CO_2 reduction. *Nat. Catal.* 3: 775–786.

56 Matsubu, J.C., Yang, V.N., and Christopher, P. (2015). Isolated metal active site concentration and stability control catalytic CO_2 reduction selectivity. *J. Am. Chem. Soc.* 137: 3076–3084.

57 Hyun, G., Song, J., Ahn, C. et al. (2020). Hierarchically porous Au nanostructures with interconnected channels for efficient mass transport in electrocatalytic CO_2 reduction. *Proc. Natl. Acad. Sci.* 117: 5680–5685.
58 Luo, W., Xie, W., Li, M. et al. (2019). 3D hierarchical porous indium catalyst for highly efficient electroreduction of CO_2. *J. Mater. Chem. A.* 7: 4505–4515.
59 Li, F., MacFarlane, D.R., and Zhang, J. (2018). Recent advances in the nanoengineering of electrocatalysts for CO_2 reduction. *Nanoscale.* 10: 6235–6260.
60 Li, X., Kou, Z., and Wang, J. (2021). Manipulating interfaces of electrocatalysts down to atomic scales: Fundamentals, strategies, and electrocatalytic applications. *Small Methods.* 5: 2001010.
61 Zhang, W., Jin, Z., and Chen, Z. (2022). Rational-designed principles for electrochemical and photoelectrochemical upgrading of CO_2 to value-added chemicals. *Adv. Sci.* 9: 2105204.
62 Chen, R., Chen, S., Wang, L. et al. (2024). Nanoscale metal particle modified single-atom catalyst: Synthesis, characterization, and application. *Adv. Mater.* 36: 2304713.
63 Li, M., Chen, X., and Lin, Z. (2024). Enhanced non-metal catalyzed CO_2 reduction on doped biphenylene. *Int. J. Hydrog. Energy.* 62: 520–531.
64 Liu, K., Cao, P., Chen, W. et al. (2022). Electrocatalysis enabled transformation of earth-abundant water, nitrogen and carbon dioxide for a sustainable future. *Mater. Adv.* 3: 1359–1400.
65 Liu, S., Yang, H., Su, X. et al. (2019). Rational design of carbon-based metal-free catalysts for electrochemical carbon dioxide reduction: A review. *J. Energy Chem.* 36: 95–105.
66 Ju, W., Bagger, A., Hao, G. et al. (2017). Understanding activity and selectivity of metal-nitrogen-doped carbon catalysts for electrochemical reduction of CO_2. *Nat. Commun.* 8: 1–9.
67 Pan, B., Zhu, X., Wu, Y. et al. (2020). Toward highly selective electrochemical CO_2 reduction using metal-free heteroatom-doped carbon. *Adv. Sci.* 7: 2001002.
68 Abbasi, P., Asadi, M., Liu, C. et al. (2017). Tailoring the edge structure of molybdenum disulfide toward electrocatalytic reduction of carbon dioxide. *ACS Nano.* 11: 453–460.
69 Besharat, F., Ahmadpoor, F., Nezafat, Z. et al. (2022). Advances in carbon nitride-based materials and their electrocatalytic applications. *ACS Catal.* 12: 5605–5660.
70 Pang, J., Bachmatiuk, A., Yin, Y. et al. (2018). Applications of phosphorene and black phosphorus in energy conversion and storage devices. *Adv. Energy Mater.* 8: 1702093.
71 Calvinho, K.U., Laursen, A.B., Yap, K.M.K. et al. (2018). Selective CO_2 reduction to C_3 and C_4 oxyhydrocarbons on nickel phosphides at overpotentials as low as 10 mV. *Energy Environ. Sci.* 11: 2550–2559.
72 Zhang, L.H., Shi, Y., Wang, Y. et al. (2020). Nanocarbon catalysts: Recent understanding regarding the active sites. *Adv. Sci.* 7: 1902126.
73 Cai, G., Yan, P., Zhang, L. et al. (2021). Metal–organic framework-based hierarchically porous materials: Synthesis and applications. *Chem. Rev.* 121: 12278–12326.

74 Cao, Y., Mao, S., Li, M. et al. (2017). Metal/porous carbon composites for heterogeneous catalysis: Old catalysts with improved performance promoted by N-doping. *ACS Catal.* 7: 8090–8112.

75 Rogers, C., Perkins, W.S., Veber, G. et al. (2017). Synergistic enhancement of electrocatalytic CO_2 reduction with gold nanoparticles embedded in functional graphene nanoribbon composite electrodes. *J. Am. Chem. Soc.* 139: 4052–4061.

76 Zhang, H., Li, J., Tan, Q. et al. (2018). Metal–organic frameworks and their derived materials as electrocatalysts and photocatalysts for CO_2 reduction: Progress, challenges, and perspectives. *Chem.–A Eur. J.* 24: 18137–18157.

77 Zheng, L., Xu, L., Gu, P. et al. (2024). Lattice engineering of noble metal-based nanomaterials via metal-nonmetal interactions for catalytic applications. *Nanoscale.* 16: 7841–7861.

78 Ye, R., Ding, J., Gong, W. et al. (2019). CO_2 hydrogenation to high-value products via heterogeneous catalysis. *Nat. Commun.* 10: 5698.

79 Lai, W., Qiao, Y., Wang, Y. et al. (2023). Stability issues in electrochemical CO_2 reduction: Recent advances in fundamental understanding and design strategies. *Adv. Mater.* 35: 2306288.

80 Popović, S., Smiljanić, M., Jovanovič, P. et al. (2020). Stability and degradation mechanisms of copper-based catalysts for electrochemical CO_2 reduction. *Angew. Chem.* 132: 14844–14854.

81 Cuesta, A., Cabello, G., Osawa, M. et al. (2012). Mechanism of the electrocatalytic oxidation of formic acid on metals. *ACS Catal.* 2: 728–738.

82 Gao, Y., Neal, L., Ding, D. et al. (2019). Recent advances in intensified ethylene production—A review. *ACS Catal.* 9: 8592–8621.

83 Adegoke, K.A., Adegoke, R.O., Ibrahim, A.O. et al. (2020). Electrocatalytic conversion of CO_2 to hydrocarbon and alcohol products: Realities and prospects of Cu based materials. *Sustain. Mater. Technol.* 25: e00200.

84 Zhang, W., Hu, Y., Ma, L. et al. (2018). Progress and perspective of electrocatalytic CO_2 reduction for renewable carbonaceous fuels and chemicals. *Adv. Sci.* 5: 1700275.

85 Nguyen, D.L.T., Kim, Y., Hwang, Y.J. et al. (2020). Progress in development of electrocatalyst for CO_2 conversion to selective CO production. *Carbon Energy.* 2: 72–98.

86 Grabow, L.C. and Mavrikakis, M. (2011). Mechanism of methanol synthesis on Cu through CO_2 and CO hydrogenation. *ACS Catal.* 1: 365–384.

87 Zhao, K. and Quan, X. (2021). Carbon-based materials for electrochemical reduction of CO_2 to C_{2+} oxygenates: Recent progress and remaining challenges. *ACS Catal.* 11: 2076–2097.

88 Vasileff, A., Xu, C., Jiao, Y. et al. (2018). Surface and interface engineering in copper-based bimetallic materials for selective CO_2 electroreduction. *Chem.* 4: 1809–1831.

89 Zhang, A., Liang, Y., Li, H. et al. (2019). Harmonizing the electronic structures of the adsorbate and catalysts for efficient CO_2 reduction. *Nano Lett.* 19: 6547–6553.

90 Dutta, A., Rahaman, M., Luedi, N.C. et al. (2016). Morphology matters: Tuning the product distribution of CO_2 electroreduction on oxide-derived Cu foam catalysts. *ACS Catal.* 6: 3804–3814.

91 Wang, H. (2023). Different distributions of multi-carbon products in CO_2 and CO electroreduction under practical reaction conditions. *Nat. Catal.* 6: 1115–1124

92 Vos, R.E., Kolmeijer, K.E., Jacobs, T.S. et al. (2023). How temperature affects the selectivity of the electrochemical CO_2 reduction on copper. *ACS Catal.* 13: 8080–8091.

93 Li, J., Kuang, Y., Zhang, X. et al. (2023). Electrochemical acetate production from high-pressure gaseous and liquid CO_2. *Nat. Catal.* 6: 1151–1163.

94 Resasco, J., Lum, Y., Clark, E. et al. (2018). Effects of anion identity and concentration on electrochemical reduction of CO_2. *ChemElectroChem.* 5: 1064–1072.

95 Wu, W., Xu, L., Lu, Q. et al. (2024). Addressing the carbonate issue: Electrocatalysts for acidic CO_2 reduction reaction. *Adv. Mater.* 37: 2312894.

96 Marcos Madrazo, A., Casado Coterillo, C., and Irabien, Á. (2019). Sustainable membrane-coated electrodes for CO_2 electroreduction to methanol in alkaline media. *ChemElectroChem.* 6: 5273–5282.

97 Grim, R.G., Huang, Z., Guarnieri, M.T. et al. (2020). Transforming the carbon economy: Challenges and opportunities in the convergence of low-cost electricity and reductive CO_2 utilization. *Energy Environ. Sci.* 13: 472–494.

98 Lai, W., Qiao, Y., Zhang, J. et al. (2022). Design strategies for markedly enhancing energy efficiency in the electrocatalytic CO_2 reduction reaction. *Energy Environ. Sci.* 15: 3603–3629.

99 Chen, C., Kotyk, J.F.K. and Sheehan, S.W. (2018). Progress toward commercial application of electrochemical carbon dioxide reduction. *Chem.* 4: 2571–2586.

100 Xu, D., Li, K., Jia, B. et al. (2023). Electrocatalytic CO_2 reduction towards industrial applications. *Carbon Energy.* 5: e230.

101 Jin, S., Hao, Z., Zhang, K. et al. (2021). Advances and challenges for the electrochemical reduction of CO_2 to CO: From fundamentals to industrialization. *Angew. Chem. Int. Ed.* 133: 20795–20816.

102 Fan, L., Xia, C., Yang, F. et al. (2020). Strategies in catalysts and electrolyzer design for electrochemical CO_2 reduction toward C_{2+} products. *Sci. Adv.* 6: eaay3111.

103 Tufa, R. A., Chanda, D., Ma, M. et al. (2020). Towards highly efficient electrochemical CO_2 reduction: Cell designs, membranes and electrocatalysts. *Appl. Energy.* 277: 115557.

104 Zhong, M., Tran, K., Min, M. et al. (2020). Accelerated discovery of CO_2 electrocatalysts using active machine learning. *Nature.* 581: 178–183.

105 Roy, D., Mandal, S.C., and Pathak, B. (2021). Machine learning-driven high-throughput screening of alloy-based catalysts for selective CO_2 hydrogenation to methanol. *ACS Appl. Mater. Interfaces.* 13: 56151–56163.

106 Ma, X., Li, Z., Achenie, L.E. et al. (2015). Machine-learning-augmented chemisorption model for CO_2 electroreduction catalyst screening. *J. Phys. Chem. Lett.* 6: 3528–3533.

107 Perry, S.C., Ponce De León, C., and Walsh, F.C. (2020). Review—The design, performance and continuing development of electrochemical reactors for clean electrosynthesis. *J. Electrochem. Soc.* 167: 155525.

108 Xi, W., Peng Y., Jiang M. et al. (2024). Electrochemical CO_2 reduction coupled with alternative oxidation reactions: Electrocatalysts, electrolytes, and electrolyzers. *Appl. Catal. B: Environ.* 341: 123291.

109 Vass, Á., Endrődi, B., and Janáky, C. (2021). Coupling electrochemical carbon dioxide conversion with value-added anode processes: An emerging paradigm. *Curr. Opin. Electrochem.* 25: 100621.

110 Spurgeon, J.M. and Kumar, B. (2018). A comparative technoeconomic analysis of pathways for commercial electrochemical CO_2 reduction to liquid products. *Energy Environ. Sci.* 11: 1536–1551.

111 Bowen, A. and Fankhauser, S. (2017). Good practice in low-carbon policy. *Trends Clim. Change Legislation*, 123–140.

112 Mo, W., Tan, X., and Ong, W. (2023). Prospective life cycle assessment bridging biochemical, thermochemical, and electrochemical CO_2 reduction toward sustainable ethanol synthesis. *ACS Sustain. Chem. Eng.* 11: 5782.

113 Jouny, M., Luc, W., and Jiao, F. (2018). General techno-economic analysis of CO_2 electrolysis systems. *Ind. Eng. Chem. Res.* 57: 2165–2177.

114 Somoza-Tornos, A., Guerra, O.J., Crow, A.M. et al. (2021). Process modeling, techno-economic assessment, and life cycle assessment of the electrochemical reduction of CO_2: A review. *IScience.* 24: 102813.

115 Pérez-Fortes, M., Schöneberger, J.C., Boulamanti, A. et al. (2016). Formic acid synthesis using CO_2 as raw material: Techno-economic and environmental evaluation and market potential. *Int. J. Hydrog. Energy.* 41: 16444–16462.

116 Zhu, P. and Wang, H. (2021). High-purity and high-concentration liquid fuels through CO_2 electroreduction. *Nat. Catal.* 4: 943–951.

117 Rumayor, M., Dominguez-Ramos, A., and Irabien, A. (2019). Environmental and economic assessment of the formic acid electrochemical manufacture using carbon dioxide: Influence of the electrode lifetime. *Sustain. Prod. Consump.* 18: 72–82.

118 Overa, S., Ko, B.H., Zhao, Y. et al. (2022). Electrochemical approaches for CO_2 conversion to chemicals: A journey toward practical applications. *Acc. Chem. Res.* 55: 638–648.

119 Artz, J., Müller, T.E., Thenert, K. et al. (2018). Sustainable conversion of carbon dioxide: An integrated review of catalysis and life cycle assessment. *Chem. Rev.* 118: 434–504.

120 Liu, M., Pang, Y., Zhang, B. et al. (2016). Enhanced electrocatalytic CO_2 reduction via field-induced reagent concentration. *Nature.* 537: 382–386.

121 Li, F., Thevenon, A., Rosas-Hernández, A. et al. (2020). Molecular tuning of CO_2-to-ethylene conversion. *Nature.* 577: 509–513.

122 Wu, Y., Jiang, Z., Lu, X. et al. (2019). Domino electroreduction of CO_2 to methanol on a molecular catalyst. *Nature.* 575: 639–642.

123 Ren, S., Joulié, D., Salvatore, D. et al. (2019). Molecular electrocatalysts can mediate fast, selective CO_2 reduction in a flow cell. *Science.* 365: 367–369.

124 Fang, W., Guo, W., Lu, R. et al. (2024). Durable CO_2 conversion in the proton-exchange membrane system. *Nature.* 626: 86–91.

125 Hou, P., Wang, X., Wang, Z. et al. (2018). Gas phase electrolysis of carbon dioxide to carbon monoxide using nickel nitride as the carbon enrichment catalyst. *ACS Appl. Mater. Interfaces*. 10: 38024–38031.

126 Cheng, Y., Hou, P., Wang, X. et al. (2022). CO_2 electrolysis system under industrially relevant conditions. *Acc. Chem. Res.* 55: 231–240.

127 Bushuyev, O.S., Luna, P.D., Dinh, C.T. et al. (2018). What should we make with CO_2 and how can we make it? *Joule*. 2: 825–832.

128 Gabardo, C.M., Seifitokaldani, A., Edwards, J.P. et al. (2018). Combined high alkalinity and pressurization enable efficient CO_2 electroreduction to CO. *Energy Environ. Sci.* 11: 2531–2539.

129 Jeanty, P., Scherer, C., Magori, E. et al. (2018). Upscaling and continuous operation of electrochemical CO_2 to CO conversion in aqueous solutions on silver gas diffusion electrodes. *J. CO_2 Util.* 24: 454–462.

130 Kaczur, J.J., Yang, H., Liu, Z. et al. (2018). Carbon dioxide and water electrolysis using new alkaline stable anion membranes. *Front. Chem.* 6: 263.

131 Wang, L., Chen, W., Zhang, D. et al. (2019). Surface strategies for catalytic CO_2 reduction: From two-dimensional materials to nanoclusters to single atoms. *Chem. Soc. Rev.* 48: 5310–5349.

132 Tang, H., Wu, L., Guo, S. et al. (2024). Metabolic engineering of yeast for the production of carbohydrate-derived foods and chemicals from C_{1-3} molecules. *Nat. Catal.* 7: 21–34.

133 Lilliestam, J., Patt, A., and Bersalli, G. (2021). The effect of carbon pricing on technological change for full energy decarbonization: A review of empirical ex-post evidence. *Wiley Interdiscip. Rev. Clim. Change*. 12: e681.

134 van den Bergh, J. and Savin, I. (2021). Impact of carbon pricing on low-carbon innovation and deep decarbonisation: Controversies and path forward. *Environ. Resour. Econ.* 80: 705–715.

12

Electrochemical Nitrogen Fixation and Conversion

Rong Zhang[1,#], Shaoce Zhang[1,#], Ying Guo[2,]*

[1]*Department of Materials Science and Engineering, City University of Hong Kong, Kowloon, Hong Kong, China*
[2]*School of Interdisciplinary Studies, Lingnan University, Tuen Mun, Hong Kong, China*

*Corresponding author: ying.guo@ln.edu.cn
#These authors contributed equally to this work.

12.1 Introduction

Nitrogen (N) is an essential element for life on Earth, serving as a fundamental building block for organisms in the form of nucleotides and proteins [1–3]. Although there is a large amount of dinitrogen (N_2) gas naturally present in the atmosphere, its chemical inertness prevents its direct artificial application unless it is converted to a more reactive species [3–5]. Historically, almost all reactive nitrogen originated from biological fixation by microbes until the development of an artificial nitrogen fixation technology invented by Fritz Haber and Carl Bosch in the early 20th century [6–8]. This process, called the Haber–Bosch (H–B) process, combines atmospheric N_2 with H_2 to produce NH_3 at relatively high pressures (20–40 MPa) and temperatures (400–600 °C) [9–11]. This H–B process is identified one of the greatest inventions of the 20th century, revolutionizing agriculture and society and playing a key role in the global population growth over the past hundred years [10]. Consequently, the global production of NH_3 (~170 Mt per year) ranks second among all chemicals worldwide, with over 80% utilized for fertilizer production and the remainder for refrigerants or chemical feedstocks. However, this artificial process is highly energy-intensive ($N_2 + 3H_2 \leftrightarrow 2NH_3$, $\Delta H^\circ_{298K} = -45.9\,\text{kJ mol}^{-1}$, $\Delta G^\circ_{298K} = -16.4\,\text{kJ mol}^{-1}$, $K_{eq} = 750$), accounting for about 2% of global energy consumption per year. It also heavily depends on fossil fuels, contributing approximately 1% of carbon emission (~420 Mt per year) related with global energy [12, 13]. Furthermore, the H–B process relies on a large centralized infrastructure, which brings capital risks of more than $1 billion, high innovation barriers, and uneven regional distribution [14–16]. Therefore, developing sustainable and energy-efficient strategies for artificial nitrogen fixation is highly desirable but presents significant challenges faced in the field of fundamental science and chemical engineering.

Electrochemical Energy Conversion via Small Molecule Catalysis, First Edition.
Edited by Zhicheng Zhang and Yuanmiao Sun.

Currently, electrochemical nitrogen fixation and conversion, including nitrogen reduction to ammonia (NRR) and nitrogen oxidation to nitrate (NOR) under ambient conditions, have garnered significant research interest and achieved remarkable progress [17–19]. NH_3, due to its high hydrogen atom proportion (17.6 wt.%) and energy density (4.32 kWh L^{-1}) for liquid NH_3 [11], can directly utilize existing liquefaction, storage, and transportation facilities [20, 21]. Besides being essential for nitrogenous fertilizer production, ammonia is considered an ideal hydrogen carrier and energy storage medium for renewable energy, as it is easier to transport than hydrogen gas [22]. Additionally, ammonia is a crucial feedstock for nitric acid production [23]. Traditionally, nitric acid is synthesized through a two-step process: artificial NH_3 production via the H–B process, followed by catalytic NH_3 oxidation through the Ostwald process under harsh conditions (400–600 °C and 15–25 MPa) [24, 25]. This two-step process is more energy-intensive than the H–B procedure alone. Consequently, novel techniques for sustainable nitric acid production, especially those powered by renewable energy, are necessary [26, 27]. The electrochemical NOR process under mild conditions offers an attractive alternative for distributed, on-site production, which is compatible with intermittent renewable energy [28, 29]. Thus, traditional high-energy consumption processes are expected to be replaced by electrochemical technology, which is also expected to become a core component of future zero-carbon green energy systems.

In recent decades, due to the rapid development of sustainable electricity, energy-related electrochemical reactions have received widespread interest, driving significant advancements in electrocatalysis technology [30–36]. However, development in nitrogen-related electrochemical reactions has been sluggish due to slow progress in designing electrochemical systems, including electrodes and electrolytes. Fortunately, recent years have seen significant advancements in electrochemical N_2 fixation and conversion, particularly for NRR since 2017 [3]. The impressive achievements in NRR have spurred interest in other related electrocatalytic N_2 transformation reactions, such as nitrogen oxidation to nitrate and nitrogen reduction to hydrazine. Various functional materials have been identified to be potential catalyst materials for these processes, including precious metals (Au, Ag, Ir, Pt, Pd, Rh, Ru) [37–42], transition metals (TMs) and their compounds (Fe, Co, Ni, Cr, Mo, Ti, V) [9, 43–48], metal-free materials (B, C, P, N) [49–52], and single-atom catalysts [53–57]. Strategies like defect/interface engineering and electrolyte optimization have been employed to enhance electrocatalytic performance [58]. Despite this progress, the field remains in its infancy and faces many challenges. Current studies often rely on trial-and-error approaches or previous experience, offering limited guidance and future-oriented insights [1, 59]. The diversity and ambiguous experimental details between research groups hinder result comparability and mutual communication. Additionally, some overlooked but influential experimental issues can invalidate performance claims, affecting reproducibility and potentially leading to misleading conclusions [60, 16]. As such, to establish a dependable framework and speed up further investigations, it is essential to define community standards and best practices, like those established in oxygen reduction reaction (ORR), oxygen evolution reaction (OER), hydrogen evolution reaction (HER), and CO_2RR.

This chapter aims to present a thorough summary of the latest progress in electrochemical N_2 fixation and conversion covering both theoretical and experimental viewpoints, focusing on the design of electrochemical systems, including electrodes, electrolytes, and configurations for electrocatalytic N_2 reduction and oxidation.

12.2 Electrocatalytic N_2 Reduction Reaction

12.2.1 Electrochemistry of NRR

N_2 reduction reaction (NRR) under ambient conditions involves a six-electron reaction process, where each N_2 molecule is reduced to produce two NH_3 molecules, or undergoes a four-electron side reaction to form N_2H_4. This multi-electron process is more complex than the four- and two-electron processes (OER/ORR and HER/HOR). Previous studies often did not explicitly state the onset potential of the NRR or used assumed nitrogen and ammonia partial pressures to determine the nonequilibrium potential. Only in a few cases where the onset potential of standard equilibrium NRR was discussed (usually referring to the gaseous NH_3 formation, based on the standard molar Gibbs free energy of formation), $\Delta_f G^o_{298K}$ ($NH_{3,g}$) $= -16.4$ kJ mol^{-1}).

$$N_{2,g} + 6H^+ + 6e^- \rightarrow 2NH_{3,g} E^0 = +0.057 \text{ V vs. RHE} \quad (12.1)$$

The high solubility of NH_3 in water makes the dissolution, hydration, and protonation of NH_3 hydrates significantly contribute to the energy, especially under acidic conditions, as noted in many studies. The relationship between the standard equilibrium potential for NRR to NH_3 or NH_4^+ and pH is shown in the H_2O-N_2 system (Figure 12.1) [3, 17, 19]. The form of NH_3 is closely related to the solution's pH: at pH > 9.25, NH_3 exists as a gas, whereas at pH < 9.25, NH_4^+ predominates. Similarly, for hydrazine, the pH boundary is 6.07. N_2 can also be reduced to hydrazine or diazene through a four-electron or two-electron process, respectively. As shown in Figure 12.1, pH affects the electrochemical potential of NRR. From a thermodynamic perspective,

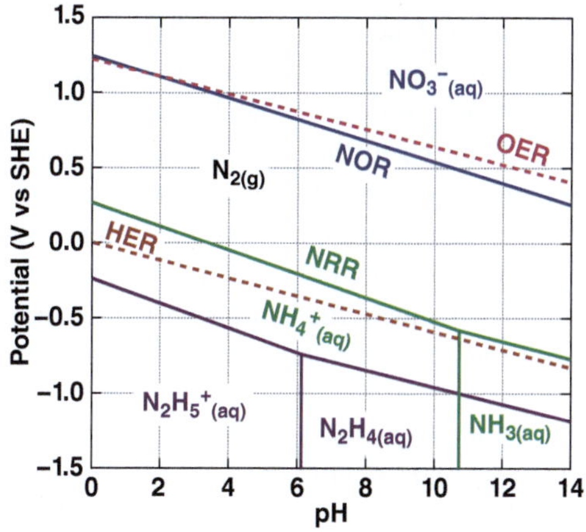

Figure 12.1 Pourbaix diagram of the N_2–H_2O system. *Source:* Reproduced from [17] / with permission of American Chemical Society.

across the entire pH range, the onset potential for NRR to ammonia is higher than that for HER. Acidic media appear more favorable for the starting potential, and the potentials for NH_3 formation and HER are very close. There is a gap between the N_2-to-NH_3 line and the H_2O-to-H_2 line where only NRR occurs, completely free from the influence of competing HER. However, under non-equilibrium conditions, especially during the initial reaction stages, the actual starting potential for HER may result in a positive shift because of the low partial pressure of H_2 gas, causing lower NH_3 efficiency. In addition, NRR to N_2H_4 requires a more negative potential, making the reaction difficult to occur, and it requires acidic conditions. The initial reaction step of NRR involves transferring of the first electron to the adsorbed dinitrogen molecule on the surface of catalyst electrode, which may or may not involve a proton. The equilibrium potentials for this step are estimated as follows [61]:

$$N_{2,g} + H^+ + e^- \rightarrow N_2H_g E^0 = -3.2 \text{ V vs. RHE} \quad (12.2)$$

$$N_{2,g} + e^- \rightarrow N_2^-, aq \quad E^0 = -4.16 \text{ V vs. RHE or} \\ E^0 = -3.37 \text{ V vs. RHE at pH} = 14 \quad (12.3)$$

The potentials apply to independent reactions that are not promoted by the catalytic surface. Nonetheless, the negative values suggest higher energy barrier inherent in the initial hydrogenation step, which makes N_2-to-NH_3 conversion more thermodynamically challenging.

12.2.2 Reaction Pathways of NRR

NRR mainly refers to three steps: (i) the N_2 adsorption, (ii) the $N \equiv N$ bond cleavage and hydrogenation, and (iii) desorption of the formed NH_3 or N-containing intermediates from the catalyst surface. The N_2 hydrogenation process experiences the process of protonation and electron transfer. Based on the sequence of hydrogenation and bond cleavage, the reaction pathways would be associated with dissociative and associative pathways, as illustrated in Figure 12.2 [62–64]. As for the dissociative pathway, the $N \equiv N$ triple bond breaks before the hydrogenation step, producing two adsorbed N atoms that are independently hydrogenated to yield NH_3. In contrast, the associative pathway involves simultaneous bond cleavage and hydrogenation.

If the nitrogen molecule coordinates end-to-end with the catalyst surface, the associative pathway can be further referred to two types: the distal and the alternative processes. In the distal pathway, the uncoordinated N atom is hydrogenated until the $N \equiv N$ bond breaks completely [65], releasing one NH_3 molecule and leaving an N atom on the surface of the catalyst for the following hydrogenation step. As for the alternative pathway, the hydrogenation reaction alternates between the two N atoms until the $N \equiv N$ triple bond is cleaved completely and two NH_3 molecules are released.

Biological catalysts or nitrogenases are able to bind N_2 strongly, and N_2 binds laterally to the surface of catalysts at both ends and undergo hydrogenation similar to the alternative pathway in enzymatic processes [63]. Notably, desorption of $*N_2H_4$ from the catalyst surface requires significant energy and typically exists as an intermediate during the reaction.

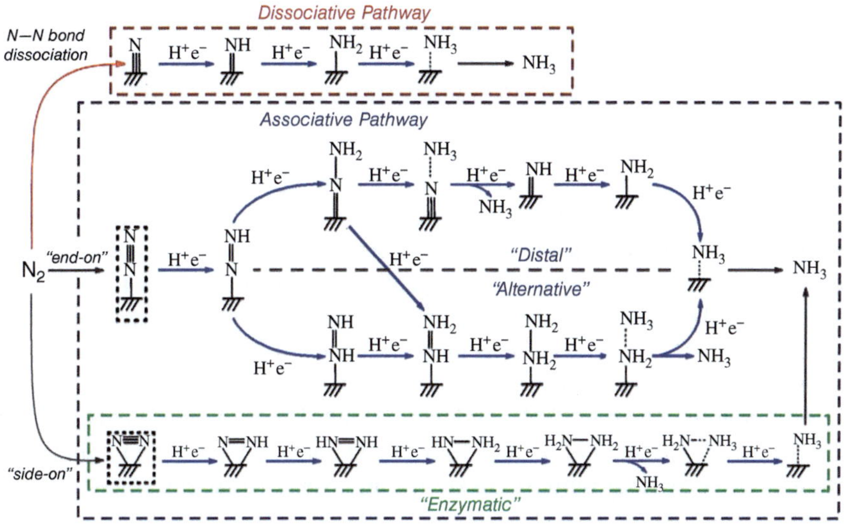

Figure 12.2 NRR pathways. *Source:* [67] / Royal Society of Chemistry.

Due to weak N_2 adsorption on catalyst surfaces, low N_2 coverage, and rapid intermediate transformation, experimentally determining the NRR mechanism is difficult. Recently, Surface Enhanced Infrared Absorption Spectroscopy (SEIAS) has been applied to detect N_2H_x ($0 \leqq x \leqq 2$) intermediate species on metallic Ru electrodes at cathodic potentials. With the cathodic potential being more negative, the coverage of N_2H_x species increases. Similar signals were also observed on Au and Pt surfaces in alkaline electrolytes [7, 66]. While some progress has been made, our understanding of the electrochemical NRR mechanism is still limited. Therefore, other in situ/operando spectroscopies that have higher sensitivity and time resolution are required to gain further insight into the NRR mechanisms.

Inspired by the catalytic oxidation of vanadium oxides [68], Skúlason and Abghoui proposed a Mars–van Krevelen (MvK) mechanism for the NRR process on transition metal nitrides (TMNs) [69–72]. Unlike traditional dissociative and associative pathways, the MvK mechanism involves the hydrogenation reaction of an N atom on the TMN catalyst surface, creating an N-vacancy [69–71]. Unlike traditional pathways, the MvK mechanism is related to the hydrogenation of nitrogen atoms on the TMN surface, creating nitrogen vacancies, which are then filled by nitrogen molecules dissolved in the electrolyte. A second nitrogen atom is subsequently reduced to a second NH_3, completing the catalytic process. After the first NH_3 is released, the MvK process would undergo three different ways: (i) gaseous N_2 dissociates directly to fill the nitrogen vacancy; (ii) N_2 binds directly to the vacancy without dissociating; and (iii) two nitrogen vacancies are created, and after releasing two NH_3 molecules, nitrogen dissociates into two nitrogen atoms to fill the dimer vacancy, regenerating the catalyst. The repair of the dimer vacancy is a non-electrochemical process that cannot be controlled by changing the bias potential. The slow kinetics of vacancy repair may lead to continuous N_2 desorption from the catalyst surface, causing the deactivation of the catalyst. This is the most important issue faced by the design and application of metal-nitride-based catalyst materials.

12.2.3 Criteria for the NRR

Establishing standardized methods for evaluating NRR catalytic performance is crucial. One major concern is the source of nitrogen in the produced ammonia. Since NRR catalysts typically generate only trace amounts of ammonia, the detected ammonia may come from external sources such as gas feed, electrolyte, electrode surface contamination, or decomposition of the nitrogen-containing catalyst itself [73, 74]. To accurately determine the nitrogen source in NRR experiments, Wang and his team suggested a standardized method, which involves initial material screening and conducting control experiments, including ^{15}N isotope labeling test, to eliminate NH_3 contamination from surrounding environment or catalyst decomposition [1, 75, 76]. Proper validation of the nitrogen source is essential for a precise evaluation of a catalyst's NRR performance. In this section, we discuss the key parameters influencing NRR performance and provide an overview of the current state, theoretical predictions, and future prospects.

Ammonia Production Rate (R_{NH_3}): The ammonia production rate per is a direct parameter for evaluating catalyst performance. The general calculation formula is as follows:

$$R_{NH_3} = \frac{n_{NH_3}}{t \times m_{cat}} \tag{12.4}$$

where n_{NH_3} is the total amount of ammonia, m_{cat} is the total catalyst mass, and t is the total time of electrolysis process. For film catalyst materials, m_{cat} can be referred to as the electrode area (typically electrochemical surface area or geometric area). However, in some metal-containing catalysts, metal elements are often considered as the active species, and the production rate is typically referred to as the mass of the active metal species rather than the total mass of the catalyst used. However, it is not an ideal universal standard for evaluating catalyst performance, as large amounts of catalysts are required for practical applications, and total mass or geometric area needs to be considered. If the active metal content is very low (<10 wt. %), it will lead to a significant overestimation of catalytic activity. Additionally, thanks to the strong interactions between the active centers and the supporter, the contribution of the support material cannot be simply ignored.

Faraday Efficiency (FE, %): FE refers to the percentage of total charge used to generate a specific reduction product. It is commonly used as a measure of the selectivity between NRR and the competing HER.

The NH_3 FE can be calculated using the following formula:

$$FE_{NH_3}(\%) = 3 \times F \frac{n_{NH_3}}{Q} \tag{12.5}$$

where 3 represents the number of electron transfers per NH_3 molecule, Q is the total charge consumed during the NRR process (C), and F is the Faraday constant (96485.3 C mol^{-1}).

For N_2H_4, the N_2H_4 FE is calculated as follows:

$$FE_{N_2H_4}(\%) = 4 \times F\frac{n_{N_2H_4}}{Q} \tag{12.6}$$

where 4 represents the number of electron transfers per N_2H_4 molecule.

The total FE of the NRR is the sum of the efficiencies for both reactions:

$$FENRR\ (\%) = FE_{NH_3}(\%) + FE_{N_2H_4}(\%) \tag{12.7}$$

Energy Efficiency (EE, %): Although EE is rarely discussed in previous reports, it is a key parameter for practical applications. It is usually expressed as the average energy cost per mole of NH_3 (kJ mol^{-1}) or the inverse, which represents the NH_3 yield per unit of energy (mol kJ^{-1}). The most advanced HB process requires about 485 kJ mol^{-1} [77], which can serve as a benchmark for assessing the efficiency of alternative approaches.

$$NH_3(g) + \frac{3}{4}\ O_2(g) \leftrightarrow \frac{3}{2}\ H_2O(l) + \frac{1}{2}N_2(g)$$

$$\Delta_c G_m^0 = -339.2\ \text{mol kJ}^{-1}$$

Assuming the standard molar Gibbs free energy change, $\Delta_c\ G_m^0$, for this reaction is the maximum energy obtainable from one mole of NH_3, the total EE of the H–B process is around 70%. Based on this, we can determine the EE of a catalytic NRR system.

The average energy input can be determined as follows:

$$E_{NH_3} = \frac{U \times Q}{1000\ \times\ n_{NH_3}} \tag{12.8}$$

where E_{NH_3} is the average energy per mole (kJ mol^{-1}), Q is the total charge consumed (C), U is the cell voltage (V), and $n_{NH_3}$is the total NH_3 amount produced (mol). In the aqueous NRR system, the anodic oxidation process is related to the OER process, whose theoretical equilibrium potential is 1.23 V vs. RHE. Since it involves the cathodic reduction reaction, we can assume an ideal, non-polarized OER at the anode, without overpotentials or kinetic limitations. Under these ideal conditions, the cell voltage can be expressed as follows:

$$U = 1.23\ V{-}\eta \tag{12.9}$$

where η represents the cathodic reduction potential during NRR (in V vs. RHE). Finally, the energy input, E_{NH_3}, can be calculated as follows:

$$E_{NH_3} = \frac{3 \times F \times (1.23 - \eta)}{1000 \times FE_{NH_3}} \tag{12.10}$$

The E_{NH_3}value then is used to calculate the NH_3 EE:

$$EE_{NH_3}(\%) = \frac{\Delta_c G_m^0}{E_{NH_3}} = \frac{1000 \times 339.2 \times FE_{NH_3}}{3 \times F \times (1.23 - \eta)} \tag{12.11}$$

The E_{NH_3} values of many reported NRR electrocatalysts can be determined using the above calculation formula. Figure 12.3a shows the EE values and NH_3

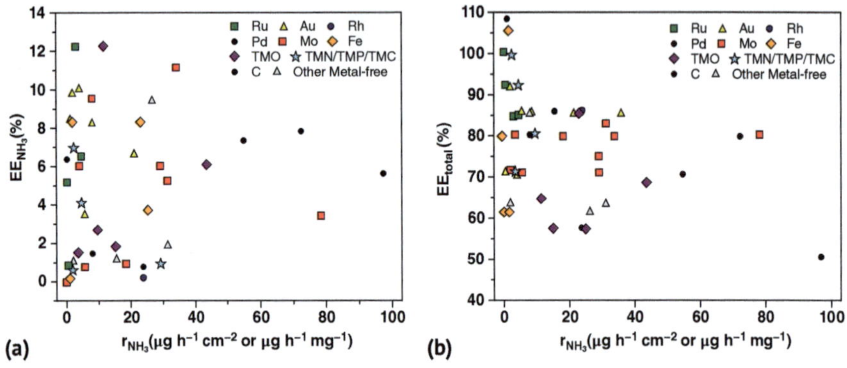

Figure 12.3 NH_3 EE of reported NRR catalysts. *Source:* [67] / Royal Society of Chemistry.

production rates of these catalysts at their optimized potentials. Each point represents an electrocatalyst, with solid markers indicating the mass-normalized mass-specific E_{NH_3} and hollow markers representing the area-normalized E_{NH_3}. It is evident that the NH_3 EE is below 13% for all reported NRR catalysts. Among them, Ru- and Pd-based catalyst materials often exhibit relatively high NH_3 EE, but their production rates are lower. Additionally, Au, Mo, and other TM-based composites show relatively better catalytic behaviors in terms of catalytic activity and EE. However, Fe-based catalyst materials, widely adopted in the industrial H–B route, have demonstrated lower NH_3 EE and E_{NH_3} values. Carbon-based materials may offer higher E_{NH_3} but they typically require very negative potentials, which results in lower NH_3 EE. VN with different morphologies (nanosheets, nanorods, and nanoparticles) exhibit varied performance, suggesting that morphology may affect the exposure of active sites, thus influencing NRR performance. For NRR to replace the commercial H–B process, NH_3 EE (~70%) and yield rate of catalysts need to be up to ~70% and ~500 mg h^{-1} mg^{-1}, respectively, but actual values of current catalysts fall far short of this standard. Note that HER is a parasitic reaction in electrochemical NRR and a major product in the reduction process. However, the above discussion does not take into account the energy required to produce H_2, so the total EE is often underestimated. Even the most efficient biological systems achieve NH_3 FE around 70–75% due to significant amounts of H_2. The ideal catalyst would have 100% FE for NRR without HER. However, to date, density functional theory (DFT) screening has not identified any most ideal materials having much higher NRR onset potential than that of the HER. Alternatively, both NRR and HER may need to be considered together. NRR catalysts that produce high NH_3 yields but have low Fe may be suitable candidates for simultaneous NH_3/H_2 production, as H_2 is a green energy source and can be easily separated from highly soluble NH_3.

Selectivity: Selectivity in electrochemical nitrogen reduction refers to the proportion of NH_3 produced relative to other nitrogen reduction products, such as hydrazine (N_2H_4). Unlike the FE, which measures the ratio of NH_3 to H_2 (HER), selectivity focuses on the nitrogen products. N_2H_4 is widely used in the aerospace sector as a deoxidizer, reducing agent, raw material, and high-energy liquid fuel.

However, it is highly toxic, and even small amounts in ammonia solutions make them unsuitable as fertilizers. Therefore, hydrazine formation should be avoided in NRR. As mentioned earlier, the desorption of N_2H_4 is often a thermodynamically hindered process during electrochemical NRR, leading to negligible N_2H_4 produced. Most reported catalysts do not generate N_2H_4, which is beneficial for distributed ammonia production systems, where the ammonia solution can be used directly without needing to remove toxic N_2H_4.

So far, only a few studies have reported N_2H_4 formation, such as using Au nanorods, with a maximum FE for N_2H_4 of just 0.163% [78]. More work is needed to develop catalysts with higher FE for N_2H_4.

Stability: Catalyst stability is a critical issue for NRR. Catalyst surfaces can be deactivated by H_2, O_2, and H_2, which block active sites for the NRR process. Water, oxygen, and hydrogen can poison the catalyst surface, reducing its ability to adsorb N_2. This poisoning can happen early in the electrolysis, causing deviations from DFT predictions and mostly promoting HER. Over extended electrolysis, hydrogen atoms produced in situ can accumulate, gradually saturating the active surface of catalysts. Another deactivation pathway is responsible for the reduction or decay of active centers, such as surface oxygen–nitrogen species in VN [79]. However, reports on catalyst stability are still not desirable. Understanding the deactivation mechanisms is important for designing highly active and stable catalysts.

12.2.4 Cathodic NRR Catalyst Materials

NRR is a complex process that involves gas–liquid–solid three-phase interfaces. The performance of NRR is influenced by factors such as gas diffusion, mass/electron transport, and intrinsic activity of catalysts. As such, it is critical for rational design of high-performance catalyst materials. Since the initial efforts in 1969 [80], due to slow reaction kinetics, with lower FE (less than 1%) and higher energy consumption than industrial and biological systems, research in the NRR field has been relatively limited [81, 82]. However, significant progress was made starting in 2016, driven by advancements in electrochemistry, nanomaterials, and characterization techniques, which improved FE and ammonia formation yields at low temperatures in aqueous media. Although the identification of suitable catalysts is not yet completed, recent reports have provided a systematic understanding and strategies for future catalyst design.

The earliest NH_3 electrosynthesis study was reported by van Tamelen and Seeley in 1969 [83], using a titanium–aluminum system containing a glyme solution with nichrome and aluminum electrodes. The Ti(II) species acted as active centers to fix and reduce N_2, forming NH_3 with a FE of 19.7%. In 1983, Sclafani et al. fabricated the first electrochemical system with Fe cathode and aqueous electrolytes under ambient conditions [82]. In 1989, Furuya and Yoshiba introduced a Fe-porphyrin-modified gas diffusion electrode, achieving an initial ammonia current efficiency of about 1.6%, though it quickly declined. They further explored various inorganic materials on gas-diffusion electrodes, but none showed significant activity or selectivity [84]. Since the turn of the century, systematic studies have been conducted to develop a comprehensive understanding of NRR electrocatalysts. Plenty of catalysts have been applied to NRR, including precious and non-precious metals, inorganic compounds, conductive polymers, carbon-based nonmetallic catalyst materials, metal-organic frameworks (MOFs), and single-atom catalysts. By tuning the chemical composition, microstructures, crystallinity, crystal facets, vacancies, and hydrophobicity of the catalysts, the catalytic activity has been significantly improved.

12.2.4.1 Metal-based NRR Catalysts

12.2.4.1.1 Precious Metal-based Catalysts

Ru-based NRR Catalysts: Early studies have proven the effectiveness of Ru-based materials for NRR. It has been found that Ru with stepped crystal surfaces adsorbs N_2 more strongly, resulting in significantly higher NRR activity compared to flat Ru surfaces [85–88]. Ru is positioned close to the peak of the volcano plot for the theoretical limits of metal surface potential (U) and N_2 adsorption energy (ΔE), which promotes N_2 dissociation and facilitates ammonia production through a low-energy associative pathway [89–91]. Moreover, Ru's vacant d-orbitals allow it to accept electron pairs from N_2, enabling the N_2 activation. Ru/C electrocatalysts, where Ru is supported in graphite, achieved an impressive NRR performance in terms of NH_3 FE and yield rate [92]. Nano-dispersed Ru materials (Ru/C) showed enhanced NRR activity [93]. Geng et al. introduced Ru^{3+} into a zeolitic imidazolate framework (ZIF), carbonized the material, and obtained single-atom Ru catalyst (Ru SAs/N-C) on nitrogen-doped carbon, achieving NH_3 yield of 120 μg h^{-1} mg^{-1} and FE of 29.6%, significantly outperforming bulk or clustered Ru catalysts (Figure 12.4a) [94]. This high catalytic performance can be ascribed to more Ru active sites and a synergistic effect.

Other studies have also linked the Ru-based catalysts to their supports. For instance, when UiO-66 was used as a support, $RuCl_3$ was carbonized to form single-atom Ru@ZrO_2/NC, which improved NRR by promoting the NRR pathway and suppressing the competing HER [56]. Wang et al. employed DFT to investigate the effect of different supports (C_2N, C_3N_4, and γ-graphene) on single-atom Ru's NRR performance (Figure 12.4b) [95]. These supports coordinate with Ru atoms, preventing aggregation and leaching while enhancing electron transfer. Simulations suggested that N_2 likely adsorbs on the active sites via the associative pathway, with T-C_3N_4 showing the highest N_2 adsorption energy and lowest NRR overpotential (Figure 12.4c). Experiments confirmed that single-atom Ru on g-C_3N_4 (SAs/g-C_3N_4) showed better NRR performance than bulk Ru, with an NH_3 production rate of 23.0 μg h^{-1} mg^{-1} and an FE of 8.3%, indicating that g-C_3N_4 modifies Ru's d-electron states and improves its Fermi level [96].

Pd-based NRR Catalysts: Pd-based catalysts are more cost-effective compared to Ru-based catalysts and they are also effective for NRR. Pd-catalyzed NRR is limited by the step of N_2 conversion to N_2H. Hydride transfer in α-Pd-H, similar to the Grotthuss mechanism, lowers the energy barrier for this reaction. In this mechanism, protons in water transfer between molecules through oxygen atoms, like a relay race [98]. Pd-based materials can absorb H from the environment, acting as intermediaries for proton transfer from H_2O to N_2. Wang et al. developed α-Pd-H on carbon black, demonstrating enhanced NRR performance. Compared to Pt and

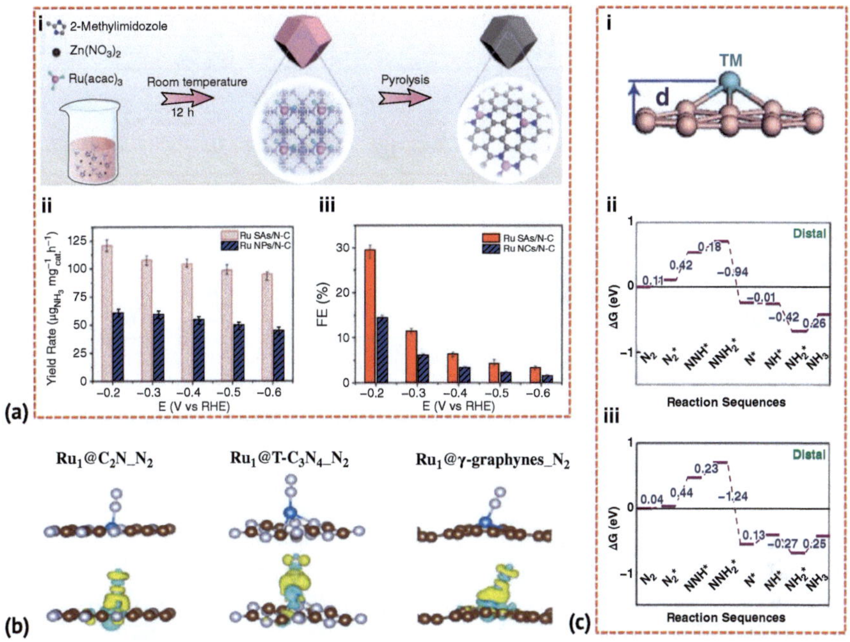

Figure 12.4 (a) The synthesis route, NH_3 yield, and FE of Ru SAs/N-C. *Source:* [94] / John Wiley & Sons. (b) The charge density difference of N_2 on different Ru-based catalysts. *Source:* [95] / John Wiley & Sons. (c) DFT results for Ru/Bα (ii: distal) and Ru/Bβ (iii: distal), respectively. *Source:* [97] / Royal Society of Chemistry.

Au, this catalyst exhibits thermodynamically favorable surface *H kinetics for the conversion of N_2 to $*N_2H$ [99]. The hydrogenation of Pd to α-Pd-H enables proton hopping, lowering the reaction energy for N_2 by 0.3 eV and promoting NRR at low overpotentials (see Figure 12.5a). Further studies on PdH catalysts reveal that porous $PdH_{0.43}$ exhibits excellent NRR performance at −150 mV (NH_3 yield rate: 20.4 μg h^{-1} mg^{-1}, NH_3 FE: 43.6%; Figure 12.5b) [100]. Isotope labeling shows that lattice hydrogen in $PdH_{0.43}$ provides the active H for the hydrogenation process of NRR. In situ characterizations and theoretical calculations indicate that Pd hydrides reduce the activation energy for the N_2-to-N_2H conversion (Figure 12.5c–e).

Au-based NRR Catalysts: Au-based catalysts are seen as highly promising for NRR and have been widely studied. The unoccupied d-orbitals can accept lone pair of electrons from N_2 molecules, while their HER activity is significantly inferior to Ru-, Pt-, and Pd-based materials, making them more favorable for NRR. Current efforts focus on manipulating the microstructure (or crystal facets) and enhancing the interaction between Au catalysts and their supports to suppress HER. For example, catalysts with fluffy structures increase the active Au site exposure. Wang and co-workers introduced a porous Au film catalyst (pAu/NF), which showed excellent performance because of the high exposure of the active Au centers [101]. Similarly, ZIFs like ZIF-8 can support Au-based porous nanomaterials, enhancing

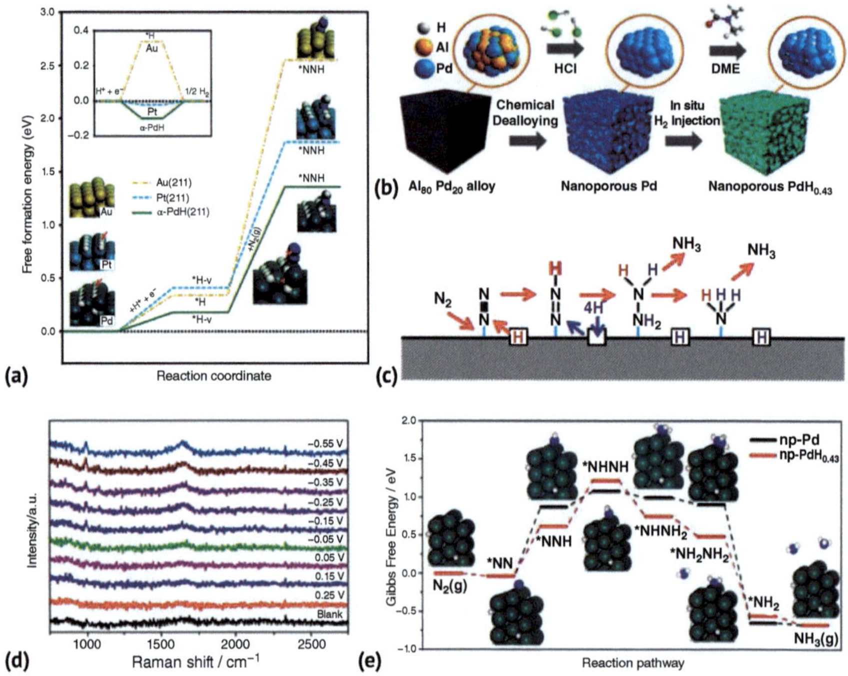

Figure 12.5 (a) DFT calculation of HER (insets) and NRR energy files for different surfaces. (b) Preparation process for $PdH_{0.43}$ catalyst, (c) the corresponding NRR pathways, and (d) in situ Raman spectra. (e) Gibbs free energy for np-Pd and np-$PdH_{0.43}$. *Source:* [99] / Springer Nature / CC BY 4.0.

NRR activity by providing abundant nanometal attachment sites and dual functions of preventing Au loss and suppressing HER. In one notable study, a Ni-Au pair catalyst achieved a Faradaic efficiency (FE) of 67.8% [102]. Theoretical studies indicate that the synergy interaction between Ni and Au species boosts the NRR process by transferring electrons from Ni to Au, creating an electron-rich state favorable for NRR (Figure 12.6a and b). Electron-rich Au catalysts have also demonstrated strong performance, such as an unique catalyst formed by depositing CoO_x on nano-Au, which adjusted the Au electronic structure and promoted the conversion from Au nanoclusters to a higher oxidation state (Figure 12.6c) [103]. Further calculations suggested that the Au oxidation state influences the adsorption N_2 molecules and reduces the reaction-free energy of NRR intermediates (Figure 12.6d). Despite these advancements, Au-based catalysts still have untapped potential for NRR. Future work should focus on designing optimized Au catalysts to leverage their HER resistance and unique d-electron properties. Further spectral studies, like those by Yao et al., are also needed to better understand the NRR mechanism on Au catalysts [66].

Other Noble Metal-based NRR Catalysts: Compared to Ru, Pd, Au, Rh, and Ag have been less explored as NRR catalysts, with most studies focusing on the influence of morphology, such as research on Ag and Rh nanosheets [41, 104–106]. Pt, despite being a top choice for HER catalysis, is seldom used for NRR due to its extremely

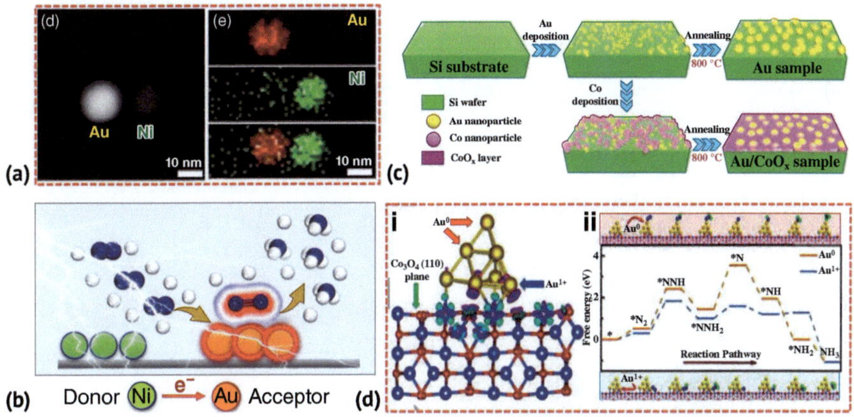

Figure 12.6 (a) HAADF-TEM and element mapping images of Au/Ni nanoparticles. (b) Illustration of electron transfer between Au–Ni couples for the NRR process. *Source:* [102] / American Chemical Society. (c) Synthesis procedure of Au/CoO_x catalysts. (d) The electron density difference of Au/CoO_x for NRR process at the Au^0 and Au^{1+} positions of Au/CoO_x samples. *Source:* [103] / John Wiley & Sons.

poor selectivity for the process [107–109]. While noble metal–based catalysts are highly effective for water splitting, ORR, and CO_2RR, their application in NRR often encounters challenges due to their shortage and high cost. Despite these limitations, noble metal–based catalysts still show promising NRR performance. Optimizing their morphology and electronic structure can help leverage their advantages while minimizing their drawbacks.

12.2.4.1.2 Non-noble Metal-based Catalysts

Alternatively, there has been growing interest in non-precious metals, especially TMs. These metal-based materials, with vacant d-orbitals, can accept lone pairs from N_2 and form M—N bonds on their surfaces, weakening the $N \equiv N$ triple bonds. This process is facilitated by p-back bonding interaction, which promotes π^* orbital formation. Fe and Mo, key cofactors in natural nitrogenase enzymes, have inspired the development of Mo- and Fe-based catalyst materials for NRR. Research is also expanding to early TM-based materials like Ti, which have also been developed. This overview highlights recent advances in the development of non-precious-based NRR catalysts, focusing on Fe-, Mo-, and other tTM-based materials.

Fe-based NRR Catalysts: Iron (Fe), as a key component of nitrogenase, is considered capable of effectively catalyzing NRR. Research of Fe-based catalyst materials has mainly focused on metallic Fe and its oxides, with activity often enhanced by the doping process. The metallic Fe in NRR catalysts remains relatively rare and typically involves atomizing Fe active sites and introducing oxygen vacancies to enhance activity. Li et al. introduced a graphene-supported Fe-N_3 catalyst [63], showing that highly polarized Fe-N_3 sites are essential for NRR and can be optimized to improve N_2 adsorption. Another notable development by Lv et al. involved creating a monoatomic Fe-based catalyst by impregnating a MOF with an Fe precursor, followed by carbonization and acid treatment. A single-atom Fe dispersion

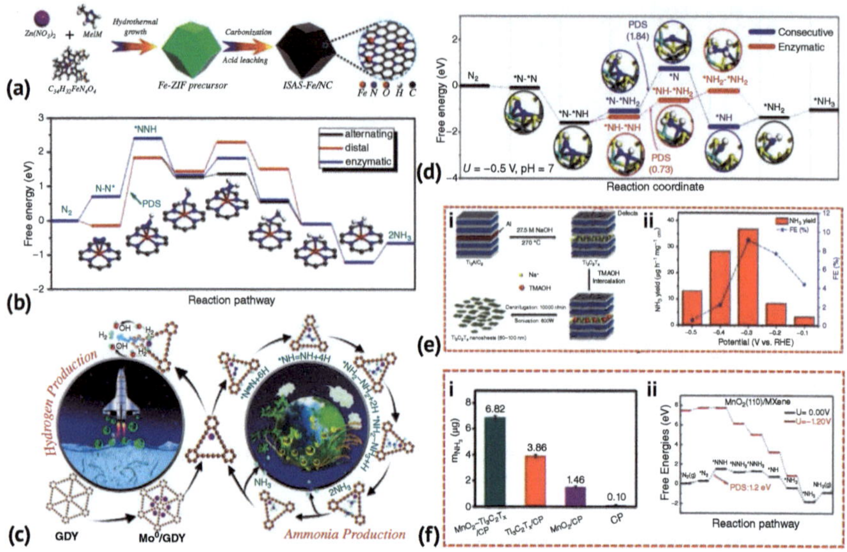

Figure 12.7 (a) Preparation route for ISAS-Fe/NC catalyst. (b) Gibbs free energy diagram over ISAS-Fe/NC. *Source:* Reproduced from [110] / with permission of Elsevier. (c) Synthesis of Mo^0/GDY and reaction pathway for HER and NRR. *Source:* Reproduced from [122] / with permission of American Chemical Society. (d) Gibbs free energy diagrams over the Mo–Sn–Sn trimer. *Source:* [125] / Royal Society of Chemistry. (e-i) Synthesis route of MXene nanosheet catalyst and (e-ii) the corresponding NH_3 yield and FE. *Source:* [126] / Royal Society of Chemistry. (f-i) NH_3 yields over MnO_2–$Ti_3C_2T_x$ heterostructure and other control catalysts and (F-ii) Gibbs free energy diagram on the surface of MnO_2(110)-MXene catalyst. *Source:* [127] / Royal Society of Chemistry.

(ISAS-Fe/NC, Figure 12.7a) built on an N-doped carbon framework demonstrated impressive NRR activity [110]. The Fe-N_4 (Figure 12.7b) was found to enhance the dispersion and stability of active Fe centers through N-Fe coordination, suppressing the HER and enhancing NRR. During electrochemical processes, Fe-based catalysts may undergo oxidation or reduction, requiring advanced testing methods to observe changes during NRR and to better understand the origin of their activity. Despite Fe being a key component of natural nitrogenase, the performance of current Fe-based catalysts in NRR remains limited, necessitating further inspiration from natural N_2 fixation mechanisms to enhance catalytic efficiency.

Mo-based NRR Catalysts: another essential component of nitrogenase, which has demonstrated superior NRR catalytic performance compared to Fe. Various Mo-based catalysts, including zero-valent Mo [53, 65, 111], Mo_xS_y [112–115], Mo_xN [116, 117], and Mo_xC [118–121], have been reported. Strategies such as exposing active sites, morphology control, doping (e.g., B, N, C, P, and S), and electronic structure modification have significantly enhanced Mo-based catalyst performance. For example, Hui et al. developed a single-atom molybdenum catalyst supported by graphdiyne (Mo^0/GDY), which exhibited excellent NRR and HER activities (Figure 12.7c) [122]. In addition, Mo single-atom catalysts on N-doped porous carbon (SA-Mo/NPC) and single-atom Mo catalysts anchored on gt-C3N4 also showed outstanding performance [53, 123]. The high specific surface area of the supports effectively dispersed Mo atoms, improving the active site exposure, and ensured rapid electron transfer. The crystal facets of Mo significantly impact its catalytic NRR activity. For instance, Wang et al. synthesized Mo nanocatalysts with (110) and (210) crystal facets and indicated that the (110) facet exhibited higher NRR activity due to its lower nitrogen adsorption energy compared to hydrogen, which is key to promoting NRR [124]. Moreover, Mo nanorod materials demonstrated at least 100 times the NRR performance of commercial Mo materials, because of the exposed Mo (110) surface. Further studies revealed that doping Mo into SnS_2 can adjust the catalyst's electronic orbitals and introduce abundant S vacancies (SVs). These would enhance the N_2 activation process over Mo-Sn-S sites and lower the energy barrier for NRR, significantly improving catalytic activity (Figure 12.7d).

Ti-based NRR Catalysts: since titanium metal has vacant d-orbitals, causing electronic interactions with Lewis bases of N_2 molecules, making it a promising candidate for NRR. Current studies on Ti-based NRR catalysts focus primarily on 2D MXene ($Ti_3C_2T_x$), TiO_2, and their derivatives [9]. MXene has demonstrated electrochemical NRR activity, with N_2 adsorbing at Ti sites within the MXene layers rather than on the surface, confirming Ti as the active site [128]. Zhao et al. reported that 2D MXene (T = F, OH) achieved NH_3 yields and FE of up to 20.4 μg h^{-1} mg^{-1} and 9.3% (at −0.4 V), respectively. Moreover, theoretical studies showed that N_2 adsorption on MXene elongates the $N \equiv N$ bond [129]. Additionally, residual fluorine (F) from HF used during synthesis may partially remain in $Ti_3C_2T_x$, reducing its conductivity and negatively affecting electrochemical processes involving electron transport. Li et al. addressed this issue by developing fluorine-free $Ti_3C_2T_x$ nanosheets, resulting in an improved NH_3 yield of 36.9 μg h^{-1} mg^{-1} (Figure 12.7e) [126]. These findings prompted further efforts to reduce residual fluorine and enhance the catalytic performance of $Ti_3C_2T_x$ [9]. Additionally, $Ti_3C_2T_x$ was shown to boost MnO_2's electron transfer efficiency [130, 131]. Researchers speculated and confirmed through experiments and simulations that MnO_2-doped $Ti_3C_2T_x$ would outperform both pristine MnO_2 and $Ti_3C_2T_x$ in NRR performance (Figure 12.7f) [127].

Other Non-noble Metal-based NRR Catalysts: Bi-based materials have also been investigated for NRR. Qiao et al. enhanced NRR activity through morphology control and electronic tuning, creating Bi nanosheets with p-electron delocalization, achieving activity 10 times higher than Bi nanoparticles [132]. Defect engineering and structural optimization further improved Bi catalysts' performance [133–135]. Hao et al. demonstrated that K^+ doping in Bi nanocrystals increased NH_3 yield rate to 200 mmol g^{-1} h^{-1}, outperforming most reported NRR catalysts by 1–2 orders of magnitude [136]. Table 12.1 summarizes the catalytic activity of non-precious metal-based catalysts for a comprehensive overview of the current status of NRR-related catalysts.

12.2.4.2 Nonmetal-based NRR Catalysts

Nonmetal-based catalysts are gaining attention due to their unique advantages. These advantages include the following: (i) Although advances in nanometal preparation have enabled the development of single-atom catalysts, limited surface exposure and low loading efficiency continue to hinder the optimal utilization of metal active

Table 12.1 The NRR activity of reported non-noble metal–based catalysts.

Catalysts	Electrolyte	Potential (vs. RHE)	NH_3 yield	FE (%)	Ref.
B-TiO_2	0.1 M Na_2SO_4	−0.8 V	14.4 μg h^{-1} mg^{-1}	3.4	[137]
V-TiO_2	0.5 M $LiClO_4$	−0.4 V	17.73 μg h^{-1} mg^{-1}	15.3	[138]
VN NSs	0.1 M HCl	−0.5 V	8.40×10^{-11} mol s^{-1} cm^{-2}	2.25	[139]
VN NPs	0.05 M H_2SO_4	−0.1 V	3.30×10^{-10} mol s^{-1} cm^{-2}	6	[79]
VN NW/CC	0.1 M HCl	−0.3 V	2.48×10^{-10} mol s^{-1} cm^{-2}	3.85	[140]
mVO_x-rGO	0.1 M Na_2SO_4	−0.35 V	18.84 μg h^{-1} mg^{-1}	16.97	[141]
Cr_3C_2@CNF	0.1 M HCl	−0.3 V	23.9 μg h^{-1} mg^{-1}	8.6	[142]
Cr_2O_3	0.1 M Na_2SO_4	−0.9 V	4.96×10^{-11} mol s^{-1} cm^{-2}	6.78	[143]
$CrO_{0.66}N_{0.56}$	1 m M H_2SO_4	2.0 V	8.9×10^{-11} mol s^{-1} cm^{-2}	6.7	[144]
Mn_3O_4	0.1 M Na_2SO_4	−0.8 V	11.6 μg h^{-1} mg^{-1}	3	[145]
CoO QD/RGO	0.1 M Na_2SO_4	−0.6 V	21.5 μg h^{-1} mg^{-1}	8.3	[146]
C@CoS@TiO_2	0.1 M Na_2SO_4	−0.55 V	8.09×10^{-10} mol s^{-1} cm^{-2}	28.6	[43]
Co–N_x–C	0.05 M H_2SO_4	−0.3 V	37.6 μg h^{-1} cm^{-2}	17.6	[147]
CoP HNC	1 M KOH	−0.4 V	5.39 μg h^{-1} cm^{-2}	7.36	[148]
CoS_2/NS-G	0.05 M H_2SO_4	−0.05 V	25.0 μg h^{-1} mg^{-1}	25.9	[149]
GDY/Co_2N	0.1 M HCl	−0.2 V	219.72 μg h^{-1} mg^{-1}	58.6	[150]
Mo–Co	0.1 M Na_2SO_4	−0.1 V	89 μg h^{-1} mg^{-1}	13.5	[151]
Fe–Ni LDH	0.1 M Na_2SO_4	−0.5 V	1.72×10^{-10} mol s^{-1} cm^{-2}	23	[152]
N-NiO/CC	0.5 M $LiClO_4$	−0.5 V	22.7 μg h^{-1} mg^{-1}	7.3	[153]
BCN/Ni	0.1 M HCl	−0.3 V	16.72 μg h^{-1} cm^{-2}	13.06	[154]
Ni	0.1 M LiCl	1.8 V	3.58×10^{-11} mol s^{-1} cm^{-2}	17.2	[155]
Porous Ni	Propanol/H_2O	—	1.54×10^{-11} mol s^{-1} cm^{-2}	0.89	[156]
Cu/PI-300	0.1 M KOH	−0.3 V	12.4 μg h^{-1} cm^{-2}	6.56	[157]
Cu-TiO_2/CP	0.5 M $LiClO_4$	−0.55 V	21.31 μg h^{-1} mg^{-1}	21.99	[158]
Nb_3O_7(OH)/CFC	0.1 M Na_2SO_4	−0.4 V	622 μg h^{-1} mg^{-1}	39.9	[159]
Nb_2O_5 nanofiber	0.1 M HCl	−0.55 V	43.6 μg h^{-1} mg^{-1}	9.26	[160]
$NbSe_2$ nanosheet	0.1 M Na_2SO_4	−0.4 V	89.5 μg h^{-1} mg^{-1}	13.9	[161]
SnO_2	0.1 M Na_2SO_4	−0.7 V	1.47×10^{-10} mol s^{-1} cm^{-2}	2.17	[162]
Sn/SnS_2	0.1 M PBS	−0.8 V	23.8 μg h^{-1} mg^{-1}	6.5	[163]
SnS_2@Ni	0.1 M Na_2SO_4	−0.5 V	9.17×10^{-10} mol s^{-1} cm^{-2}	10.8	[164]
LaF	0.5 M $LiClO_4$	−0.45 V	55.9 μg h^{-1} mg^{-1}	16	[165]
$LaFeO_3$	0.1 M HCl	−0.55 V	18.59 μg h^{-1} mg^{-1}	8.77	[166]
2D W_2N_3	0.1 M KOH	−0.2 V	11.66 μg h^{-1} mg^{-1}	11.67	[167]
$W_{18}O_{49}$	0.1 M Na_2SO_4	−0.2 V	5.3 μg h^{-1} mg^{-1}	12.1	[168]
PC/Sb/$SbPO_4$	0.1 M HCl	−0.25 V	25 μg h^{-1} mg^{-1}	31	[169]
Sb_2Te_3	0.1 M KOH	0 V	13.8 μg h^{-1} cm^{-2}	27.7	[170]

Table 12.1 (Continued)

Catalysts	Electrolyte	Potential (vs. RHE)	NH_3 yield	FE (%)	Ref.
Sc/NC	0.1 M HCl	−0.1 V	20.40 $\mu g\ h^{-1}\ cm^{-2}$	11.2	[171]
Y/NC	0.1 M HCl	−0.1 V	23.2 $\mu g\ h^{-1}\ cm^{-2}$	12.1	[171]
Ta_2O_5	0.1 M HCl	−0.7 V	15.9 $\mu g\ h^{-1}\ mg^{-1}$	8.9	[172]
Re_2MnS_6	0.1 M Na_2SO_4	−0.3 V	15 $\mu g\ h^{-1}\ mg^{-1}$	14	[173]
Fe-ReS_2	0.1 M Na_2SO_4	−0.2 V	80.4 $\mu g\ h^{-1}\ mg^{-1}$	12.3	[174]
DyF_3	0.1 M Na_2SO_4	−0.45 V	10.9 $\mu g\ h^{-1}\ mg^{-1}$	8.8	[175]

sites. Nonmetal-based catalysts are gaining attention in the field of NRR due to their unique advantages. These advantages include the following: (i) Although advances in nanometal preparation have enabled the development of single-atom catalysts, limited surface exposure and low loading efficiency would continue to hinder the optimal utilization of metal active sites. (ii) TMs with d-orbitals have ability to effectively adsorb N_2 and form M—N bonds, weakening the N ≡ N bonds. However, they can also generate M-H intermediates, which favor the HER process over NRR [64, 176]. (iii) Many metal-based catalysts are prone to oxidation or reduction, which severely limits their recyclability. While the consumption of metals can be managed on a laboratory scale, the use of valuable metal resources becomes economically unfeasible on an industrial scale. To overcome these challenges, researchers have developed nonmetal-based catalysts, including polymers and carbon-based materials. These catalysts often feature high surface areas, doping effects, structural defects (e.g., vacancies), and distinctive electronic properties, making them highly promising for NRR. This section highlights advancements in nonmetal-based NRR catalysts, with an emphasis on carbon-, boron-, and phosphorus-based materials.

12.2.4.2.1 *Carbon-based NRR Catalysts*

C-based materials have also been essential in electrocatalysis, either as active catalysts or carriers for hybrid materials. Their significance stems from features such as porous or layered structures, excellent conductivity, versatile surface engineering options (e.g., defect introduction or heteroatom doping), and low cost. Surface defects typically act as active species for the N_2 adsorption, while heteroatom doping (e.g., N, B, P) can modify the bandgap, charge density, and spin density, boosting NRR performance [177]. Generally, pure carbon materials show limited NRR activity. For instance, Li et al. demonstrated that CC-450 with rich defects, created through thermal treatment of commercial carbon cloth (CC), outperformed defect-free CC in NRR activity [178]. Furthermore, high electron-affinity heteroatoms can increase the positive charge density on nearby carbon atoms, further enhancing NRR efficiency.

N-doped Carbon-based NRR Catalysts: N-doping promotes charge transfer and increases positive charge accumulation on adjacent carbon atoms, enhancing N_2 adsorption. For example, pyrolyzing cicada sloughs with $ZnCl_2$ produced a nitrogen-rich, micro/mesoporous carbon-based catalysts (NCF-900) with

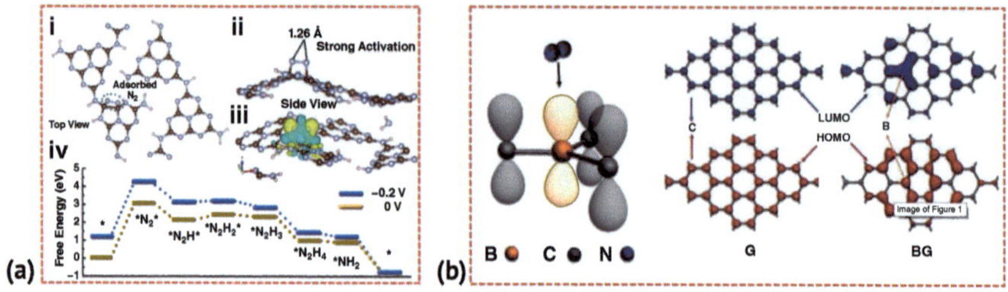

Figure 12.8 (a) The structure of PCN with nitrogen vacancy (a-i) and the adsorbed structure with N_2 (a-ii and a-iii) and free energy map of NRR (a-iv). *Source:* [117] / John Wiley & Sons. (b) Schematic diagram and LUMO (blue) and HOMO (red) of B-doped graphene and the undoped graphene. *Source:* Reproduced from [51] / with permission of Elsevier.

impressive NRR activity due to its high nitrogen content and porous microstructure [177, 179]. Similarly, Lv et al. prepared nitrogen-vacancy-rich polymeric carbon nitride (PCN-NV4) by pyrolyzing melamine polycondensates, achieving significant NRR performance. DFT calculations indicated that dual-terminal sites near nitrogen vacancies (two carbons linked to an imidazole group) enhanced interfacial electron transfer, promoted N_2 adsorption, and elongated the N—N bond (Figure 12.8a). Compared to PCN without nitrogen vacancies, PCN-NV4 demonstrated superior NRR activity, highlighting the importance of engineering nitrogen-vacancy defects. To prevent false positives due to nitrogen contamination, rigorous contamination control and $^{15}N_2$ isotopic labeling are crucial.

Boron-doped Carbon-based NRR Catalysts: Boron (B) atoms, known for their electron deficiency and Lewis acidity, can effectively interact with Lewis-basic N_2 molecules, making them attractive for NRR applications. Boron-doped carbon materials have shown significant potential as NRR catalysts [51, 180, 181]. Zheng et al. designed boron-doped graphene (BC_3) and validated its outstanding NRR performance through both theoretical and experimental analyses. The improved activity was linked to the following: (i) the preservation of graphene's sp^2 hybridization and 2D planar conjugated structure post-doping, providing a large surface area and exposing active sites; (ii) boron's electron deficiency, which enhances N_2 attachment via lone pair interactions; and (iii) boron's ability to chemically adsorb negatively polarized oxygen atoms, thereby suppressing HER and boosting NRR selectivity [182]. In addition, B's lower electronegativity (2.04) compared to carbon (2.55) causes significant differentiation of electron density, altering the stability of carbon rings in boron-doped graphene (Figure 12.8b). The highest occupied molecular orbital (HOMO) and lowest unoccupied molecular orbital (LUMO) of non-doped and B-doped graphene indicate that the electron density is rearranged due to the doping effect, disrupting the even distribution of electrons in graphene. By combining ultrasonic treatment with thermal processing, commercial bulk B_4C was converted into boron-doped graphene quantum dots (BGQD), which achieved an ammonia yield rate of 28.6 $\mu g\ h^{-1}\ mg^{-1}$, outperforming boron carbide nanosheets [181]. While these quantum dots are believed to enhance charge transfer and N_2 adsorption, further studies are needed to clarify the underlying mechanisms.

Oxygen Group Element–doped Carbon-based NRR Catalyst: Beyond B and N doping, elements like O, S, and P also play crucial roles in tuning the electronic configurations of C-based materials, thereby improving their electrochemical performance [31, 183–185]. For example, O doping can create O-Li active states, influencing NRR efficiency [186]. Similarly, S doping has demonstrated the effectiveness of enhancing the NRR activity of C-based materials with S quantum dot-graphene composites achieving an NH_3 yield of 28.56 μg h^{-1} mg^{-1} [187]. The combined effects of S/N co-doping can further elevate graphene's catalytic performance [184]. Tian et al. demonstrated that S/N co-doped graphene catalyst (NSG) outperformed both undoped and single-element-doped graphene in NRR activity [187]. DFT studies revealed that co-doping introduced more defects, facilitated electron transfer, and enabled NSG to adsorb N_2 molecules and subsequently weaken the N—N bonds. S/N co-doped carbon can be prepared by pyrolyzing carbon cloth with ammonium persulfate treatment, which also acts as the N and S source. The resulting material (CC-APS-800) showed significant NRR activity, owing to the abundance of defects and heteroatoms. Moreover, O-doped carbon materials, including amorphous O-doped nanosheets (NS), hollow O-doped nanotubes, and O-doped graphene, have also considered to be promising NRR catalysts [188–190].

12.2.4.2.2 Boron-based NRR Catalysts

Boron (B), due to its electron deficiency and low electronegativity, significantly enhances its ability to facilitate electron transfer for NRR process. Studies have shown that B-based nonmetallic catalysts are highly efficient catalysts, with both experimental and theoretical evidence supporting their activity [52, 191–195]. Sun et al. developed B_4C NSs and achieved an NH_3 yield of 26.57 μg h^{-1} mg^{-1} and FE of 15.95% (Figure 12.9a). Theoretical calculations identified the $*NH_2—NH_2$ to $*NH_2—NH_3$ transition as the rate-limiting step in NRR on the B_4C (110) surface [195]. The proposed mechanism involves N_2 being captured by distal adsorption and binding to B at both ends of B_4C through enzymatic pathway. The conversion steps from $*N \equiv N$ to $*NH—*NH_2*$ and then to $*NH_2—NH_2$ are easily implemented, with the rate-limiting step having a barrier of only 0.34 eV (Figure 12.9b). Although the energy barrier for NH_3 desorption ($*NH_3 \rightarrow NH_3$) is 1.73 eV, the total free energy of 2.7 eV renders the process thermodynamically favorable.

Another study shows that 2D boron nanosheets (BNS) can selectively enhance NRR [52]. When N_2 adsorbs onto three adjacent boron atoms (3B), the $N \equiv N$ bond is elongated from 1.112 to 1.132 Å, enabling a distal associative mechanism (Figure 12.9c). Additionally, charge accumulation at the interface of the 3B sites and N_2 molecule results in a transfer of 0.38 e^- to N_2 (Figure 12.9d). While NH_3 desorption poses challenges, higher NH_3 coverage lowers the desorption energy barrier, aiding its release from the catalyst surface (Figure 12.9e, f). Moreover, hydrogenation of the B(104) surface enhances NRR while inhibiting HER. The overlap between boron's 2p orbitals and $*N_2$'s 2p orbitals, supported by σ-donation and π-back bonding, further underscores the catalytic potential of boron-based materials in NRR (Figure 12.9g). These findings highlight the mechanisms behind boron-based catalysis and offer strategies for efficient ammonia synthesis.

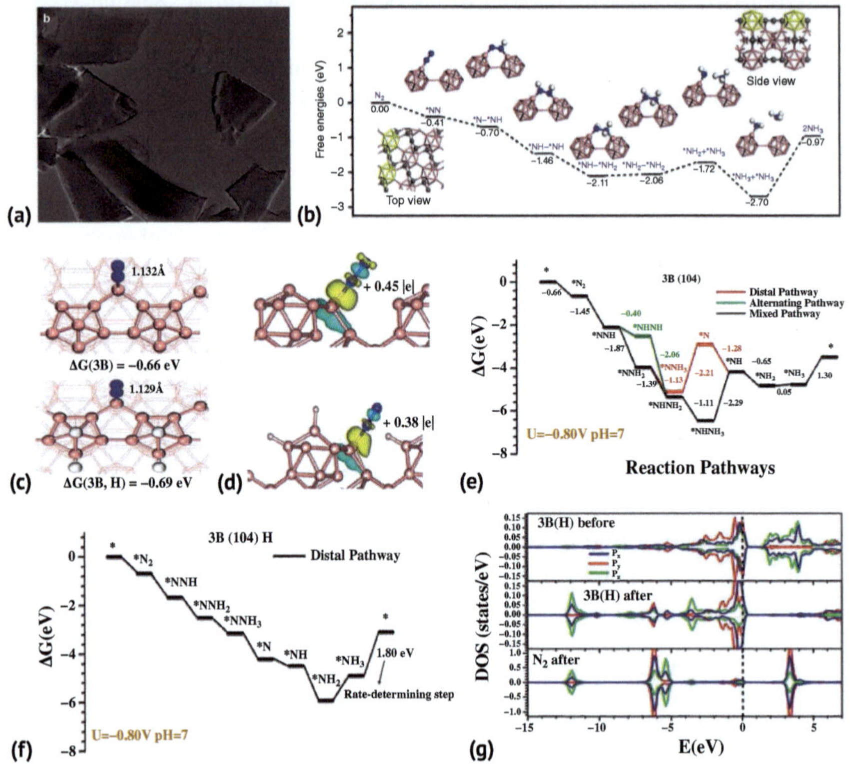

Figure 12.9 (a) SEM image of B_4C nanosheets and (b) Free energy profile of NRR over B_4C. *Source:* [195] / Springer Nature / CC BY 4.0. (c) The optimal structure of N_2 adsorbed on different B-based catalysts, (d) The electron density difference, and (e, f) Free energy profile for the different NRR pathways over B(104) surface. (g) DOS plots of 3B(H). (c–g) *Source:* Reproduced from [52] / with permission of American Chemical Society.

12.2.4.2.3 Phosphorous-based NRR Catalysts

Phosphorus (P)-based materials, though less explored in NRR, show substantial promise as catalysts. Phosphorus and nitrogen, both in the same periodic table group, share similar valence electron configurations (P: $3s^23p^3$; N: $2s^22p^3$), giving them compatible chemical properties [196]. Black phosphorus can be transformed into monolayers, due to its layered crystalline structure, resulting in abundant active sites for the N_2 adsorption [197, 198]. Recent studies confirm the high NRR activity of certain P-based materials [199–202]. For example, Wang et al. used fully exfoliated black P nanosheets (FL-BP NSs) as NRR catalysts, delivering an NH_3 yield of 31.37 μg h^{-1} mg^{-1} [202]. Theoretical calculations revealed that FL-BP NSs' asymmetric electronic distribution is key to their activity. HOMO and LUMO electron densities, concentrated at zigzag edges, enable N_2 conversion to NH_3 via an associative mechanism involving substitutional hydrogenation.

12.2.5 Challenges and Prospects of the NRR

Despite decades of research, NRR still encounters considerable scientific and technical obstacles. One major issue is the extremely low ammonia yield, which raises concerns about the validity of nitrogen sources used in experiments. Recently, researchers have increasingly stressed the necessity of carefully verifying nitrogen sources during the NRR process. Ensuring the reliability of experimental data is of utmost importance. Achieving high ammonia yields (approximately mg $mmol^{-1}$ or more) necessitates the adoption of stringent verification methods, such as blank/argon controls and isotopic labeling, to trace the origin of the produced NH_3 product. Early studies conducted before 2017 often neglected isotopic labeling experiments and frequently employed nitrates or other nitrogen-containing precursors [203], leading to skepticism about the credibility of their findings. Even studies that utilized ^{15}N labeling frequently failed to purify commercially obtained $^{15}N_2$ gas beforehand, which could introduce contamination, and most only conducted qualitative analyses [78, 135]. Future work must adopt stricter protocols to eliminate potential nitrogen contamination. The use of airtight systems combined with quantitative ^{15}N measurements can significantly reduce the chances of false positives.

12.3 Electrocatalytic N_2 Oxidation Reaction

12.3.1 Development of Nitric Acid Production

Nitric acid (HNO_3) plays a vital role in industrial and agricultural applications [204, 205]. The earliest industrial production of HNO_3 was achieved in the early 20th century using the Birkeland–Eyde (B–E) process (Figure 12.10) [206–209]. This process employs thermal plasma to react atmospheric N_2 and O_2, forming nitrogen oxides (NO_x), which are then absorbed by water to produce nitric acid. The following reactions outline the process (Eqs. (12–14)).

$$N_2 + O_2 \leftrightarrow 2NO \ (\textit{plasma catalysis}) \tag{12.12}$$

$$2NO + O_2 \rightarrow 2NO_2 \ (\textit{thermal catalysis}) \tag{12.13}$$

$$3NO_2 + H_2O \rightarrow 2HNO_3 + NO \tag{12.14}$$

While the theoretical energy demand to fix 907 kg of nitrogen is 1628 kWh, the actual energy input to the arc furnace reaches 61 000 kWh, with only about 3% of the energy utilized for nitrogen oxidation [209]. Efforts to enhance the efficiency of the B–E process have included recovering waste heat from processed gases and operating the furnace at elevated pressures [207]. However, these enhancements have not provided a substantial economic advantage compared to the combined H–B process and Ostwald oxidation method [210]. Despite its high efficiency, the H–B process is energy-intensive and generates significant carbon emissions due to its reliance on natural gas or coal as a hydrogen source [11, 211–213]. The Ostwald

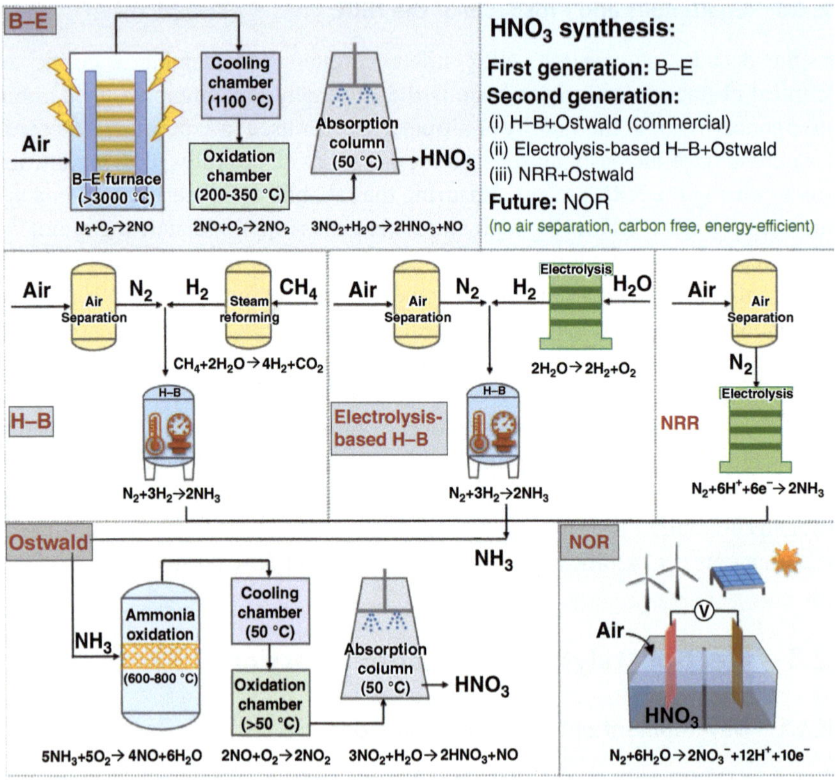

Figure 12.10 The different pathways for HNO_3 synthesis. *Source:* [19] / John Wiley & Sons.

oxidation method further converts the ammonia produced by the H–B process into nitrogen oxides via the complex catalytic oxidation process under extreme working conditions (15–25 MPa, 400–600 °C) [214, 215]. While this method is widely used for synthesizing HNO_3, it also has notable shortcomings, including high energy consumption and substantial CO_2 emissions.

Recent advancements have highlighted the potential of renewable energy-powered electrocatalytic water splitting, combined with the H–B process, for synthesizing green ammonia [11]. Trammo, Inc., in partnership with an Australian energy company, has initiated a project in Chile aimed at producing 1 million tons of green ammonia annually [19]. This approach is expected to drastically lower fossil fuel usage and carbon emissions compared to traditional H–B methods. A novel strategy for artificial ammonia synthesis differs from the H–B process [213, 216] Nevertheless, conventional HNO_3 production methods typically involve multiple stages, making direct electrocatalytic oxidation of nitrogen in water an attractive alternative [217, 218]. This process can be described by an electrochemical reaction (Eq. (10)), with its standard potential referenced to the RHE.

$$N_2 + 6H_2O \rightarrow 2NO_3^- + 12H^+ + 10e^- \quad (12.15)$$

$$E^0 = 1.24 \text{ V vs. RHE}$$

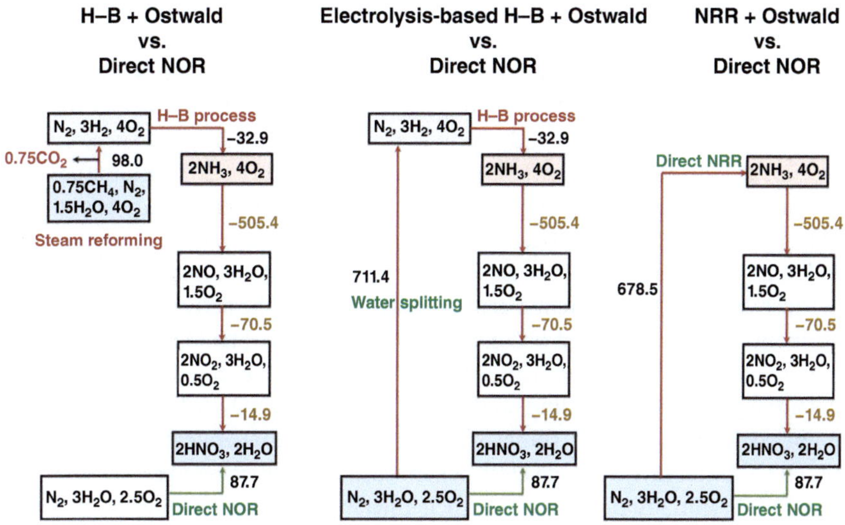

Figure 12.11 Some important thermochemical data related to different reaction routes. *Source:* [19] / John Wiley & Sons.

Unlike nitrogenase-inspired artificial NRR, no biological systems are known to directly oxidize N_2 and O_2 into aqueous HNO_3. This explains why direct nitrogen oxidation, despite its potential to replace conventional methods, remains relatively underdeveloped [3]. Electrocatalytic nitrogen oxidation reaction (NOR) presents distinct advantages. First, in comparison to energy-intensive H–B and Ostwald method—where H source is derived from water electrolysis or steam reforming—NOR offers significant reductions in energy consumption and carbon footprint. Moreover, NOR enables decentralized, on-demand production of HNO_3, reducing challenges associated with storage and transportation. Second, while combining NRR-based ammonia production with the Ostwald process bypasses the H–B step and appears eco-friendly, it is thermodynamically inefficient [219]. The multistep nature of NRR followed by oxidation results in substantial energy losses as heat, as illustrated in Figure 12.11. Furthermore, the NRR process demands additional energy input for air separation, whereas NOR can directly utilize air as the feedstock. From the Pourbaix diagram for the N_2–H_2O system, at pH levels above 1.3, the 10-electron NOR pathway is thermodynamically more favorable than the competing 4eORR [3]. This indicates that achieving high selectivity for HNO_3 production under appropriate conditions is achievable. Therefore, the development of advanced catalyst materials for nitrate and HNO_3 production is crucial to realizing energy-efficient, sustainable, and carbon-neutral production pathways.

12.3.2 Electrochemistry of NOR

The electrocatalytic NOR is a relatively recent area of exploration, and its mechanisms remain largely unclear. Most proposed pathways are based on theoretical calculations, with limited experimental validation of intermediates, highlighting a critical need for further study. Advanced in situ characterization techniques could offer valuable insights into these mechanisms [220], presenting both opportunities and challenges. This section reviews reference mechanisms for NOR, aiming to support the development of more efficient catalysts for future use.

Current NOR catalysts mainly include precious metals like Pt, Ru, Rh, and Pd [24, 25, 27, 28, 39], as well as TMs such as Fe and Co [221–223]. These studies provide an initial understanding of the field, focusing on effective catalysts and potential reaction mechanisms. Despite differences in pathways among catalysts, most follow a "+OH^- and $-e^-$" process. Similar to NRR, NOR involves three main stages: (1) adsorption of N_2, where molecules are absorbed to the active surface of the catalysts in "side-on" or "end-on" configurations without breaking the N ≡ N bond [62]. (2) *Nitrogen Activation:* This step involves continuous deprotonation, with two potential pathways depending on the final products (HNO_3 or NO_x). One pathway, followed by $ZnFe_{0.4}Co_{1.6}O_4$ spinel oxide, is purely electrochemical, producing HNO_3 [221]. The other involves both electrochemical and chemical redox processes, where NO/NO_2 is initially produced electrochemically and then chemically oxidized to HNO_3 [24, 26]. The OER competes with the NOR process, necessitating a catalyst that can balance the N_2 activation and OER competition. For example, Ru-doped TiO_2 adjusts OER performance through RuO_2 [224]. (3) *Desorption of Products:* The final step is the desorption of products from the catalyst surface. Nørskov et al. outlined the same pathway on TiO_2 and IrO_2 surfaces, combining electrochemical and chemical steps [26]. Here, *OH generated from water oxidation reacts with N_2 to form N_2O, followed by further reactions. Competing processes like OER can be partially suppressed under specific conditions [225]. Figure 12.12 illustrates reported bond-breaking mechanisms and electrocatalytic pathways, offering a guide for future research.

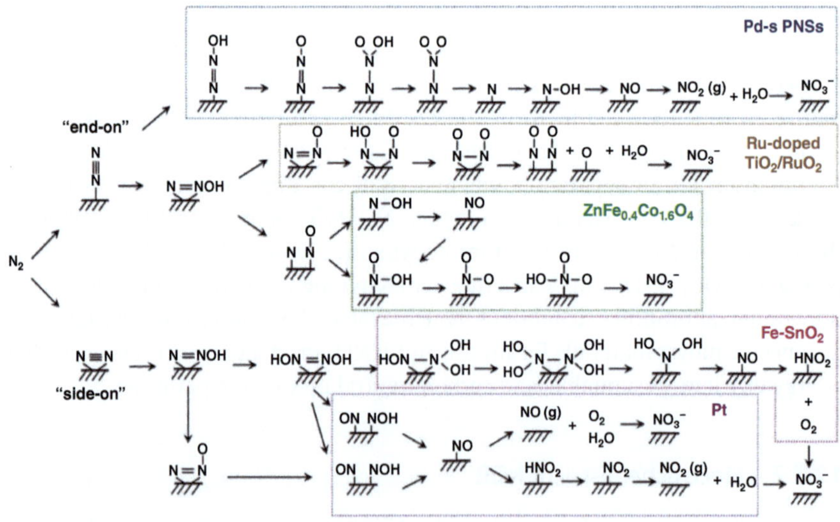

Figure 12.12 The possible NOR pathways based on previously reported NOR. *Source:* [19] / John Wiley & Sons.

12.3.3 Anodic NOR Catalyst Materials

NOR is an endothermic process with reduced entropy [216–218], similar to cathodic NRR. It involves N_2 activation, *OH adsorption, and multielectron transfer at the solid–liquid–gas interface. The strong $N \equiv N$ triple bond and high cleavage energy (410 kJ mol^{-1}) [26, 62–64, 219, 220, 226, 227] make NOR kinetically slow, particularly for the 10-electron transfer step, which is less efficient than the 4-electron transfer in OER [224]. Furthermore, the NOR process competes with the parasitic OER over a wide range of oxidation potentials, resulting in poor NOR performance, although OER can also promote NOR by providing active oxygen intermediates. Therefore, highly active and selective electrocatalysts are crucial for improving NOR performance.

Recent advancements in NOR catalysts have focused on noble metals like Pt, Ru, and Pd, along with their compounds [228]. For example, Yan and colleagues developed a catalyst by over-doping Ru into TiO_2 particles, creating additional RuO_2 and enhancing electrochemical conversion of N_2 to NO* (Figure 12.13a) [229]. However, the high cost and limited availability of these materials hinder large-scale applications [27]. TMs such as Ti, Fe, and Co show promise for N_2 activation due to their partially filled d-orbitals, which form strong metal—N bonds [24, 25, 28, 221, 230]. TiO_2-based materials, in particular, are promising due to their abundance, high OER overpotential, and thermal stability [26, 219, 222]. However, pristine TiO_2 offers few active sites for N_2 adsorption. Doping with heteroatoms has been shown to improve performance by enhancing nitrogen adsorption, reducing activation barriers, and adjusting oxygen vacancies, thereby increasing NRR efficiency. For instance, doping TiO_2 with atoms such as Fe, Mo, and Ru can significantly enhance NOR performance [26, 225].

The effectiveness of N_2 activation depends largely on the number of unpaired electrons occupied in d-orbitals of the coordinating metal atom. Ding et al. prepared catalysts such as Fe/SnO_2, Fe/TiO_2, Ru/Mn_3O_4, and RuO_2, achieving impressive NOR behaviors in terms of yield rate and FE [222, 224, 230]. They systematically investigated the catalytic mechanism from the perspectives of electronic structures,

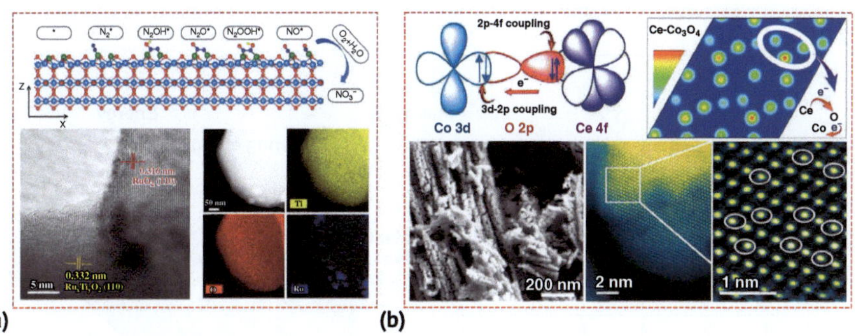

Figure 12.13 (a) The proposed NOR pathways, the TEM image, and element mapping of the Ru/TiO_2 catalyst. *Source:* [24] / John Wiley & Sons. (b) The schematic diagram for the orbital interaction and SEM and TEM images of the $Ce\text{-}Co_3O_4$ catalyst. *Source:* [29] / John Wiley & Sons.

coordination environments, and orbitals of the active sites. Orbital hybridization, which increases the number of unpaired electrons occupied in d-orbitals, has proven effective [204, 231]. However, conventional 3d–2p and 3d–3d orbital hybridization method often fails because of their small energy differences between electronic orbitals. However, incorporating 4f and 5f elements for electronic modulation offers a promising approach to boost charge transfer between metal atoms.

Cerium (Ce), with its unique 4f orbital, actively participates in chemical bonding by donating electrons to coordinated elements. Among lanthanides, Ce uniquely transitions between Ce^{3+} ($4f^1$) and Ce^{4+} (4f), attributed to the stability of its 4f configurations. Yan et al. proposed an orbital hybridization strategy by creating a Ce 4f–O 2p–Co 3d structure in Ce-doped Co_3O_4 (Ce–Co_3O_4). This catalyst achieved an HNO_3 yield of 24.76 μg h^{-1} mg^{-1} and a FE of 31.93%, significantly outperforming pure Co_3O_4 (Figure 12.13b) [29].

12.3.4 Summary and Challenges of the NOR

Electrochemical NOR to produce nitrate or nitric acid is an emerging area that warrants significant interest and development. Currently, no benchmark catalysts with exceptional performance exist, necessitating clear objectives for future studies. A techno-economic analysis (TEA) evaluated catalytic performance requirements, electricity costs, and the market price of HNO_3. Existing materials struggle to compete with traditional methods such as the H–B and Ostwald processes, highlighting the urgent need for more efficient catalysts, particularly those operating at lower potentials. Key performance targets and existing challenges are illustrated in Figure 12.14a.

The low N_2 solubility in water limits its conversion efficiency. To overcome this, increasing the concentration of dissolved N_2 in the solution or designing nanomaterials with strong nitrogen adsorption capabilities is critical. While fluorinated ionic liquids and alcohols can increase the N_2 solubility, their high costs and complex post-processing make them unsuitable for large-scale use. Alternatively, nanomaterials with tailored structures, such as pointed or porous designs or

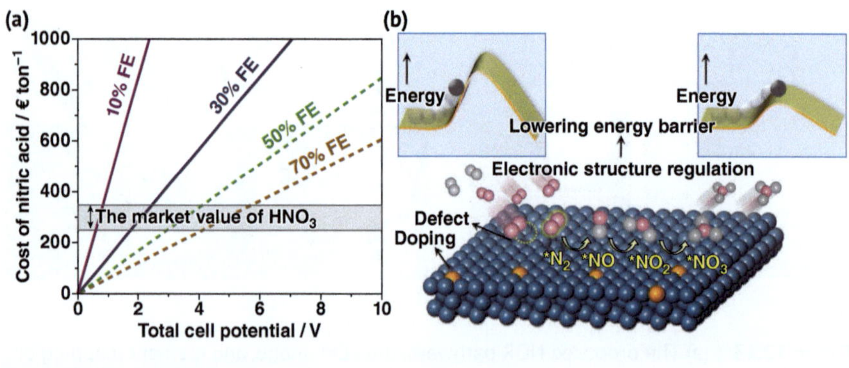

Figure 12.14 (a) The predicted cost for the HNO_3 production from the NOR process and a comparison with the market value of HNO_3. *Source:* [232] / Royal Society of Chemistry / CC BY 3.0. (b) Schematic illustration to enhance the catalytic NOR activity. *Source:* [19] / John Wiley & Sons.

gas-affinitive surfaces, could enrich nitrogen at catalyst surfaces. Flow cells also hold promise for improving nitrogen mass transfer, increasing N_2 concentration at reaction interfaces, and extending retention time during NOR tests. However, further research is needed to optimize flow cell systems, including their NOR kinetics, thermodynamics, and fluid dynamics.

Theoretical modeling can facilitate the discovery of efficient NOR catalysts by identifying activity descriptors and experimental optimization strategies. While Nørskov and colleagues explored possible NOR mechanisms and its competition with OER at the molecular level, their findings were limited to specific catalysts [26]. Therefore, systematic theoretical simulations of NOR catalysts are needed. Catalysts exposed to acidic or alkaline environments under positive potentials may undergo surface evolution. Identifying the actual active species through in situ characterizations can help design efficient catalysts. Insights from other well-established electrocatalytic oxidation systems could inspire the development of novel NOR catalysts. Introducing features such as O vacancies and dopants can modify the electronic structure and atomic configuration of catalyst materials, reducing energy barriers and improving the binding energies of nitrogen intermediates (Figure 12.14b). Developing catalysts that perform efficiently at lower overpotentials is crucial to avoid excessive catalyst decomposition.

The reaction mechanism of electrochemical NOR remains unclear. Advanced spectroscopy techniques are needed to study the reaction pathway and local reaction environment near the electrode, providing corroboration for DFT calculations and confirming the actual conversion of nitrogen. In situ FTIR, Raman, XPS, and online DEMS are useful tools for identifying NOR intermediates and confirming the formation of NO_3^-, NO_2^-, and other nitrogen oxides [66, 233–235].

Adjusting the pH of the electrolyte can significantly impact NOR. Alkaline electrolytes are particularly effective for NOR, as the reaction is more favorable in basic media, and theoretical potentials decrease with increasing pH, as shown in the N_2–H_2O Pourbaix diagram. However, product recovery must be considered. In acidic or neutral media with high salt concentrations, distillation can be used to collect nitric acid. In alkaline media, nitrate produced through this process can be directly employed to manufacture KNO_3 fertilizer without further purification. Practical process designs must carefully account for real-world application scenarios. Additionally, HNO_3 or nitrates generated from NOR can serve as nitrogen sources for electrocatalytic C—N coupling in CO_2 reduction, enabling the synthesis of organic nitrogen compounds such as urea, acetamide, and amines [236–238].

12.4 Conclusion and Outlook

The global nitrogen cycle is critical for maintaining ecological balance, but human activities have significantly disrupted it. Electrochemical nitrogen fixation methods, including nitrogen reduction (NRR) and nitrogen oxidation (NOR), offer sustainable pathways to mitigate these impacts. These processes, powered by clean energy sources and characterized by low costs, represent the future evolution of the Earth's nitrogen cycle. This chapter provides an in-depth analysis of NRR and NOR, discussing thermodynamic considerations, reaction mechanisms, and recent advancements in catalyst materials. Despite significant progress in enhancing catalyst activity and selectivity for ammonia and nitric acid production, the artificial N_2 fixation process still faces competition from HER and OER. To address these challenges and improve electrochemical nitrogen fixation, several key issues need to be considered:

1. *Computational Studies*: Computational methods can identify active centers for nitrogen activation on catalyst surfaces and reveal reaction pathways with minimal energy requirements. DFT-guided synthesis can optimize defective catalyst surfaces, creating more active centers or species and reducing energy barriers and overpotential.
2. *Electrolyte Optimization*: Selecting the appropriate electrolytes is vital for enhancing the activity and selectivity of NRR and NOR processes. Stable aqueous electrolytes are particularly desirable as they suppress parasitic HER and OER reactions, thereby improving overall electrochemical nitrogen fixation performance.
3. *Metal-free Catalysts*: Although earth-abundant transition-metal catalysts have garnered considerable interest, they frequently face challenges related to HER or OER. Metal-free catalyst materials, such as C-, B-, and P-based electrocatalysts, present promising alternatives. These materials can enhance reaction efficiency by facilitating electron transfer between N2 and active sites, forming nonmetal—N bonds, and improving N_2 fixation performance.
4. *Mechanistic Understanding*: The current understanding of electrochemical nitrogen fixation mechanisms, especially NOR, remains incomplete, hindering the development of artificial N_2 fixation. In situ techniques, including optical methods, X-ray-based techniques, and electron microscopy, provide valuable insights into the changes occurring at the electrode/electrolyte interface during reactions. These insights can deepen our understanding of the mechanisms and promote the development of efficient catalytic systems for NRR and NOR.

By addressing these key issues, we can advance the field of electrochemical nitrogen fixation and contribute to a more sustainable and balanced nitrogen cycle.

Acknowledgment

This work was supported by the start-up fund from Lingnan University.

References

1 Andersen, S.Z., Čolić, V., Yang, S. et al. (2019). A rigorous electrochemical ammonia synthesis protocol with quantitative isotope measurements. *Nature* 570 (7762): 504–508.

2 Ashida, Y., Arashiba, K., Nakajima, K. et al. (2019). Molybdenum-catalysed ammonia production with samarium diiodide and alcohols or water. *Nature* 568 (7753): 536–540.

3 Chen, J.G., Crooks, R.M., Seefeldt, L.C. et al. (2018). Beyond fossil fuel–driven nitrogen transformations. *Science* 360 (6391): eaar6611.

4 Fu, X., Pedersen, J.B., Zhou, Y. et al. (2023). Continuous-flow electrosynthesis of ammonia by nitrogen reduction and hydrogen oxidation. *Science* 379 (6633): 707–712.

5 Li, K., Andersen, S.Z., Statt, M.J. et al. (2021). Enhancement of lithium-mediated ammonia synthesis by addition of oxygen. *Science* 374 (6575): 1593–1597.

6 Hu, L., Xing, Z., and Feng, X. (2020). Understanding the electrocatalytic interface for ambient ammonia synthesis, *ACS Energy Lett.* 5 (2): 430–436.

7 Yao, Y., Wang, H., Yuan, X.-z. et al. (2019). Electrochemical nitrogen reduction reaction on ruthenium. *ACS Energy Lett.* 4 (6): 1336–1341.

8 Zheng, G., Yan, J.-M., and Yu, G. (2019). Nitrogen reduction reaction. *Small Methods* 3 (6): 1900070.

9 Guo, Y., Wang, T., Yang, Q. et al. (2020). Highly efficient electrochemical reduction of nitrogen to ammonia on surface termination modified $Ti_3C_2T_x$ MXene nanosheets. *ACS Nano* 14 (7): 9089–9097.

10 Kyriakou, V., Garagounis, I., Vourros, A. et al. (2020). An electrochemical Haber–Bosch process. *Joule* 4 (1): 142–158.

11 MacFarlane, D.R., Cherepanov, P.V., Choi, J. et al. (2020). A roadmap to the ammonia economy. *Joule* 4 (6): 1186–1205.

12 Qing, G., Ghazfar, R., Jackowski, S.T. et al. (2020). Recent advances and challenges of electrocatalytic n_2 reduction to ammonia. *Chem. Rev.* 12: 5437–5516.

13 van der Ham, C.J.M., Koper, M.T.M., and Hetterscheid, D.G.H. (2014). Challenges in reduction of dinitrogen by proton and electron transfer. *Chem. Soc. Rev.* 43 (15): 5183–5191.

14 Johnsen, K., Ryu, H.J., Grace, J.R. et al. (2016). Introduction to ammonia production. *Chem. Eng. Sci.* 61 (4): 1195–1202.

15 Medford, A.J. and Hatzell, M.C. (2017). Photon-driven nitrogen fixation: current progress, thermodynamic considerations, and future outlook. *ACS Catal.* 7 (4): 2624–2643.

16 Zhao, X., Hu, G., Chen, G.-F. et al. (2021). Comprehensive understanding of the thriving ambient electrochemical nitrogen reduction reaction. *Adv. Mater.* 33 (33): 2007650.

17 Guo, Y., Zhang, S., Zhang, R. et al. (2022). Electrochemical nitrate production via nitrogen oxidation with atomically dispersed Fe on N-doped carbon nanosheets. *ACS Nano* 16 (1): 655–663.

18 Rosca, V., Duca, M., de Groot, M.T. et al. (2009). Nitrogen cycle electrocatalysis. *Chem. Rev.* 109 (6): 2209–2244.

19 Wang, Y., Li, T., Yu, Y. et al. (2022). Electrochemical synthesis of nitric acid from nitrogen oxidation. *Angew. Chem. Int. Ed.* 61 (12): e202115409.

20 Zhang, R., Guo, Y., Zhang, S. et al. (2022). Efficient ammonia electrosynthesis and energy conversion through a Zn-nitrate battery by iron doping engineered nickel phosphide catalyst. *Adv. Energy Mater.* 12 (13): 2103872.

21 Zhang, S., Chen, D., Chen, P. et al. (2024). Concurrent mechanisms of hot electrons and interfacial water molecule ordering in plasmon-enhanced nitrogen fixation. *Adv. Mater.* 36 (15): 2310776.

22 Zhang, R., Zhang, Y., Xiao, B. et al. (2024). Phase engineering of high-entropy alloy for enhanced electrocatalytic nitrate reduction to ammonia. *Angew. Chem. Int. Ed.* 63 (35): e202407589.

23 Zhang, R., Zhang, S., Guo, Y. et al. (2022). A Zn–nitrite battery as an energy-output electrocatalytic system for high-efficiency ammonia synthesis using carbon-doped cobalt oxide nanotubes. *Energy Environ. Sci.* 15 (7): 3024–3032.

24 Kuang, M., Wang, Y., Fang, W. et al. (2020). Efficient nitrate synthesis via ambient nitrogen oxidation with Ru-doped TiO_2/RuO_2 electrocatalysts. *Adv. Mater.* 32 (26): 2002189.

25 Li, T., Han, S., Wang, C. et al. (2021). Ru-doped Pd nanoparticles for nitrogen electrooxidation to nitrate. *ACS Catal.* 11 (22): 14032–14037.

26 Anand, M., Abraham, C.S., and Nørskov, J.K. (2021). Electrochemical oxidation of molecular nitrogen to nitric acid – towards a molecular level understanding of the challenges. *Chem. Sci.* 12 (18): 6442–6448.

27 Wang, Y., Yu, Y., Jia, R. et al. (2019). Electrochemical synthesis of nitric acid from air and ammonia through waste utilization. *Natl. Sci. Rev.* 6 (4): 730–738.

28 Han, S., Wang, C., Wang, Y. et al. (2021). Electrosynthesis of nitrate via the oxidation of nitrogen on tensile-strained palladium porous nanosheets. *Angew. Chem. Int. Ed.* 60 (9): 4474–4478.

29 Li, S., Wang, X., Chi, X. et al. (2023). Boosted N_2 activation through 4f–2p–3d orbital hybridization for efficient nitrate electrosynthesis. *Adv. Funct. Mater.* 33 (49): 2306098.

30 Guo, Y., Zhang, R., Zhang, S. et al. (2022). Recent advances in Zn–CO_2 batteries for the co-production of electricity and carbonaceous fuels. *Nano Mater. Sci.*

31 Jiao, Y., Zheng, Y., Jaroniec, M. et al. (2015). Design of electrocatalysts for oxygen- and hydrogen-involving energy conversion reactions. *Chem. Soc. Rev.* 44 (8): 2060–2086.

32 Li, C.W. and Kanan, M.W. (2012). CO_2 reduction at low overpotential on Cu electrodes resulting from the reduction of thick Cu_2O films. *J. Am. Chem. Soc.* 134 (17): 7231–7234.

33 Liu, X., Jiao, Y., Zheng, Y. et al. (2019). Building up a picture of the electrocatalytic nitrogen reduction activity of transition metal single-atom catalysts. *J. Am. Chem. Soc.* 141 (24): 9664–9672.

34 Xia, W., Mahmood, A., Liang, Z. et al. (2016). Earth-abundant nanomaterials for oxygen reduction. *Angew. Chem. Int. Ed.* 55 (8): 2650–2676.

35 Xia, W., Mahmood, A., Zou, R. et al. (2015). Metal–organic frameworks and their derived nanostructures for electrochemical energy storage and conversion. *Energy Environ. Sci.* 8 (7): 1837–1866.

36 Zhang, L., Zhao, Z.-J., and Gong, J. (2017). Nanostructured materials for heterogeneous electrocatalytic CO_2 reduction and their related reaction mechanisms. *Angew. Chem. Int. Ed.* 56 (38): 11326–11353.

37 Guo, Y., Liu, J., Yang, Q. et al. (2021). Regulating nitrogenous adsorption and desorption on Pd clusters by the acetylene linkages of hydrogen substituted graphdiyne for efficient electrocatalytic ammonia synthesis. *Nano Energy* 86: 106099.

38 Li, S.-J., Bao, D., Shi, M.-M. et al. (2017). Amorphizing of Au nanoparticles by CeO_x–RGO hybrid support towards highly efficient electrocatalyst for N_2 reduction under ambient conditions. *Adv. Mater.* 29 (33): 1700001.

39 Li, T., Han, S., Cheng, C. et al. (2022). Sulfate-enabled nitrate synthesis from nitrogen electrooxidation on a rhodium electrocatalyst. *Angew. Chem. Int. Ed.* 61 (26): e202204541.

40 Liu, D., Zhang, G., Ji, Q. et al. (2019). Synergistic electrocatalytic nitrogen reduction enabled by confinement of nanosized Au particles onto a two-dimensional Ti_3C_2 substrate. *ACS Appl. Mater. Interfaces* 11 (29): 25758–25765.

41 Zhang, M., Ma, Z., Zhou, S. et al. (2024). Surface engineering on Ag-decorated Co_3O_4 electrocatalysts for boosting nitrate reduction to ammonia. *ACS Catal.* 14 (15): 11231–11242.

42 Zhang, Y., Zhao, Y., Chen, Z. et al. (2018). Electrochemical reduction of nitrate via Cu/Ni composite cathode paired with Ir-Ru/Ti anode: high efficiency and N_2 selectivity. *Electrochimica Acta* 291: 151–160.

43 Liu, Y.-T., Chen, X., Yu, J. et al. (2019). Carbon-nanoplated CoS@TiO_2 nanofibrous membrane: an interface-engineered heterojunction for high-efficiency electrocatalytic nitrogen reduction. *Angew. Chem. Int. Ed.* 58 (52): 18903–18907.

44 Su, C.-Y., Cheng, H., Li, W. et al. (2017). Atomic modulation of FeCo–nitrogen–carbon bifunctional oxygen electrodes for rechargeable and flexible all-solid-state zinc–air battery. *Adv. Energy Mater.* 7 (13): 1602420.

45 Tan, H., Li, Y., Kim, J. et al. (2018). Sub-50 nm iron–nitrogen-doped hollow carbon sphere-encapsulated iron carbide nanoparticles as efficient oxygen reduction catalysts. *Adv. Sci.* 5 (7): 1800120.

46 Wang, H.-B., Wang, J.-Q., Zhang, R. et al. (2020). Bionic design of a Mo(IV)-doped FeS_2 catalyst for electroreduction of dinitrogen to ammonia. *ACS Catal.* 10 (9): 4914–4921.

47 Zhang, R., Ren, X., Shi, X. et al. (2018). Enabling effective electrocatalytic N_2 conversion to NH_3 by the TiO_2 nanosheets array under ambient conditions. *ACS Appl. Mater. Interfaces* 10 (34): 28251–28255.

48 Zhang, Y., Qiu, W., Ma, Y. et al. (2018). High-performance electrohydrogenation of N_2 to NH_3 catalyzed by multishelled hollow Cr_2O_3 microspheres under ambient conditions. *ACS Catal.* 8 (9): 8540–8544.

49 Liu, S., Wang, M., Qian, T. et al. (2019). Facilitating nitrogen accessibility to boron-rich covalent organic frameworks via electrochemical excitation for efficient nitrogen fixation. *Nat. Commun.* 10 (1): 3898.

50 Mukherjee, S., Cullen, D.A., Karakalos, S. et al. (2018). Metal-organic framework-derived nitrogen-doped highly disordered carbon for electrochemical ammonia synthesis using N_2 and H_2O in alkaline electrolytes. *Nano Energy* 48: 217–226.

51 Yu, X., Han, P., Wei, Z. et al. (2018). Boron-doped graphene for electrocatalytic N_2 reduction. *Joule* 2 (8): 1610–1622.

52 Zhang, X., Wu, T., Wang, H. et al. (2019). Boron nanosheet: an elemental two-dimensional (2D) material for ambient electrocatalytic N_2-to-NH_3 fixation in neutral media. *ACS Catal.* 9 (5): 4609–4615.

53 Han, L., Liu, X., Chen, J. et al. (2019). Atomically dispersed molybdenum catalysts for efficient ambient nitrogen fixation. *Angew. Chem. Int. Ed.* 58 (8): 2321–2325.

54 Huang, B., Li, N., Ong, W.-J. et al. (2019). Single atom-supported MXene: how single-atomic-site catalysts tune the high activity and selectivity of electrochemical nitrogen fixation. *J. Mater. Chem. A* 7 (48): 27620–27631.

55 Li, R. and Wang, D. (2022). Superiority of dual-atom catalysts in electrocatalysis: one step further than single-atom catalysts. *Adv. Energy Mater.* 12 (9): 2103564.

56 Tao, H., Choi, C., Ding, L.-X. et al. (2019). Nitrogen fixation by Ru single-atom electrocatalytic reduction. *Chem* 5 (1): 204–214.

57 Zhang, R., Jiao, L., Yang, W. et al. (2019). Single-atom catalysts templated by metal–organic frameworks for electrochemical nitrogen reduction. *J. Mater. Chem. A* 7 (46): 26371–26377.

58 Ren, Y., Yu, C., Tan, X. et al. (2021). Strategies to suppress hydrogen evolution for highly selective electrocatalytic nitrogen reduction: challenges and perspectives. *Energy Environ. Sci.* 14(3).

59 Tang, C. and Qiao, S.-Z. (2019). True or false in electrochemical nitrogen reduction. *Joule* 3 (7): 1573–1575.

60 Seh, Z.W., Kibsgaard, J., Dickens, C.F. et al. (2017). Combining theory and experiment in electrocatalysis: insights into materials design. *Science* 355 (6321): eaad4998.

61 Bauer, N. (1960). Theoretical pathways for the reduction of N_2 molecules in aqueous media: thermodynamics of $N_2H_n{}^1$. *J. Phy. Chem.* 64 (7): 833–837.

62 Cui, X., Tang, C., and Zhang, Q. (2018). A review of electrocatalytic reduction of dinitrogen to ammonia under ambient conditions. *Adv. Energy Mater.* 8 (22): 1800369.

63 Li, X.-F., Li, Q.-K., Cheng, J. et al. (2016). Conversion of dinitrogen to ammonia by FeN_3-embedded graphene. *J. Am. Chem. Soc.* 138 (28): 8706–8709.

64 Skúlason, E., Bligaard, T., Gudmundsdóttir, S. et al. (2012). A theoretical evaluation of possible transition metal electro-catalysts for N_2 reduction. *Phys. Chem. Chem. Phys.* 14 (3): 1235–1245.

65 Schrock, R.R. (2008). Catalytic reduction of dinitrogen to ammonia by molybdenum: theory versus experiment. *Angew. Chem. Int. Ed.* 47 (30): 5512–5522.

66 Yao, Y., Zhu, S., Wang, H. et al. (2018). A spectroscopic study on the nitrogen electrochemical reduction reaction on gold and platinum surfaces. *J. Am. Chem. Soc.* 140 (4): 1496–1501.

67 Guo, W., Zhang, K., Liang, Z. et al. (2019). Electrochemical nitrogen fixation and utilization: theories, advanced catalyst materials and system design. *Chem. Soc. Rev.* 48 (24): 5658–5716.

68 Mars, P. and van Krevelen, D.W. (1954). Oxidations carried out by means of vanadium oxide catalysts. *Chem. Engineer. Sci.* 3: 41–59.

69 Abghoui, Y., Garden, A.L., Hlynsson, V.F. et al. (2015). Enabling electrochemical reduction of nitrogen to ammonia at ambient conditions through rational catalyst design. *Phys. Chem. Chem. Phys.* 17 (7): 4909–4918.

70 Abghoui, Y., Garden, A.L., Howalt, J.G. et al. (2016). Electroreduction of N_2 to ammonia at ambient conditions on mononitrides of Zr, Nb, Cr, and V: A DFT guide for experiments. *ACS Catal.* 6 (2): 635–646.

71 Abghoui, Y. and Skúlason, E. (2017). Electrochemical synthesis of ammonia via Mars-van Krevelen mechanism on the (111) facets of group III–VII transition metal mononitrides. *Catal. Today* 286: 78–84.

72 Abghoui, Y. and Skúlason, E. (2017). Onset potentials for different reaction mechanisms of nitrogen activation to ammonia on transition metal nitride electro-catalysts. *Catal. Today* 286: 69–77.

73 Greenlee, L.F., Renner, J.N., and Foster, S.L. (2018). The use of controls for consistent and accurate measurements of electrocatalytic ammonia synthesis from dinitrogen. *ACS Catal.* 8 (9): 7820–7827.

74 MacLaughlin, C. (2019). Role for standardization in electrocatalytic ammonia synthesis: a conversation with Leo Liu, Lauren Greenlee, and Douglas MacFarlane. *ACS Energy Lett.* 4 (6): 1432–1436.

75 Chen, G.-F., Ren, S., Zhang, L. et al. (2019). Nitrogen reduction reactions: advances in electrocatalytic N_2 reduction—strategies to tackle the selectivity challenge. *Small Methods* 3 (6): 1970016.

76 Tang, C., and Qiao, S.-Z. (2019). How to explore ambient electrocatalytic nitrogen reduction reliably and insightfully. *Chem. Soc. Rev.* 48 (12): 3166–3180.

77 Bielawa, H., Hinrichsen, O., Birkner, A. et al. (2001). The ammonia-synthesis catalyst of the next generation: barium-promoted oxide-supported ruthenium. *Angew. Chem. Int. Ed.* 40 (6): 1061–1063.

78 Bao, D., Zhang, Q., Meng, F.-L. et al. (2017). Electrochemical reduction of N_2 under ambient conditions for artificial N_2 fixation and renewable energy storage using N_2/NH_3 cycle. *Adv. Mater.* 29 (3): 1604799.

79 Yang, X., Nash, J., Anibal, J. et al. (2018). Mechanistic insights into electrochemical nitrogen reduction reaction on vanadium nitride nanoparticles. *J. Am. Chem. Soc.* 140 (41): 13387–13391.

80 Van Tamelen, E.E. and Seeley, D.A. (1969). Catalytic fixation of molecular nitrogen by electrolytic and chemical reduction. *J. Am. Chem. Soc.* 91 (18): 5194–5194.

81 Howalt, J.G., Bligaard, T., Rossmeisl, J. et al. (2013). DFT based study of transition metal nano-clusters for electrochemical NH_3 production. *Phys. Chem. Chem. Phys.* 15 (20): 7785–7795.

82 Sclafani, A., Augugliaro, V., and Schiavello, M. (1983). Dinitrogen electrochemical reduction to ammonia over iron cathode in aqueous medium. *J. Electrochem. Soc.* 130 (3): 734.

83 Burgess, B.K. and Lowe, D.J. (1996). Mechanism of molybdenum nitrogenase. *Chem. Sci.* 96 (7): 2983–3012.

84 Furuya, N. and Yoshiba, H. (1990). Electroreduction of nitrogen to ammonia on gas-diffusion electrodes loaded with inorganic catalyst. *J. Electroanal. Chem. Interfaces Electrochem.* 291 (1): 269–272.

85 Dahl, S., Törnqvist, E., and Chorkendorff, I. (2000). Dissociative adsorption of N_2 on Ru(0001): a surface reaction totally dominated by steps. *J. Catal.* 192 (2): 381–390.

86 Hellman, A., Baerends, E.J., Biczysko, M. et al. (2006). Predicting catalysis: understanding ammonia synthesis from first-principles calculations. *J. Phy. Chem. B* 110 (36): 17719–17735.

87 Logadóttir, Á. and Nørskov, J.K. (2003). Ammonia synthesis over a Ru(0001) surface studied by density functional calculations. *J. Catal.* 220 (2): 273–279.

88 Montoya, J.H., Tsai, C., Vojvodic, A. et al. (2015). The challenge of electrochemical ammonia synthesis: a new perspective on the role of nitrogen scaling relations. *ChemSusChem* 8 (13): 2180–2186.

89 Antonietti, M. and Oschatz, M. (2018). The concept of "noble, heteroatom-doped carbons," their directed synthesis by electronic band control of carbonization, and applications in catalysis and energy materials. *Adv. Mater.* 30 (21): 1706836.

90 Garden, A.L. and Skúlason, E. (2015). The mechanism of industrial ammonia synthesis revisited: calculations of the role of the associative mechanism. *J. Phy. Chem. C* 119 (47): 26554–26559.

91 Singh, A.R., Rohr, B.A., Statt, M.J. et al. (2019). Strategies toward selective electrochemical ammonia synthesis. *ACS Catal.* 9 (9): 8316–8324.

92 Kordali, V., Kyriacou, G., and Lambrou, C. (2000). Electrochemical synthesis of ammonia at atmospheric pressure and low temperature in a solid polymer electrolyte cell. *Chem. Commun.* (17): 1673–1674.

93 Wang, D., Azofra, L.M., Harb, M. et al. (2018). Energy-efficient nitrogen reduction to ammonia at low overpotential in aqueous electrolyte under ambient conditions. *ChemSusChem* 11 (19): 3416–3422.

94 Geng, Z., Liu, Y., Kong, X. et al. (2018). Achieving a record-high yield rate of 120.9 for N_2 electrochemical reduction over Ru single-atom catalysts. *Adv. Mater.* 30 (40): 1803498.

95 Cao, Y., Gao, Y., Zhou, H. et al. (2018). Highly efficient ammonia synthesis electrocatalyst: single Ru atom on naturally nanoporous carbon materials. *Adv. Theory Simul.* 1 (5): 1800018.

96 Yu, B., Li, H., White, J. et al. (2020). Tuning the catalytic preference of ruthenium catalysts for nitrogen reduction by atomic dispersion. *Adv. Fun. Mater.* 30 (6): 1905665.

97 Liu, C., Li, Q., Zhang, J. et al. (2019). Conversion of dinitrogen to ammonia on Ru atoms supported on boron sheets: a DFT study. *J. Mater. Chem. A* 7 (9): 4771–4776.

98 Agmon, N. (1995). The Grotthuss mechanism. *Chem. Phy. Lett.* 244 (5): 456–462.

99 Wang, J., Yu, L., Hu, L. et al. (2018). Ambient ammonia synthesis via palladium-catalyzed electrohydrogenation of dinitrogen at low overpotential. *Nat. Commun.* 9 (1): 1795.

100 Xu, W., Fan, G., Chen, J. et al. (2020). Nanoporous palladium hydride for electrocatalytic N_2 reduction under ambient conditions. *Angew. Chem. Int. Ed.* 59 (9): 3511–3516.

101 Wang, H., Yu, H., Wang, Z. et al. (2019). Electrochemical fabrication of porous Au film on Ni foam for nitrogen reduction to ammonia. *Small* 15 (6): 1804769.

102 Xue, Z.-H., Zhang, S.-N., Lin, Y.-X. et al. (2019). Electrochemical reduction of N_2 into NH_3 by donor–acceptor couples of Ni and Au nanoparticles with a 67.8% faradaic efficiency. *J. Am. Chem. Soc.* 141 (38): 14976–14980.

103 Zheng, J., Lyu, Y., Qiao, M. et al. (2019). Tuning the electron localization of gold enables the control of nitrogen-to-ammonia fixation. *Angew. Chem. Int. Ed.* 58 (51): 18604–18609.

104 Huang, H., Xia, L., Shi, X. et al. (2018). Ag nanosheets for efficient electrocatalytic N_2 fixation to NH_3 under ambient conditions. *Chem. Commun.* 54 (81): 11427–11430.

105 Kugler, K., Luhn, M., Schramm, J.A. et al. (2015). Galvanic deposition of Rh and Ru on randomly structured Ti felts for the electrochemical NH_3 synthesis. *Phys. Chem. Chem. Phys.* 17 (5): 3768–3782.

106 Liu, H.-M., Han, S.-H., Zhao, Y. et al. (2018). Surfactant-free atomically ultrathin rhodium nanosheet nanoassemblies for efficient nitrogen electroreduction. *J. Mater. Chem. A* 6 (7): 3211–3217.

107 Lan, R., Irvine, J.T.S., and Tao, S. (2013). Synthesis of ammonia directly from air and water at ambient temperature and pressure. *Sci. Rep.* 3 (1): 1145.

108 Lan, R. and Tao, S. (2013). Electrochemical synthesis of ammonia directly from air and water using a $Li^+/H^+/NH_4^+$ mixed conducting electrolyte. *RSC Adv.* 3 (39): 18016–18021.

109 Mao, Y.-J., Wei, L., Zhao, X.-S. et al. (2019). Excavated cubic platinum–iridium alloy nanocrystals with high-index facets as highly efficient electrocatalysts in N_2 fixation to NH_3. *Chem. Commun.* 55 (63): 9335–9338.

110 Lv, F., Zhao, S., Guo, R. et al. (2019). Nitrogen-coordinated single Fe sites for efficient electrocatalytic N_2 fixation in neutral media. *Nano Energy* 61: 420–427.

111 Cui, Q., Qin, G., Wang, W. et al. (2019). Mo-based 2D MOF as a highly efficient electrocatalyst for reduction of N_2 to NH_3: a density functional theory study. *J. Mater. Chem. A* 7 (24): 14510–14518.

112 Li, X., Li, T., Ma, Y. et al. (2018). Boosted electrocatalytic N_2 reduction to NH_3 by defect-rich MoS_2 nanoflower. *Adv. Energy Mater.* 8 (30): 1801357.

113 Li, X., Ren, X., Liu, X. et al. (2019). A MoS_2 nanosheet–reduced graphene oxide hybrid: an efficient electrocatalyst for electrocatalytic N_2 reduction to NH_3 under ambient conditions. *J. Mater. Chem. A* 7 (6): 2524–2528.

114 Zeng, L., Chen, S., van der Zalm, J. et al. (2019). Sulfur vacancy-rich N-doped MoS_2 nanoflowers for highly boosting electrocatalytic N_2 fixation to NH_3 under ambient conditions. *Chem. Commun.* 55 (51): 7386–7389.

115 Zhang, L., Ji, X., Ren, X. et al. (2018). Electrochemical ammonia synthesis via nitrogen reduction reaction on a MoS_2 catalyst: theoretical and experimental studies. *Adv. Mater.* 30 (28): 1800191.

116 Ren, X., Cui, G., Chen, L. et al. (2018). Electrochemical N_2 fixation to NH_3 under ambient conditions: Mo_2N nanorod as a highly efficient and selective catalyst. *Chem. Commun.* 54 (61): 8474–8477.

117 Zhang, L., Ji, X., Ren, X. et al. (2018). Efficient electrochemical N_2 reduction to NH_3 on MoN nanosheets array under ambient conditions. *ACS Sustain. Chem. Eng.* 6 (8): 9550–9554.

118 Ba, K., Wang, G., Ye, T. et al. (2020). Single faceted two-dimensional Mo_2C electrocatalyst for highly efficient nitrogen fixation. *ACS Catal.* 10 (14): 7864–7870.

119 Cheng, H., Ding, L.-X., Chen, G.-F. et al. (2018). Molybdenum carbide nanodots enable efficient electrocatalytic nitrogen fixation under ambient conditions. *Adv. Mater.* 30 (46): 1803694.

120 Liu, Y., Zhu, X., Zhang, Q. et al. (2020). Engineering Mo/Mo_2C/MoC hetero-interfaces for enhanced electrocatalytic nitrogen reduction. *J. Mater. Chem. A* 8 (18): 8920–8926.

121 Wu, H.B., Xia, B.Y., Yu, L. et al. (2015). Porous molybdenum carbide nano-octahedrons synthesized via confined carburization in metal-organic frameworks for efficient hydrogen production. *Nat. Commun.* 6 (1): 6512.

122 Hui, L., Xue, Y., Yu, H. et al. (2019). Highly efficient and selective generation of ammonia and hydrogen on a graphdiyne-based catalyst. *J. Am. Chem. Soc.* 141 (27): 10677–10683.

123 Ling, C., Bai, X., Ouyang, Y. et al. (2018). Single molybdenum atom anchored on N-doped carbon as a promising electrocatalyst for nitrogen reduction into ammonia at ambient conditions. *J. Phys. Chem. C* 122 (29): 16842–16847.

124 Yang, D., Chen, T., and Wang, Z. (2017). Electrochemical reduction of aqueous nitrogen (N_2) at a low overpotential on (110)-oriented Mo nanofilm. *J. Mater. Chem. A* 5 (36): 18967–18971.

125 Chu, K., Wang, J., Liu, Y.-p. et al. (2020). Mo-doped SnS_2 with enriched S-vacancies for highly efficient electrocatalytic N2 reduction: the critical role of the Mo–Sn–Sn trimer. *J. Mater. Chem. A* 8 (15): 7117–7124.

126 Li, T., Yan, X., Huang, L. et al. (2019). Fluorine-free $Ti_3C_2T_x$ (T = O, OH) nanosheets (~50–100 nm) for nitrogen fixation under ambient conditions. *J. Mater. Chem. A* 7 (24): 14462–14465.

127 Kong, W., Gong, F., Zhou, Q. et al. (2019). An MnO_2–$Ti_3C_2T_x$ MXene nanohybrid: an efficient and durable electrocatalyst toward artificial N_2 fixation to NH_3 under ambient conditions. *J. Mater. Chem. A* 7 (32): 18823–18827.

128 Luo, Y., Chen, G.-F., Ding, L. et al. (2019). Efficient electrocatalytic N_2 fixation with MXene under ambient conditions. *Joule* 3 (1): 279–289.

129 Zhao, J., Zhang, L., Xie, X.-Y. et al. (2018). $Ti_3C_2T_x$ (T = F, OH) MXene nanosheets: conductive 2D catalysts for ambient electrohydrogenation of N_2 to NH_3. *J. Mater. Chem. A* 6 (47): 24031–24035.

130 Fang, Y., Liu, Z., Han, J. et al. (2019). High-performance electrocatalytic conversion of N_2 to NH_3 using oxygen-vacancy-rich TiO_2 in situ grown on Ti_3C_2T MXene. *Adv. Energy Mater.* 9 (16): 1803406.

131 Rakhi, R.B., Ahmed, B., Anjum, D. et al. (2016). Direct chemical synthesis of MnO_2 nanowhiskers on transition-metal carbide surfaces for supercapacitor applications. *ACS Appl. Mater. Interfaces* 8 (29): 18806–18814.

132 Li, L., Tang, C., Xia, B. et al. (2019). Two-dimensional mosaic bismuth nanosheets for highly selective ambient electrocatalytic nitrogen reduction. *ACS Catal.* 9 (4): 2902–2908.

133 Lv, C., Yan, C., Chen, G. et al. (2018). An amorphous noble-metal-free electrocatalyst that enables nitrogen fixation under ambient conditions. *Angew. Chem. Int. Ed.* 57 (21): 6073–6076.

134 Wang, F., Lv, X., Zhu, X. et al. (2020). Bi nanodendrites for efficient electrocatalytic N_2 fixation to NH_3 under ambient conditions. *Chem. Commun.* 56 (14): 2107–2110.

135 Wang, Y., Shi, M.-m., Bao, D. et al. (2019). Generating defect-rich bismuth for enhancing the rate of nitrogen electroreduction to ammonia. *Angew. Chem. Int. Ed.* 58 (28): 9464–9469.

136 Hao, Y.-C., Guo, Y., Chen, L.-W. et al. (2019). Promoting nitrogen electroreduction to ammonia with bismuth nanocrystals and potassium cations in water. *Nat. Catal.* 2 (5): 448–456.

137 Wang, Y., Jia, K., Pan, Q. et al. (2019). Boron-doped TiO_2 for efficient electrocatalytic N_2 fixation to NH_3 at ambient conditions. *ACS Sustain. Chem. Eng.* 7 (1): 117–122.

138 Wu, T., Kong, W., Zhang, Y. et al. (2019). Greatly enhanced electrocatalytic N_2 reduction on TiO_2 via V doping. *Small Methods* 3 (11): 1900356.

139 Zhang, R., Zhang, Y., Ren, X. et al. (2018). High-efficiency electrosynthesis of ammonia with high selectivity under ambient conditions enabled by VN nanosheet array. *ACS Sustain. Chem. Eng.* 6 (8): 9545–9549.

140 Zhang, X., Kong, R.-M., Du, H. et al. (2018). Highly efficient electrochemical ammonia synthesis via nitrogen reduction reactions on a VN nanowire array under ambient conditions. *Chem. Commun.* 54 (42): 5323–5325.

141 Fang, W., Zhao, J., Wu, T. et al. (2020). Hydrophilic engineering of VO_x-based nanosheets for ambient electrochemical ammonia synthesis at neutral pH. *J. Mater. Chem. A* 8 (12): 5913–5918.

142 Zhang, Y., Qiu, W., Ma, Y. et al. (2018). High-performance electrohydrogenation of N_2 to NH_3 catalyzed by multishelled hollow Cr2O3 microspheres under ambient conditions. *ACS Catalysis* 8 (9): 8540–8544.

143 Du, H., Guo, X., Kong, R.-M. et al. (2018). Cr_2O_3 nanofiber: a high-performance electrocatalyst toward artificial N_2 fixation to NH_3 under ambient conditions. *Chem. Commun.* 54 (91): 12848–12851.

144 Yao, Y., Feng, Q., Zhu, S. et al. (2019). Chromium oxynitride electrocatalysts for electrochemical synthesis of ammonia under ambient conditions. *Small Methods* 3 (6): 1800324.

145 Wu, X., Xia, L., Wang, Y. et al. (2018). Mn3O4 nanocube: an efficient electrocatalyst toward artificial N_2 fixation to NH3. *Small* 14 (48): 1803111.

146 Chu, K., Liu, Y.-p., Li, Y.-b. et al. (2019). Efficient electrocatalytic N_2 reduction on CoO quantum dots. *J. Mater. Chem. A* 7 (9): 4389–4394.

147 Qin, M., Li, X., Gan, G. et al. (2020). Boosting electrocatalytic nitrogen fixation with Co–N_3 site-decorated porous carbon. *ACS Sustain. Chem. Eng.* 8 (35): 13430–13439.

148 Guo, W., Liang, Z., Zhao, J. et al. (2018). Hierarchical cobalt phosphide hollow nanocages toward electrocatalytic ammonia synthesis under ambient pressure and room temperature. *Small Methods* 2 (12): 1800204.

149 Chen, P., Zhang, N., Wang, S. et al. (2019). Interfacial engineering of cobalt sulfide/graphene hybrids for highly efficient ammonia electrosynthesis. *P. Natl. Acad. Sci.* 116 (14): 6635–6640.

150 Fang, Y., Xue, Y., Li, Y. et al. (2020). Graphdiyne interface engineering: highly active and selective ammonia synthesis. *Angew. Chem. Int. Ed.* 59 (31): 13021–13027.

151 Zhang, Y., Hu, J., Zhang, C. et al. (2020). Bimetallic Mo–Co nanoparticles anchored on nitrogen-doped carbon for enhanced electrochemical nitrogen fixation. *J. Mater. Chem. A* 8 (18): 9091–9098.

152 Liu, Y.-T., Tang, L., Dai, J. et al. (2020). Promoted electrocatalytic nitrogen fixation in Fe-Ni layered double hydroxide arrays coupled to carbon nanofibers: the role of phosphorus doping. *Angew. Chem. Int. Ed.* 59 (32): 13623–13627.

153 Wang, X.-h., Wang, J., Li, Y.-b. et al. (2019). Nitrogen-doped NiO nanosheet array for boosted electrocatalytic N2 reduction. *ChemCatChem* 11 (18): 4529–4536.

154 Zhao, X., Yang, Z., Kuklin, A.V. et al. (2020). BCN-encapsulated nano-nickel synergistically promotes ambient electrochemical dinitrogen reduction. *ACS Appl. Mater. Interfaces* 12 (28): 31419–31430.

155 Kim, K., Yoo, C.-Y., Kim, J.-N. et al. (2016). Electrochemical synthesis of ammonia from water and nitrogen in ethylenediamine under ambient temperature and pressure. *J. Electrochem. Soc.* 163 (14): F1523.

156 Kim, K., Lee, N., Yoo, C.-Y. et al. (2016). Communication—electrochemical reduction of nitrogen to ammonia in 2-propanol under ambient temperature and pressure. *J. Electrochem. Soc.* 163 (7): F610.

157 Lin, Y.-X., Zhang, S.-N., Xue, Z.-H. et al. (2019). Boosting selective nitrogen reduction to ammonia on electron-deficient copper nanoparticles. *Nat. Commun* 10 (1): 4380.

158 Wu, T., Zhao, H., Zhu, X. et al. (2020). Identifying the origin of Ti_3^+ activity toward enhanced electrocatalytic N_2 reduction over TiO_2 nanoparticles modulated by mixed-valent copper. *Adv. Mater.* 32 (30): 2000299.

159 Wu, T., Han, M., Zhu, X. et al. (2019). Experimental and theoretical understanding on electrochemical activation and inactivation processes of $Nb_3O_7(OH)$ for ambient electrosynthesis of NH_3. *J. Mater. Chem. A* 7 (28): 16969–16978.

160 Han, J., Liu, Z., Ma, Y. et al. (2018). Ambient N2 fixation to NH3 at ambient conditions: using Nb2O5 nanofiber as a high-performance electrocatalyst. *Nano Energy* 52: 264–270.

161 Wang, Y., Chen, A., Lai, S. et al. (2020). Self-supported $NbSe_2$ nanosheet arrays for highly efficient ammonia electrosynthesis under ambient conditions. *J. Catal.* 381: 78–83.

162 Zhang, L., Ren, X., Luo, Y. et al. (2018). Ambient NH_3 synthesis via electrochemical reduction of N_2 over cubic sub-micron SnO_2 particles. *Chem. Commun.* 54 (92): 12966–12969.

163 Li, P., Fu, W., Zhuang, P., et al. (2019). Amorphous Sn/crystalline SnS2 nanosheets via in situ electrochemical reduction methodology for highly efficient ambient N2 fixation. *Small* 15 (40): 1902535.

164 Chen, X., Liu, Y.-T., Ma, C. et al. (2019). Self-organized growth of flower-like SnS_2 and forest-like ZnS nanoarrays on nickel foam for synergistic superiority in electrochemical ammonia synthesis. *J. Mater. Chem. A* 7 (39): 22235–22241.

165 Li, P., Liu, Z., Wu, T. et al. (2019). Ambient electrocatalytic N_2 reduction to NH_3 by metal fluorides. *J. Mater. Chem. A* 7 (30): 17761–17765.

166 Li, C., Ma, D., Mou, S. et al. (2020). Porous $LaFeO_3$ nanofiber with oxygen vacancies as an efficient electrocatalyst for N_2 conversion to NH_3 under ambient conditions. *J. Energy Chem.* 50: 402–408.

167 Jin, H., Li, L., Liu, X. et al. (2019). Nitrogen vacancies on 2D layered W_2N_3: a stable and efficient active site for nitrogen reduction reaction. *Adv. Mater.* 31 (32): 1902709.

168 Yang, M., Huo, R., Shen, H. et al. (2020). Metal-tuned $W_{18}O_{49}$ for efficient electrocatalytic N_2 reduction. *ACS Sustain. Chem. Eng.* 8 (7): 2957–2963.

169 Liu, X., Jang, H., Li, P. et al. (2019). Antimony-based composites loaded on phosphorus-doped carbon for boosting faradaic efficiency of the electrochemical nitrogen reduction reaction. *Angew. Chem. Int. Ed.* 58 (38): 13329–13334.

170 Han, Y., Cai, W., Wu, X. et al. (2020). Electrocatalytic nitrogen fixation on metal tellurides boosted by multiple promoted-synergetic effects of telluride. *Cell. Rep. Phys. Sci.* 1 (11): 100232.

171 Liu, J., Kong, X., Zheng, L. et al. (2020). Rare earth single-atom catalysts for nitrogen and carbon dioxide reduction. *ACS Nano* 14 (1): 1093–1101.

172 Fu, W., Zhuang, P., OliverLam Chee, M. et al. (2019). Oxygen vacancies in Ta_2O_5 nanorods for highly efficient electrocatalytic N_2 reduction to NH_3 under ambient conditions. *ACS Sustain. Chem. Eng.* 7 (10): 9622–9628.

173 Fu, Y., Li, T., Zhou, G. et al. (2020). Dual-metal-driven selective pathway of nitrogen reduction in orderly atomic-hybridized Re2MnS6 ultrathin nanosheets. *Nano Letters* 20 (7): 4960–4967.

174 Lai, F., Chen, N., Ye, X. et al. (2020). Refining energy levels in ReS_2 nanosheets by low-valent transition-metal doping for dual-boosted electrochemical ammonia/hydrogen production. *Adv. Fun. Mater.* 30 (11): 1907376.

175 Li, Y., Li, T., Zhu, X. et al. (2020). DyF_3: an efficient electrocatalyst for N_2 fixation to NH_3 under ambient conditions. *Chem. Asian J.* 15 (4): 487–489.

176 Zheng, Y., Jiao, Y., Jaroniec, M. et al. (2015). Advancing the electrochemistry of the hydrogen-evolution reaction through combining experiment and theory. *Angew. Chem. Int. Ed.* 54 (1): 52–65.

177 Lv, C., Qian, Y., Yan, C. et al. (2018). Defect engineering metal-free polymeric carbon nitride electrocatalyst for effective nitrogen fixation under ambient conditions. *Angew. Chem. Int. Ed.* 57 (32): 10246–10250.

178 Li, W., Wu, T., Zhang, S. et al. (2018). Nitrogen-free commercial carbon cloth with rich defects for electrocatalytic ammonia synthesis under ambient conditions. *Chem. Commun.* 54 (79): 11188–11191.

179 Yang, X., Li, K., Cheng, D. et al. (2018). Nitrogen-doped porous carbon: highly efficient trifunctional electrocatalyst for oxygen reversible catalysis and nitrogen reduction reaction. *J. Mater. Chem. A* 6 (17): 7762–7769.

180 Chen, C., Yan, D., Wang, Y. et al. (2019). B-N pairs enriched defective carbon nanosheets for ammonia synthesis with high efficiency. *Small* 15 (7): 1805029.

181 Qiu, W.-B., Luo, Y.-X., Liang, R.-P. et al. (2019). B_4C nanosheets decorated with in situ-derived boron-doped graphene quantum dots for high-efficiency ambient N_2 fixation. *Chem. Commun.* 55 (51): 7406–7409.

182 Jiao, Y., Zheng, Y., Jaroniec, M. et al. (2014). Origin of the electrocatalytic oxygen reduction activity of graphene-based catalysts: a roadmap to achieve the best performance. *J. Am. Chem. Soc.* 136 (11): 4394–4403.

183 Hu, C. and Dai, L. (2016). Carbon-based metal-free catalysts for electrocatalysis beyond the ORR. *Angew. Chem. Int. Ed.* 55 (39): 11736–11758.

184 Wang, X., Sun, G., Routh, P. et al. (2014). Heteroatom-doped graphene materials: syntheses, properties and applications. *Chem. Soc. Rev.* 43 (20): 7067–7098.

185 Yeh, M.-H., Leu, Y.-A., Chiang, W.-H. et al. (2018). Boron-doped carbon nanotubes as metal-free electrocatalyst for dye-sensitized solar cells: Heteroatom doping level effect on tri-iodide reduction reaction. *J. Power Sources* 375: 29–36.

186 Chen, G.-F., Cao, X., Wu, S. et al. (2017). Ammonia electrosynthesis with high selectivity under ambient conditions via a Li^+ incorporation strategy. *J. Am. Chem. Soc.* 139 (29): 9771–9774.

187 Tian, Y., Xu, D., Chu, K. et al. (2019). Metal-free N, S co-doped graphene for efficient and durable nitrogen reduction reaction. *J. Mater. Sci.* 54 (12): 9088–9097.

188 Huang, H., Xia, L., Cao, R. et al. (2019). A biomass-derived carbon-based electrocatalyst for efficient N_2 fixation to NH_3 under ambient conditions. *Chem. Eur. J.* 25 (8): 1914–1917.

189 Wang, T., Xia, L., Yang, J.-J. et al. (2019). Electrocatalytic N_2-to-NH_3 conversion using oxygen-doped graphene: experimental and theoretical studies. *Chem. Commun.* 55 (52): 7502–7505.

190 Wu, T., Li, P., Wang, H. et al. (2019). Biomass-derived oxygen-doped hollow carbon microtubes for electrocatalytic N_2-to-NH_3 fixation under ambient conditions. *Chem. Commun.* 55 (18): 2684–2687.

191 Légaré, M.-A., Bélanger-Chabot, G., Dewhurst, R.D. et al. (2018). Nitrogen fixation and reduction at boron. *Science* 359 (6378): 896–900.

192 Légaré, M.-A., Courtemanche, M.-A., Rochette, É. et al. (2015). Metal-free catalytic C–H bond activation and borylation of heteroarenes. *Science* 349 (6247): 513–516.

193 Liu, C., Li, Q., Wu, C. et al. (2019). Single-boron catalysts for nitrogen reduction reaction. *J. Am. Chem. Soc.* 141 (7): 2884–2888.

194 Liu, C., Li, Q., Zhang, J. et al. (2018). Theoretical evaluation of possible 2D boron monolayer in N_2 electrochemical conversion into ammonia. *J. Phys. Chem. C* 122 (44): 25268–25273.

195 Qiu, W., Xie, X.-Y., Qiu, J. et al. (2018). High-performance artificial nitrogen fixation at ambient conditions using a metal-free electrocatalyst. *Nat. Commun.* 9 (1): 3485.

196 Boukhvalov, D.W. (2015). The atomic and electronic structure of nitrogen- and boron-doped phosphorene. *Phys. Chem. Chem. Phys.* 17 (40): 27210–27216.

197 Guo, Z., Zhang, H., Lu, S. et al. (2015). From black phosphorus to phosphorene: basic solvent exfoliation, evolution of Raman scattering, and applications to ultrafast photonics. *Adv. Fun. Mater.* 25 (45): 6996–7002.

198 Li, L., Yu, Y., Ye, G.J. et al. (2014). Black phosphorus field-effect transistors. *Nat. Nanotech.* 9 (5): 372–377.

199 Liu, Q., Zhang, X., Wang, J. et al. (2020). Crystalline red phosphorus nanoribbons: large-scale synthesis and electrochemical nitrogen fixation. *Angew. Chem. Int. Ed.* 59 (34): 14383–14387.

200 Liu, Y.-T., Li, D., Yu, J. et al. (2019). Stable confinement of black phosphorus quantum dots on black tin oxide nanotubes: a robust, double-active electrocatalyst toward efficient nitrogen fixation. *Angew. Chem. Int. Ed.* 58 (46): 16439–16444.

201 Xu, G., Li, H., Bati, A.S.R. et al. (2020). Nitrogen-doped phosphorene for electrocatalytic ammonia synthesis. *J. Mater. Chem. A* 8 (31): 15875–15883.

202 Zhang, L., Ding, L.-X., Chen, G.-F. et al. (2019). Ammonia synthesis under ambient conditions: selective electroreduction of dinitrogen to ammonia on black phosphorus nanosheets. *Angew. Chem. Int. Ed.* 58 (9): 2612–2616.

203 Suryanto, B.H.R., Du, H.-L., Wang, D. et al. (2019). Challenges and prospects in the catalysis of electroreduction of nitrogen to ammonia. *Nat. Catal.* 2 (4): 290–296.

204 Wiesler, F. (1998). Comparative assessment of the efficacy of various nitrogen fertilizers. *J. Crop Prod.* 1 (2): 81–114.

205 Yao, B., Cao, J., Zhao, C. et al. (2011). Influence of ammonium and nitrate supply on growth, nitrate reductase activity and N-use efficiency in a natural hybrid pine and its parents. *J. Plant Ecol.* 4 (4): 275–282.

206 Birkeland, K. (1906). On the oxidation of atmospheric nitrogen in electric arcs. *Trans. Faraday Soc.* 2 (December): 98–116.

207 Cherkasov, N., Ibhadon, A.O., and Fitzpatrick, P. (2015). A review of the existing and alternative methods for greener nitrogen fixation. *Chem. Eng. Process. Process Intensif.* 90: 24–33.

208 Eyde, S. (1912). Oxidation of atmospheric nitrogen and development of resulting industries in Norway. *J. Ind. Eng. Chem.* 4 (10): 771–774.

209 Patil, B.S., Wang, Q., Hessel, V. et al. (2015). Plasma N_2-fixation: 1900–2014. *Catal. Today* 256: 49–66.

210 Smith, C., Hill, A.K., and Torrente-Murciano, L. (2020). Current and future role of Haber–Bosch ammonia in a carbon-free energy landscape. *Energy Environ. Sci.* 13 (2): 331–344.

211 Guo, C., Ran, J., Vasileff, A. et al. (2018). Rational design of electrocatalysts and photo(electro)catalysts for nitrogen reduction to ammonia (NH_3) under ambient conditions. *Energy Environ. Sci.* 11 (1): 45–56.

212 van Langevelde, P.H., Katsounaros, I., and Koper, M.T.M. (2021). Electrocatalytic nitrate reduction for sustainable ammonia production. *Joule* 5 (2): 290–294.

213 Yao, D., Tang, C., Li, L. et al. (2020). In situ fragmented bismuth nanoparticles for electrocatalytic nitrogen reduction. *Adv. Energy Mater.* 10 (10): 2001289.

214 Liu, Y., Cheng, M., He, Z. et al. (2019). Pothole-rich ultrathin WO_3 nanosheets that trigger $N \equiv N$ bond activation of nitrogen for direct nitrate photosynthesis. *Angew. Chem. Int. Ed.* 58 (3): 731–735.

215 Research, G.V. (2020). Nitric Acid Market Size, Share & Trends Analysis Report By Application (Fertilizers, Nitrobenzene, Adipic Acid, Toluene Di-isocyanate, Nitoroclorobenzene), By Region, And Segment Forecasts, 2023–2030, https://www.grandviewresearch.com/industry-analysis/nitric-acid-market.

216 Wang, S., Ichihara, F., Pang, H. et al. (2018). Nitrogen fixation reaction derived from nanostructured catalytic materials. *Adv. Fun. Mater.* 28 (50): 1803309.

217 Comer, B.M., Fuentes, P., Dimkpa, C.O. et al. (2019). Prospects and challenges for solar fertilizers. *Joule* 3 (7): 1578–1605.

218 Iriawan, H., Andersen, S.Z., Zhang, X. et al. (2021). Methods for nitrogen activation by reduction and oxidation. *Nat. Rev. Methods Primers* 1 (1): 56.

219 Comer, B.M. and Medford, A.J. (2018). Analysis of photocatalytic nitrogen fixation on rutile TiO_2(110). *ACS Sustain. Chem. Eng.* 6 (4): 4648–4660.

220 Wang, Y., Wang, C., Li, M. et al. (2021). Nitrate electroreduction: mechanism insight, in situ characterization, performance evaluation, and challenges. *Chem. Soc. Rev.* 50 (12): 6720–6733.

221 Dai, C., Sun, Y., Chen, G. et al. (2020). Electrochemical oxidation of nitrogen towards direct nitrate production on spinel oxides. *Angew. Chem. Int. Ed.* 59 (24): 9418–9422.

222 Guo, M., Fang, L., Zhang, L. et al. (2023). Pulsed electrocatalysis enabling high overall nitrogen fixation performance for atomically dispersed Fe on TiO_2. *Angew. Chem. Int. Ed.* 62 (13): e202217635.

223 Zheng, X., Wu, H., Gao, Y. et al. (2024). Controllable assembly of highly oxidized cobalt on graphdiyne surface for efficient conversion of nitrogen into nitric acid. *Angew. Chem. Int. Ed.* 63 (9): e202316723.

224 Nie, Z., Zhang, L., Ding, X. et al. (2022). Catalytic kinetics regulation for enhanced electrochemical nitrogen oxidation by Ru-nanoclusters-coupled Mn_3O_4 catalysts decorated with atomically dispersed Ru atoms. *Adv. Mater.* 34 (14): 2108180.

225 Wan, H., Bagger, A., and Rossmeisl, J. (2022). Limitations of electrochemical nitrogen oxidation toward nitrate. *J. Phys. Chem. Lett.* 13 (38): 8928–8934.

226 Li, L., Tang, C., Jin, H. et al. (2021). Main-group elements boost electrochemical nitrogen fixation. *Chem* 7 (12): 3232–3255.

227 Shipman, M.A. and Symes, M.D. (2017). Recent progress towards the electrosynthesis of ammonia from sustainable resources. *Catal. Today* 286: 57–68.

228 Gruber, N. and Galloway, J.N. (2008). An earth-system perspective of the global nitrogen cycle. *Nature* 451 (7176): 293–296.

229 Rayleigh, L. (1897). XIII.—Observations on the oxidation of nitrogen gas. *J. Chem. Soc., Trans.* 71 (0): 181–186.

230 Zhang, L., Cong, M., Ding, X. et al. (2020). A Janus Fe-SnO_2 catalyst that enables bifunctional electrochemical nitrogen fixation. *Angew. Chem. Int. Ed.* 59 (27): 10888–10893.

231 Galloway, J.N., Townsend, A.R., Erisman, J.W. et al. (2008). Transformation of the nitrogen cycle: recent trends, questions, and potential solutions. *Science* 320 (5878): 889–892.

232 Rouwenhorst, K.H.R., Jardali, F., Bogaerts, A. et al. (2021). From the Birkeland–Eyde process towards energy-efficient plasma-based NO_X synthesis: a techno-economic analysis. *Energy Environ. Sci.* 14 (5): 2520–2534.

233 Li, T., Han, S., Wang, Y. et al. (2023). A spectroscopic study on nitrogen electrooxidation to nitrate. *Angew. Chem. Int. Ed.* 62 (19): e202217411.

234 Timoshenko, J. and Roldan Cuenya, B. (2021). In situ/operando electrocatalyst characterization by X-ray absorption spectroscopy. *Chem. Sci.* 121 (2): 882–961.

235 Wu, J., Wang, S., Ji, R. et al. (2024). In situ characterization techniques for electrochemical nitrogen reduction reaction. *ACS Nano* 18 (32): 20934–20956.

236 Chen, C., Zhu, X., Wen, X. et al. (2020). Coupling N_2 and CO_2 in H_2O to synthesize urea under ambient conditions. *Nat. Chem.* 12 (8): 717–724.

237 Lv, C., Zhong, L., Liu, H. et al. (2021). Selective electrocatalytic synthesis of urea with nitrate and carbon dioxide. *Nat. Sustain.* 4 (10): 868–876.

238 Xing, P., Wei, S., Zhang, Y. et al. (2023). Electrochemical co-reduction of N_2 and CO_2 to urea using Bi_2S_3 nanorods anchored to N-doped reduced graphene oxide. *ACS Appl. Mater. Interfaces* 15 (18): 22101–22111.

13

Recent Progress in Electrochemical C—N Coupling Reactions

Chu Zhang[1], Chunyuan Feng[2], Lixiang Zhong[2,], Chade Lv[1,*]*

[1] *MIIT Key Laboratory of Critical Materials Technology for New Energy Conversion and Storage, State Key Laboratory of Space Power-Sources, School of Chemistry and Chemical Engineering, Harbin Institute of Technology, Harbin, China*
[2] *School of Physics, Beijing Institute of Technology, Beijing, China*

*Corresponding authors: zhonglx@bit.edu.cn; lv.chade@hit.edu.cn

13.1 Introduction

The industrial synthesis of chemicals is usually carried out under harsh conditions and with high energy consumption, exacerbating the energy crisis and environmental problems [1]. Driven by renewable electricity, electrocatalysis can lower the energy barrier of the reaction, enabling efficient, green synthesis of chemicals under milder conditions [2]. Currently, the resource utilization of abundant gas molecules through electrocatalytic activation and fixation is a hot topic in electrocatalytic research [3, 4]. Through the electrocatalytic CO_2 reduction reactions (CO_2RR), CO_2 could couple with renewable energy to be effectively converted into value-added products with zero or even negative carbon footprint [5–7]. Although a wide variety of single carbon (C_1) and multi-carbon (C_{2+}) products containing C—H, C—C, and/or C—O bonds can be selectively synthesized by CO_2RR, further adding value and expanding the range of products is limited in systems relying on CO_2 and H_2O (or different proton sources) as reactants [8, 9]. An alternative approach is bonding nitrogen atoms through C—N coupling rather than bonding carbon, hydrogen, or oxygen through electrocatalytic CO_2RR [10]. The construction of C—N bond through electrocatalytic coupling of N_2 and CO_2 is achievable to synthesize organonitrogen chemicals (such as urea, amides, amines, and nitriles, etc.), which are significant chemicals in the field of chemical synthesis, pharmaceutical chemistry, agriculture, and the aerospace industry. More importantly, compared with the current industrial synthesis routes, the electrocatalytic C—N coupling reaction can offer several notable benefits: First, using CO_2 as a C source might facilitate carbon neutrality and promote economic development; second, using other nitrogen-containing chemicals could replace NH_3 as N sources. As the industrial synthesis of organic nitrogen compounds consumes more than 50% of

Electrochemical Energy Conversion via Small Molecule Catalysis, First Edition.
Edited by Zhicheng Zhang and Yuanmiao Sun.

global NH_3 production, which still relies on the energy-intensive and fossil fuel-based Haber–Bosch process. Directly utilizing N_2, the most abundant and cheapest N source on the earth, is desirable for C—N coupling, while using NO_2^-, NO_3^- as N sources can convert serious environmental pollutants into reactants to realize "transforming waste into treasure." [11]. Finally, electrochemical reaction is a more intriguing and sustainable process. In sharp contrast with the traditional C—N coupling approach dominated by high energy consumption, high pollution, high temperature, and high pressure, the electrochemical reactions can mainly rely on green electricity originating from the intermittent renewable energy sources, such as wind turbines and photovoltaic techniques.

Up till now, converting CO_2 and N-containing compounds into value-added chemicals has been mainly studied as individual electrochemical reduction reactions, namely CO_2RR, N_2 reduction reactions (N_2RR), and nitrate/nitrite/nitric oxide reduction to ammonia (NRA) reactions. Although it has been demonstrated that the multifunctional bond properties of N make it possible to establish C—N bonds simultaneously with CO_2 electroreduction, guiding CO_2 and N precursors to the ideal C—N product requires innovative design and refined reaction pathway control. In recent years, there has been a lot of studies on the formation of C—N bonds and the synthesis of amine products by co-reduction of CO_2 and N-containing chemicals.

In this chapter, we summarize the basic principle, product detection, and test method of electrocatalytic C—N coupling reaction systematically and analyze the research progress of electrosynthesis of urea by C—N coupling reaction between CO_2 and small N-containing molecules. Finally, we discuss the advanced characterization methods for exploring the reaction mechanism of C—N coupling. Through the basic introduction of the electrocatalytic C—N coupling reaction, we let more readers know its research progress to arouse their interest in exploring more research on C—N coupling electrosynthesis.

13.2 Milestones in Electrocatalytic C—N Coupling

Over the past decades, significant effort has been devoted to exploring efficient routes of electrochemical co-reduction of CO_2 and N-containing species to synthesize organonitrogen compounds. Jiao and coworkers creatively introduced ammonia as an additional reactant in the CO reduction reaction, achieving a maximum Faradaic efficiency (FE) of nearly 40% for acetamide [12]. The C—N bond formed through the nucleophilic attack of ammonia on *C=C=O intermediates. By combining electrocatalytic processes with organic reactions, the investigation and application of C—N bond coupling have expanded. In 1963, Rapson and Bird successfully synthesized glycine by the electroreduction of oxalic acid and nitric acid in the same cell [13]. The electrochemical interaction between glyoxylic acid and hydroxylamine enabled C—N coupling. Research on electrocatalytic C—N coupling for synthetic urea can be traced back to 1995 when Furuya and coworkers synthesized urea through the co-reduction of CO_2 and NO_2^-/NO_3^- using Cu-loaded gas-diffusion electrodes [14]. In addition to NO_3^-/NO_2^- as an N source, some researchers have also proposed N_2 as an alternative N source, particularly for the electrochemical co-reduction of CO_2 and N_2 for urea synthesis. In 2016, Köleli and colleagues validated this concept, achieving urea synthesis on polyaniline- or polypyrrole-coated platinum electrodes under a total pressure of 60 bar (30 bar N_2 + 30 bar CO_2) in a 0.1 M Li_2SO_4/0.03 M H^+ solution [15]. Due to the low solubility of N_2 in the electrolyte and the intense competitive hydrogen evolution reaction (HER) in the aqueous solution, the catalytic performance of the electrochemical co-reduction of CO_2 and N_2 is limited, and the search for additional N-containing products is necessary. In 2022, Zhang and colleagues reported the electrochemical conversion of CO_2 and NO to urea on ZnO nanosheet catalysts with NO contaminants as N sources and H_2O as hydrogen sources under environmental conditions [16].

In addition to synthetic urea, methylamine, a product of C—N coupling, is a crucial ingredient in the synthesis of valuable chemicals, which is typically produced by high-temperature and high-pressure reactions between methanol and ammonia. In 2021, Wang et al. achieved the direct electrosynthesis of methylamine by co-electrolysis of CO_2 and NO_3^-, reducing hydroxylamine with nitric acid and condensing formaldehyde with CO_2 [17].

In recent years, on the basis of the above pioneering research, significant work has been done in the field of electrocatalytic co-reduction of CO_2 and N-containing compounds. However, the existing catalysts still fall short of meeting the requirements for large-scale commercial applications due to their low activity and selectivity in urea electrosynthesis. Additionally, there is limited research on the progress of electrocatalytic urea synthesis and its reaction mechanism. Therefore, it is essential to systematically summarize the development history, research progress, and reaction mechanisms of electrochemical C—N coupling. This will provide valuable insights for researchers to design efficient electrocatalysts and enhance electrochemical synthesis performance.

13.3 Fundamentals of Electrocatalytic C—N Coupling

Electrochemical C—N coupling reactions have emerged as a new "green synthetic strategy," which is extremely significant for synthesizing organonitrogen compounds with vital applications in energy, agriculture, medicine, and chemical industry. Electrochemical C—N coupling involves a co-electrolysis system that includes carbon dioxide and N-containing species (N_2, NO, NO_2^-, and NO_3^-). The principle is based on the interaction between CO_2 and N-containing intermediates to generate C—N bonds and produce target coupling products on the same electrode. In contrast to conventional industrial methods that predominantly rely on fossil fuels, the electrochemical co-reduction of CO_2 and N-containing species not only makes the synthesis process more energy-efficient, environmentally friendly, and practical, but also provides more possibilities for synthesizing high-value products and expands product categories.

As a new research field, it faces challenges due to an immature research system and inconsistent evaluation standards. Therefore, it is essential to establish systematic research methods and strict evaluation criteria. In this section, we provide a detailed introduction to electrocatalytic C—N coupling reactions from three perspectives: product detection technology, key evaluation indices, and electrochemical reactor configuration.

13.3.1 Detection Techniques

In the electrochemical co-reduction process involving CO_2 and nitrogen-containing species (such as N_2, NO, NO_2^-, and NO_3^-), there are highly competitive reactions, including the independent CO_2RR and the nitrogen-containing species reduction (NO_xRR), as well as HER. To accurately evaluate catalytic performance, accurate qualitative and quantitative analysis of various products (both liquid and gas phase products) generated during the reaction process is essential. The analytical methods used must exhibit extremely high sensitivity, reliability, accuracy, and repeatability to minimize the impact of interfering factors that could affect the accurate determination of the products, such as the duration of electrolysis, pH of the solution, and cleaning agents.

1. **Liquid-phase products.** Recently, the synthesis of urea through electrocatalytic C—N coupling reactions has garnered significant attention from researchers, making the accurate detection of urea products critically important. Currently, well-established methods include the diacetylmonoxime method, the urease method, ^{1}H nuclear magnetic resonance (^{1}H NMR) spectroscopy, and high-performance liquid chromatography (HPLC). This summary comprehensively reviews the detection mechanisms, advantages, and disadvantages of these urea determination methods, with a particular focus on potential sources of interference.

   The diacetylmonoxime method depends on the chromogenic reaction between diacetylmonoxime and urea, where the concentration of urea is directly proportional to the absorbance of the chromogenic product measured by UV-VIS spectroscopy. In this method, diacetyl imide is first converted to a diacetyl group, which then reacts with urea in the presence of ferric chloride, thiosemicarbazone, and an acidic environment. The resulting red compound reaches its maximum absorbance at 525 nm, allowing urea concentration to be quantitatively determined based on the linear relationship between concentration and absorbance [18]. However, NO_2^- can react with the chromogenic agents used in this method, interfering with accurate urea detection [16]. Therefore, the diacetylmonoxime method is unsuitable for quantifying urea when NO_2^- is a byproduct or when NO_2^-/NO_3^- are used as nitrogen sources for co-reduction with CO_2. In addition, reducing agents, high-concentration electrolytes, and NH_3 can also affect the accuracy of the diacetyl monoxime method.

   The urease method is an indirect method for quantifying urea and involves a two-step reaction process. Urea is first decomposed into NH_3 and CO_2 by urease. The NH_3 concentration is then quantitatively analyzed using the indophenol blue method. Urea concentration is indirectly determined by the difference between the NH_3 concentration after urease decomposition and the initial NH_3 concentration. Interfering factors include conditions during urease decomposition that can affect the accuracy of urea detection, such as enzyme concentration, reaction time, temperature, etc. Additionally, NO_2^-, electrolyte pH, and specific metal ions can influence the color reaction of the indophenol blue method, thereby interfering with the quantification of urea [19].

Compared to the spectrophotometric methods mentioned above, ^{1}H NMR spectroscopy is considered a more reliable method for urea detection due to its lower interference and better detection limit. For ^{1}H NMR measurement, the sample solution is first mixed with a deuterated reagent and an internal standard, then placed into an NMR tube to obtain the ^{1}H NMR spectrum [20]. Urea can be accurately detected and distinguished from other potential disturbances based on the specific chemical environment of its hydrogen atoms in the ^{1}H NMR spectrum. The concentration of urea is calculated by comparing the integral areas of the internal standard's characteristic peak and urea's characteristic peak. Additionally, ^{1}H NMR is useful for isotope labeling measurements. For instance, ^{15}N-labeled urea shows characteristic peaks at approximately 5.65 and 5.55 ppm, which can be differentiated from the single peak of ^{14}N-containing urea at around 5.6 ppm [21]. However, the quality of ^{1}H NMR data depends on the equipment, operator experience, and the amount of data accumulation. Bruker 800 MHz equipment is known for its quick and accurate detection of urea, but it is more expensive and less widely available.

HPLC offers advantages such as low detection limit, high sensitivity, and fast analysis speed. It is not affected by various C/N-containing byproducts, metal ions, or other potential interferences in the detection of urea. Additionally, in isotope labeling experiments, HPLC coupled with mass spectrometry (HPLC-MS) can detect ^{15}N-labeled urea ($m/z=63$) and distinguish urea containing ^{13}C ($m/z=62$), providing strong evidence for the source of urea. Compared to 1H NMR, HPLC is more convenient and easier to operate.

In summary, the widely used diacetylmonoxime and urease methods, which rely on color reactions, are susceptible to interference from byproducts or changes in reaction conditions, affecting the accuracy of urea detection. Therefore, it is recommended that the results from these methods should be verified using ^{1}H NMR or HPLC.

2. **Gas-phase products.** Gas chromatography is commonly used for both quantitative and qualitative analysis of gas-phase products. The electrocatalytic process and the analytical procedure can occur simultaneously. Chemicals are identified and detected as they exit the column. The size of the peak (either peak height or peak area) indicates the quantity of the component, while the time at which the peak appears, known as the retention time, is used to characterize each component.

Thermal conductivity detectors (TCD) and flame ionization detectors (FID) are commonly used in electrochemical CO_2 reduction reaction (ECRR) systems to detect gas products. FID is particularly suited for investigating organic compounds as it can identify hydrocarbons produced by the reaction but cannot detect H_2O or CO_2. This type of destructive mass detector operates with a flame created by burning hydrogen and air. The electrical signal generated by the ion stream of an organic compound entering the flame correlates with the amount of the compound, allowing for quantitative analysis based on the signal intensity. FID is effective for detecting nearly all volatile organic compounds, especially alkanes from C_1 to C_6. In contrast, the TCD is a nondestructive, concentration-based detector that measures the thermal conductivity of different gases to identify components in a sample. It converts thermal signals into electrical signals for quantitative analysis of organic, inorganic, and elemental substances.

3. **Isotope labeling experiments.** To eliminate interference from C/N-containing pollutants present in the environment or introduced during the electrocatalytic process, it is essential to verify the origin of the products through isotope labeling experiments. This method not only accurately assesses the synthesis performance of electrocatalysts but also helps deduce the plausible C—N coupling reaction pathway. The carbon and nitrogen sources used in the electrochemical C—N coupling reaction should exclusively contain ^{15}N and ^{13}C. Various techniques, such as infrared spectroscopy (IR), mass spectrometry (MS), and NMR spectroscopy, can be used to distinguish ^{15}N and ^{13}C from possible contaminants like ^{14}N and ^{12}C. While ^{1}H NMR can effectively identify N-sources due to the different chemical shifts between ^{14}N and ^{15}N, it is not suitable for ^{13}C isotope labeling experiments. For the carbon source, it is necessary to determine whether the CO produced during the reaction originates from the ^{13}C source by analyzing the position, size, and shape of the C peak in mass spectrometry (GC-MS) measurements.

13.3.2 Key Evaluation Index

To initially assess catalytic effectiveness in electrosynthesis, key parameters such as activity, kinetics, and stability should be measured. The following summarizes several important evaluation metrics:

1. **Overpotential (η).** Overpotential is defined as the difference between the equilibrium potential (E^0) and the applied potential (E_{cat}) required to drive a specific electrochemical reaction. The practical process often requires a higher applied potential to overcome kinetic barriers caused by issues such as high activation energy and low energy efficiency. Therefore, overpotential (η) can be expressed by the following equation:

$$\eta = E^0 - E_{cat} \tag{13.1}$$

A smaller η, indicating less extra energy required for the same current density, suggests that the catalyst can efficiently catalyze the reaction at a lower potential.

2. **Faradaic efficiency.** The FE of the target product directly represents the selectivity of the catalyst in a reaction system with multiple reaction routes and competitive reactions. A higher FE indicates greater selectivity of the catalyst, making it more effective in producing a specific product. FE is calculated as the ratio of the charge used to generate the target product to the total charge passing through the electrode.

FE can be calculated by Eq. (13.2):

$$FE = \frac{znF}{Q} \tag{13.2}$$

where z is the number of transfer electrons needed for generating one target product molecule, n is the number of moles of the target product (mol), F is the

Faradaic constant (96500 C mol^{-1}), and Q is the total amount of charge transferred in the electrochemical reaction.

3. **Current density (j).** Current density refers to the amount of electric current passing through a unit area of an electrode and is a crucial metric for evaluating catalyst performance in electrochemical reactions. It includes both total current density and partial current density for specific products. Partial current density is determined by multiplying the total current by the FE of the desired product at the corresponding potential. This measure indicates the production rate of the target product and is an important index for industrial applications. Therefore, understanding and optimizing current density can significantly enhance the efficiency and selectivity of electrochemical processes, making large-scale production more feasible.
4. **Stability.** Stability is a crucial parameter for assessing the potential practical application of a catalyst. Two electrochemical measurements are commonly used to evaluate stability: linear sweep voltammetry (LSV) or cyclic voltammetry (CV) and galvanostatic or potentiostatic electrolysis. In the voltammetry approach, stability is assessed by comparing changes in overpotential before and after a certain number of cycles within a specific potential range using CV or LSV. Smaller changes in overpotential indicate higher catalyst stability. The galvanostatic or potentiostatic method involves long-term testing at a constant current density or potential to monitor changes in potential or current density over time. If the current density or potential remains stable over an extended period, it indicates good catalyst stability.
5. **Tafel Plot.** The Tafel slope is derived from the overpotential with the logarithm of the partial current density of the target product. This concept is derived from Tafel's empirical formula (Eq. 13.3):

$$\eta = a + b \times \log|i| \tag{13.3}$$

   where η is the overpotential, the intercept a is a constant which subject to the properties of the electrode material and solution, i is the partial current density of the target product, and b is the Tafel slope. In addition, there is another new method to determine b by plotting the logarithm R_{ct} versus η, where R_{ct} is the charge transfer resistance measured by electrochemical impedance spectroscopy (EIS) [22]. The method is suitable for electrocatalysts with poor conductivity. The Tafel slope represents the intrinsic properties of the electrocatalyst and is closely related to the rate of the electrocatalytic reaction. The size of the Tafel slope reflects how voltage affects the partial current density of a specific product. A smaller Tafel slope indicates that the current density increases more rapidly with each unit increase in voltage. In other words, a smaller Tafel slope means that less η is needed to achieve the same increment in current density, suggesting that the catalyst has faster electron transfer kinetics. Therefore, a high-performance electrocatalyst should have a small Tafel slope.
6. **Yield rate.** The yield of the target product is expressed in terms of its chemical yield and current efficiency. Current efficiency is the ratio of the coulombs used to form the target product to the total number of coulombs consumed during the entire reaction process. Current efficiency can decrease due to the

presence of other byproducts. The yield rate for the target product is calculated as follows (Eq. (13.4)):

$$r = \frac{CV}{tA} \tag{13.4}$$

where C is the measured concentration of a certain product, V is the volume of electrolyte, t is the reaction time, A can be the geometric area, electrochemical specific surface area, or the weight of the catalyst (contribution of catalyst per milligram to yield rate of product).

7. **Turnover frequency.** The recent definition of turnover frequency (TOF) is the total number of molecules converted into the desired product per unit of time per catalytic site. The most common method for estimating TOF is based on the following equation (Eq. 13.5):

$$\mathrm{TOF} = \frac{j/NF}{m_{cat} w/M} \tag{13.5}$$

where j is the current density for a certain product, N is the number of electrons that need to be transferred to form the product, m_{cat} is the catalyst mass of the working electrode, w is metal loading in the catalyst, and M is the atomic mass of the metal. However, for most solid-state catalysts, particularly complex ones, obtaining an exact TOF value is challenging because not all surface atoms in the catalyst are catalytically active or equally accessible. Consequently, this method may not effectively compare catalytic activity between different catalysts, especially in similar systems.

13.3.3 Reactors

The electrochemical reactor is the most fundamental equipment for researching C—N coupling reactions and significantly influences electrocatalytic properties since the configuration of the electrochemical cell and the structure of the working electrode directly affect mass diffusion, charge transfer, and gas transformation, which in turn impact reaction kinetics [23]. Therefore, rational design and optimization of the electrochemical reactor for C—N coupling reactions are crucial strategies for achieving efficient electrocatalytic synthesis. Currently, three typical types of reactors are suitable for C—N electrochemical coupling reactions: H-type cells, flow cells, and membrane electrode assembly (MEA) cells (Figure 13.1).

Among different electrochemical cells, the H-type cell is the most widely used due to its simplicity in operation and its ability to prevent interference between the cathode and anode reactions [24]. The H-type cell consists of a cathode chamber and an anode chamber separated by a proton or ion-exchange membrane (Figure 13.1a). During electrochemical measurements, the working electrode is fully immersed in the electrolyte, and the feeding gases must diffuse from the bulk electrolyte to the electrode surface to participate in the reaction. However, a major drawback of the H-type cell is its poor gas mass transfer. For example, the low solubility of CO_2/N_2 in aqueous solutions can hinder gas conversion and mass diffusion, resulting in low current density and slower reaction kinetics.

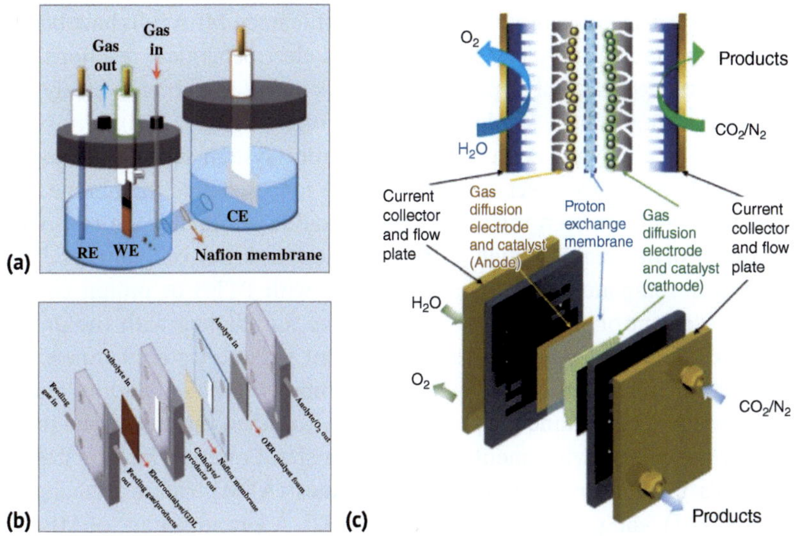

Figure 13.1 Schematic illustration of (a) H cell (WE: working electrode; CE: counter electrode; RE: reference electrode) and (b) flow cell (GDE: gas diffusion electrode). *Source:* Reproduced from [25] / with permission of American Chemical Society. (c) MEA cell for CN electrocatalysis. *Source:* Reproduced from [26] / with permission of Elsevier.

To address the limitations of gas transformation and mass diffusion, the flow cell with a gas diffusion electrode (GDE) was developed to enhance C—N coupling performance. Unlike the traditional H-type cell, the flow cell features a gaseous cathode compartment, a catholyte compartment, and an anode compartment, with the catholyte and anolyte compartments separated by a proton or ion exchange membrane (Figure 13.1b). The electrolyte is circulated through the cathode and anode chambers by a peristaltic pump, improving the mass transfer of reactants and products. The GDE, with the catalyst deposited on a gas diffusion layer (GDL) made of a carbon fiber substrate and a polytetrafluoroethylene (PTFE) hydrophobic porous carbon layer [27], allows direct exposure of the GDL and catalyst layers to both the gas phase and the electrolyte [28]. The GDL is crucial in the flow cell, as it facilitates effective gas diffusion to the electrocatalyst surface, overcoming the sluggish gas diffusion seen in H-type cells. This increases the local gas concentration, forming a catalyst–electrolyte-gas three-phase boundary that accelerates gas conversion and improves reaction rates [29]. The porous hydrophobic substrate of the GDE provides a gas channel and limits the availability of local protons, thereby reducing competitive HER. Although the GDL itself does not participate directly in the electrochemical reactions, it significantly enhances electrode performance. For the electrochemical co-reduction of CO_2 and NO_2^-/NO_3^-, the flow cell improves CO_2 diffusion, accelerates NO_2^-/NO_3^- transfer, and facilitates product release, making it more effective for high current density C—N coupling reactions than the H-type cell. However, the flow cell has two main disadvantages: higher resistance due to the separate compartments and increased costs from the heavy use of electrolytes, which limits its large-scale industrial application.

To reduce energy consumption and enhance efficiency, MEA cells have been developed and are increasingly popular for various electrochemical reactions [30, 31]. The MEA cell consists of a proton or ion exchange membrane, catalyst layers, and GDL arranged in a sandwich-like structure (Figure 13.1c). Unlike H-type and flow cells, the MEA cell uses a two-electrode system where the cathode and anode catalysts are in direct contact with the ion exchange membrane, with no liquid electrolyte between them. This configuration eliminates the resistance of a liquid electrolyte, enabling high current densities at lower cell voltages. The GDL is a critical component in MEA, often made with PTFE or proton conductors. Proper treatment of the catalytic layer ensures full contact with the proton and electronic conductors while providing sufficient gas and water channels, expanding the three-dimensional reaction area, and improving overall performance. Since the catalyst and membrane are pressed directly together, the membrane influences the interface microenvironment and mass transfer. Common membranes used in MEA cells include anion exchange membranes (AEM), cation exchange membranes (CEM), and porous solid electrolytes [32]. A key advantage of MEA cells is that the GDE is attached directly to the ion exchange membrane, allowing for efficient collection of gas/liquid products and reducing the need for flow between the catalyst layer and membrane. This membrane-based design significantly reduces ohmic resistance during the reaction, increasing current density [30].

In general, the electrocatalytic reactor is crucial for enhancing electrocatalysis performance. An ideal reactor not only facilitates efficient electron and ion transfer and controlled gas diffusion but also prevents the mixing of anolyte and catholyte, which is essential for scaling up from laboratory to industrial production. The MEA cell is particularly suitable for electrocatalytic C—N coupling reactions, accommodating various reaction conditions such as cathode coupling and generating intermediates at both cathode and anode through redox mediators. By considering the coupling method and substrate properties, the reactor design can be optimized to achieve efficient C—N coupling.

13.4 Electrocatalytic C—N Coupling for Urea Synthesis

Urea, a highly efficient nitrogen fertilizer, is widely used in agriculture and industry and is one of the most important and valuable products in electrochemical C—N coupling reactions. Traditionally, urea is synthesized on an industrial scale through a direct reaction between CO_2 and NH_3 using the Haber–Bosch process, which requires extremely high temperatures and pressures, resulting in high energy consumption and a significant carbon footprint [33]. Recently, there has been growing interest in using CO_2 and various nitrogen-containing small molecules (such as N_2, NO, NO_2^-, and NO_3^-) as feedstocks to synthesize urea under more environmentally friendly conditions. This approach aims to transform environmental pollutants and greenhouse gases into high-value chemicals. However, achieving high selectivity and yield for target products in electrochemical C—N coupling reactions remains a significant challenge due to the presence of various competitive reactions.

This section focuses on the electrocatalytic coupling reactions for the direct synthesis of urea from CO_2 and nitrogen sources under environmental conditions. It aims to enhance our understanding of the reaction mechanisms and key steps involved, as well as to optimize the reaction system for improved performance.

13.4.1 C—N Coupling with CO_2 and N_2

Using N_2 as the nitrogen source for urea synthesis has notable advantages, limitations, and mechanistic implications. N_2 is the most abundant nitrogen source on Earth and can be directly obtained from the atmosphere. Urea electrosynthesis from CO_2 and N_2 not only reduces material costs but also eliminates the need for the energy-intensive Haber–Bosch process. However, similar to the electrochemical reduction of N_2 to NH_3 [34, 35], using N_2 in C—N coupling reactions faces the challenge of sluggish reduction kinetics, which reduces the rate of urea production. Köleli et al. made a pioneering contribution to the electrochemical synthesis of urea by co-reducing CO_2 and N_2 in 2016 [15]. Since then, research has focused on improving urea yield and FE through optimized catalyst design and synthesis. For instance, Wang and colleagues developed PdCu alloy nanoparticles on TiO_2 nanosheets with oxygen-rich vacancies as an electrocatalyst for C—N coupling (Figure 13.2a) [36]. The reaction involved four stages: adsorption of N_2 and CO_2, reduction of $*CO_2$ to *CO, coupling of $*N_2$ and *CO to form *NCON, and finally hydrogenation of *NCON to urea. The Pd_1Cu_1/TiO_2-400 catalyst enhanced the chemisorption of N_2 and CO_2 and improved their coadsorption, resulting in efficient urea production. In flow cells, this catalyst achieved a urea formation rate of 3.36 mmol g^{-1} h^{-1} and an FE of 8.92% at −0.4 V versus reversible hydrogen electrode (RHE), outperforming the H-type cell and demonstrating that flow devices enhance reaction rate and selectivity (Figure 13.2b).

However, it is important to note that the competitive reduction of CO_2 and N_2 at the same active sites can lead to reduced C—N coupling efficiency and lower urea selectivity. This issue can be addressed by designing electrocatalysts with dual-active sites through various strategies, such as single-site configurations, artificial frustrated Lewis pairs (FLPs), or heterostructures.

Wang and coworkers synthesized the Pd_1-TiO_2 single-atom catalyst (SAC) and Pd_1Cu_1-TiO_2 using vacancy-anchorage strategy. The synergistic interaction between dual-atom Pd_1Cu_1 and oxygen vacancy (V_O) active centers plays a vital role in promoting the coupling of C–N, which exhibited higher electrocatalytic performance than Pd_1Cu_1-TiO_2, with a urea production rate of 166.67 mol_{urea} $mol_{Pd}{}^{-1}$ h^{-1} and FE of 22.54 % in H-type cell. Then, copper phthalocyanine nanotubes (CuPc NTs) have been reported to possess multiple active sites (metal centers or various coordination structures with nitrogen atoms), making them highly efficient electrocatalysts (Figure 13.2c). They exhibited a urea yield of 143.47 µg h^{-1} $mg^{-1}{}_{cat}$ and FE of 12.99% at −0.6 V versus RHE (Figure 13.2d) [18]. Theoretical calculations suggested that the pyridinic C—N_1 and Cu centers were responsible for forming C—N bonds, with pyridinic C—N_1 facilitating the co-reduction of N_2 to NN* and Cu centers reducing CO_2 to *CO.

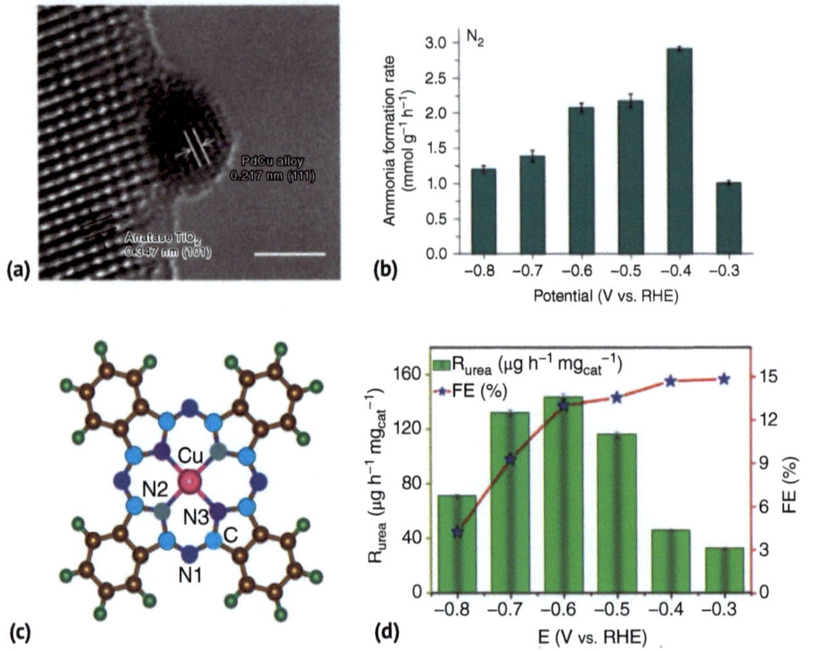

Figure 13.2 (a) High-resolution TEM image of Pd_1Cu_1/TiO-400. The scale bar is 2 nm. (b) Ammonia synthesis with N_2 as the feeding gas at various potentials. *Source:* Reproduced from [36] / with permission of Springer Nature. (c) The model of CuPc and possible active sites denoted with different color. (d) Urea yield rate and FE with N_2 and CO_2 as feeding gas at different potentials for CuPc NTs. *Source:* Reproduced from [18] / with permission of John Wiley & Sons.

Recently, FLPs have also been studied for electrocatalytic C—N coupling. These pairs consist of steric Lewis acids and Lewis bases, which uniquely allows them to chemically absorb and activate two gas molecules simultaneously at both the acid and base sites. For instance, Zhang's group designed FLPs using flower-like nickel borate ($Ni_3(BO_3)_2$), where surface hydroxyl groups acted as Lewis bases and adjacent Ni sites served as Lewis acids (Figure 13.3a) [37]. This design enhanced the directional trapping of CO_2 and N_2, leading to a remarkable urea yield and FE for the $Ni_3(BO_3)_2$-150 catalyst (9.7 mmol h^{-1} and 20.36% at −0.4 V vs. RHE, respectively) (Figure 13.3b). Specifically, the frustrated Lewis pairs promoted CO_2 chemisorption by capturing C atoms on the Lewis base sites. The Lewis acid and base sites then worked together to activate the adsorbed CO_2, facilitating the formation of *CO intermediates. When N_2 is activated to *NN*, it spontaneously coupled with CO. Finally, the *NCON* intermediate was formed through the "S-orbital carbonylation" strategy. Similarly, InOOH nanocapsules (NCs), with an in site acting as an electron-deficient Lewis acid and an In-OH group as an electron-rich Lewis base, also exhibited excellent electrocatalytic urea synthesis properties [38]. They not only facilitated the chemisorption of N_2 and CO_2 through electronic interactions but also enhanced the orbital matching between intermediates *NN* and CO, promoting the C—N coupling process to generate *NCON*. Thus, InOOH FLPs achieved a urea yield of 6.85 mmol h^{-1} g^{-1} and an FE of 20.97%. The use of FLPs underscored a crucial factor in the urea electrosynthesis process and offered a new pathway for developing novel and efficient electrocatalysts of the co-reduction of CO_2 and N_2.

Apart from FLPs, Mott–Schottky structures, which consist of a metal–semiconductor heterojunction, have been reported as efficient catalysts for C–N coupling. Zhang et al. constructed Bi-$BiVO_4$, an electrocatalyst with a heterojunction structure, which achieved a urea FE of 12.55% and a yield rate of 5.91 mmol h^{-1} g^{-1} at −0.4 V versus RHE (Figure 13.3c, d) [39]. The superior catalytic performance was primarily due to the unique electron interactions within the hybrid structure. The electron-rich N atoms in N_2 and the electron-deficient C atoms in CO_2 are preferentially adsorbed in the electrophilic and nucleophilic regions of the Bi/$BiVO_4$ electrocatalyst, respectively. The adsorbed N_2 then promoted the reduction of CO_2 to form CO and participated in the subsequent reactions. Yuan and colleagues also

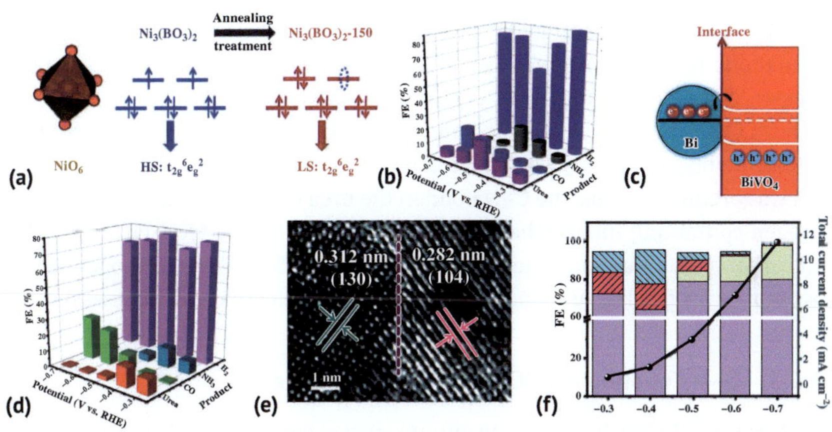

Figure 13.3 (a) Schematic illustration of the spin-state regulation of pristine $Ni_3(BO_3)_2$ after annealing treatment. (b) Corresponding product distribution of H_2, CO, NH_3, and urea with N_2 and CO_2 as the feed gas at various potentials for the $Ni_3(BO_3)_2$-150 catalysts. *Source:* Reproduced from [37] / with permission of Royal Society of Chemistry. (c) Schematic illustration of the charge transfer process in Bi-$BiVO_4$. (d) The corresponding product distribution of H_2 (purple), CO (cyan–blue), NH_3 (blue), and urea (red) with N_2 and CO_2 as the feeding gas at various potentials for Bi-$BiVO_4$ hybrids. *Source:* Reproduced from [39] / with permission of John Wiley & Sons. (e) High-resolution TEM image of $BiFeO_3$/$BiVO_4$ hybrids. *Source:* Reproduced from [40] / with permission of Royal Society of Chemistry. (f) the corresponding product distribution of H_2 (purple color), CO (cyan–blue color), NH_3 (red color), and urea (blue color) for urea production with N_2 and CO_2 as the feed gas at various potentials for $BiFeO_3$/$BiVO_4$ hybrids. *Source:* [40] / Royal Society of Chemistry / CC BY 3.0.

demonstrated that the perovskite $BiFeO_3/BiVO_4$ hybrid catalyst, which benefited from local charge redistribution in the heterogeneous structure, thus delivered a urea yield rate of 4.94 mmol h^{-1} g^{-1} with an FE of 17.18% at −0.4 V versus RHE in 0.1 M $KHCO_3$ (Figures 13.3e and f) [40].

Although significant progress has been made in using N_2 as a nitrogen source for urea formation through C—N coupling, its inherent chemical inertness still limits industrial applications. From the perspective of "converting waste into valuable products" through renewable energy, utilizing other nitrogen sources (such as NO, NO_2^-, and NO_3^-) along with CO_2 for electrocatalytic co-reduction to produce high-value organic nitrogen compounds presents a promising strategy.

13.4.2 C—N Coupling with CO_2 and Nitric Oxide

Nitric oxide (NO), a pollutant gas emitted from the combustion of fossil fuels in vehicles, factories, and power plants, can be electrochemically reduced and used as a nitrogen source for the C—N coupling reaction. This approach addresses environmental concerns while also producing high-value products [41, 42]. Zhang's group prepared Zn nanobelts (Zn NBs) through the in situ electrochemical reduction of ZnO nanosheets [43]. Electrochemical tests showed that the Zn NB catalyst achieved a urea yield of 15.13 mmol h^{-1} g^{-1} and an FE of 11.26% at −0.92 V versus RHE in NO- and CO_2-saturated 0.2 M KOH electrolyte (Figures 13.4a and b). DFT calculations revealed that urea generation proceeds through a 10-step reduction process: *NO →*NHO →*NHOH → (equal to *HNOH)→*NH_2OH→*HN_2→*NH_3, while *COOH and *CO intermediates achieved CO_2-to-CO conversion (Figure 13.4c). Notably, when CO_2 was replaced by CO or NO was substituted by NH_4^+, no urea was produced. Thus, the C–N bond in the urea originated from the coupling of nitrogen-containing intermediates (*NO, *NHO, *NHOH, *NH_2OH, and *NH_2) and carbon-containing intermediates (*COOH and *CO). This work opens up new possibilities for finding suitable nitrogen sources for electrocatalytic urea synthesis.

13.4.3 C—N Coupling with CO_2 and Nitrate/Nitrite

The N=O bond (204 kJ mol^{-1}) in nitrate/nitrite has a lower dissociation energy compared to the N ≡ N bond in N_2 (941 kJ mol^{-1}), making nitrate or nitrite as a more favorable nitrogen source [44, 45]. Recently, researchers have explored the reduction of nitrate or nitrite (NO_3RR/NO_2RR) to avoid the nitrogen fixation reaction [46]. One approach is to pair NO_2^- or NO_3^- with CO_2 to drive electrocatalytic urea synthesis. However, reducing NO_2^-/NO_3^- and CO_2 generates numerous intermediate products, which can lead to the formation of unwanted byproducts like NH_3 and CO [47, 48]. Designing complex electrocatalysts is crucial for enhancing urea production. This section summarizes the research progress on several typical electrocatalysts and suggests directions for the simultaneous reduction of CO_2 and NO_2^-/NO_3^- in urea electrosynthesis.

The first use of nitrite as an N source coupling with CO_2 to synthesize electrosynthesis of urea can be traced back to 1995, when Shibata's group synthesized Cu-loaded GDE [41]. Inspired by this work, the researchers designed and

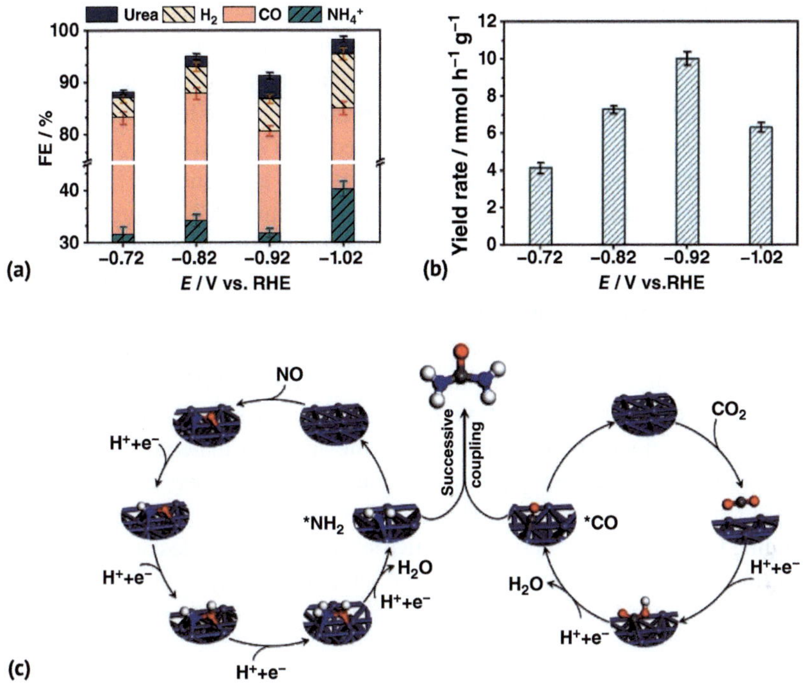

Figure 13.4 (a) Product distribution and (b) urea yield rate over Zn NBs at different applied potentials. (c) Schematic illustration of urea formation mechanism on the surface of Zn using NO and CO_2 as feed gas. *Source:* Reproduced from [43] / with permission of American Chemical Society.

synthesized a series of highly efficient electrocatalysts for C—N coupling reactions. For example, doping nanoparticles with a second metal can regulate the electronic structure and optimize the adsorption energy of the reactants, thus achieving highly selective synthesis of specific products. Feng and coworkers reported Te-Pd nanocrystal catalysts, prepared by doping Te in Pd nanocrystals, for urea electrosynthesis of CO_2 and NO_2^-, which achieved nearly 12.2% FE and 88.7% N atom efficiency at −1.1 V versus RHE, which was much higher than pure Pd nanocrystals [49]. The introduction of Te in Pd significantly promoted the conversion of CO_2 to *CO and NO_2 to *NH_2, and further accelerated the coupling of *CO and *NH_2 to produce urea. Similarly, He's group designed novel carbon-supported indium-doped bismuth nanoparticles for the urea synthesis through the co-reduction of NO_3^- and CO_2, achieving an average FE of 20.31% and a urea yield of 606.38 μg mg^{-1} h^{-1} [50].

Similar to designing electrocatalysts for N_2 and CO_2 co-reduction, creating dual active sites in the electrocatalyst is crucial for efficient urea synthesis from CO_2 and NO_2^-/NO_3^-. For example, using alloys can integrate different metal active sites, alter the electronic structure of the metals, and thereby adjust the adsorption energy of various reactants and intermediates. As Zhao and co-workers reported, abricating PdCu alloying nanoparticles (NP) anchored on carbonized bacterial cellulose (CBC) achieved superior electrocatalytic activity toward urea synthesis with CO_2 and NO_3^-, affording a remarkable urea yield rate of 763.8 ± 42.8 mg h^{-1} mg_{cat}^{-1}

at −0.50 V versus. RHE and an exceptional FE of 69.1 ± 3.8% at −0.40 V versus RHE (Figure 13.5a) [51]. The alloying catalyst provided Pd and Cu dual active sites with favored internal electron transferability, enabling the generation of key intermediates (*NO_2 and *CO_2) to facilitate C—N coupling reaction for urea synthesis (Figure 13.5b). Like that, Sun et al. synthesized a $CuWO_4$ catalyst with Cu and W primary bimetallic active sites, which optimized the adsorption of *NO_2 and *CO intermediates (Figures 13.5c and d) [52]. Liu et al. found that the properties of the catalyst surface significantly affected molecular adsorption and the activity of urea electrosynthesis. By constructing bimetallic electrocatalysts, such as Cu-Bi and Cu-Sn, they discovered that a negatively charged C-bond surface promoted the formation of *COOH intermediates, which facilitated the subsequent C—N coupling process. In contrast, an O-bond configuration (*OCHO) hindered further non-electrochemical steps, resulting in poor urea synthesis activity [54].

Atomic scale modulation, which reduces the size of a material to the scale of a single atom, can impart unique physical and chemical properties to the material. SACs, in particular, have garnered significant attention due to their advantages, including high dispersion, optimized active sites, and adjustable coordination structures. Inspired by the wide application of Cu-based catalysts in CO_2RR, Leverett et al. used Cu SACs to electrosynthesize urea and investigated the effect of

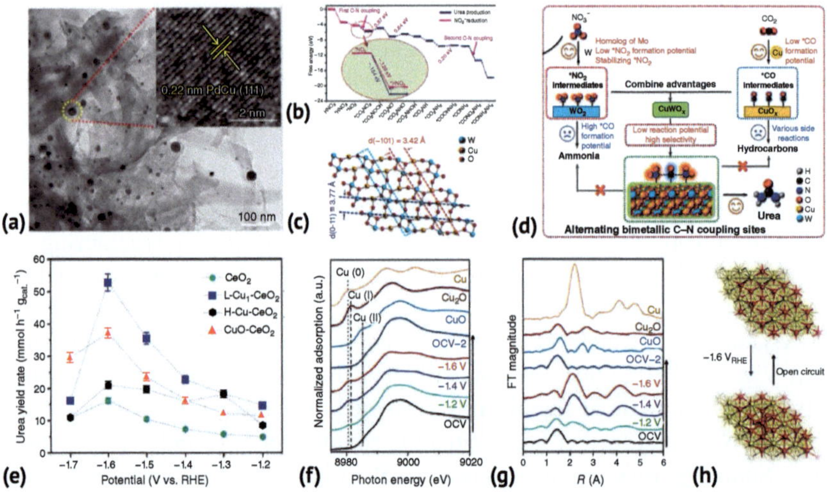

Figure 13.5 (a) TEM images of the PdCu/CBC. *Source:* Ref. [51], Royal Society of Chemistry, CC BY-NC 3.0. (b) Free-energy diagram for urea production of the PdCu/CBC. *Source:* Reproduced from [51] / with permission of Royal Society of Chemistry. (c) Atomic structure of $CuWO_4$ (111) facet. (d) Schematic illustration of design strategy for bioinspired alternating bimetallic sites of $CuWO_4$ catalyst for urea electrosynthesis. *Source:* [52] / Springer nature / CC BY 4.0. (e) The urea yield rates of CeO_2, L-Cu_1−CeO_2, H-Cu−CeO_2, and CuO−CeO_2 catalysts at different applied potentials. (f) Cu K-edge XANES spectra of Cu_1−CeO_2 recorded at different cathodic potentials during C—N coupling. (g) Cu R-space EXAFS spectra of Cu_1−CeO_2 recorded at different cathodic potentials during C—N coupling. (h) Schematic diagram of reconstitution of copper single atoms to clusters suggested by the operando XAS measurements. *Source:* Reproduced from [53] / with permission of John Wiley & Sons.

coordination structure on C—N coupling properties [55]. The Cu SACs with Cu-N_4-C coordination structure exhibited excellent urea yield rate of 4.3 nmol s^{-1} cm^{-2} at −1.0 V versus RHE and the maximum FE of 28% at −0.9 V versus RHE. DFT calculations revealed that Cu-N_{4-x}-C_x structure was more favorable for C—N coupling of *CO and *NH_2. The report on using SACs for urea electrosynthesis highlighted that the relationship between the coordination structure and catalytic performance requires further investigation. Recently, Wang and colleagues loaded single copper atoms onto CeO_2 (Cu-CeO_2) using a wet impregnation and calcination method to study the dynamic reconfiguration during urea production [53]. The Cu-CeO_2 electrocatalyst exhibited a high urea yield rate of 52.84 mmol h^{-1} $g_{cat.}^{-1}$ at −1.6 V versus RHE by co-reducing CO_2 and NO_3^- (Figure 13.5e). However, in situ characterization results showed that single copper atoms transform into Cu clusters (Cu_4) under reducing potential, which were the actual active sites for the C—N coupling reaction (Figures 13.5f and g). DFT calculations indicated that modifying Cu clusters on CeO_2 enhanced the adsorption of *NO and facilitated the subsequent C—N coupling between *NO and *CO, forming the *OCNO intermediate (Figure 13.5h). This modification lowered the energy barrier of the key reaction step for urea production.

Given that electrocatalytic urea synthesis involves different reactants, including CO_2 and nitrogen-containing species, dual-atom catalysts (DACs) can offer advantages over SACs by providing multiple active sites. This enables the synergistic reduction of CO_2 and NO_2^-/NO_3^-. For example, Pan et al. synthesized Ru and Co dual single atoms anchored on N-doped carbon ($CoRuN_6$), which exhibited a high urea yield rate of 8.98 mmol h^{-1} g^{-1} and an FE of 25.31% at −0.6 V versus RHE through the co-reduction of CO_2 and NO_3^- (Figures 13.6a and b) [56]. DFT calculations showed that the Co and Ru sites in $CoRuN_6$ served as active sites for the adsorption of NO_2^- and CO_2, respectively, to form *CO_2NO_2 intermediates in the C—N coupling process. The excellent catalytic performance arose from the synergistic interaction between Co and Ru dual atoms. Compared with RuN_4 and CoN_4 structures, $CoRuN_6$ effectively reduced the energy barrier of the two-step C—N coupling process, thus promoting electrochemical urea synthesis. For atomic-level catalysts, the efficiency of C—N coupling electrochemical synthesis of urea can be further enhanced by selecting appropriate metal species, adjusting the coordination structure, and optimizing the geometry.

Additionally, vacancies in catalysts play a crucial role in improving electrocatalytic performance. In particular, oxygen vacancies can enhance the adsorption of oxygen-containing reactants and stabilize key intermediates. Zhang's group prepared carbon cloth-supported ZnO porous nanosheets and created oxygen vacancies through an electrochemical reduction process (Figure 13.6c) [57]. The oxygen vacancy-rich ZnO (ZnO—V_O) catalysts achieved a maximum urea yield rate of 16.56 mmol h^{-1} cm^{-2} and the highest FE of 23.26% at −0.79 V versus RHE when co-reducing CO_2 and NO_2^-. In situ attenuated total reflectance-Fourier transform infrared spectroscopy (ATR-FTIR) analysis revealed the reaction mechanism: NO_2^- was absorbed on the V_O site and converted into *NH intermediate. The oxygen atom of CO_2 also occupied another V_O and was hydrogenated into *COOH intermediate. C—N coupling process occurred between *NH_2 and *COOH for urea synthesis. Besides co-reduction of CO_2 and NO_2^-, oxygen vacancies are also beneficial in the

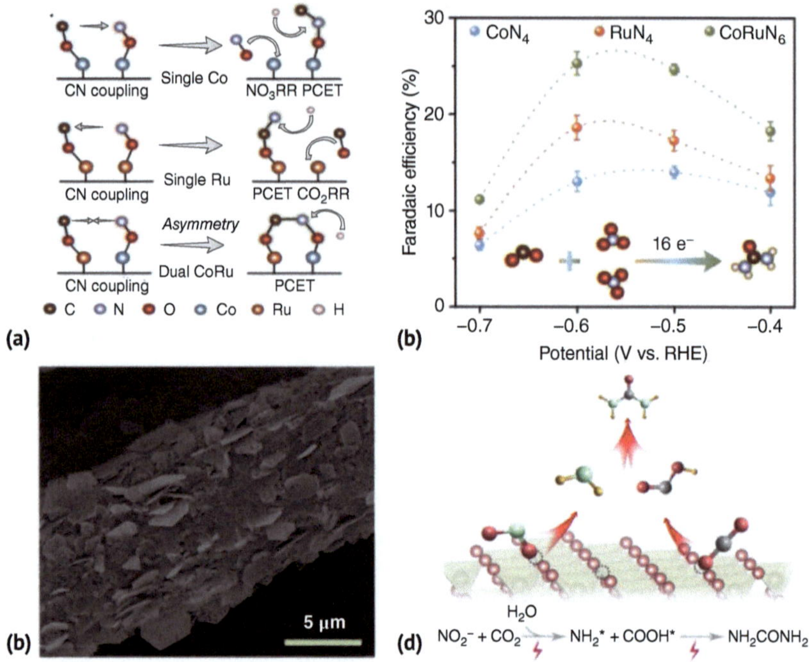

Figure 13.6 (a) Schematic illustration of C—N coupling process on single and dual sites. (b) FE_{Urea} for different catalysts at different potentials. *Source:* Reproduced from [56] / with permission of Elsevier. (c) High-magnification SEM images of ZnO—N porous nanosheets. *Source:* [57] / CC BY-NC-ND 4.0. (d) Schematic illustration and reaction pathways of urea formation over ZnO—V. *Source:* Reproduced from [57] / with permission of Elsevier.

electrochemical coupling of CO_2 and NO_3^- (Figure 13.6d). Wang and coworkers developed oxygen-rich vacancy CeO_2 nanoparticles, where the oxygen vacancies stabilized *NO intermediates and facilitated the C—N coupling process to produce the important *OCNO intermediates [16].

The heterogeneous structure can integrate the functions of different components, making it well-suited for designing C—N coupling electrocatalysts. Li's group prepared a heterostructure composed of mixed Co-Ni oxides and graphdiyne (Co-NiO$_x$@GDY) [58]. The Co-NiO$_x$@GDY heterostructure, with its abundant grain boundaries, dislocations, and disordered domains, provides numerous active sites for the co-reduction of CO_2 and NO_2^-. Thanks to this advantageous heterointerface, the Co-NiO$_x$@GDY electrocatalyst exhibited a high urea yield rate of 913.2 μg $h^{-1} mg_{cat.}^{-1}$ and an improved FE of 64.3% at −0.7 V vs. RHE, outperforming comparative electrocatalysts with single components.

To sum up, the rational design and preparation of catalysts are crucial for the electrocatalytic synthesis of urea. This chapter primarily discusses the types of catalysts used for the co-reduction of various nitrogen sources and CO_2. It aims to provide readers with an understanding of the current research progress in electrocatalytic urea synthesis and to generate interest in this field.

13.4.4 C—N Coupling with CO and NH_3

In addition, N_2 and NO_x are common sources of nitrogen in the process of urea electrosynthesis. However, the direct activation of N_2 to create C—N bond compounds is usually slow and nonselective due to the strong triple bond in N_2 and the efficient HER [59]. If NO_3^- or NO_2^- are used as reactants in urea production, it would involve an eight or six-electron reduction per nitrogen atom [60, 61]. To improve the energy and electron efficiency of C—N bond formation, it is important to minimize unnecessary changes in the oxidation state of nitrogen. Following this logical sequence, using NH_3 as the N source can help avoid fluctuations in the N oxidation state. The reaction of NH_3 with a carbon source in the anode compartment could also suppress the HER. Therefore, the oxidative coupling between NH_3 and a carbon source to produce urea may be a more efficient process in terms of energy, as long as ammonia synthesis remains the most cost-effective method for N_2 activation [62].

Recently, Xiong et al. introduced a new approach to urea synthesis by electrocatalytic oxidative coupling between CO and NH_3 on Pt catalysts (Figure 13.7a),

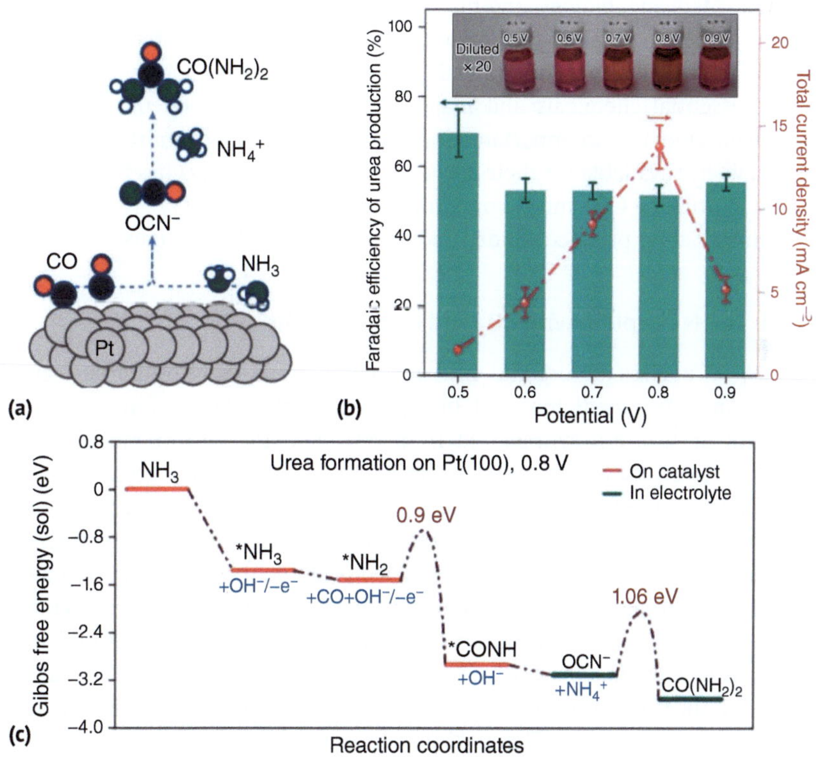

Figure 13.7 (a) A schematic depicting cyanate and urea formation from the electro-oxidative coupling reaction of CO and NH_3 on the Pt catalyst. (b) Three separate measurements performed on three different 70 wt.% Pt/C electrodes. The inserted photograph shows the post-reaction electrolytes after DAMO assays. (c) Solvation (sol)-corrected Gibbs free energy diagram for urea formation on Pt (100) at 0.8 V.
Source: Reproduced from [63] / with permission of Springer Nature.

achieving a high selectivity of around 70% for urea and maintaining over 50% selectivity across a wide range of potentials. The electrocatalytic C—N bond formation rate reached up to 100 mmol h^{-1} g^{-1} (Figure 13.7b) [63]. Through electrochemical, in situ spectroscopic, and computational studies, they revealed that the coupling of CO and NH_3 on Pt involved two proton-coupled electron transfer (PCET) steps for the formation of (iso)cyanic acid (HNCO), followed by a neutralization reaction to produce cyanate, which then reacted with NH_4^+ in the electrolyte to form urea. This pathway not only offers a promising method for converting CO, a significant byproduct of CO_2RR, into valuable urea, but also provides high electron efficiency and low energy consumption compared to current industrial methods for urea production and the coupling of CO_2 and nitrate (Figure 13.7c).

13.5 Electrocatalytic C—N Coupling for Amide Synthesis

Amides play a crucial role in chemistry and have various applications in organic synthesis, medicine, biology, and material sciences [64, 65]. Their strong hydrogen bonding ability, found in many amides, impacts the structure of biomolecules, particularly proteins and pharmaceuticals. Aromatic amides such as nylon and aramides are essential chemicals and materials [66–68]. Therefore, the formation of amide bonds remains an important aspect of organic and biochemistry, with various methodologies available for their creation [69]. Recently, there has been a surge of interest in utilizing CO_2 and N-containing small molecules (NO_2^-, NO_3^-, NH_3, etc.) as raw materials to produce amides under environmental conditions.

13.5.1 C—N Coupling with CO_2 and Nitrate/Nitrite

Amines, as a class of organic nitrogen compounds, have important applications in industry, medicine, and biological activities. Most of the industrially important amines are synthesized from NH_3 and the corresponding alcohols, while the electrocatalytic co-reduction of CO_2 and small N-containing molecules for the synthesis of amines remains largely unexplored. Recently, Wang's research group successfully synthesized methylamine through the electrocatalytic co-reduction of CO_2 and NO_3^- under aqueous ambient conditions using a cobalt β-tetraaminophthalocyanine molecular catalyst supported on carbon nanotubes (CoPC—NH_2/CNT) [17]. The optimized FE for methylamine production was 13%, with a partial current density of 3.4 mA cm^{-2} at −0.92 V versus RHE. It was worth noting that the catalytic performance could be maintained for 16 h with an average FE(methylamine) of 12% (Figure 13.8a). In the entire reaction, the formation of each methylamine molecule involved the transfer of 14 electrons and 15 protons, which was an eight-step catalytic cascade process achieved through the coupling of two reaction intermediates near the catalyst surface (Figure 13.8b). Meanwhile, the key C—N bond formation step was a nucleophilic attack between formaldehyde and hydroxylamine intermediates along the electrocatalytic CO_2RR pathway to ethanol and the NO_3RR pathway to NH_3, respectively. This C—N coupling

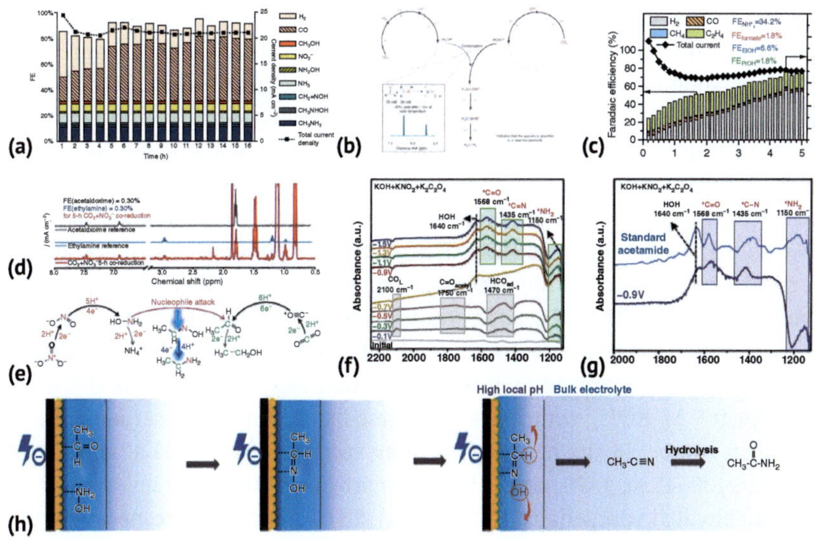

Figure 13.8 (a) Product distribution and total current density during 16 hours electrolysis at −0.94 V versus RHE. (b)The proposed reaction pathway of the eight-step cascade electrosynthesis of methylamine from CO_2 and NO_3^- catalyzed by CoPc-NH_2 /CNT. *Source:* Reproduced from [17] / with permission of Springer Nature. (c) Product distribution and total current density for a 5-hour co-reduction electrolysis; the average liquid product FEs were measured after the electrolysis. (d) ^1H NMR spectrum of the post-electrolysis solution corresponding to (c). (e) Proposed reaction pathway to form acetaldoxime and ethylamine from electrochemical co-reduction of CO_2 and NO_3^-. *Source:* Reproduced from [70] / with permission of Elsevier. (f) Potential-dependent, attenuated total reflection infrared (ATR-IR) spectra on Cu nanoparticles with oxalate as the C source and NO_2^- as the N source in 0.5 M KOH. (g) Full-range spectra on Cu nanoparticles with oxalate as the C source and NO_2^- (dark blue) and standard acetamide (light blue) at −0.9 V versus RHE. (h) A schematic that depicts acetamide formation from acetaldehyde and hydroxylamine on Cu nanoparticles during alkaline electrolysis. *Source:* Reproduced from [71] / with permission of John Wiley & Sons.

reaction first produced formaloxime, which was then reduced to methylamine by an electrochemical process.

Lately, the researchers have further exploited the amine–aldehyde coupling reaction, using the key C—N coupling of amines and aldehydes to synthesize more complex ethylamines. Specifically, ethylamine was successfully synthesized by the electrocatalytic co-reduction of CO_2 and nitrate (NO_3^-) using a CuO-derived copper nanoparticle catalyst, which was a 20-electron, 21-proton reduction cascade (Figure 13.8e) [70]. Similar to the process of methylamine synthesis, acetaldoxime was identified as a key intermediate in the spontaneous condensation of acetaldehyde and hydroxylamine intermediates, which are produced along the electrocatalytic reduction of CO_2 to ethanol and NO_3^- to ammonia (NH_3), respectively. This hypothesis was tested by a series of controlled electrolysis experiments with different carbon or nitrogen sources, and the spontaneous chemical reaction between acetaldehyde and hydroxylamine was confirmed (Figure 13.8d). After 5 h of potentiostatic electrolysis at −1.0 V versus RHE, the FE for ethylamine was only 0.15%, and the FE for acetaldoxime was 0.30%, which might be limited by several competing processes, in particular the self-reduction of acetaldehyde and hydroxylamine (Figure 13.8c).

Amides, such as formamide and acetamide, could also be produced from CO_2 and NO_2^- or NO_3^- through several electron transfer steps. Very recently, Ma's research group achieved the electrochemical co-reduction of CO_2 and NO_2^- to acetamide through Cu catalyst in an alkaline environment using a GDE [71]. Under 0.5 M KOH, within the wide potential range detected for acetamide, the highest FE was 20% at −1.4 V versus RHE. The partial current densities of acetamide on Cu nanoparticles reached approximately 20 mA cm^{-2} and 80 mA cm^{-2} at −1.4 V versus RHE (0.5 M KOH) and −1.6 V versus RHE (1.0 M KOH), respectively. Subsequently, following attenuated total reflection infrared (ATR-IR) and electrochemical co-reduction tests, it was found that NO_2^- was used as the nitrogen source in 0.5 M KOH, with oxalate intermediates instead of carbon dioxide as the carbon source. In the negative scanning spectrum from −0.1 V to −0.7 V, the bands at 2100, 1750, 1470, and 1150 cm^{-1} correspond to the stretching vibrations of acetaldehyde, hydrogenated CO (HCO_{ad}), *NH2, and C=O, respectively, which implied the formation of acetaldehyde and hydroxylamine (NH_2OH) intermediates. Notably, the presence of *C=O (1568 cm^{-1}) and C—N (1435 cm^{-1}) at − 0.9 V indicated the formation of acetamide (Figure 13.8f). Based on the ATR-IR results, the reaction pathway from CO_2 to acetaldehyde (CH_3CHO) was proposed (Figure 13.8g). The reaction pathway of acetaldehyde and hydroxylamine to acetamide on Cu nanoparticles was further analyzed by DFT calculations. It should be mentioned that in the vacuum model, acetaldoxime cannot form acetonitrile (ACN), which actually occurred at the copper–water interface (Figure 13.8h). Therefore, a complex liquid/electrode interface was constructed to simulate the reaction processes ($48H_2O$-Cu(100) and $48H_2O$-K-Cu(100)). The results showed that the adsorption of ACN on the $48H_2O$-Cu(100) surface was stable. However, the "non-adsorption phenomenon" was observed on the $48H_2O$-K-Cu(100) surface, which was almost linear. The adsorption energy difference between the two was −0.7 eV, indicating that ACN molecules tend to desorb into the solvents at negative potentials. This suggested that the reaction intermediate ACN would immediately desorb and diffuse away from the catalyst surface in the alkaline electrolyte, promoting its hydrolysis into acetamide. In summary, the researchers confirmed that acetaldehyde and hydroxylamine (NH_2OH) were the key intermediates in the reaction of acetamide production by the co-reduction of CO_2 and NO_2^-. It further performed a nucleophilic addition reaction to produce acetaldoxime, which was then dehydrated to ACN and hydrolyzed to acetamide.

13.5.2 C—N Coupling with CO and NO

The use of N-heterocyclic carbene was a simple strategy to alter the catalyst surface and weaken the repulsive dipole–dipole interaction [72]. Owing to its excellent conductivity, high surface area, and tunable surface functionalities effective for the ECRR and electrochemical nitrogen reduction reaction (ENRR), Mo_2NO_2 with an V_O (Cu_2@v-Mo_2NO_2) and two anchored Cu atoms was chosen as the model catalyst for the C—N coupling reaction from CO and NO [73, 74].

The analysis of the electronic configuration revealed that carbene functionalization played a crucial role in C—N bond formation by disrupting the electronic configuration between the Cu pairs, leading to charge redistribution in the adsorbed intermediates. Specifically, the d-orbital states that fully overlapped experienced a significant splitting upon carbene ligation. After the surface functionalization with carbene, the N atom acquired a negative charge while the C atom retained its positive charge. This resulted in a shift from repulsion to attraction between the N and C atoms, causing a decrease in the C*CO – N*NO distance and confirming the weakening of dipole–dipole repulsion. Additionally, the reduction of dipole–dipole repulsion between neighboring carbonaceous and nitrogenous intermediates was shown to facilitate C—N bond formation, with the most stable intermediate being identified as the one leading to the production of methylamine on the designed catalyst. While the *CO intermediate was converted to *CHOH and the *NO intermediate was reduced to *NH in this reaction pathway, structural optimization revealed that the *NH intermediate autonomously migrated to a nearby Cu site and connected with the *CHOH intermediate to generate *NH – CHOH. This electrochemical step in C–N coupling had a low kinetic barrier and was highly thermodynamically exothermic, indicating that surface carbene functionalization effectively reduced dipole–dipole repulsion to enhance the C—N coupling reaction.

13.5.3 C—N Coupling with CO_2 and NH_3

Li and his colleagues utilized commercially available Cu or CuO particles as electrocatalysts, using CO_2 as the carbon source to create formamide and acetamide as the primary C—N products. They achieved a maximum FE of 0.4% for formamide and 10% for acetamide (at −0.58 V vs. RHE, corresponding to partial current densities of 0.2 and 2.2 mA cm^{-2}, respectively) [75]. Through IR analysis, they identified new vibrational bands associated with the reaction between CO_2 and NH_3. In the presence of both CO_2 and NH_3, distinct positive and negative bands appeared, indicating the presence of carbonates and ammonium salts. These spectral characteristics aligned with those of *COO and *COOH intermediates observed in earlier studies, bolstering the evidence for the formation of formamide and acetamide. However, despite the production of acetamide and formamide, the catalytic selectivity and activity experienced a notable decrease. Recently, Li et al. developed an electrosynthesis method that coupled CO_2 and NH_3 at the gas–liquid–solid boundary to form small molecules with C—N bonds [76]. Specifically, they successfully synthesized formamide and acetamide by introducing gas-phase CO_2 and liquid-phase NH_3 onto solid copper catalysts (Figure 13.9a), with an initial current of about 0 V versus RHE, reaching 100 mA cm^{-2} at −1.0 V versus RHE (Figure 13.9b). When the local current density was between 0.1 and 1.2 mA cm^{-2}, formamide and acetamide were formed, and the FE was about 1% (Figure 13.9c). Two reaction pathways were proposed (Figure 13.9d):

1. The electrosynthesis pathways of formic acid and formamide were linked together on the Cu surface due to the presence of a common intermediate. The formation of C—N products involved the nucleophilic attack of carbon atoms

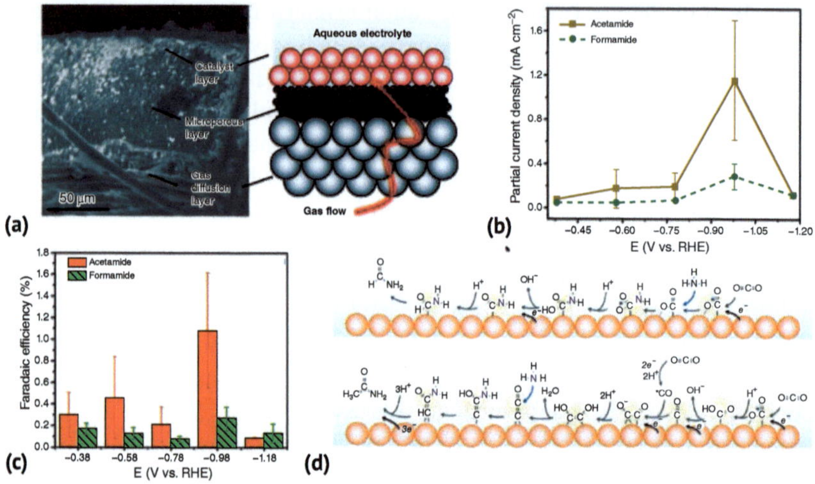

Figure 13.9 (a) The CuO catalyst particles were drop-cast onto a gas-diffusion electrode, which was characterized through scanning electron microscopy. *Source:* [76] / Royal Society of Chemistry / CC BY-NC 3.0. (b, c) In the gas diffusion electrode cell with a 1 M KOH electrolyte, 6SCCM CO_2 flow, and the optimized concentration of NH_3, formamides were quantified and their partial current densities and Faradaic efficiencies for CuO catalysts. (d) Plausible surface reaction pathways in the electrosynthetic process of formamide and acetamide generation. *Source:* Reproduced from [76] / with permission of Royal Society of Chemistry.

on ammonia nitrogen atoms, and the exposed carbon species might be coupled with ammonia to produce formamide, which may be μ_2, —C, —O combined with $^*CO_2^-$. There would then be a competition between hydrogenation of this species to produce formate or a nucleophilic attack to eventually form formamide.

2. The synthesis of acetamide might have the same reaction pathway as acetate and therefore required the presence of C_2 intermediates. A branch of this step may also occur with $^*CO_2^-$ intermediates, which were converted to *CO. The coupling of CO molecules was considered to be a key step in the formation of C_2 products (acetate, ethanol, and ethylene). DFT analysis of CO and NH_3 synthesis of acetamide suggests that *CCO intermediates are possible candidates for this process, with the carbon in between going to get nucleophilic attack from NH_3.

At present, the electrocatalytic co-reduction of CO_2 and nitrogen-containing molecules to synthesize complex nitrogen compounds is still facing serious challenges. Most importantly, the efficiency of C—N coupling is unsatisfactory in most existing cases compared to the side reaction, i.e., the self-reduction of the active C and N intermediates responsible for C—N coupling. For example, the co-reduction electrosynthesis of ethylamine from CO_2 and NO_3^- has a very low yield because the C—N coupling step is hardly competitive with the self-reduction of the acetaldehyde and hydroxylamine intermediates. Therefore, more efforts should be made to study the mechanism, especially the kinetic analysis of the C—N coupling step and the self-reduction reaction of the coupling intermediates. This will improve the efficiency of electrocatalytic coupling and provide new guidance for the design of efficient catalysts.

13.6 Mechanistic Understanding of Electrocatalytic C—N Coupling

Understanding the C—N coupling mechanism between CO_2 and different nitrogen sources is the key to designing and synthesizing efficient catalysts for electrochemical urea synthesis. At present, the exploration of the C—N coupling mechanism mainly relies on DFT calculations and in situ characterization. However, due to the complexity of the reaction pathway, the diversity of intermediates, and the differences in electrocatalysts, the accurate verification of the urea synthesis mechanism is still a challenge. Therefore, this section will review the characterization techniques for exploring the reaction mechanism in C—N coupling and further reveal the urea synthesis pathway.

13.6.1 Theoretical Calculations

The electrochemical process of the electrocatalytic C—N coupling reaction and competitive reactions (CO_2RR and NO_xRR) involves many intermediate reduction products. Therefore, the combination of structural and performance characterization, as well as theoretical calculations, provides a favorable way to further explore the reaction pathway.

Many studies have claimed that the coupling of $*NH_2$ and $*CO_2$ intermediates can effectively promote the formation of C—N bonds. Under the condition of NO_2^- and CO_2 co-reduction, the Te-Pd nanocrystal (Te-doped Pd NPs) catalyst achieved a 12% FE for urea production at −1.1V versus RHE, which was higher than the 4.2% FE(urea) obtained by Pd nanocrystals [49]. The experimental data showed that the reason for the improvement in urea synthesis efficiency was Te doping, which not only optimized the adsorption of CO_2 and CO, but also promoted the reduction of NO_3^- to NH_3 and inhibited the generation of N_2. In addition, DFT calculations indicated that Te doping effectively reduced the formation energy of *COOH and enhanced the adsorption strength of CO on the surface during the CO_2RR (Figure 13.10a). In the NO_2^- reduction reaction, *NO intermediates were more likely to hydrogenate to form NH_3 than undergo N—N coupling to form N_2 (Figure 13.10b). The formation of $*CONH_2$ by *CO and $*NH_2$ intermediates showed a lower kinetic energy barrier in the Te-Pd nanocrystals catalyst, which was more conducive to the formation of C—N bonds. Shen et al. further confirmed whether the N-containing intermediate in the coupling reaction during electrocatalytic urea synthesis was $*NH_2$ or $*NH_2OH$, through DFT calculations and experimental study (Figure 13.10c) [60]. This work considered two coupling pathways of $*NH_2OH$ and *CO intermediates in the electrochemical co-reduction of CO_2 and NO_3^- to form urea: the stepwise coupling pathway ($*NH_2OH + *CO \rightarrow *NH_2 + *CO \rightarrow *OCNH_2$)

and the one-step coordinated coupling pathway (*NH_2OH + *CO → *$OCNH_2$) (Figure 13.10d). The first step of the cascade coupling pathway, the decomposition reaction of *NH_2OH → *NH_2, had a high activation energy barrier of 1.65 eV, indicating that the decomposition of *NH_2OH into *NH_2 was extremely difficult. Even though the activation energy barrier of the subsequent reaction *NH_2 + *CO → *$CONH_2$ was pretty low, the coupling reaction was unlikely to produce urea through a stepwise pathway. In contrast, the activation energy barrier of the one-step cooperative coupling pathway was only 0.47 eV, which was an exothermic reaction (−1.29 eV), revealing that this direct coupling reaction is favorable from a kinetic perspective. Additionally, the N—OH bond (2.457 Å) and C—N bond (1.365 Å) in the transition state of *NH_2OH coupled with *CO on AuPd nanoalloys were longer than the bond lengths of *NH_2OH (1.449 Å) and *$CONH_2$ (1.347 Å), indicating that the formation of the C—N bond in the transition state and the breaking of the N—OH bond occur simultaneously, which was characteristic of a one-step co-coupling reaction (Figures 13.10e and f).

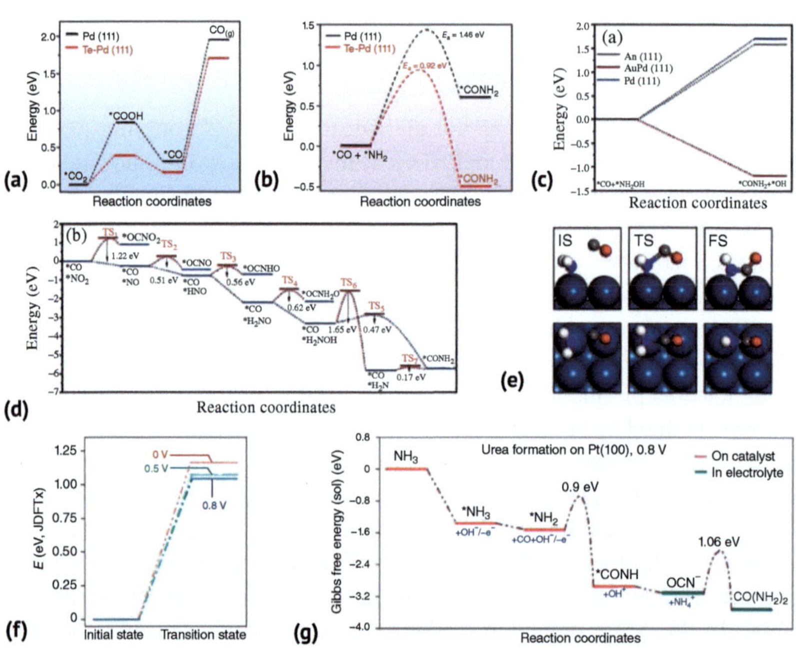

Figure 13.10 (a) Free energy profiles of *CO_2 reduction to CO on pure Pd (111) and Te-doped Pd (111) surface, respectively. The "*" represents the active site. (b) Diagram of free energy changes and activation barriers of *$CONH_2$ formation on the Te-Pd (111) surface and pure Pd (111) surface, respectively. *Source:* Reproduced from [49] / with permission of American Chemical Society. (c) Free energies for *CO + *NH_2OH → *$CONH_2$ on Au, AuPd, and Pd. (d) The C—N coupling pathway on AuPd. *Source:* Reproduced from [60] / with permission of Elsevier. (e) The geometry of initial state (IS), transition state (TS), and final state (FS) for the C—N coupling process from DFT calculations. (f) The calculated energy barriers of the C—N coupling process at different applied potentials. (g) Solvation (sol)-corrected Gibbs free energy diagram for urea formation on Pt (100) at 0.8 V. *Source:* Reproduced from [63] / with permission of Springer Nature.

Yu and co-workers used a well-defined single-faceted $In(OH)_3$ catalyst and a defective engineered InOOH catalyst with oxygen vacancies (V_O-InOOH) to electrocatalyze the C—N coupling to urea and further clarified the reaction pathway of the C—N coupling process combined with DFT calculations [77]. First of all, the absence of C-based byproducts such as CO in the electrocatalytic co-reduction of CO_2 and NO_3^- indicated that C—N coupling occurs in the early stages of the CO_2RR. According to the DFT calculations, the kinetic barrier of $*CO_2$ coupling with $*NO_2$ to generate $*CO_2NO_2$ was smaller than that of $*NO_2$ hydrogenation to generate $*HNO_2$. Subsequently, the $*CO_2NO_2$ intermediate underwent a PCET reaction to produce $*CO_2NH_2$, which then combined with a second $*NO_2$ intermediate to form urea. Notably, the necessary N-containing intermediate in this reaction pathway was $*NO_2$ instead of $*NH_2$ or $*NH_2OH$.

13.6.2 Operando Studies

Advanced in situ characterization techniques are necessary to reveal the catalytic mechanism and reaction pathway of the C—N coupling process. These techniques include operando synchrotron-radiation Fourier transform infrared spectroscopy (SR-FTIR), in situ diffuse reflectance infrared Fourier transform spectroscopy (DRIFTS), and in situ Raman vibrational absorption spectra. The application of these techniques not only provides insights into the reaction mechanism, but also indirectly promotes the rational design of catalysts.

In situ DRIFTS is a suitable technique for studying the mechanism of C—N coupling reactions, as it can provide useful information about reaction intermediates by tracking the changes in substance concentrations or the formation of new species during the reaction. Some previous studies based on in situ experiments have proposed that the C—N coupling occurs in the early stages of the co-reduction of CO_2 and NO_x. To verify this hypothesis, Zhang's group performed an electrocatalytic co-reduction of CO_2 and nitrite using an oxygen-deficient zinc oxide (ZnO—V) catalyst [57]. They found that the primary electric reduction product of nitrite was NH_3, while the primary reduction product of CO_2 was CO. When CO_2 was replaced by CO or nitrite was replaced by NH_3, urea was not produced, indicating that the formation of urea originates from intermediates formed before the production of CO and NH_3. To further investigate the reaction mechanism, the researchers used DRIFTS to monitor the intermediates during the single reduction of CO_2 and NO_2^-, as well as their co-reduction. The DRIFTS spectra showed that in the CO_2RR, *COOH intermediates were observed at 1360 and 1210 cm^{-1}, and their intensity increased with decreasing potential. In the NO_2^- reduction process, a peak at 1100 cm^{-1} was attributed to $*NH_3$ intermediates (Figure 13.11a). During the electrocatalytic co-reduction of CO_2 and NO_2^-, a C—N bond signal was observed at 1440 cm^{-1}, but no *COOH intermediate was detected at 1360 cm^{-1}, suggesting that the *COOH intermediate was consumed to form urea. Therefore, the authors propose that the C—N coupling step occurs between the *COOH and * NH_2 intermediates. Additionally, the differential electrochemical mass spectrometry (DEMS) results showed that when both CO_2 and NO_2^- were present, the signal intensities of both *CO and $*NH_3$ were reduced, confirming a competitive

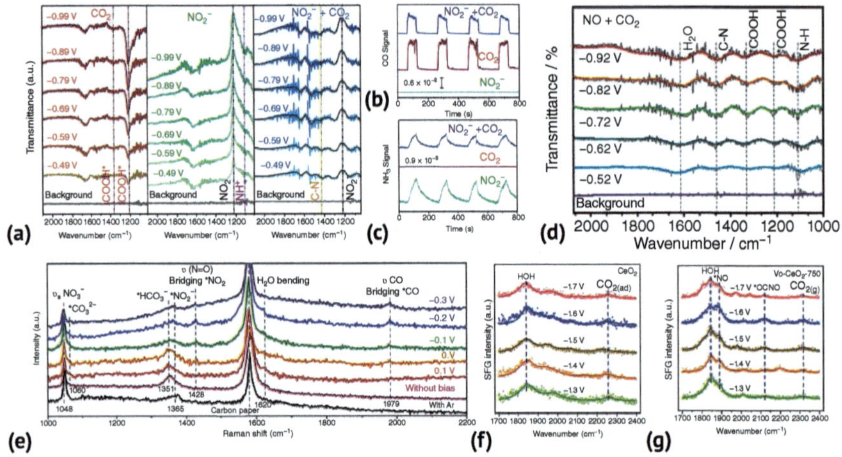

Figure 13.11 (a) In situ ART-FTIR spectra of ZnO—V under CO_2, $NaNO_2$, and both. Online DEMS spectra of CO (b) and NH_3. (c) Signals over ZnO—V. *Source:* Reproduced from [57] / with permission of Elsevier. (d) In situ ATR-FTIR spectra over Zn NBs using NO + CO_2 as the feed gas. *Source:* [43] / with permission of American Chemical Society. (e) In situ Raman spectra of $CuWO_4$ in 0.1M KNO_3 with CO_2 bubbling at different applied potentials or Ar bubbling at open circuit state. *Source:* [52] / Springer nature / CC BY 4.0. (f, g) SFG signals of intermediate species on pristine CeO_2 and Vo-CeO_2-750. *Source:* Reproduced from [16] / with permission of American Chemical Society.

relationship between the CO_2RR and NO_2RR reduction on the catalyst surface (Figure 13.11b and c).

Similarly, Huang et al. used in situ ATR-FTIR spectroscopy to systematically study the intermediates involved in the co-reduction of CO and NO for urea synthesis by ZnNB catalysis and further deduced the reaction process [43]. ATR-FTIR can obtain structural information about the organic composition of the sample through the reflected signal on the surface of the sample material and can further determine the type and concentration of the intermediates based on the position and intensity of the vibrational peaks, making it a suitable technique for tracking changes in the C—N coupling reaction intermediates (Figure 13.11d). They used ATR-FTIR and online DEMS tests to provide detailed intermediate information. The DEMS results showed that HNO, NH_2OH, N_2, and NH_3 signals were periodically detected when a voltage was applied to the working electrode during the electrocatalytic reduction of NO, thus confirming the presence of *NHO, *NH_2OH, *N_2, and *NH3 intermediates. In the CO_2RR experiment, only the CO signal was detected. In contrast, NH_2OH, HNO, NH_3, CO, and N_2 signals could be detected during the electrocatalytic co-reduction of NO and CO_2. Meanwhile, the in situ ATR-FTIR test revealed a new C—N bond vibration peak at 1465 cm^{-1}, which proved that C—N coupling had occurred. With the decrease of applied potential, the vibration intensity of *COOH, *N – H, and *C – N increased gradually. According to the detected changes in intermediate substances and chemical bonds, the reaction pathway of NO reduction was *NO → *NHO → *NHOH → *NH_2OH → *NH_2 → *NH_3, and the reaction pathway of CO_2RR was *CO_2 → *COOH → *CO. By further combining in situ characterization spectra with thermodynamic calculations, the authors concluded that in the electrocatalytic co-reduction of NO and CO_2, $^*NH_2OH$ was first spontaneously reduced to $^*NH_2$, which then couples with *CO to form the C—N bond.

In addition to the $^*NH_2$ intermediate, $^*NO_x$ intermediates without being hydrogenated could also be involved in the C—N coupling process. Sun's group used in situ Raman spectroscopy combined with theoretical calculations, confirming that the combination of stable intermediates $^*NO_2$ and *CO reduces the reaction barrier and increases the possibility of C—N coupling, resulting in low overpotential and high FE for the $CuWO_4$ catalyst to synthesize urea through C—N coupling [52]. In situ Raman spectroscopy is often used to detect intermediates in the reaction process. The characteristic vibration peak position can determine the intermediate types, and the linear proportional relationship between the vibration intensity and the analyte concentration can determine the content of the intermediate. This differs from in situ IR in the scanning range and is a complement to the infrared spectrum. In this work, in situ Raman spectroscopy was used to monitor the tensile vibration of the intermediates produced on the $CuWO_4$ surface under working conditions. Compared to the Ar-saturated condition, in the Raman spectrum of the CO_2-saturated open-circuit state, in addition to the NO_3^- (1048 and 1365 cm^{-1}) vibration peaks, two new peaks appeared at 1060 and 1352 cm^{-1}, corresponding to * CO_3^{2-} and $^*HCO_3^-$, respectively. Meanwhile, *CO (1979 cm^{-1}) on the bridge sites could be observed when the potential reached −0.1 V versus RHE, and the peak intensity of $^*CO_3^{2-}$ and $^*HCO_3^-$ gradually decreased with the increase of applied potential (potential range from −0.1 to −0.3 V vs. RHE), showing that the production of *CO consumed $^*CO_3^{2-}$ and $^*HCO_3^-$. Interestingly, at the optimal potential of −0.2 V versus RHE, the $^*NO_2$ and *CO intermediates on the $CuWO_4$ surface had the highest coverage, indicating that they were essential intermediates in urea electrosynthesis. The DFT further concluded that the rate-determining step is the C−N coupling between the *CO and $^*NO_2$ intermediates (Figure 13.11c).

Wei et al. reported a new urea production mechanism for the electrocatalytic co-reduction of CO_2 and NO_3^- through a CeO_2 nanorod catalyst rich in oxygen vacancies [16]. Moreover, in situ ultraviolet-visible-near-infrared (UV-Vis-NIR) vibrational spectroscopy was used to determine the type and concentration of intermediate products, which was determined by the change in the position or intensity of the absorption peak in the absorption spectrum at the operating potential. The in situ UV-Vis-NIR vibrational spectrum analysis showed that V_O-CeO_2-750 exhibited obvious peaks of *NO and *OCNO intermediates at 1888 cm^{-1} and 2120 cm^{-1}, respectively, and the SFG (sum-frequency generation) intensity of *OCNO gradually increased with the negative shift of potential, reaching a maximum at a potential of −1.6 V versus RHE (Figure 13.11f). The variation of the characteristic peak intensity of *OCNO at different potentials was consistent with the variation of urea yield, so it was concluded that *OCNO intermediates participated in urea synthesis as C−N coupling intermediates (Figure 13.11g). Based on the DFT calculations, the authors proposed that the V_O first adsorbed the *NO intermediate, and then the C—N coupling reaction occurred with *CO to produce *OCNO. The chemical process of C—N coupling of V_O adsorbing *NO intermediates was more favorable than the reaction of PCET hydrogenation of *NO. Both the UV-Vis-NIR vibrational spectroscopy and the DFT calculations supported that the V_O promoted the formation of the key intermediate *OCNO by using *NO as the coupling intermediate, which effectively promoted the electrochemical synthesis of urea. UV-Vis-NIR vibrational spectroscopy can provide information about the type and concentration of intermediates, and when combined with other characterization methods, it can effectively analyze the dynamic changes in the catalytic process and help understand the reaction mechanism of complex reactions.

13.6.3 Other Advanced Techniques

DEMS is a modern electrochemical field measurement method developed by combining electrochemical and MS techniques. DEMS can detect the volatile gas products and kinetic parameters in the electrochemical reaction, as well as the properties and structure of intermediates. When multiple electrode reaction products are formed, DEMS can simultaneously determine the change of Faradaic current of each product with electrode potential or time. Yu's group used DEMS to confirm the critical role of *CO and $*NH_2$ intermediates in the C—N coupling process in the Zn@Cu nanowires catalyst (Figure 13.12) [78]. The DEMS results showed that both the CO and NH_2 signal strengths decreased when CO_2 and NO_3^- existed simultaneously, indicating a competitive relationship between CO_2 and nitrate reduction. This initially confirmed that $*NH_2$ and *CO participated in the C—N coupling process as intermediates. Subsequent theoretical calculations showed that electron transfer from Zn to Cu promoted the formation of the key intermediates

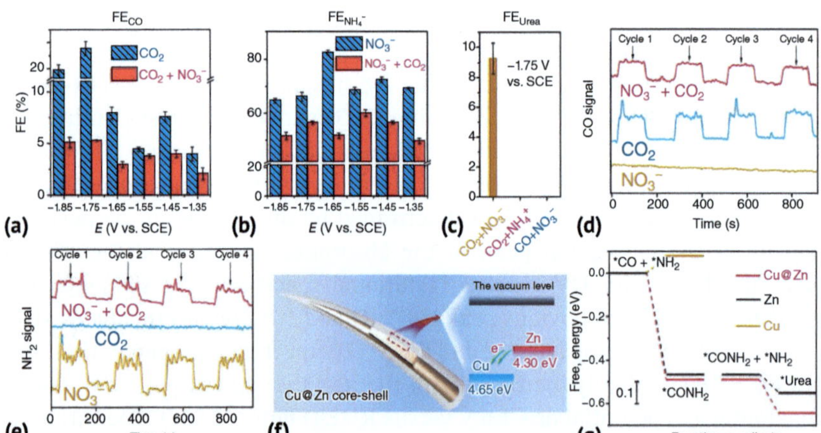

Figure 13.12 (a, b) FE_{CO} FE_{NH4+} and for the electrochemical reduction of CO_2 and the mixture (NO_3^- + CO_2) over Cu@Zn, respectively. (c) FE_{urea} for the electroreduction of mixtures (CO_2 + NO_3^-), (CO_2 + NH_4^+), and (CO + NO_3^-) over Cu@Zn. (d, e) Online DEMS spectra of CO and NH_2 signals over Cu@Zn. (f) Work function analysis of core–shell structured Cu@Zn. (g) C—N coupling on Cu@Zn, pure Zn, and pure Cu surfaces. "*" represents the active site. *Source:* Reproduced from [78] / with permission of American Chemical Society.

(*CO and *NH_2) and reduced the coupling energy barrier of *CO and *NH_2, ultimately realizing the efficient catalysis of the C—N coupling reaction. In summary, DEMS is a powerful technique that can provide valuable information about the volatile products, kinetics, and intermediates in electrochemical reactions. By combining DEMS with theoretical calculations, the critical role of key intermediates in complex reaction mechanisms can be elucidated.

13.7 Summary

In conclusion, this chapter mainly introduces the latest research on the electrosynthesis of high-value organic nitrogen compounds through electrochemical C—N coupling reactions using abundant CO_2 and various nitrogen-containing molecules (N_2, NO, NO_2^-, NO_3^-, etc.). Specifically, the electrosynthesis of urea not only produces high-value urea cleanly and sustainably under environmental conditions but also contributes to artificial carbon neutrality and nitrogen fixation. Furthermore, it offers a promising and sustainable approach for using renewable energy to drive the synthesis of valuable chemicals. The chapter details the basic principles of electrochemical C—N coupling reactions, including detection methods for target products, performance evaluation parameters, and various electrocatalytic reactors (H-type cell, flow cell, and MEA). This information will help readers better evaluate catalytic performance and understand reactor configurations to optimize reactor details and improve catalytic efficiency. Additionally, it discusses the design principles and research progress of electrocatalysts for CO_2 and nitrogen source coupling to synthesize urea. However, due to the complex, multi-step nature of electrocatalytic urea synthesis, identifying the reaction mechanism experimentally remains challenging. Therefore, the chapter summarizes advanced in situ characterization techniques and theoretical calculation methods to explore the reaction mechanism, aiming to enhance readers' understanding of the process and guide the rational design of efficient catalysts.

Acknowledgment

C.L. acknowledges the National Natural Science Foundation of China (52302231 and 22479034), and Key Research & Development Plan of Heilongjiang Province (2024ZXJ03C06). L.Z. acknowledges the National Natural Science Foundation of China (22202008) and Beijing Institute of Technology Research Fund Program for Young Scholars (XSQD-202215006 and RCPT-6120220009).

References

1 Service, R.F. (2014). New recipe produces ammonia from air, water, and sunlight. *Science* 345 (6197): 610–610.
2 Soloveichik, G. (2019). Electrochemical synthesis of ammonia as a potential alternative to the Haber–Bosch process. *Nat. Catal.* 2 (5): 377–380.

3 Chen, C., Yan, D., Wang, Y. et al. (2019). B-N pairs enriched defective carbon nanosheets for ammonia synthesis with high efficiency. *Small* 15 (7): 1805029.

4 Zhu, D.D., Liu, J.L., and Qiao, S.Z. (2016). Recent advances in inorganic heterogeneous electrocatalysts for reduction of carbon dioxide. *Adv. Mater.* 28 (18): 3423–3452.

5 Kuhl, K.P., Hatsukade, T., Cave, E.R. et al. (2014). Electrocatalytic conversion of carbon dioxide to methane and methanol on transition metal surfaces. *J. Am. Chem. Soc.* 136 (40): 14107–14113.

6 Lee, C.W., Yang, K.D., Nam, D.H. et al. (2018). Defining a materials database for the design of copper binary alloy catalysts for electrochemical CO_2 conversion. *Adv. Mater.* 30 (42): 1704717.

7 Sun, Z., Ma, T., Tao, H. et al. (2017). Fundamentals and challenges of electrochemical CO_2 reduction using two-dimensional materials. *Chem* 3 (4): 560–587.

8 Chang, X., Malkani, A., Yang, X. et al. (2020). Mechanistic insights into electroreductive C–C coupling between CO and acetaldehyde into multicarbon products. *J. Am. Chem. Soc.* 142 (6): 2975–2983.

9 Ting, L.R.L., García-Muelas, R., Martín, A.J. et al. (2020). Electrochemical reduction of carbon dioxide to 1-butanol on oxide-derived copper. *Angew. Chem. Int. Ed.* 59 (47): 21072–21079.

10 Tang, C., Zheng, Y., Jaroniec, M. et al. (2021). Electrocatalytic refinery for sustainable production of fuels and chemicals. *Angew. Chem. Int. Ed.* 60 (36): 19572–19590.

11 McIsaac, G.F., David, M.B., Gertner, G.Z. et al. (2001). Nitrate flux in the Mississippi River. *Nature* 414 (6860): 166–167.

12 Jouny, M., Lv, J.-J., Cheng, T. et al. (2019). Formation of carbon–nitrogen bonds in carbon monoxide electrolysis. *Nat. Chem.* 11 (9): 846–851.

13 Rapson, H.D.C. and Bird, A.E. (2007). The electrochemical preparation of glycine. *J. Appl. Chem.* 13 (6): 233–239.

14 Shibata, M., Yoshida, K., and Furuya, N. (1995). Electrochemical synthesis of urea on reduction of carbon dioxide with nitrate and nitrite ions using Cu-loaded gas-diffusion electrode. *J. Electroanal. Chem.* 387: 143—145.

15 Kayan, D.B. and Köleli, F. (2016). Simultaneous electrocatalytic reduction of dinitrogen and carbon dioxide on conducting polymer electrodes. *Appl. Catal. B-Environ. Energy* 181: 88–93.

16 Wei, X., Wen, X., Liu, Y. et al. (2022). Oxygen vacancy-mediated selective C–N coupling toward electrocatalytic urea synthesis. *J. Am. Chem. Soc.* 144 (26): 11530–11535.

17 Wu, Y., Jiang, Z., Lin, Z. et al. (2021). Direct electrosynthesis of methylamine from carbon dioxide and nitrate. *Nat. Sustain.* 4 (8): 725–730.

18 Mukherjee, J., Paul, S., Adalder, A. et al. (2022). Understanding the site-selective electrocatalytic co-reduction mechanism for green urea synthesis using copper phthalocyanine nanotubes. *Adv. Funct. Mater.* 32 (31): 2200882.

19 Zhao, Y., Shi, R., Bian, X. et al. (2019). Ammonia detection methods in photocatalytic and electrocatalytic experiments: how to improve the reliability of NH_3 production rates? *Adv. Sci.* 6 (8): 1802109.

20 Li, P., Jin, Z., Fang, Z. et al. (2020). A surface-strained and geometry-tailored nanoreactor that promotes ammonia electrosynthesis. *Angew. Chem. Int. Ed.* 59 (50): 22610–22616.

21 Li, D., Xu, N., Zhao, Y. et al. (2022). A reliable and precise protocol for urea quantification in photo/electrocatalysis. *Small Methods* 6 (9): 2200561.

22 Vrubel, H., Moehl, T., Grätzel, M. et al. (2013). Revealing and accelerating slow electron transport in amorphous molybdenum sulphide particles for hydrogen evolution reaction. *Chem. Commun.* 49 (79): 8985–8987.

23 Nguyen, T.N. and Dinh, C.-T. (2020). Gas diffusion electrode design for electrochemical carbon dioxide reduction. *Chem. Soc. Rev.* 49 (21): 7488–7504.

24 Burdyny, T. and Smith, W.A. (2019). CO_2 reduction on gas-diffusion electrodes and why catalytic performance must be assessed at commercially-relevant conditions. *Energ. Environ. Sci.* 12 (5): 1442–1453.

25 Mei, Z., Zhou, Y., Lv, W. et al. (2022). Recent progress in electrocatalytic urea synthesis under ambient conditions. *ACS Sustainable Chem. Eng.* 10 (38): 12477–12496.

26 Ge, L., Rabiee, H., Li, M. et al. (2022). Electrochemical CO_2 reduction in membrane-electrode assemblies. *Chem* 8 (3): 663–692.

27 Rabiee, H., Ge, L., Zhang, X. et al. (2021). Gas diffusion electrodes (GDEs) for electrochemical reduction of carbon dioxide, carbon monoxide, and dinitrogen to value-added products: a review. *Energ. Environ. Sci.* 14 (4): 1959–2008.

28 Hu, L., Xing, Z., and Feng, X. (2020). Understanding the electrocatalytic interface for ambient ammonia synthesis. *ACS Energy Lett.* 5 (2): 430–436.

29 Niu, Z.-Z., Chi, L.-P., Liu, R. et al. (2021). Rigorous assessment of CO_2 electroreduction products in a flow cell. *Energ. Environ. Sci.* 14 (8): 4169–4176.

30 Weng, L.-C., Bell, A.T., and Weber, A.Z. (2019). Towards membrane-electrode assembly systems for CO_2 reduction: a modeling study. *Energ. Environ. Sci.* 12 (6): 1950–1968.

31 Lazaridis, T., Stühmeier, B.M., Gasteiger, H.A. et al. (2022). Capabilities and limitations of rotating disk electrodes versus membrane electrode assemblies in the investigation of electrocatalysts. *Nat. Catal.* 5 (5): 363–373.

32 Zhang, Z., Huang, X., Chen, Z. et al. (2023). Membrane electrode assembly for electrocatalytic CO_2 reduction: principle and application. *Angew. Chem. Int. Ed.* 62 (28): e202302789.

33 Alfian, M. and Purwanto, W.W. (2019). Multi-objective optimization of green urea production. *Energ. Environ. Sci.* 7 (2): 292–304.

34 Yao, Y., Zhu, S., Wang, H. et al. (2018). A spectroscopic study on the nitrogen electrochemical reduction reaction on gold and platinum surfaces. *J. Am. Chem. Soc.* 140 (4): 1496–1501.

35 Wang, Y., Shi, M., Bao, D. et al. (2019). Generating defect-rich bismuth for enhancing the rate of nitrogen electroreduction to ammonia. *Angew. Chem. Int. Ed.* 58 (28): 9464–9469.

36 Chen, C., Zhu, X., Wen, X. et al. (2020). Coupling N_2 and CO_2 in H_2O to synthesize urea under ambient conditions. *Nat. Chem.* 12 (8): 717–724.

37 Yuan, M., Chen, J., Xu, Y. et al. (2021). Highly selective electroreduction of N_2 and CO_2 to urea over artificial frustrated Lewis pairs. *Energ. Environ. Sci.* 14 (12): 6605–6615.

38 Yuan, M., Zhang, H., Xu, Y. et al. (2022). Artificial frustrated Lewis pairs facilitating the electrochemical N_2 and CO_2 conversion to urea. *Chem. Catal.* 2 (2): 309–320.

39 Yuan, M., Chen, J., Bai, Y. et al. (2021). Unveiling electrochemical urea synthesis by co-activation of CO_2 and N_2 with Mott–Schottky heterostructure catalysts. *Angew. Chem. Int. Ed.* 60 (19): 10910–10918.

40 Yuan, M., Chen, J., Bai, Y. et al. (2021). Electrochemical C–N coupling with perovskite hybrids toward efficient urea synthesis. *Chem. Sci.* 12 (17): 6048–6058.

41 Wan, H., Wang, X., Tan, L. et al. (2023). Electrochemical synthesis of urea: co-reduction of nitric oxide and carbon monoxide. *ACS Catal.* 13 (3): 1926–1933.

42 Xian, J., Li, S., Su, H. et al. (2023). Electrocatalytic synthesis of essential amino acids from nitric oxide using atomically dispersed Fe on N-doped carbon. *Angew. Chem. Int. Ed.* 62 (26): e202302789.

43 Huang, Y., Yang, R., Wang, C. et al. (2021). Direct electrosynthesis of urea from carbon dioxide and nitric oxide. *ACS Energy Lett.* 7 (1): 284–291.

44 Zhu, X., Fan, X., Lin, H. et al. (2023). Highly efficient electroenzymatic cascade reduction reaction for the conversion of nitrite to ammonia. *Adv. Energy Mater.* 13 (20): 2300669.

45 Fang, L., Wang, S., Song, C. et al. (2022). Enhanced nitrate reduction reaction via efficient intermediate nitrite conversion on tunable Cu_xNi_y/NC electrocatalysts. *J. Hazard. Mater.* 421: 126628.

46 Sedlmair, C. (2003). Elementary steps of NOx adsorption and surface reaction on a commercial storage–reduction catalyst. *J. Catal.* 214 (2): 308–316.

47 Garcia-Segura, S., Lanzarini-Lopes, M., Hristovski, K. et al. (2018). Electrocatalytic reduction of nitrate: Fundamentals to full-scale water treatment applications. *Appl. Catal. B-Environ. Energy* 236: 546–568.

48 Li, X., Wu, X., Lv, X. et al. (2022). Recent advances in metal-based electrocatalysts with hetero-interfaces for CO_2 reduction reaction. *Chem Catal.* 2 (2): 262–291.

49 Feng, Y., Yang, H., Zhang, Y. et al. (2020). Te-Doped Pd nanocrystal for electrochemical urea production by efficiently coupling carbon dioxide reduction with nitrite reduction. *Nano Lett.* 20 (11): 8282–8289.

50 Mao, Y., Jiang, Y., Gou, Q. et al. (2024). Indium-activated bismuth-based catalysts for efficient electrocatalytic synthesis of urea. *Appl. Catal. B-Environ. Energy* 340: 123189.

51 Zhang, S., Geng, J., Zhao, Z. et al. (2023). High-efficiency electrosynthesis of urea over bacterial cellulose regulated Pd–Cu bimetallic catalyst. *EES Catalysis* 1 (1): 45–53.

52 Zhao, Y., Ding, Y., Li, W. et al. (2023). Efficient urea electrosynthesis from carbon dioxide and nitrate via alternating Cu–W bimetallic C–N coupling sites. *Nat. Commun.* 14 (1): 4491.

53 Wei, X., Liu, Y., Zhu, X. et al. (2023). Dynamic reconstitution between copper single atoms and clusters for electrocatalytic urea synthesis. *Adv. Mater.* 35 (18): 2300020.

54 Liu, Y., Tu, X., Wei, X. et al. (2023). C-Bound or O-Bound Surface: Which one boosts electrocatalytic urea synthesis? *Angew. Chem. Int. Ed.* 62 (19): e202302789.

55 Leverett, J., Tran-Phu, T., Yuwono, J.A. et al. (2022). Tuning the coordination structure of Cu–N–C single atom catalysts for simultaneous electrochemical reduction of CO_2 and NO_3– to urea. *Adv. Energy Mater.* 12 (32): 2201500.

56 Liu, C., Tong, H., Wang, P. et al. (2023). The asymmetric orbital hybridization in single-atom-dimers for urea synthesis by optimizing the C–N coupling reaction pathway. *Appl. Catal. B-Environ. Energy* 336: 122917.

57 Meng, N., Huang, Y., Liu, Y. et al. (2021). Electrosynthesis of urea from nitrite and CO_2 over oxygen vacancy-rich ZnO porous nanosheets. *Cell Rep. Phys. Sci.* 2 (3): 100378.

58 Zhang, D., Xue, Y., Zheng, X. et al. (2023). Multi-heterointerfaces for selective and efficient urea production. *Nati. Sci. Rev.* 10 (2): nwac209.

59 Li, Z., Zhou, P., Zhou, M. et al. (2023). Synergistic electrocatalysis of crystal facet and O-vacancy for enhancive urea synthesis from nitrate and CO_2. *Appl. Catal. B-Environ. Engery* 338: 1222962.

60 Wang, H., Jiang, Y., Li, S. et al. (2022). Realizing efficient C–N coupling via electrochemical co-reduction of CO_2 and NO_3– on AuPd nanoalloy to form urea: Key C–N coupling intermediates. *Appl. Catal. B-Environ. Energy* 318: 121819.

61 Jiao, D., Dong, Y., Cui, X. et al. (2023). Boosting the efficiency of urea synthesis via cooperative electroreduction of N_2 and CO_2 on MoP. *J. Mater. Chem. A* 11 (1): 232–240.

62 Lv, Z., Zhou, S., Zhao, L. et al. (2023). Coactivation of multiphase reactants for the electrosynthesis of urea. *Adv. Energy Mater.* 13 (25): 2300946.

63 Xiong, H., Yu, P., Chen, K. et al. (2024). Urea synthesis via electrocatalytic oxidative coupling of CO with NH_3 on Pt. *Nat. Catal.* 7 (7): 785–795.

64 Wang, X. (2019). Challenges and outlook for catalytic direct amidation reactions. *Nat. Catal.* 2 (2): 98–102.

65 Lundberg, H., Tinnis, F., Selander, N. et al. (2014). Catalytic amide formation from non-activated carboxylic acids and amines. *Chem. Soc. Rev.* 43 (8): 2714–2742.

66 Sarak, S., Khobragade, T.P., Jeon, H. et al. (2021). One-pot biocatalytic synthesis of nylon monomers from cyclohexanol using Escherichia coli-based concurrent cascade consortia. *Green Chem.* 23 (23): 9447–9453.

67 Zhang, D.-W., Zhao, X., Hou, J.-L. et al. (2012). Aromatic amide foldamers: structures, properties, and functions. *Chem. Rev.* 112 (10): 5271–5316.

68 Yokozawa, T., Ogawa, M., Sekino, A. et al. (2002). Chain-growth polycondensation for well-defined aramide synthesis of unprecedented block copolymer containing aramide with low polydispersity. *J. Am. Chem. Soc.* 124 (51): 15158–15159.

69 Gnanaprakasam, B. and Milstein, D. (2011). Synthesis of amides from esters and amines with liberation of H2 under neutral conditions. *J. Am. Chem. Soc.* 133 (6): 1682–1685.

70 Tao, Z., Wu, Y., Wu, Z. et al. (2022). Cascade electrocatalytic reduction of carbon dioxide and nitrate to ethylamine. *J. Energy Chem.* 65: 367–370.

71 Kuang, S., Xiao, T., Chi, H. et al. (2024). Acetamide electrosynthesis from CO_2 and nitrite in water. *Angew. Chem. Int. Ed.* 63 (9): e202316772.

72 Peng, X., Zeng, L., Wang, D. et al. (2023). Electrochemical C–N coupling of CO_2 and nitrogenous small molecules for the electrosynthesis of organonitrogen compounds. *Chem. Soc. Rev.* 52 (6): 2193–2237.

73 Chen, H., Handoko, A.D., Wang, T. et al. (2020). Defect-enhanced CO_2 reduction catalytic performance in O-terminated MXenes. *ChemSusChem* 13 (21): 5690–5698.

74 Luo, Y., Chen, G.-F., Ding, L. et al. (2019). Efficient electrocatalytic N_2 fixation with MXene under ambient conditions. *Joule* 3 (1): 279–289.

75 Chernyshova, I.V., Somasundaran, P., and Ponnurangam, S. (2018). On the origin of the elusive first intermediate of CO_2 electroreduction. *P. Nati. A. Sci.* 115 (40): E9261–E9270.

76 Li, J. and Kornienko, N. (2022). Electrochemically driven C–N bond formation from CO_2 and ammonia at the triple-phase boundary. *Chem. Sci.* 13 (14): 3957–3964.

77 Lv, C., Zhong, L., Liu, H. et al. (2021). Selective electrocatalytic synthesis of urea with nitrate and carbon dioxide. *Nat. Sustain.* 4 (10): 868–876.

78 Meng, N., Ma, X., Wang, C. et al. (2022). Oxide-derived core–shell Cu@Zn nanowires for urea electrosynthesis from carbon dioxide and nitrate in water. *ACS Nano*. 16 (6): 9095–9104.

14

Electrochemical Fluorination

Meng Li[1,#], Chuang Fan[2,#], Gengtao Fu[1,]*

[1] *Jiangsu Key Laboratory of New Power Batteries, Jiangsu Collaborative Innovation Center of Biomedical Functional Materials, School of Chemistry and Materials Science, Nanjing Normal University, Nanjing, China*
[2] *Anhui Key Laboratory of Low-Temperature Co-Fired Material, School of Chemical and Materials Engineering, Huainan Normal University, China*

*Corresponding author: gengtaofu@njnu.edu.cn
#These authors contributed equally to this work.

14.1 Introduction

The high electronegativity and small atomic radius of fluorine atoms enable their incorporation into organic compounds, resulting in significant changes to physical and chemical properties, such as enhanced bioactivity, metabolic stability and lipophilicity [1–4]. These attributes make fluorine-containing compounds highly desirable in pharmaceuticals, agrochemicals, and material sciences. However, naturally occurring organofluorine compounds are scarce, necessitating the development of efficient and safe fluorination methods for organic synthesis, which remains a crucial challenge in modern organic chemistry.

Conventional fluorination strategies for organic compounds typically proceed through the use of monomeric fluorine, the CoF_3-mediated hydrocarbon radical fluorination [5], deoxyfluorination [6, 7], electrophilic fluorination [8, 9], nucleophilic fluorination of functionalized substrates [10, 11], or the addition of fluorine to olefins/alkynes to produce organofluorine compounds [12, 13]. While these methods have proven effective, they often suffer from significant limitations, including harsh reaction conditions, reliance on costly or hazardous fluorinating agents, and challenges in achieving high selectivity [14, 15]. To address these challenges, electrochemical fluorination (ECF) has emerged as a promising alternative, offering safer and more environmentally friendly pathways for fluorination [16]. By utilizing electrical energy to generate reactive fluorine species, ECF avoids the need for traditional hazardous fluorine reagents. Moreover, the ability to fine-tune key electrochemical parameters, such as current density and electrode potential, enhances the precision and efficiency of the process. As a versatile tool in both academic and industrial applications, ECF not only advances organic synthesis

Electrochemical Energy Conversion via Small Molecule Catalysis, First Edition.
Edited by Zhicheng Zhang and Yuanmiao Sun.

but also contributes to the development of sustainable energy materials [17]. Over recent years, significant efforts have been devoted to advancing the field of ECF, including catalyst and electrode materials design [18], reaction mechanism exploration [19], and selective fluorination strategies [20]. In general, ECF processes go through two different reaction paths: direct anodic fluorination and indirect electrochemical fluorination [19, 21, 22]. In direct anodic fluorination, fluoride ions (F^-) are oxidized at the anode to produce highly reactive fluorine gas (F_2) or other high-valent metal fluoride species, which then react with the target molecules. In contrast, indirect fluorination involves electrochemical generation of intermediate fluorinating reagents, such as select fluor or other fluorine-containing compounds, to mediate the fluorination reaction. However, both of these two pathways face several challenges. Reaction selectivity and controllability of the fluorination process remain critical hurdles, particularly in complex molecular systems. The fluorination reaction involves highly reactive fluorine species, which often results in undesired substrate fluorination or other side reactions, such as defluorination, fluoride polymerization, etc., leading to uncontrolled product distributions. Another significant challenge lies in identifying suitable electrolytes and solvents to optimize reaction efficiency and selectivity [23, 24]. Achieving scalability for industrial applications also demands cost-effective, safe, and efficient solutions. Finally, integrating ECF into existing synthetic methodologies and extending its applicability to complex molecular systems remain areas requiring further research [25, 26].

In this chapter, we aim to provide a comprehensive overview of recent advances in ECF, with a particular focus on electrochemical fluorination of various organic compounds, highlighting strategies to enhance reaction selectivity and achieve precise control over the fluorine source. Furthermore, the critical roles of electrodes and electrolytes in ECF are also reviewed, underscoring the significance of the reaction medium in determining efficiency and selectivity. Notable advancements include electrochemically driven fluoroalkylation and selective fluorination of olefins, which leverage the generation of fluorine radicals or ions under optimized electrolytic conditions to enable precise and selective fluorination under mild, environmentally friendly conditions. Ionic liquids and alkali metal fluorides (e.g., KF and CsF) have emerged as promising electrolytes due to their high conductivity, stability, and ease of handling. These attributes significantly enhance reaction selectivity and efficiency while addressing key challenges such as anodic passivation and the weak nucleophilicity of fluoride ions. By analyzing recent progress and addressing ongoing challenges, this account provides valuable insights into the evolving landscape of ECF research. These findings are expected to propel the development of more efficient, selective, and sustainable fluorination methodologies, thereby advancing the fields of organic synthesis and energy storage technologies.

14.2 Electrochemically Driven Fluoroalkylation

14.2.1 Electrochemical Trifluoromethylation

The trifluoromethyl (CF_3) functional group occupies a crucial role in pharmaceuticals, agrochemicals, and materials science due to its exceptional ability to modulate molecular properties. The electron-withdrawing nature of the CF_3 group alters molecular charge distribution, thereby enhancing interactions with target biomolecules and improving bioactivity. Electrochemical trifluoromethylation has emerged as an innovative and sustainable strategy for incorporating CF_3 groups into molecular frameworks. This method leverages electrochemical techniques to generate CF_3 radicals in situ, providing a more efficient and environmentally friendly alternative to traditional approaches. By eliminating the need for catalysts, this approach adheres to green chemistry principles, minimizing dependence on costly and potentially hazardous reagents. Through precise control of electrolyte composition, electrode materials, and reaction parameters, electrochemical trifluoromethylation achieves high efficiency and selectivity under mild conditions. Recent advancements have further expanded the versatility of this technique, integrating it with complementary reaction paradigms to enable the rapid synthesis of complex, functionalized molecules, thus broadening its synthetic applicability.

In 2014, Baran pioneered an electrochemical method for the trifluoromethylation of *N*-heteroarenes using $Zn(SO_2CF_3)_2$ (ZnTFMS) (Figure 14.1a) [27]. This approach offers significant advantages over peroxide-initiated trifluoromethylation, enabling precise control of radical generation through modulation of current and voltage. Such control minimizes side reactions, enhances selectivity, and improves yields, particularly with less reactive heterocyclic substrates. Current influence on ZnTFMS consumption, product formation rates, and selectivity underscores the benefits of controlled radical generation. Zeng et al. introduced a complementary strategy using bromide ions as mediators for the Minisci-type trifluoromethylation of electron-deficient heterocycles [28]. By finely tuning CF_3 radical concentrations, this method minimizes side reactions and increases yields. The process involves anodic oxidation of bromide to Br_2, which reacts with $Zn(SO_2CF_3)_2$ to form intermediates such as sulfonyl hypobromite or sulfonyl bromide. These intermediates undergo cathodic reduction or homolytic cleavage to produce CF_3 radicals, facilitating their addition to heterocycles (Figure 14.1b). Performed in a simple undivided cell without the need for costly supporting electrolytes, this method demonstrates high efficiency and broad applicability for functionalizing complex substrates. Yu et al. developed a metal-free electrochemical trifluoromethylation protocol employing trifluoromethyl thianthrenium triflate ($TT{-}CF_3{}^+OTf^-$) as the CF_3 source (Figure 14.1c) [29]. This green methodology eliminates the need for external oxidants or catalysts, significantly reducing by-product formation and streamlining the reaction process and exhibits excellent substrate compatibility and high yields. Mechanistically, anodic oxidation of $TT{-}CF_3{}^+OTf^-$ generates CF_3 radicals, which react with electron-deficient heterocycles. Subsequent single-electron oxidation and deprotonation produce the trifluoromethylated products. Cantillo et al. achieved C—H trifluoromethylation of arenes and heteroarenes by generating CF_3 radicals via cathodic reduction (Figure 14.1d) [30]. The reaction employs a trifluoromethyltriem complex, formed in situ from triflyl chloride and triethylamine, as the CF_3 source. Conducted on graphite electrodes, the reaction achieves exceptional selectivity, with triethylamine effectively suppressing side reactions. Mechanistic studies reveal that single-electron reduction of the complex at the cathode produces a neutral radical, which decomposes to release CF_3 radicals. These radicals react with aromatic

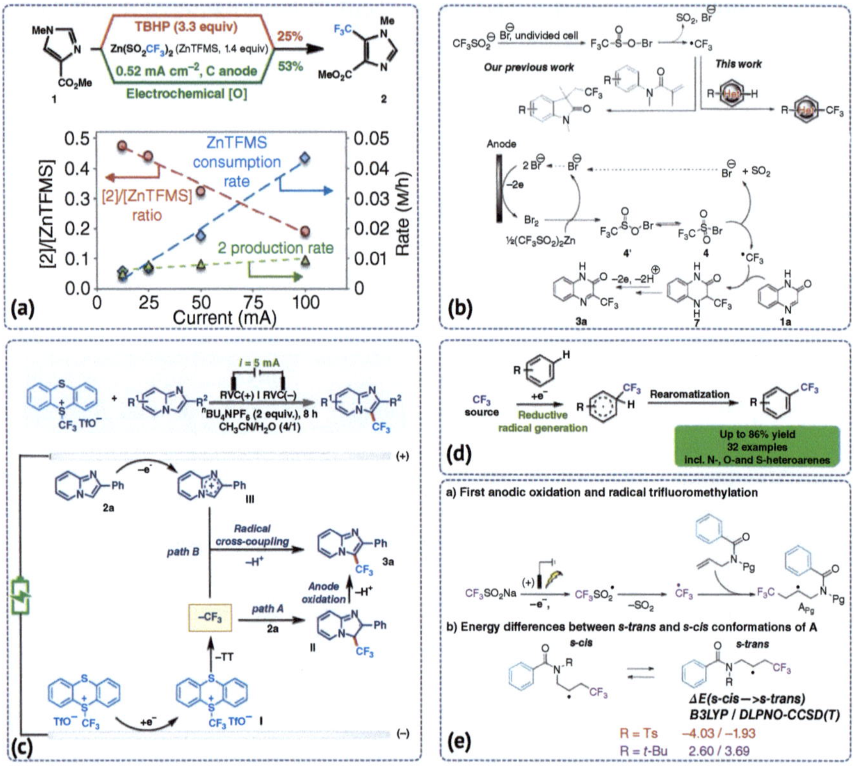

Figure 14.1 Electrochemical trifluoromethylation. (a) Radical C—H functionalization of heteroarenes. *Source:* [27] /with permission of John Wiley & Sons. (b) Electrochemical trifluoromethylation of organic molecules and a proposed mechanism for the trifluoromethylation of quinoxalinone. *Source:* [28] / with permission of Royal Society of Chemistry. (c) Metal-free electrochemical trifluoromethylation of imidazole-fused heterocycles with trifluoromethyl thianthrenium triflate and a proposed mechanism. *Source:* [29] / with permission of John Wilrey & Sons. (d) Strategy for arene trifluoromethylation and a proposed mechanism. *Source:* [30] / with permission of American Chemical Society. (e) Selective generation of a β-amino radical on benzamide for decarbonylative Truce–Smiles rearrangement and mechanistic studies. *Source:* [31] / with permission of John Wiley & Sons.

substrates, forming intermediates that rearomatize through hydrogen abstraction or single-electron transfer followed by deprotonation. Claraz et al. recently developed an electrochemical method for 1,4-aryl migration in N-allylbenzamides via radical fluoromethylation (Figure 14.1e) [31]. High selectivity was achieved through the use of an *N*-tert-butyl protecting group, which also prevented 6-endo cyclization. Mechanistic insights from DFT calculations and cyclic voltammetry revealed that fluoromethyl radicals, generated via anodic oxidation, add to the enamide double bond to form a β-amino radical intermediate. This intermediate undergoes 1,4-aryl migration through ipso-attack, yielding spirocyclic products. The method offers a novel pathway for synthesizing fluorine-containing β-arylethylamine derivatives, with broad applicability and significant industrial potential.

14.2.2 Electrochemical Difluoromethylation

In tandem with advancements in electrochemical trifluoromethylation, electrochemical difluoromethylation has gained prominence as a pivotal area of research. The difluoromethyl group (CF_2H) is highly regarded in drug design and molecular engineering due to its distinctive chemical properties. Serving as a lipophilic isostere for functional groups such as methanol, thiol, hydroxamic acid, or amides, CF_2H enables versatile structural modifications. Additionally, its weak acidity facilitates hydrogen bonding interactions, thereby improving the binding affinity and selectivity of bioactive compounds. These characteristics highlight the transformative role of CF_2H in pharmaceutical development and related fields.

Xu et al. synthesized a stable and user-friendly difluoromethylating reagent, $CF_2HSO_2NHNHBoc$, through a straightforward one-step reaction with commercially available difluoromethanesulfonyl chloride and tert-butyl carbazate [32]. This reagent undergoes ferrocene-catalyzed electrochemical oxidation to generate difluoromethyl (CF_2H) radicals, which react with alkyne substrates to form dibenzazepine frameworks (Figure 14.2a). Additionally, it enables efficient difluoromethylation of alkenes and enynes, constructing nitrogen-containing heterocycles with yields up to 70% under electrolysis at 70 °C. Mechanistic studies indicated that CF_2H radicals initially add to alkynes, forming vinyl radical intermediates that subsequently undergo homolytic aromatic substitution to yield the final products. Xu et al. further extended the application of $CF_2HSO_2NHNHBoc$ for the 1,2-hydroxydifluoromethylation and C—H difluoromethylation of electron-deficient alkenes, such as acrylamides, without relying on transition-metal catalysts or external oxidants (Figure 14.2b) [33]. In this method, a reticulated vitreous carbon (RVC) anode and a platinum cathode were used in a mixed solvent system, where sodium difluoromethanesulfinate underwent anodic oxidation to generate CF_2H radicals. These radicals react with acrylamides to form α-carbonyl carbocation intermediates, which subsequently react with water or eliminated protons, yielding the desired difluoromethylated products. Complementing these approaches, Poisson et al. developed an electrochemical method for difluoromethylation of electron-rich olefins, such as enamides and styrene derivatives, leveraging anodically generated CF_2H radicals (Figure 14.2c) [34]. This method achieved yields ranging from 23% to 87% across diverse substituted substrates under optimized conditions. It is proved that anodic oxidation of sodium difluoromethanesulfinate generates CF_2H radicals, which react with olefins to form stable alkyl radical intermediates that undergo further anodic oxidation or solvent-mediated nucleophilic addition to yield the final products.

Zhang et al. developed a scalable electrochemical oxidation-promoted method for *N*-ortho-selective difluoromethylation of heterocyclic *N*-oxides using sodium difluoromethanesulfinate (HCF_2SO_2Na) as the CF_2H source (Figure 14.2d) [35]. Conducted under constant current, this approach achieved high yields for quinoline and isoquinoline *N*-oxides. The HCF_2SO_2Na underwent anodic oxidation to generate CF_2H radicals, which selectively added to the *N*-oxides, followed by hydrogen elimination and electron transfer to deliver the final products with high efficiency and selectivity. Similarly, He et al. introduced an electrochemical oxidative radical cascade reaction for synthesizing difluoromethylated benzoxazines in an undivided

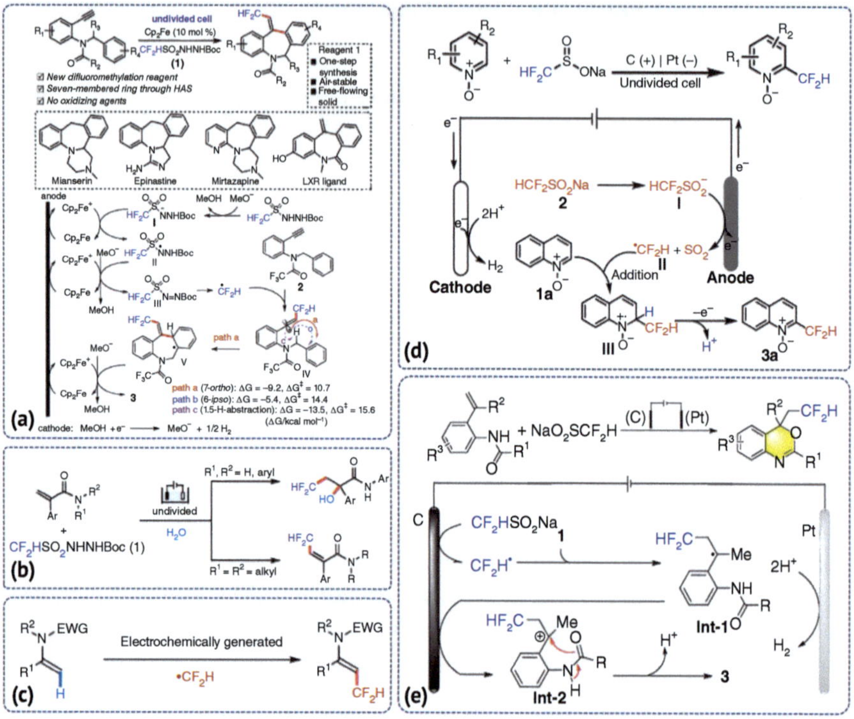

Figure 14.2 Electrochemical difluoromethylation. (a) Radical difluoromethylation of alkynes and a proposed mechanism. *Source:* [32] / with permission of American Chemistry Society. (b) Difluoromethylation of electron-deficient acrylamides and a proposed mechanism. *Source:* [33] with permission of John Wiley & Sons. (c) Electrochemical difluoromethylation of electron-rich olefins and a proposed mechanism. *Source:* [34] / with permission of American Chemistry Society. (d) Difluoromethylation of the aromatic heterocycle and a proposed mechanism. *Source:* [35] / with permission of American Chemistry Society. (e) Synthesis of benzoxazines and a proposed mechanism. *Source:* [36] / with permission of Royal Society of Chemistry.

electrochemical cell, notably eliminating the need for transition-metal catalysts or external oxidants (Figure 14.2e) [36]. Using sodium difluoromethanesulfinate as the CF_2H source, this method transformed olefinic amides into benzoxazines with up to 87% yield. The process involved anodic oxidation of sodium difluoromethanesulfinate to produce CF_2H radicals, which reacted with olefinic amides to form benzyl radical intermediates, and subsequently oxidation to benzyl cations, which underwent intramolecular nucleophilic attack, while proton reduction at the cathode generated hydrogen gas as a by-product.

14.3 Electrochemical Selective Fluorination of Alkenes

The electrochemical selective fluorination of alkenes, as a class of readily available and cost-effective feedstocks, has emerged as an attractive approach for obtaining valuable organic intermediates with significant applications in pharmaceuticals, agrochemicals, and materials science [37, 38]. This approach leverages the principles of electrochemistry to incorporate fluorine atoms into alkene molecules with exceptional precision and efficiency, offering significant advantages over conventional chemical fluorination methods. By enabling enhanced control over reaction selectivity and reducing dependence on hazardous reagents, electrochemical fluorination provides a sustainable and versatile alternative [39]. The ability to achieve regioselectivity, stereoselectivity, and chemoselective orientation in alkene fluorination highlights its potential as a pivotal tool in synthetic organic chemistry. The underlying mechanisms are influenced by various factors, including substrate properties, applied voltage, and reaction conditions. This section offers a comprehensive overview of the regioselectivity, stereoselectivity, and chemoselective orientation observed in these transformations, providing insights into the mechanistic principles that govern reaction outcomes. By addressing these critical aspects, this discussion aims to inform future advancements in electrochemical fluorination research.

14.3.1 Regioselectivity of Alkenes

Alkenes, characterized by their carbon–carbon double bond, display unique reactivity patterns in electrochemical fluorination, enabling precise incorporation of fluorine. The double bond plays a pivotal role in modulating electronic distribution within the molecule, which influences both the reaction pathway and the site of fluorination. This influence arises from the localized π-electron density of the double bond, which enhances reactivity and facilitates selective interactions during fluorination. A key feature of alkene fluorination is its regioselectivity, where fluorine addition predominantly occurs at nucleophilicity sites. This selective behavior arises from the electrophilic nature of the fluorine species, which preferentially interacts with nucleophilic regions of the molecule. The ability to predict and control regioselective outcomes has rendered alkenes essential in designing advanced strategies for synthesizing fluorinated compounds, especially in pharmaceuticals and agrochemicals. For example, Vasile Dinoiu et al. [40] achieved high regioselectivity through an EC_NEC_N mechanism involving radical cation intermediates (Figure 14.3a). They demonstrated that the electron-withdrawing substituents, such as F and CF_3, improved reaction yields, while electron-donating groups like CH_3 and CH_3O resulted in lower yields due to reduced anodic potentials of the substrates. Wigman et al. [3] reported the formation of vinyl carbocations from alkenyl boronic esters and boronates *via* sequential single-electron oxidation (Figure 14.3b). These carbocations were selectively trapped by nucleophilic fluoride reagents once formed, leading to the production of fully substituted vinyl fluorides. Their findings highlighted that strongly electron-donating substituents significantly enhanced the vinyl fluoride yields. Similarly, Luan et al. [34] explored the electrochemical difluoromethylation of electron-rich olefins, such as enamides and styrene derivatives (Figure 14.3c). They reported that the fluorination reaction with enamides containing an electron-donating group, including naphthyl or methoxy, proceeded well with good efficiencies. Additionally, the nucleophilicity of the fluorine source is a crucial factor for the regioselectivity during the electrochemical fluorination

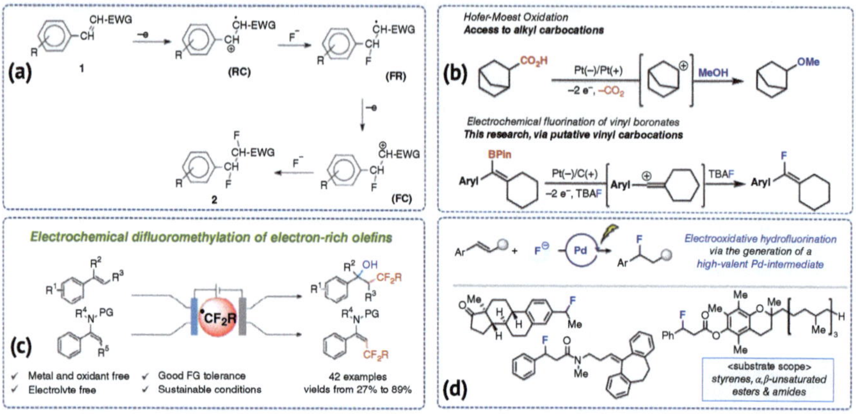

Figure 14.3 (a) An EC_NEC_N (electrochemical–chemical–electrochemical–chemical) mechanism. *Source:* [40] / with permission of Elsevier. (b) Previous work: Hofer–Moest electrochemical oxidation and this research: electrochemical fluorination of alkenyl boronates via vinyl carbocations. *Source:* Ref. [3], with permission of John Wiley & Sons. (c) The mechanism of electrochemical difluoromethylation of electron-rich olefins. *Source:* Ref. [34], with permission of American Chemical Society. (d) Palladium-catalyzed hydrofluorination of alkenes. *Source:* Ref. [41], with permission of American Chemical Society.

of alkenes. Mandal et al. [41] developed an electrocatalytic hydrofluorination of aryl-substituted alkenes with $Et_3N{\cdot}3HF$ as the nucleophilic fluorine source (Figure 14.3d). The anodic oxidation can finely tune the oxidation potential to selectively generate high-valent organopalladium intermediates, facilitating subsequent C—F bond formation.

14.3.2 Stereoselective of Alkenes

Stereoselective fluorination in electrochemical fluorination marks a notable advancement in modern synthetic chemistry, enabling the precise incorporation of fluorine atoms into alkenes with meticulous control over the three-dimensional arrangement of atoms. This selectivity arises from the interplay between the substrate's inherent stereoelectronic properties and the reaction conditions, collectively determining the stereochemical outcome. For instance, Doobary et al. [42] employed an "ex-cell" approach to generate hypervalent iodine reagents as mediators, pioneering a novel strategy for the vicinal difluorination of alkenes through electrochemical methods (Figure 14.4a). Remarkably, both *cis* and *trans* isomers of internal alkenes were suitable substrates, retaining their stereochemical features in the products with a diastereoselectivity ratios ranging from 5 : 1 to 9 : 1.

14.3.3 Chemoselectivity of Alkenes

Chemoselectivity, defined as the preferential reaction at a specific functional group within a molecule containing multiple reactive sites, is a fundamental principle

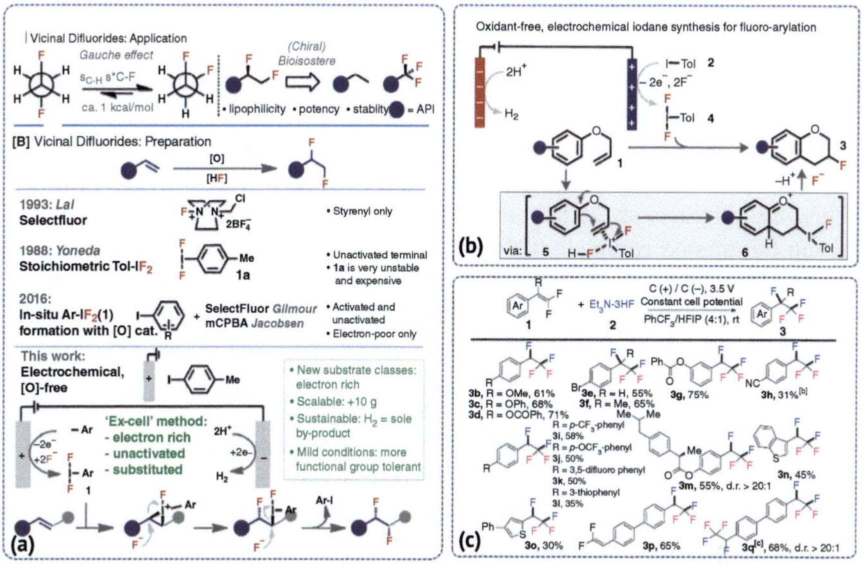

Figure 14.4 (a) Significance and application of vicinal difluorides and methods to prepare vicinal difluorides. *Source:* Ref. [42], John Wiley & Sons, CC BY 4.0. (b) Electrochemical generation of an iodane mediator for sustainable and scalable synthesis of 3-fluorochromanes. *Source:* Ref. [43], with permission American Chemical Society. (c) Overview of methodologies for fluorination difunctionalization of *gem*-difluorostyrenes. *Source:* Ref. [44], with permission of Royal Society of Chemistry.

in synthetic chemistry. In the context of electrochemical fluorination of alkenes, achieving high chemoselectivity is particularly critical due to the inherent reactivity of alkenes, which often feature double bonds, alkyl chains, and additional functional groups as potential reaction sites. For instance, Doobary et al. [43] utilized the electrochemically generated hypervalent iodine species to mediate the alkene fluoroarylation of phenolic ethers, achieving high yields and selectivity for electron-deficient substrates (Figure 14.4b). This method employed electrochemically in situ generated difluoro-λ^3-tolyl iodane, eliminating the need for external oxidants and enhancing sustainability and scalability. Chemoselectivity remained excellent for most substrates with oxidation potentials between 1.2 and 1.8 V, though it diminished for those with oxidation potentials exceeding 1.8 V. Similarly, Wen et al. [44] presented a direct electrochemical method for the selective difluorination, azo-fluorination, and fluoro-alkoxylation of gem-difluorostyrenes (Figure 14.4c). Using both $Et_3N{\cdot}3HF$ and azole as nucleophiles, the protocol selectively functionalized both ends of the gem-difluorostyrene, yielding 1-(1,1,2-trifluoro-2-arylethyl)-1H-azole as a single isomer with high chemo- and regioselectivity.

The selective electrochemical fluorination of alkenes delivers exceptional precision and efficiency in fluorine incorporation, providing unparalleled control over regioselectivity, stereoselectivity, and chemoselectivity. By leveraging specialized catalysts and finely tuned reaction conditions, this method ensures that fluorination occurs at targeted molecular sites with desired spatial configuration. A major advantage of electrochemical fluorination lies in its environmentally friendly approach, employing electricity as a sustainable energy source and reducing dependency on hazardous reagents typically required in conventional fluorination methods. Additionally, the mild reaction conditions inherent to electrochemical techniques mitigate substrate degradation and limit undesired side reactions, enhancing both the process efficiency and the integrity of sensitive compounds. This versatility makes electrochemical fluorination an invaluable tool for synthesizing fluorinated compounds essential in pharmaceuticals, agrochemicals, and advanced materials. By harmonizing precision, efficiency, and sustainability, this approach addresses the increasing demand for more environmentally friendly selective fluorination strategies in both industrial and academic settings.

14.4 Ionic Liquids-facilitated Electrofluorination

Recently, task-specific ionic liquids (ILs) have garnered significant attention as catalysts or promoters in diverse chemical transformations, owing to their ability to markedly enhance the reactivity and selectivity in nucleophilic substitution reactions [45]. ILs-facilitated fluorination represents one of the most dynamic and promising frontiers in contemporary organic synthesis. Over the past few years, several studies have highlighted the role of ionic liquids in assisting fluorination reactions [23, 24], with electrochemical approaches providing a reliable and efficient platform for selective fluorination [46, 47]. The selectivity and efficiency of fluorinated products are significantly influenced by the choice of solvents and supporting fluoride salts [48]. Through the integration of a diverse array of cations and anions with unique structural features, it is feasible to custom-design the chemical properties of task-specific ILs to suit specific transformation requirements. Despite these advancements, the potential versatility and application scope of structurally diverse task-specific ionic liquids in selective fluorination remain to be fully explored.

14.4.1 Electrochemical Fluorination in Amine-HF Composite ILs

The process of electrochemical fluorination, utilizing anhydrous hydrogen fluoride (HF) as the reaction medium and a nickel anode, stands as a well-established technique that has found widespread application in the synthesis and industrial production of perfluorinated compounds [49]. However, the development of selective electrochemical fluorination has been constrained by several critical challenges. These include limited reaction selectivity, which often leads to undesired by-products; the low nucleophilicity of fluoride ions, which reduces their reactivity toward electrophilic centers; and anode passivation, where a non-conductive layer on the anode surface suppresses current flow and hinders efficiency. Moreover, HF itself is highly hazardous and challenging to handle due to its gaseous state at room temperature. To address these issues, poly(HF) salt ionic liquids, such as 70% HF/pyridine (Olah's reagent), Et_3N–nHF ($n = 3 - 5$), and Et_4NF–nHF ($n = 3 - 5$), have been explored extensively as nucleophilic fluorine sources [50]. These ILs, known for their nonflammable nature, offer significantly lower viscosity and higher ionic conductivity compared to conventional ionic liquids like 1-ethyl-3-methylimidazolium trifluoromethanesulfonate. For example, Et_3N–3HF exhibits an ionic conductivity of 32.6 mS cm^{-1}, while Et_4NF–4HF reaches 99 mS cm^{-1}. Additionally, the poly(HF) salt ILs with a high HF content demonstrate anodic stability up to 3 V vs. SCE or even higher, making them suitable for diverse electrochemical applications. Using these ILs, researchers have reported the successful electrochemical fluorination of various substrates, including aromatics, heteroaromatics, α-arylacetonitrile, cyclic ketones, and cyclic unsaturated ethers [45, 51, 52]. Despite their advantages, these organic base-containing ionic liquids also present limitations. For instance, they may lower the acidity of the reaction environment, adversely affecting acid-catalyzed reactions. Furthermore, organic bases like pyridine can interact with metal catalysts, forming stable metal-pyridine complexes that sequester metal centers, reducing catalytic efficiency and altering the electronic properties of the metal. Such interactions may compromise the performance and selectivity of catalytic cycles, highlighting the need for careful optimization in their application [53].

HF can establish stable complexes through robust hydrogen bonding interactions when paired with carefully selected hydrogen-bond acceptors (HBAs) that are non-basic, non-nucleophilic, and weakly coordinating properties toward cationic metals [55]. These HBAs minimally interfere with metal cations, thereby preserving the catalytic activity while providing a controlled reaction environment conducive to specific transformations. Such HF complexes have emerged as promising nucleophilic fluorination reagents. A good HBA forms a liquid solution upon complexing with HF; however, common organic bases or neutral H-bond acceptors can reduce the system's acidity. To address this challenge, quite recently, Xu et al. developed 1,3-dimethyl-3,4,5,6-tetrahydro-2(1*H*)-pyrimidinone (DMPU)/HF complex, which demonstrated superior performance for monofluorination of alkenes (Figure 14.5a) [54]. DMPU ($pK_{BHX} = 2.82$) is a better HBA than pyridine ($pK_{BHX} = 1.86$) or Et_3N($pK_{BHX} = 1.98$) and simultaneously less basic according to Laurence's database [55], enabling higher acidity in the DMPU/HF complex compared to pyridine/HF and Et_3N/HF complexes. Based on this, in the monohydrofluorination of acceptor-substituted terminal alkyne (1m), Xu et al. observed a notable shift in regioselectivity [54]. The uncatalyzed HF addition to 1 m favored the Michael addition product 4, whereas the DMPU/HF/Au-1 system directed the reaction towards 2 m as the sole isomer (Figure 14.5b). This finding underscores the significant influence of the DMPU/HF/Au-1 catalyst on regioselectivity. Despite its effectiveness in alkyne hydrofluorination, the system proved less suitable for alkene hydrofluorination. To overcome these limitations, researchers proposed a multifunctional activation strategy using potassium bisulfate ($KHSO_4$) as an HBA. The bisulfate anion exhibits a bifunctional behavior: its —OH terminus is strongly acidic ($pK_a = 1.9$) and its ionic $-O^-K^+$ terminus is a potent HBA. This dual functionality enables $KHSO_4$ to form an extensive hydrogen bond network with HF, condensing gaseous HF into a stable liquid phase and enhancing its nucleophilicity (Figure 14.5c) [53]. This hydrogen bonding interaction achieves two aims: (i) it condenses gaseous HF and forms a stable liquid at room temperature; (ii) it enhances the nucleophilicity of HF [56]. The high HF affinity of $KHSO_4$ could be rationalized using Hunter and co-workers' recently reported H-bond scale (parameter β), where the hydrogen bonding basicity

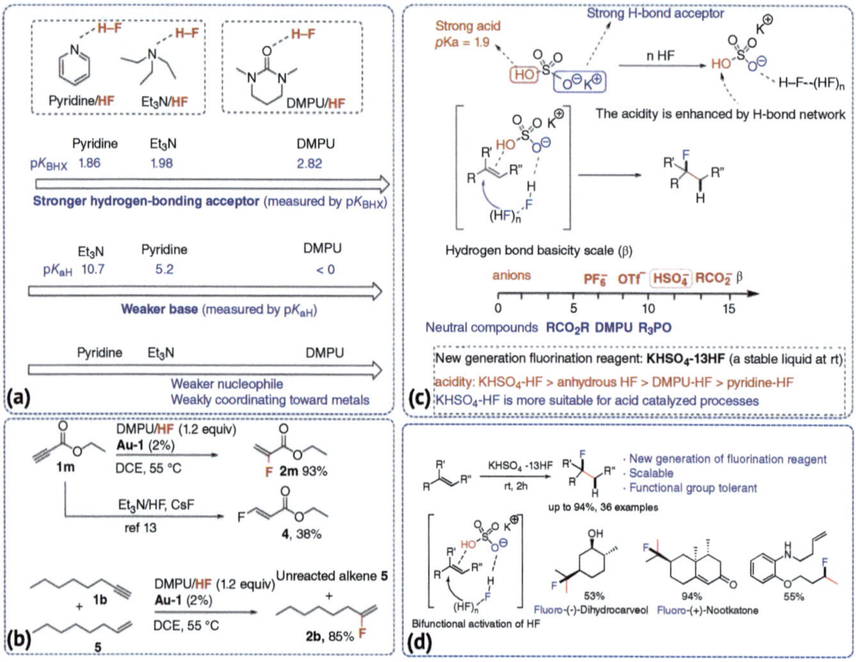

Figure 14.5 (a) Comparison of DMPU/HF with pyridine/HF and Et_3N/HF. *Source:* Ref. [54], with permission of American Chemical Society. (b) The selectivity of the DMPU/HF/Au-1 fluorination system. *Source:* Ref. [54], with permission of American Chemical Society. (c) Bifunctional activation strategy for hydrofluorination of alkenes. *Source:* Ref. [53], with permission of American Chemical Society. (d) Hydrofluorination of alkenes via $KHSO_4$-13HF complex. *Source:* Ref. [53], with permission of American Chemical Society.

of anionic species like HSO_4^- is significantly higher than that of neutral HBAs, including DMPU and even R_3PO [57]. Additionally, $KHSO_4$ is inexpensive, readily available, and easily removable in aqueous workups. Building on this, Lu et al. developed the $KHSO_4$-13HF reagent (HF 68 wt/wt%), which is cost-effective, safe to handle, and highly efficient for hydrofluorination of various highly functionalized alkenes (Figure 14.5d) [53]. The high efficiency of this system highlights the critical role of activating HF with a strongly acidic HBA, making it a promising next-generation nucleophilic fluorination reagent.

14.4.2 Electrochemical Fluorination in Using Mediators in ILs

Although the electrochemical fluorination of organic compounds in organic solvents in HF-based ILs electrolytic systems has been extensively investigated, it frequently encounters severe anode passivation. To address this issue, solvent-free electrochemical fluorination in neat HF-based ILs electrolytic systems has been developed. However, anode passivation persists in these systems, depending on the substrates used. A promising solution to this challenge is the use of indirect electrolysis employing mediators. Mediators provide several advantages, including reducing overpotential, enhancing selectivity control, and mitigating anode passivation. For instance, recent developments include the first electrochemical application of task-specific ionic liquids (TSILs) containing iodoarene moieties as mediators (Figure 14.6a) [58]. These mediators facilitate electron transfer, preventing the formation of a non-conductive film on the anode surface. By ensuring uninterrupted and efficient electron transfer, mediators enable smooth fluorination reactions, enhancing the yield and selectivity of target products. This approach has enabled volatile organic solvent-free indirect electrochemical fluorination of various organic compounds using reusable iodoarene mediators with ionic liquid moieties. Additionally, task-specific ionic liquids with triarylamine and iodoarene functionalities, as well as polymer-supported iodobenzenes, have been developed to improve further efficiency. For example, Figure 14.6b illustrates how these mediators significantly enhance electrochemical fluorination reactions in undivided cells, while remaining intact in ionic liquids after product extraction, allowing for reuse in subsequent cycles [16]. Moreover, the predominant reduction of protons at the cathode prevents the reduction of hypervalent iodobenzene intermediates generated at the anode, ensuring consistent mediator performance throughout the process.

It is worth noting that a polymer-supported iodobenzene (PSIB) mediator has also proven effective for indirect anodic fluorination in HF salts. In this case, the iodobenzene moiety pendent from the solid polymer support could not be directly oxidized. Therefore, a dual mediator system was employed. As shown in Figure 14.6c, the process begins with the electrooxidation of Cl^- to Cl^+, which reacted with the

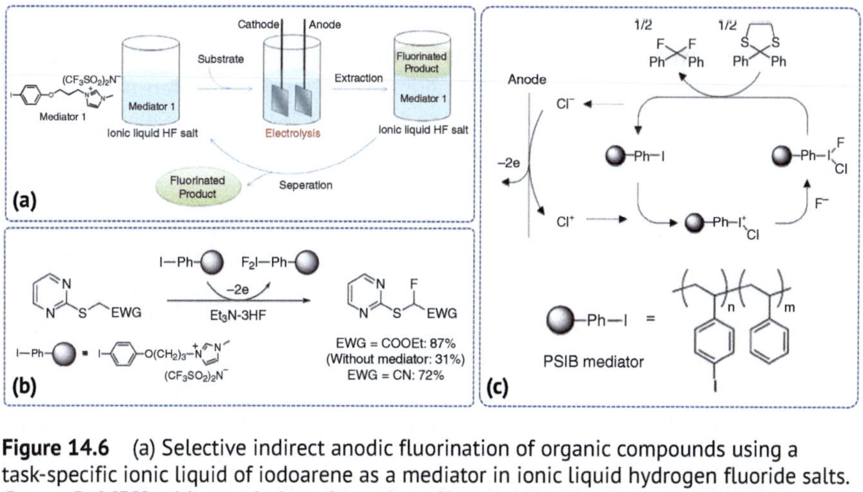

Figure 14.6 (a) Selective indirect anodic fluorination of organic compounds using a task-specific ionic liquid of iodoarene as a mediator in ionic liquid hydrogen fluoride salts. *Source:* Ref. [58], with permission of American Chemical Society. (b) Anodic fluorination using a recyclable task-specific iodobenzene mediator in Et_3N-3HF. *Source:* Ref. [16], with permission of American Chemical Society. (c) Reaction pathway for the double mediatory anodic fluorination. *Source:* Ref. [59], with permission of John Wiley & Sons.

iodobenzene moiety to form $PhI^{+}Cl$, and subsequently captures a fluoride ion to generate the hypervalent [(chloro)(fluoro)iodo]benzene moiety [59]. This hypervalent iodine moiety intermediate oxidizes the substrate, regenerating the original PSIB mediator. Remarkably, the PSIB mediator retained its functionality and facilitated good yields (86–79%) of the fluorinated products across 10 reaction cycles. These findings underscore the efficacy of the mediator system in enabling the indirect electrochemical fluorination of substrates with oxidation potentials lower than that of iodoarene.

The expanding application of ILs further highlights their versatility in the synthesis of fluorinated compounds. These unique solvents, with properties that can be finely tuned for specific synthetic objectives, represent a transformative alternative to traditional volatile and hazardous organic solvents. ILs have demonstrated significant utility as catalysts, reaction media, and promoters in fluorine functionalization processes, providing numerous advantages. Their non-volatility and thermal stability ensure safer handling and reduced environmental impact, aligning with green chemistry principles. In addition to these environmental benefits, IL-based methodologies often feature key advantages, such as the reusability of ILs, which reduces waste and lowers costs in industrial contexts. The ability to perform reactions under mild conditions enhances safety, limits side reactions, and contributes to higher yields and purer products. This combination of environmental, economic, and operational benefits makes ILs a compelling choice for advancing fluorination processes.

14.5 Electrochemical Fluorination Using Alkali Metal Fluorides

The fluorination of organic compounds traditionally relies on HF or its ammonium derivatives, such as $Et_3N{\cdot}n HF$ and $Et_4NF{\cdot}n HF$ (n = 3–5), which serve as both fluoride sources and supporting electrolytes [60, 61]. While electrochemical fluorination offers a safer alternative to conventional methods that use hazardous fluorinating agents, the use of HF and its derivatives is limited by their high cost, toxicity, and corrosiveness. These challenges have driven the exploration for simpler, less hazardous fluoride sources [16]. Among the promising alternatives, inorganic fluoride salts, particularly alkali metal fluorides (MFs), have gained attention due to their stability, cost-effectiveness, and ease of handling. These properties make MFs attractive as nucleophilic fluorinating agents and supporting electrolytes in chemical and electrochemical applications. However, the poor solubility and low nucleophilicity of MFs in organic solvents present significant barriers to their broader adoption [62]. To overcome these limitations, phase-transfer catalysts, including crown ethers and quaternary ammonium or phosphonium salts, are frequently employed. These catalysts enhance the solubility and nucleophilicity of MFs by shifting the equilibrium of reaction (1) toward production of reactive fluoride ions and reducing the Coulombic interactions:

$$MF \rightleftharpoons M^{+} + F^{-} \quad (1) \qquad M = Li, Na, K, Cs$$

Fuchigami and Inagi's research teams have pioneered innovative methodologies for anodic fluorination. In 2007, Fuchigami's group introduced a novel anodic C—H fluorination method employing cation exchange between solid-supported acids with terminal SO_3H group to promote the dissociation of MFs in organic solvents (Figure 14.7a) [63]. Their study demonstrated that the fluorination yield was significantly influenced by the selectivity sequence of Amberlyst 15Dry ($Li^+ < H^+ < Na^+ < K^+ < Cs^+$), which governs the concentration and nucleophilicity of F^- in the system. Furthermore, the system generates 2,6-lutidine·HF in situ, a compound that acts both as a fluorinating agent and a supporting electrolyte, eliminating the direct handling of HF. Building on this concept, the group developed an enhanced system by integrating anodic fluorination with a poly(ethyleneglycol) [Mn ≈ 200; PEG 200]/MF (KF or CsF) (Figure 14.7b) [1]. Here, PEG 200 coordinates with metal cations to facilitate MF dissolution in MeCN, enabling electrochemical fluorination and desulfurization reactions (Figure 14.7c).

In contrast, Inagi's team further advanced the field by employing a split bipolar electrode (s-BPE) system within a similar PEG/MF setup, enabling the fluorination of substrates like triphenylmethane and triphenylphosphine in the presence of low MF concentrations [64]. This system also facilitates the monitoring of

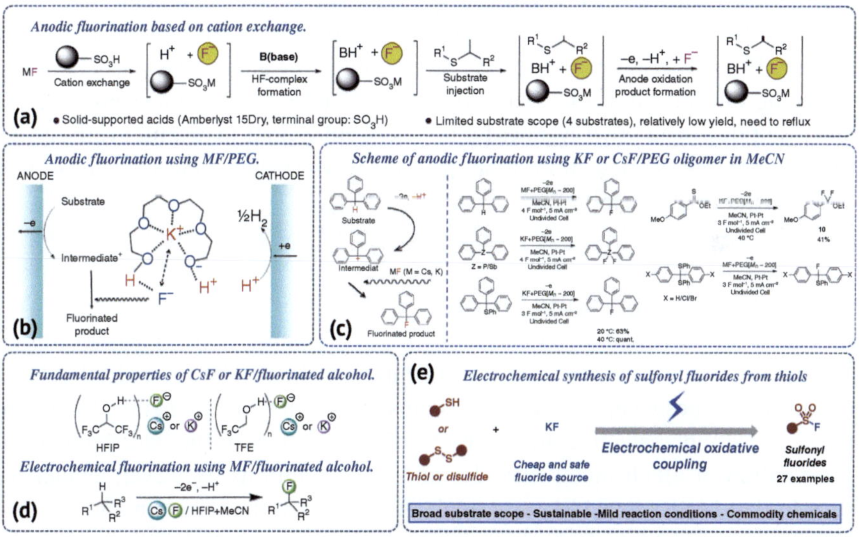

Figure 14.7 (a) Schematic representation of the anodic C—H fluorination method, which relies on cation exchange between solid-supported acids with terminal SO_3H groups and alkali metal fluorides. *Source:* Ref. [63], with permission of John Wiley & Sons.
(b) Poly(ethylene glycol) with two terminal hydroxy groups serves as a multifunctional additive for anodic fluorination using KF. *Source:* [1] / with permission of John Wiley & Sons.
(c) Anodic fluorination using KF or CsF/PEG oligomer in MeCN.
Source: [1] / with permission of John Wiley & Sons. (d) Anodic fluorination using KF or CsF/fluorinated alcohols. *Source:* Ref. [25], with permission of American Chemical Society.
(e) Electrochemical synthesis of sulfonyl fluorides from thiols/KF. *Source:* Ref. [65], with permission of American Chemical Society.

the corresponding current and charge throughout the process. In a subsequent study, the Inagi group employed weakly coordinating and oxidation-resistant fluorinated alcohols, such as 1,1,1,3,3,3-hexafluoroisopropanol (HFIP) and 3,3,3-trifluoroethanol (TFE), as alternatives to oxidizable aliphatic alcohols [25]. These fluorinated alcohols facilitate the formation of hydrogen bonds with the fluorine anion of MF (M = Cs,K), thereby offering a new reaction medium for electrochemical fluorination (Figure 14.7d). Additionally, Laudadio et al. [65] utilized widely available thiols and disulfides as starting materials, in combination with the inexpensive, abundant, and safe fluorine source KF, to successfully synthesize sulfonyl fluorides via electrochemical methods (Figure 14.7e). The proposed reaction mechanism involves the formation of radical cations from the oxidation of disulfides, which subsequently react with nucleophilic fluorides to yield the corresponding sulfonyl fluorides. This approach demonstrates mild reaction conditions and a broad substrate scope, encompassing various alkyl, benzylic, aryl, and heteroaryl thiols and disulfides.

In summary, the exploration of electrochemical fluorination using alkali metal fluorides presents offers a promising and viable alternative to traditional fluorination methods. Researchers like Fuchigami and Inagi have made significant strides in overcoming the challenges of solubility and reactivity associated with these fluoride sources, employing phase-transfer catalysts and innovative electrochemical systems to enhance efficiency and selectivity. These advancements contribute to the development of greener methodologies, reducing reliance on hazardous substances such as HF. Furthermore, the ongoing investigation into alternative reaction media, including fluorinated alcohols, and the strategic utilization of inexpensive thiols and disulfides, opens new avenues for broader applications in organic synthesis. As the field progresses, these innovations pave the way for safer, more sustainable, and versatile fluorination strategies.

14.6 Conclusion and Outlook

In conclusion, the electrochemical fluorination (ECF) plays a crucial role in chemical synthesis across several domains, including electrochemically driven fluoroalkylation, selective fluorination of alkenes, and the electrochemical fluorination of alkali metal fluorides and ionic solutions. However, challenges remain that hinder its widespread application. Firstly, the selection and stability of corrosion-resistant electrode materials are critical factors in ensuring the efficiency and longevity of the electrochemical fluorination process. Secondly, reaction selectivity poses a significant challenge due to the potential for side reactions in organic synthesis. Finally, a deeper understanding of the reaction mechanisms and kinetics is essential for optimizing reaction conditions and improving product selectivity. Insights from previous studies emphasize the importance of interdisciplinary approaches in refining electrochemical fluorination techniques, ultimately expanding their applicability in both academic research and industrial settings.

Acknowledgment

The authors are grateful for the support from the National and Local Joint Engineering Research Center of Biomedical Functional Materials and a project sponsored by the Priority Academic Program Development of Jiangsu Higher Education Institutions.

References

1 Sawamura, T., Takahashi, K., Inagi, S. et al. (2012). Electrochemical fluorination using alkali-metal fluorides. *Angew. Chem. Int. Ed.* 51 (18): 4413.

2 Chang, X., Zhang, J., Zhang, Q. et al. (2020). Merging electrosynthesis and bifunctional squaramide catalysis in the asymmetric detrifluoroacetylative alkylation reactions. *Angew. Chem. Int. Ed.* 59 (42): 18500.

3 Wigman, B., Lee, W., Wei, W. et al. (2022). Electrochemical fluorination of vinyl boronates through donor-stabilized vinyl carbocation intermediates. *Angew. Chem. Int. Ed.* 61 (12): e202113972.

4 Jian Hu, X., Ping Zheng, Y., Wei Li, Z. et al. (2024). Artificial LiF-rich interface enabled by in situ electrochemical fluorination for stable lithium-metal batteries. *Angew. Chem. Int. Ed.* 63 (12): e202319600.

5 Cheng, Y.-X., Yang, X.-G., Du, F.-H. et al. (2024). Direct trifluoromethylthiolation of terminal alkynes mediated by a hypervalent trifluoromethylthio-iodine(iii) reagent; boosting effect of fluorinated alcohol. *Green Chem.* 26 (10): 5914.

6 Li, L., Ni, C., Wang, F. et al. (2016). Deoxyfluorination of alcohols with 3,3-difluoro-1,2-diarylcyclopropenes. *Nat. Commun.* 7 (1): 13320.

7 Goldberg, N.W., Shen, X., Li, J. et al. (2016). AlkylFluor: deoxyfluorination of alcohols. *Org. Lett.* 18 (23): 6102.

8 Timofeeva, D.S., Ofial, A.R., and Mayr, H. (2018). Kinetics of electrophilic fluorinations of enamines and carbanions: comparison of the fluorinating power of N–F reagents. *J. Am. Chem. Soc.* 140 (36): 11474.

9 Yamamoto, K., Li, J., Garber, J.A.O. et al. (2018). Palladium-catalysed electrophilic aromatic C–H fluorination. *Nature* 554 (7693): 511.

10 Leibler, I.N.-M., Gandhi, S.S., Tekle-Smith, M.A. et al. (2023). Strategies for nucleophilic C(sp3)–(radio)fluorination. *J. Am. Chem. Soc.* 145 (18): 9928.

11 Adler, P., Teskey, C.J., Kaiser, D. et al. (2019). α-Fluorination of carbonyls with nucleophilic fluorine. *Nat. Chem.* 11 (4): 329.

12 Liu, Q., Shen, X., Ni, C. et al. (2017). Stereoselective carbonyl olefination with fluorosulfoximines: facile access to Z or E terminal monofluoroalkenes. *Angew. Chem. Int. Ed.* 56 (2): 619.

13 Dong, X., Jiang, W., Hua, D. et al. (2021). Radical-mediated vicinal addition of alkoxysulfonyl/fluorosulfonyl and trifluoromethyl groups to aryl alkyl alkynes. *Chem. Sci.* 12 (35): 11762.

14 Neumann, C.N. and Ritter, T. (2016). U can fluorinate unactivated bonds. *Nat. Chem.* 8 (9): 822.

15 Munoz, S.B., Krishnamurti, V., Barrio, P. et al. (2018). Direct difluorination–hydroxylation, trifluorination, and C(sp2)–H fluorination of enamides. *Org. Lett.* 20 (4): 1042.

16 Fuchigami, T. and Inagi, S. (2020). Recent advances in electrochemical systems for selective fluorination of organic compounds. *Acc. Chem. Res.* 53 (2): 322.

17 Thadathil, D.A., Varghese, A., and Radhakrishnan, K.V. (2021). Review—Electrochemical strategies for selective fluorination of organic compounds. *J. Electrochem. Soc.* 168 (7): 075503.

18 Tarantino, G. and Hammond, C. (2020). Catalytic C(sp3)–F bond formation: recent achievements and pertaining challenges. *Green Chem.* 22 (16): 5195.

19 Zou, Z., Zhang, W., Wang, Y. et al. (2021). Recent advances in electrochemically driven radical fluorination and fluoroalkylation. *Org. Chem. Front.* 8 (11): 2786.

20 Hintz, H., Bower, J., Tang, J. et al. (2023). Copper-catalyzed electrochemical C–H fluorination. *Chem. Catal.* 3 (1): 100491.

21 de Rooij, D.M.R. (2003). Electrochemical methods: fundamentals and applications. *Anti-Corr. Meth. Mater.* 50 (5).

22 Box, J.R., Avanthay, M.E., Poole, D.L. et al. (2023). Electronically ambivalent hydrodefluorination of aryl-CF_3 groups enabled by electrochemical deep-reduction on a Ni cathode. *Angew. Chem. Int. Ed.* 62 (12): e202218195.

23 Chatterjee, R. and Dandela, R. (2023). Recent advances in ionic liquid mediated selective fluorination of organic compounds. *J. Fluor. Chem.* 268: 110133.

24 Borodkin, G.I. and Shubin, V.G. (2022). Electrophilic fluorination of heterocyclic compounds with NF reagents in unconventional media. *Chem. Heterocycl. Comp.* 58 (2): 84.

25 Shida, N., Takenaka, H., Gotou, A. et al. (2021). Alkali metal fluorides in fluorinated alcohols: fundamental properties and applications to electrochemical fluorination. *J. Org. Chem.* 86 (22): 16128.

26 Watanabe, N. (1980). Electrochemical process: application to fluorine chemistry. *Journal of Fluor. Chem.* 16 (6): 499.

27 O'Brien, A.G., Maruyama, A., Inokuma, Y. et al. (2014). Radical C-H functionalization of heteroarenes under electrochemical control. *Angew. Chem. Int. Ed.* 53 (44): 11868.

28 Dou, G.-Y., Jiang, Y.-Y., Xu, K. et al. (2019). Electrochemical Minisci-type trifluoromethylation of electron-deficient heterocycles mediated by bromide ions. *Org. Chem. Front.* 6 (14): 2392.

29 Ge, C., Qiao, L., Zhang, Y. et al. (2024). Metal-free electrochemical trifluoromethylation of imidazole-fused heterocycles with trifluoromethyl thianthrenium triflate. *Chin. J. Chem.* 42 (15): 1679.

30 Jud, W., Maljuric, S., Kappe, C.O. et al. (2019). Cathodic C-H trifluoromethylation of arenes and heteroarenes enabled by an in situ-generated triflyltriethylammonium complex. *Org. Lett.* 21 (19): 7970.

31 Derat, E., Masson, G., and Claraz, A. (2024). Electrochemically-driven 1,4-aryl migration via radical fluoromethylation of N-allylbenzamides: a straightforward access to functionalized beta-arylethylamines. *Angew. Chem. Int. Ed.* 63 (29): e202406017.

32 Xiong, P., Xu, H.H., Song, J. et al. (2018). Electrochemical difluoromethylarylation of alkynes. *J. Am. Chem. Soc.* 140 (7): 2460.

33 Xu, H.H., Song, J., and Xu, H.C. (2019). Electrochemical difluoromethylation of electron-deficient alkenes. *ChemSusChem* 12 (13): 3060.

34 Luan, S., Castanheiro, T., and Poisson, T. (2023). Electrochemical difluoromethylation of electron-rich olefins. *Org. Lett.* 25 (10): 1678.

35 Zhang, D., Cai, J., Du, J. et al. (2022). Electrochemical-oxidation-promoted direct N-ortho-selective difluoromethylation of heterocyclic N-oxides. *Org. Lett.* 24 (7): 1434.

36 Chen, X., Jiang, J., Huang, X.-J. et al. (2023). Electrochemical oxidative radical cascade reactions for the synthesis of difluoromethylated benzoxazines. *Org. Chem. Front.* 10 (15): 3898.

37 Wang, Z.-X., Livingstone, K., Hümpel, C. et al. (2023). Regioselective, catalytic 1,1-difluorination of enynes. *Nat. Chem.* 15 (11): 1515.

38 Lee, C., Kim, M., Han, S. et al. (2024). Nickel-catalyzed hydrofluorination in unactivated alkenes: regio- and enantioselective C–F bond formation. *J. Am. Chem. Soc.* 146 (13): 9375.

39 Noel, M., Suryanarayanan, V., and Chellammal, S. (1997). A review of recent developments in the selective electrochemical fluorination of organic compounds. *J. Fluor. Chem.* 83 (1): 31.

40 Dinoiu, V., Kanno, K., Fukuhara, T. et al. (2005). Selective anodic fluorination of electrophilic alkenes. *J. Fluor. Chem.* 126 (5): 753.

41 Mandal, A., Jang, J., Yang, B. et al. (2023). Palladium-catalyzed electrooxidative hydrofluorination of aryl-substituted alkenes with a nucleophilic fluorine source. *Org. Lett.* 25 (1): 195.

42 Doobary, S., Sedikides, A.T., Caldora, H.P. et al. (2020). Electrochemical vicinal difluorination of alkenes: scalable and amenable to electron-rich substrates. *Angew. Chem. Int. Ed.* 59 (3): 1155.

43 Doobary, S., Poole, D.L., and Lennox, A.J.J. (2021). Intramolecular alkene fluoroarylation of phenolic ethers enabled by electrochemically generated iodane. *J. Org. Chem.* 86 (22): 16095.

44 Wen, L., Li, B., Zou, Z. et al. (2024). Direct electrochemical difluorination and azo-fluorination of gem-difluorostyrenes. *Org. Chem. Front.* 11 (1): 142.

45 Fuchigami, T. and Tajima, T. (2005). Highly selective electrochemical fluorination of organic compounds in ionic liquids. *J. Fluor. Chem.* 126 (2): 181.

46 Khan, Z.U.H., Kong, D., Chen, Y. et al. (2015). Ionic liquids based fluorination of organic compounds using electrochemical method. *J. Industr. Eng. Chem.* 31: 26.

47 Sartori, P. and Ignat'ev, N. (1998). The actual state of our knowledge about mechanism of electrochemical fluorination in anhydrous hydrogen fluoride (Simons process). *J. Fluor. Chem.* 87 (2): 157.

48 Ishii, H., Yamada, N., and Fuchigami, T. (2001). Electrolytic partial fluorination of organic compound. Part: 53 Highly regioselective anodic mono- and difluorination of 4-arylthio-1,3-dioxolan-2-ones. A marked solvent effect on fluorinated product selectivity. *Tetrahedron* 57 (44): 9067.

49 Simons, J.H. (1986). The seven ages of fluorine chemistry. *J. Fluor. Chem.* 32 (1): 7.

50 Yoneda, N. (1991). The combination of hydrogen fluoride with organic bases as fluorination agents. *Tetrahedron* 47 (29): 5329.

51 Ilayaraja, N. and Noel, M. (2009). A comparative study of anodic fluorination of N-alkyl and N,N-dialkyl phenylacetamides in Et3N·4HF medium. *J. Electroanal. Chem.* 632 (1): 45.

52 Chen, S.-Q., Hatakeyama, T., Fukuhara, T. et al. (1997). Electrochemical fluorination of aliphatic aldehydes and cyclic ketones using Et_3N-5HF electrolyte. *Electrochim. Acta* 42 (13): 1951.

53 Lu, Z., Zeng, X., Hammond, G.B. et al. (2017). Widely applicable hydrofluorination of alkenes via bifunctional activation of hydrogen fluoride. *J. Am. Chem. Soc.* 139 (50): 18202.

54 Okoromoba, O.E., Han, J., Hammond, G.B. et al. (2014). Designer HF-based fluorination reagent: highly regioselective synthesis of fluoroalkenes and gem-difluoromethylene compounds from alkynes. *J. Am. Chem. Soc.* 136 (41): 14381.

55 Laurence, C., Brameld, K.A., Graton, J. et al. (2009). The pKBHX database: toward a better understanding of hydrogen-bond basicity for medicinal chemists. *J. Med. Chem.* 52 (14): 4073.

56 Liang, S., Hammond, G.B., and Xu, B. (2017). Hydrogen bonding: regulator for nucleophilic fluorination. *Chem.: A Europ. J.* 23 (71): 17850.

57 Pike, S.J., Hutchinson, J.J., and Hunter, C.A. (2017). H-bond acceptor parameters for anions. *J. Am. Chem. Soc.* 139 (19): 6700.

58 Sawamura, T., Kuribayashi, S., Inagi, S. et al. (2010). Use of task-specific ionic liquid for selective electrocatalytic fluorination. *Org. Lett.* 12 (3): 644.

59 Sawamura, T., Kuribayashi, S., Inagi, S. et al. (2010). Recyclable polymer-supported iodobenzene-mediated electrocatalytic fluorination in ionic liquid. *Adv. Synth. Catal.* 352 (16): 2757.

60 Ignat'ev, N.V. (2017). Electrochemical fluorination. In: *Modern Synthesis Processes and Reactivity of Fluorinated Compounds* (eds. H. Groult, F.R. Leroux, and A. Tressaud), 71–123. Amsterdam: Elsevier.

61 Li, B., Liao, X., Wen, L. et al. (2024). Electrochemically direct fluorination functionalization of styrenes with different fluorine *Source:* access to fluoroalkyl derivatives. *J. Org. Chem.* 89 (13): 9440.

62 Wen, L., Zou, Z., Zhou, N. et al. (2024). Electrochemical fluorination functionalization of gem-difluoroalkenes with CsF as a fluorine *Source:* access to fluoroalkyl building blocks. *Org. Lett.* 26 (1): 241.

63 Tajima, T., Nakajima, A., Doi, Y. et al. (2007). Anodic fluorination based on cation exchange between alkali-metal fluorides and solid-supported acids. *Angew. Chem. Int. Ed.* 46 (19): 3550.

64 Miyamoto, K., Nishiyama, H., Tomita, I. et al. (2018). Development of a split bipolar electrode system for electrochemical fluorination of triphenylmethane. *ChemElectroChem* 6 (1): 97.

65 Laudadio, G., Bartolomeu, A.A., Verwijlen, L. et al. (2019). Sulfonyl fluoride synthesis through electrochemical oxidative coupling of thiols and potassium fluoride. *J. Am. Chem. Soc.* 141 (30): 11832.

15

Electrochemical Polymerization: Synthesis of Functional Films for Energy Devices

Yansong Zhou[1,2,], Shuangyin Wang[1,2,*]*

[1] *State Key Laboratory of Chemo and Biosensing, College of Chemistry and Chemical Engineering, Advanced Catalytic Engineering Research Center of the Ministry of Education, Hunan University, Changsha, P. R. China*
[2] *Greater Bay Area Institute for Innovation, Hunan University, Guangzhou, P. R. China*

*Corresponding authors: yansongzhou@huu.edu.cn; shuangyinwang@hnu.edu.cn

15.1 Introduction

The electrochemical polymerization method is recognized as one of the most rapid and reliable conventional synthetic techniques for producing high-quality polymer films [1]. Unlike traditional chemical polymerization, which often involves hazardous chemicals, electrochemical polymerization employs an appropriate electric field to facilitate electron transfer between monomers and electrodes, thereby generating active species essential for the polymerization process. With the application of an electric field, the molecular polymerization can be meticulously controlled to tailor the properties of the resulting film. Furthermore, this technique enables the precise deposition of functional organic layers onto designated areas, significantly minimizing positional errors in constructing modern high-precision devices [2, 3].

Electrochemical polymerization has found extensive applications in synthesizing second-generation conductive polymers [4] such as polypyrrole [5], polyethylenedioxythiophene [6, 7], poly(arylene-vinylene) [8], and polyaniline [9–11], along with their derivatives. Additionally, thin films composed of microporous polymer networks featuring multifunctional carbazole- and thiophene-based tectons have also been generated through electrochemical methods [12]. Currently, advancements in electrochemical polymerization concepts are being explored to synthesize emerging functional materials, including hydrogen-bonded organic frameworks [13], supramolecular materials [14], and covalent organic frameworks [15] among others – extending beyond the mere synthesis of polymers. The resultant functional films have demonstrated efficacy as active layers or modifiers in energy-related research domains encompassing optical applications, energy storage solutions, and energy conversion technologies.

Electrochemical Energy Conversion via Small Molecule Catalysis, First Edition.
Edited by Zhicheng Zhang and Yuanmiao Sun.

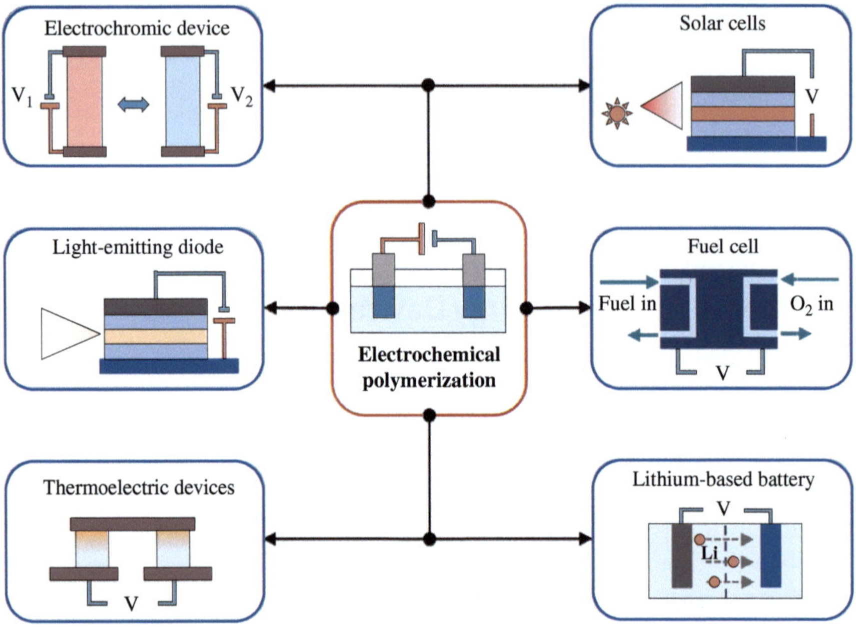

Figure 15.1 Applications of electrochemical polymerization method in fabricating functional films in energy devices.

In this chapter, we commence with a discussion of the fundamental principles underlying electrochemical polymerization before summarizing recent technological developments in this area. We will review three advanced systems: an electrochemical microflow reactor, a single-electrode electrochemical system, and a streaming potential-driven electrochemical polymerization system. Subsequently, we will focus on energy devices fabricated via electrochemical polymerization – including, but not limited to, electrochromic devices; organic light-emitting diodes; thermoelectric devices; solar cells; fuel cells; lithium-based batteries; and electrochromic batteries – all based on recently published research findings (see Figure 15.1). In the final part of this chapter, we emphasize the role of electrochemical polymerization in fabricating next-generation electronic and energy devices.

15.2 General Principles of Electrochemical Polymerization

15.2.1 Definition

Electrochemical polymerization is defined as the polymerization reaction that occurs when a monomer undergoes oxidation, reduction, or decomposition into free radicals or ions at the electrode surface through electrolysis within an electrolytic cell containing an appropriate electrolyte. This process may also be referred to as electrolytic polymerization, electropolymerization, or electroinitiated polymerization.

15.2.2 Characteristics of Electrochemical Polymerization

Electrochemical polymerization exhibits the following several key characteristics:

1. Simple equipment and ease of control over reaction conditions allow for the precise regulation of the thickness of the resulting polymer film;
2. The resulting polymer films demonstrate uniform thickness with high reproducibility;
3. Functional films with diverse structures and properties can be obtained;
4. A wide variety of conductive polymers can be synthesized;
5. Doping can occur simultaneously during monomer polymerization.

15.2.3 Types of Electrochemical Polymerization

Electrochemical polymerizations can be classified in two ways based on either the chain growth process or the nature of the polymerization reaction itself:

1. Based on chain growth processes, electrochemical reactions are categorized into electrochemical polycondensation and electrochemical polyaddition reactions.

   Polycondensation involves low-molecular compounds possessing two or more reactive functional groups interacting to form macromolecules through a stepwise growth mechanism. In contrast, polyaddition entails forming macromolecules via repeated additions starting from alkene monomers or cyclic monomers, characterized by a one-chain growth process where monomeric molecules react with active centers along the growing macromolecular chain. Depending on the nature of these active centers, polyaddition reactions are further divided into three categories: free radical polyaddition, cationic polymerization, and anionic polymerization.
2. Based on whether they occur at cathodes or anodes, these processes are classified as cathodic or anodic polymerizations.

   The most extensively studied method is anodic polymerization – conceptually akin to oxidative chemical processes but conducted without external oxidants – and it has been widely employed in synthesizing conductive polymers. Conversely, cathodic polarization – which utilizes electrons as catalysts for constructing polymers – remains relatively underexplored despite its potential for fabricating conjugated polymers on metal surfaces to create multilayer heterostructures suitable for organic optoelectronic devices [16, 17].

15.2.4 Mechanisms of Electrochemical Polymerization

Diaz et al. were pioneers in proposing mechanisms for electrochemically induced formation pathways leading to conductive polymers [18, 19]. In their model concerning pyrrole's behavior during this process, Diaz and colleagues posited that, following initial oxidation, electrochemically activated dimer formation occurs at α-positions, followed by proton elimination yielding aromatic neutral dimers. Furthermore, the subsequent steps involve additional interactions between both dimer radical cations alongside unreacted monomers post-electrooxidation.

This foundational mechanism remains widely accepted today, yet it has been recognized as overly simplistic given certain experimental observations [20, 21]. With advancements in analytical techniques such as in situ spectroscopy coupled with scanning probe microscopy, researchers have gained deeper insights into deposition mechanisms occurring on electrodes [22]. Currently, two primary models dominate discussions of these phenomena: radical-radical coupling mechanisms vs. those involving substrate interactions.

1. Radical–radical polymerizations
   The polymerization reaction is a stepwise growth process between the association between free radical-ions and free radical-ions, and the monomer itself. The polymerization products are reacted by electrodes together into intermediates of the polymerization reaction for polymerization.
2. Radical–substrate mechanism
   Characterized by sequential chains formed through interactions between radicals/ions where substrates could represent either low molecular weight precursors (monomers), or higher-weight neutral counterparts (polymers).

15.2.5 Stages of Electrochemical Polymerization

The overall progression in any given instance follows four distinct stages:

1. Formation of oligomer regions initiated by electric field-induced redox events, generating concentrated radicals near the surfaces of electrodes;
2. Nucleation commences once the concentration of interface oligomers reach critical thresholds;
3. Chain growth involves the polymerization of oligomers into long-chain structures;
4. Formation of a cross-linked film occurs through further polymerization processes involving the long-chain structures via multi-dimensional growth pathways.

15.2.6 Setup, Electrolytes, and Electrochemical Methods

The electrolytic cells commonly employed in electrochemical polymerization typically consist of either two-electrode or three-electrode systems.

The electrolytes utilized for electrochemical polymerization generally comprise three components: solvents, supporting electrolytes, and organic monomers. While water can serve as a solvent, organic solvents are predominantly used; common examples include dimethylformamide, tetrahydrofuran, acetic acid, acetonitrile, acetone, chloroform, dimethyl sulfoxide, and dichloroethane. To ensure electrical conductivity within the solution, it is often necessary to incorporate supporting electrolytes such as ammonium salts, potassium salts, sodium salts, and lithium salts among others.

Due to the typically low conductivity of electropolymerization products, potentiostatic methods are primarily employed for electrolysis; consequently, the rate of electropolymerization tends to decrease with prolonged electrolysis time.

However, alternative techniques, such as potentiometric scanning methods, galvanostatic electrolysis methods, rectangular wave electrolysis methods, and alternating current electrolysis methods also exist.

15.3 Advanced Electrochemical Polymerization Technology

Conventional electrochemical polymerization experiments are typically performed within two- or three-electrode electrochemical systems. With the rapid advancements in materials science and electrochemistry, substantial progress has been achieved in organic electrochemical technology, including the development of sophisticated electrochemical reactors (e.g., microflow electrochemical reactors [23, 24]), innovative electrochemical systems (e.g., bipolar electrochemistry [25, 26]), and novel technologies akin to lithium-ion batteries [27]. Furthermore, advanced methods for electrochemical polymerization have emerged concurrently with the evolution of organic electrochemical technology.

Electrochemical microflow reactors featuring parallel electrode configurations play a crucial role in both fundamental research and industrial applications. The electrochemical polymerization conducted within these microflow reactors facilitates enhanced monomer conversion and narrower molecular weight distribution, attributable to their unique configurations [28] (See Figure 15.2a). Mizuno et al. reported the pioneering synthesis of soluble π-conjugated polymers, specifically poly(3-hexylthiophene), via electrochemical polymerization in a flow microreactor [29]. Notably, the molecular weight and distribution of the synthesized poly(3-hexylthiophene) can be precisely controlled within this synthetic system. Additionally, another class of π-conjugated polymers, polyphenylene derived from 1,4-bis(trimethylsilyl)benzene, was effectively synthesized using the flow microreactor [30]. Similarly, the molecular weight of the synthesized polyphenylene can be accurately regulated by optimizing the electrochemical polymerization conditions in the microreactor.

A newly developed single-electrode electrochemical system has demonstrated significant success in the high-throughput electrosynthesis of gradient films (See Figure 15.2b). In this system, a single conductive electrode, typically indium tin oxide-coated glass, is employed for electrochemical measurements based on a resistance-induced potential gradient. An insulating sticker with multiple apertures, affixed to the aforementioned electrode surface creates independent microelectrochemical cells, thereby enabling high-throughput electrosynthesis [31, 32]. Utilizing this single-electrode electrochemical system, gradient polypyrrole films have been synthesized at high throughput through pyrrole electropolymerization [33].

An advanced, electric power-free electrosynthesis method has been developed to facilitate the electrochemical polymerization process (See Figure 15.2c). This system is constructed on the principle of streaming potential-driven bipolar electrochemistry [34]. The electrode reactions occur when the streaming potential, generated by a laminar flow of electrolyte within the microchannel, is sufficiently large to

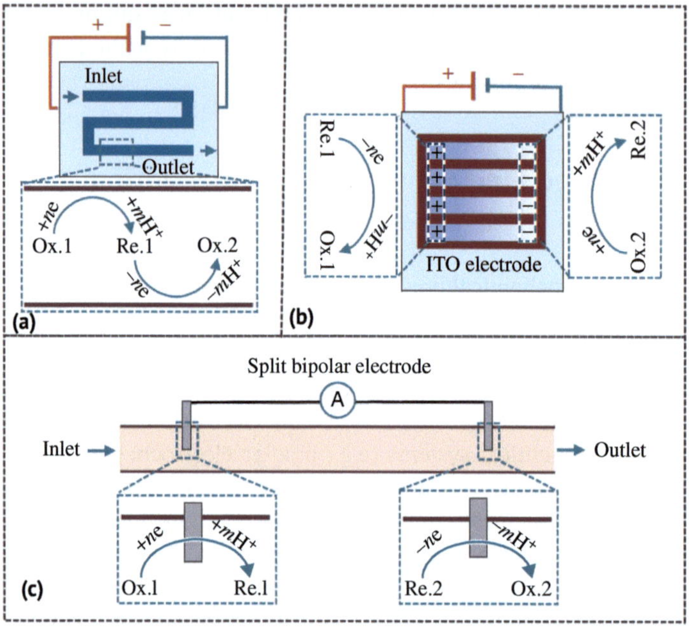

Figure 15.2 Illustrations of the advanced electrochemical polymerization technology. (a) Electrochemical polymerization in the electrochemical microflow reactor. (b) Single-electrode electrochemical system. (c) Streaming potential-driven electrochemical polymerization system.

overcome the difference between anodic and cathodic potentials of the redox species present in the electrolyte. Building on this concept, Iwai et al. introduced the first instance of streaming potential-driven electrochemical oxidative polymerization of aromatic monomers without requiring electric power [35].

15.4 Electrochemical Polymerization in the Fabrication of Functional Films

Over the past few decades, electrochemical polymerization has evolved to synthesize conductive polymers and advanced functional materials. This section discusses the role of electrochemical polymerization in fabricating functional materials for various applications, including optical devices (e.g., electrochromic devices, light-emitting diodes), energy conversion devices (e.g., thermoelectric devices, solar cells, fuel cells), and energy storage systems (e.g., lithium-ion batteries).

15.4.1 Electrochemical Polymerization-fabricated Polymers in Optical Devices

The electrochromic device, which utilizes electrochromic materials that exhibit reversible transmittance or color change properties in response to external electrical stimuli, has garnered renewed interest due to its extensive potential applications in solar-attenuated windows, optical displays, and other areas [36–38]. Compared to transition metal oxides [39] and viologens [40], organic conjugated polymer-based electrochromic devices demonstrate a variety of colors, rapid switching capabilities (sub-second switching times), high optical contrast in both the visible and near-infrared (NIR) regions, as well as excellent cycle stability [41, 42]. The synthesis of high-quality electrochromic polymer films through appropriate methods is crucial for enhancing their electrochromic performance.

The electrochemical polymerization method was initially employed to fabricate electrochromic polymer films and continues to be the most widely used approach for producing conjugated polymer-based electrochromic devices [43]. This method has been extensively utilized in synthesizing both single-component [4, 44, 45] and multi-electroactive (e.g., D-A-D type [46]) electrochromic polymers. Additionally, the electrochemical polymerization technique offers distinct advantages in fabricating nanocomposite electrodes, such as polyaniline/indium tin oxide [10], polyaniline/cerium(IV) [11], PEDOT-graphene oxide [47], and polyaniline/carbon quantum dots [48], for use in electrochromic devices. Furthermore, this method has also been applied to synthesize newly developed electrochromic materials, including covalent organic frameworks. Wang et al. employed an in situ electrochemical interfacial polymerization strategy to rapidly prepare electrochromic covalent organic framework films at room temperature. The resulting materials exhibited stable polymorphic electrochromic performance [15].

Organic light-emitting diodes have garnered increasing research interest due to their significant application potential, particularly in flexible display devices. The electrochemical polymerization method for fabricating light-emitting organic thin films offers several advantages, including selective patterning and precise deposition, as well as a unique boundary effect at microelectrode arrays that facilitates the development of high-resolution displays [49]. Ma et al. developed an electrochemical polymerization technique based on the cross-linking of carbazole to prepare the active layer of organic light-emitting diodes [50–54]. This method has demonstrated its efficacy in fabricating multilayer light-emitting diodes and displays [55].

15.4.2 Electrochemical Polymerization Fabricated Polymers in Energy Conversion Devices

Thermoelectric devices directly convert unused waste heat into electricity, presenting significant potential for constructing a sustainable society [56]. Organic thermoelectric materials have recently emerged as viable alternatives to inorganic materials, offering reasonable thermoelectric performance, lower cost, reduced toxicity, and improved mechanical properties compared to their inorganic counterparts [57–59]. Electrochemical polymerization provides a facile and effective approach for preparing large-area, flexible organic thermoelectric devices [60–62]. In addition to synthesizing pure organic thermoelectric materials (e.g., poly(3,4-ethylenedioxythiophene) [60], poly(nickel-ethylenetetrathiolate) [63], poly (3,4-ethylenedioxythiophene): poly(4-styrenesulfonate)/polythiophenes/polythiophenes [1]), electrochemical polymerization methods have also been employed to fabricate hybrid thermoelectric films that include both organic-organic and inorganic-organic combinations (e.g., bismuth-antimony-telluride alloy/poly(3,4-ethylenedioxythiophene [64]), multiwalled carbon nanotubes/poly(3,4-ethylenedioxythiophene) [65, 66], single-walled carbon nanotubes/polyaniline [67]) with superior thermoelectric performance by leveraging the advantages of each component.

Organic solar cells offer several advantages, including lightweight construction, low fabrication costs, and flexibility, compared to conventional silicon-based photovoltaic cells. The electrochemical polymerization method has long been employed to synthesize photoactive layer materials. For instance, the first polyisothianaphthene was prepared through electrochemical polymerization [68], and early soluble and processable conjugated poly(thiophenes) were primarily synthesized using this method [69]. Currently, electrochemical polymerization remains a popular approach for preparing photoactive materials in organic solar cells, such as π-conjugated microporous polymer films [70] and poly(3,4-ethylenedioxythiophene) [71]. In addition to synthesizing the active layers of organic solar cells [72, 73], the electrochemical polymerization method has also been applied to produce key components, such as hole-transporting layers and electron extraction layers [74] in perovskite solar cells.

Polymer electrolyte membrane fuel cells convert chemical energy directly into electricity. The microstructure of the electrode significantly influences the local mass transport within the membrane electrode assembly, thereby determining the overall performance of the fuel cell. Xia et al. reported the fabrication of vertically aligned polypyrrole nanowires through electrochemical polymerization for direct methanol fuel cell applications [75]. The effects of both anionic soft templates and electric fields induced conglomerations that facilitated vertical alignment in response to the electric field. Furthermore, polypyrrole nanowires can be tuned from disordered to well-ordered, vertically aligned structures by simply adjusting electrochemical deposition parameters, such as potential and temperature [76]. Electrochemically polymerized polypyrrole films have also been demonstrated to be efficient cathode materials in photo-fuel cells [77]. Additionally, other polymers, such as hydroquinone, can be electrochemically deposited onto conductive substrates for microbial fuel cell applications [78].

15.4.3 Electrochemical Polymerization Fabricated Polymers in Energy Storage Devices

Batteries are among the most widely used energy storage devices for various applications today. Among them, metal-ion batteries have garnered increased attention due to their potential to meet the future substantial demand for large-scale energy storage [79]. Electrodeposited polymers have demonstrated extensive applications in this field [80, 81].

In the most extensively studied lithium-ion batteries, irreversible structural instabilities and severe interfacial issues in both anode and cathode materials lead to rapid capacity fading. Compositing electrode materials with conductive polymers has been demonstrated as an effective solution to mitigate capacity loss. In particular, electrochemical polymerization offers a more facile in situ approach for compositing electrode materials with polymers compared to alternative methods such as oxidative polymerization [82, 83]. This strategy has proven effective in enhancing the stability of both anode (e.g., polypyrrole/SnO_2 [84], polyacrylonitrile/TiO_2 [85], polyaniline/Cu [86]) and cathode materials (e.g., polyaniline/$LiNi_{0.8}Co_{0.1}Mn_{0.1}O_2$ [82], poly(3,4-ethylenedioxythiophene)/$LiFePO_4$ [83]). The conductive polymers themselves have also been identified as viable candidates for use as cathodes in lithium-ion batteries; for instance, electrochemically polymerized 4,4'-triazole-9-yl-triphenylamine [87], 1,5-diaminoanthraquinone [88], copolymers of thiophene and 3,4-ethylenedioxythiophene [28], and copolymers of poly(3,4-ethylenedioxythiophene) with hydroquinone [89] can serve as promising cathode materials exhibiting excellent performance. Furthermore, electrochemically fabricated polymers have also been applied as promising electrolytes in lithium-ion batteries [90]. Electrochemically polymerized conductive polymers have shown unique functionalities in other lithium-based battery systems including Li—S [91, 92], and Li—O_2 batteries [93].

The electrochromic battery, a newly developed energy storage technology, is a single electrochromic energy storage device that possesses both energy storage capabilities and electrochromic characteristics [94]. The electrochromic process, which involves the reversible storing and releasing of charge carriers, imparts battery-like features [95]. A self-rechargeable electrochromic battery has been constructed using an electrodeposited p-toluenesulfonate-doped polypyrrole film [96]. However, the thin active material films required for achieving fast-switching speeds in electrochromic devices conflict with the demands of high-capacity batteries [97]. Recently, there has been significant interest in integrating electrochromic devices with rechargeable aqueous metal-ion batteries to fabricate hybrid electrochromic batteries. A notable example is a hybrid electrochromic aqueous zinc battery featuring an electrodeposited one-dimensional iron terpyri dyl coordination polymer cathode, which demonstrates outstanding performance characterized by high color contrast, rapid response time, and excellent endurance [98]. Additionally, the tunable transmittance of these electrochromic devices offers opportunities for fabricating transparent batteries. Notably, substantial progress has been made in flexible electrochromic zinc-ion batteries; numerous all-in-one flexible designs utilizing electrodeposited organic electrochromic cathodes have been reported, including Zn//self-doped polyaniline [99] and Zn@Ni@AgNFs//PANi/HTE[9].

15.5 Concluding Remarks

The electrochemical polymerization method has been demonstrated to be a facile and effective approach for precisely depositing polymer films onto electrode surfaces for molecular electronic devices. This method has also been extended to synthesize newly developed functional materials, such as covalent organic frameworks, beyond traditional polymers. Furthermore, the electrochemical polymerization technique has shown significant potential in fabricating modern energy devices. However, it presents challenges in preparing homogeneous films over large areas, particularly on substrates with relatively high resistance due to the non-uniform distribution of the electric field [100]. Additionally, film properties are largely influenced by monomer identity along with other electrochemical polymerization parameters such as electrolyte types and methods used; this results in a lack of universal correlation between polymerization parameters and film characteristics [4, 43, 101]. While the low throughput of the electrochemical polymerization method may lead to its replacement by alternative solution-processable coating and printing techniques like spray casting, its advantages in fabricating functional layers allow for precise control over film parameters including composition, thickness, and other properties. Moreover, it is well-suited for forming multi-layer films, making it one of the preferred methods for constructing precision electronics and highly efficient energy devices. It is foreseeable that this method will play an irreplaceable role in the next generation of organic or hybrid electronic devices and energy systems.

Acknowledgment

This work was supported by the National Key R&D Program of China (2023YFA1508300) and National Natural Science Foundation of China (22425021).

References

1 Shi, H., Liu, C., Xu, J., et al. (2013). Facile fabrication of PEDOT:PSS/polythiophenes bilayered nanofilms on pure organic electrodes and their thermoelectric performance. *ACS Appl. Mater. Interf.* 5 (24): 12811.

2 Yang, S., Lapsley, M.I., Cao, B., et al. (2012). Large-scale fabrication of three-dimensional surface patterns using template-defined electrochemical deposition. *Adv. Funct. Mater.* 23 (6): 720.

3 Hou, S., Wang, S., Yu, Z.T.F., et al. (2008). A hydrodynamically focused stream as a dynamic template for site-specific electrochemical micropatterning of conducting polymers. *Angew. Chem. Int. Ed.* 47 (6): 1072.

4 Heinze, J., Frontana-Uribe, B.A., Ludwigs, S. (2010). Electrochemistry of conducting polymers – persistent models and new concepts. *Chem. Rev.* 110 (8): 4724.

5 Bocchetta, P., Frattini, D., Tagliente, M., et al. (2020). Electrochemical deposition of polypyrrole nanostructures for energy applications: a review. *Curr. Nanosci.* 16 (4): 462.

6 Nie, S., Li, Z., Yao, Y., et al. (2021). Progress in synthesis of conductive polymer poly(3,4-ethylenedioxythiophene). *Front. Chem.* 9: 803509.

7 Fenoy, G.E., Azzaroni, O., Knoll, W., et al. (2021). Functionalization strategies of PEDOT and PEDOT:PSS films for organic bioelectronics applications. *Chemosensors* 9 (8): 212.

8 Fink, J.K. In *High Performance Polymers* (Second Edition), DOI: https://doi.org/10.1016/B978-0-323-31222-6.00003-0 (Ed: J. K. Fink), William Andrew Publishing 2014, p. 71.

9 Singh, S.B., Tran, D.T., Jeong, K.U., et al. (2021). A flexible and transparent zinc-nanofiber network electrode for wearable electrochromic, rechargeable Zn-ion battery. *Small* 18 (5): 2104462.
10 Yilmaz, P., Magni, M., Martinez, S., et al. (2020). Spectrally selective PANI/ITO nanocomposite electrodes for energy-efficient dual band electrochromic windows. *ACS Appl. Mater. Interf.* 3 (4): 3779.
11 Wang, B., Xu, G., Song, S., et al. (2022). Flexible, infrared adjustable, thermal radiation control device based on electro-emissive PANI/Ce4+ thin films inspired by chameleon. *Chem. Eng. J.* 445: 136819.
12 Palma-Cando, A., and Scherf, U. (2016). Electrochemically generated thin films of microporous polymer networks: synthesis, properties, and applications. *Macromol. Chem. Phys.* 217 (7): 827.
13 Feng, J.f., Liu, T.F., and Cao, R. (2020). An electrochromic hydrogen-bonded organic framework film. *Angew. Chem. Int. Ed.* 59 (50): 22392.
14 Ellis, T.K., Galerne, M., Armao, J.J., et al. (2018). Supramolecular electropolymerization. *Angew. Chem. Int. Ed.* 57 (48): 15749.
15 Wang, Y.A., Wu, Q., Wang, X., et al. (2024). In situ electrochemical interfacial polymerization for covalent organic frameworks with tunable electrochromism. *Angew. Chem. Int. Ed.* 63 (46): e202413071.
16 Wang, X., Wu, J., Liu, H., et al. (2023). Cathodic polymerization through electrochemical dehalogenation. *Macromol.* 56 (24): 10198.
17 Zhang, C., Li, Y., Li, H., et al. (2021). Overview of electric-field-induced deposition technology in fabricating organic thin films. *J. Mater. Chem. C* 9 (2): 374.
18 Diaz, A.F., Castillo, J.I., Logan, J.A., et al. (1981). Electrochemistry of conducting polypyrrole films. *J. Electroanalyt. Chem. Interf. Electrochem.* 129 (1–2): 115.
19 Genies, E.M., Bidan, G., and Diaz, A.F. (1983). Spectroelectrochemical study of polypyrrole films. *J. Electroanalyt. Chem. Interf. Electrochem.* 149 (1–2): 101.
20 Heinze, J., John, H., Dietrich, M., et al. (2001). σ-"Dimers" – key intermediates and products during generation and redox switching of conjugated oligomers and polymers. *Synth. Metals* 119 (1–3): 49.
21 Downard, A.J. and Pletcher, D. (1986). The influence of water on the electrodeposition of polypyrrole in acetonitrile. *J. Electroanalyt. Chem. Interf. Electrochem.* 206 (1–2): 139.
22 Bone, J.M. and Jenkins, J.L. (2023). Understanding polymer electrodeposition and conducting polymer modified electrodes using electrochemistry, spectroscopy, and scanning probe microscopy. *J. Chem. Educat.* 100 (10): 4062.
23 Elsherbini, M. and Wirth, T. (2019). Electroorganic synthesis under flow conditions. *Acc. Chem. Res.* 52 (12): 3287.
24 Noël, T., Cao, Y., and Laudadio, G. (2019). The fundamentals behind the use of flow reactors in electrochemistry. *Acc. Chem. Res.* 52 (10): 2858.
25 Shida, N., Zhou, Y., and Inagi, S. (2019). Bipolar electrochemistry: a powerful tool for electrifying functional material synthesis. *Acc. Chem. Res.* 52 (9): 2598.
26 Villani, E. and Inagi, S. (2022). *Electrochemistry in Organic Synthesis*. Thieme. DOI: 10.1055/sos-SD-236-00125.
27 Peters, B.K., Rodriguez, K.X., Reisberg, S.H., et al. (2019). Scalable and safe synthetic organic electroreduction inspired by Li-ion battery chemistry. *Science* 363(6429): 838.

28 Atobe, M., Tateno, H., and Matsumura, Y. (2017). Applications of flow microreactors in electrosynthetic processes. *Chem. Rev.* 118(9): 4541.

29 Mizuno, M., Tateno, H., Matsumura, Y., et al. (2017). Synthesis and molecular weight control of poly(3-hexylthiophene) using electrochemical polymerization in a flow microreactor. *React. Chem. Eng.* 2(5): 642.

30 Nakamura, Y., Tanaka, K., Matsumura, Y., et al. (2020). Flow electrosynthesis and molecular weight control of polyphenylene deriving from 1,4-bis(trimethylsilyl) benzene: effect of a silyl substituent on the coupling position. *Electrochemistry* 88(4): 336.

31 Gao, W., Muzyka, K., Ma, X., et al. (2018). A single-electrode electrochemical system for multiplex electrochemiluminescence analysis based on a resistance induced potential difference. *Chem. Sci.* 9(16): 3911.

32 Du, F., Dong, Z., Guan, Y., et al. (2022). Single-electrode electrochemical system for the visual and high-throughput electrochemiluminescence immunoassay. *Analyt. Chem.* 94(4): 2189.

33 Shi, Y., Villani, E., Chen, Y., et al. (2022). High-throughput electrosynthesis of gradient polypyrrole film using a single-electrode electrochemical system. *Analyt. Chem.* 95(2): 1532.

34 Mavré, F., Chow, K.-F., Sheridan, E., et al. (2009). A theoretical and experimental framework for understanding electrogenerated chemiluminescence (ECL) emission at bipolar electrodes. *Analyt. Chem.* 81(15): 6218.

35 Iwai, S., Suzuki, T., Sakagami, H., et al. (2022). Electropolymerization without an electric power supply. *Commun. Chem.* 5(1): 66.

36 Yen, H.-J. and Liou, G.-S. (2018). Recent advances in triphenylamine-based electrochromic derivatives and polymers. *Polym. Chem.* 9(22): 3001.

37 Feng, F., Guo, S., Ma, D., et al. (2023). An overview of electrochromic devices with electrolytes containing viologens. *Solar Energy Mater. Solar Cells* 254.

38 Tan, X.Y., Wang, H., Kim, T.G., et al. (2020). Composite Materials: Applications in Engineering, Biomedicine and Food Science. DOI: https://doi.org/10.1007/978-3-030-45489-0_9, Chapter 9, p. 213.

39 Rao, T., Zhou, Y., Jiang, J., et al. (2022). Low dimensional transition metal oxide towards advanced electrochromic devices. *Nano Energy* 100: 107479.

40 Madasamy, K., Velayutham, D., Suryanarayanan, V., et al. (2019). Viologen-based electrochromic materials and devices. *J. Mater. Chem. C* 7(16): 4622.

41 Neo, W.T., Ye, Q., Chua, S.-J., et al. (2016). Conjugated polymer-based electrochromics: materials, device fabrication and application prospects. *J. Mater. Chem. C* 4(31): 7364.

42 Lin, K., Wu, C., Ming, S., et al. (2025). Electrochromic conjugated polymers: The relationship of gradient band gap and electrochromic performance. *Dyes Pigm.* 232: 112504.

43 Jensen, J., Hösel, M., Dyer, A.L., et al. (2015). Development and manufacture of polymer-based electrochromic devices. *Adv. Funct. Mater.* 25(14): 2073.

44 Chen, Y., Niu, C., Wang, L., et al. (2024). Multi-pattern polyaniline electrochromic device by controllable three-dimensional movement of ions. *Opt. Mater.* 147: 114605.

45 Fonseca, S.M., Moreira, T., Parola, A.J., et al. (2017). PEDOT electrodeposition on oriented mesoporous silica templates for electrochromic devices. *Solar Energy Mater. Solar Cells* 159: 94.

46 Fu, W., Chen, H., Han, Y., et al. (2021). Electropolymerization of D–A–D type monomers consisting of triphenylamine and substituted quinoxaline moieties for electrochromic devices. *New J. Chem.* 45(40): 19082.

47 Xiong, S., Zhang, J., Wu, B., et al. (2020). Electrochemical preparation of covalently bonded PEDOT-graphene oxide composite electrochromic materials using thiophene-2-methylanine as bridging group. *ChemistrySelect* 5(39): 12206.

48 Qin, T., Deng, L., Zhang, N., et al. (2024). Flexible polyaniline/carbon quantum dots films for enhanced electrochromic performance. *Ceram. Int.* 50(24): 53254.

49 Liu, L. (2024). Solution-Processed Organic Light-emitting Devices. DOI: https://doi.org/10.1016/b978-0-323-95146-3.00004-5, p. 477.

50 Wang, R., Zhang, D., Xiong, Y., et al. (2018). TFT-directed electroplating of Rgb luminescent films without a vacuum or mask toward a full-color AMOLED pixel matrix. *ACS Appl. Mater. Interf.* 10(21): 17519.

51 Gu, C., Fei, T., Lv, Y., et al. (2010). Color-stable white electroluminescence based on a cross-linked network film prepared by electrochemical copolymerization. *Adv. Mater.* 22(24): 2702.

52 Li, M., Tang, S., Shen, F., et al. (2006). Highly luminescent network films from electrochemical deposition of peripheral carbazole functionalized fluorene oligomer and their applications for light-emitting diodes. *Chem. Commun.* 32: 3393.

53 Gu, C., Tang, S., Yang, B., et al. (2009). Almost completely dedoped electrochemically deposited luminescent films exhibiting excellent LED performance. *Electrochim. Acta* 54(27): 7006.

54 Lv, Y., Yao, L., Gu, C., et al. (2013). Cross linked luminescent films via electropolymerization of multifunctional precursors for highly efficient electroluminescence. *Polym. Chem.* 4(6): 2090.

55 Liang, Y., Liu, C., Zhao, M., et al. (2020). Organic electropolymerized multilayers for light-emitting diodes and displays. *ACS Appl. Mater. Interf.* 12(18): 20714.

56 Feng, M., Lv, S., Deng, J., et al. (2023). An overview of environmental energy harvesting by thermoelectric generators. *Renew. Sustain. Energy Rev.* 187.

57 Petsagkourakis, I., Tybrandt, K., Crispin, X., et al. (2018). Thermoelectric materials and applications for energy harvesting power generation. *Sci. Technol. Adv. Mater.* 19(1): 836.

58 Russ, B., Glaudell, A., Urban, J.J., et al. (2016). Organic thermoelectric materials for energy harvesting and temperature control. *Nat. Rev. Mater.* 1(10): 16050.

59 Massetti, M., Jiao, F., Ferguson, A.J., et al. (2021). Unconventional thermoelectric materials for energy harvesting and sensing applications. *Chem. Rev.* 121(20): 12465.

60 Ye, G., Xu, J., Ma, X., et al. (2017). One-step electrodeposition of free-standing flexible conducting PEDOT derivative film and its electrochemical capacitive and thermoelectric performance. *Electrochim. Acta* 224: 125.

61 Zhao, Y., Liu, L., Zhang, F., et al. (2021). Advances in organic thermoelectric materials and devices for smart applications. *SmartMat* 2(4): 426.

62 Serrano-Claumarchirant, J.F., Nasiri, M.A., Cho, C., et al. (2023). Textile-based thermoelectric generator produced via electrochemical polymerization. *Adv. Mater. Interf.* 10(8): 2202105.

63 Sun, Y., Qiu, L., Tang, L., et al. (2016). Flexible n-type high-performance thermoelectric thin films of poly(nickel-ethylenetetrathiolate) prepared by an electrochemical method. *Adv. Mater.* 28(17): 3351.

64 Cho, S., Lim, J., and Seo, Y. (2023). Fabrication of bismuth–antimony–telluride alloy/poly(3,4-ethylenedioxythiophene) hybrid films for thermoelectric applications at room temperature by a simple electrochemical process. *Chem. Mater.* 35(8): 3196.

65 Culebras, M., Cho, C., Krecker, M., et al. (2017). High thermoelectric power factor organic thin films through combination of nanotube multilayer assembly and electrochemical polymerization. *ACS Appl. Mater. Interf.* 9(7): 6306.

66 Serrano-Claumarchirant, J.F., Brotons-Alcázar, I., Culebras, M., et al. (2020). Electrochemical synthesis of an organic thermoelectric power generator. *ACS Appl. Mater. Interf.* 12(41): 46348.

67 Yin, S., Lu, W., Wu, X., et al. (2021). Enhancing thermoelectric performance of polyaniline/single-walled carbon nanotube composites via dimethyl sulfoxide-mediated electropolymerization. *ACS Appl. Mater. Interf.* 13(3): 3930.

68 Wudl, F., Kobayashi, M., and Heeger, A.J. (2002). Poly(isothianaphthene). *The J. Organ. Chem.* 49(18): 3382.

69 Roncali, J. (2002). Conjugated poly(thiophenes): synthesis, functionalization, and applications. *Chem. Rev.* 92(4): 711.

70 Gu, C., Huang, N., Chen, Y., et al. (2015). π-conjugated microporous polymer films: designed synthesis, conducting properties, and photoenergy conversions. *Angew. Chem. Int. Ed.* 127(46): 13798.

71 Carbas, B.B., Gulen, M., Tolu, M.C., et al. (2017). Hydrogen sulphate-based ionic liquid-assisted electro-polymerization of PEDOT catalyst material for high-efficiency photoelectrochemical solar cells. *Sci. Rep.* 7(1): 11672.

72 Liu, Z., Xu, Y., Li, A., et al. (2024). Efficient and stable perovskite solar cells through electrochemically polymerized porphyrin-based hole-transporting materials. *J. Mater. Chem. C* 12(32): 12282.

73 Mabrouk, S., Gurung, A., Bahrami, B., et al. (2022). Electrochemically prepared polyaniline as an alternative to poly(3,4-ethylenedioxythiophene)-poly(styrenesulfonate) for inverted perovskite solar cells. *ACS Appl. Mater. Interf.* 5(8): 9351.

74 Suárez, M.B., Aranda, C., Macor, L., et al. (2018). Perovskite solar cells with versatile electropolymerized fullerene as electron extraction layer. *Electrochim. Acta* 292: 697.

75 Xia, Z., Wang, S., Li, Y., et al. (2013). Vertically oriented polypyrrole nanowire arrays on Pd-plated Nafion® membrane and its application in direct methanol fuel cells. *J. Mater. Chem. A* 1(3): 491.

76 Xia, Z., Wang, S., Jiang, L., et al. (2014). Controllable synthesis of vertically aligned polypyrrole nanowires as advanced electrode support for fuel cells. *J. Power Sourc.* 256: 125.

77 Li, K., Zhang, H., Tang, T., et al. (2016). Facile electrochemical polymerization of polypyrrole film applied as cathode material in dual rotating disk photo fuel cell. *J. Power Sourc.* 324: 368.

78 Wang, G. and Feng, C. (2017). Electrochemical polymerization of hydroquinone on graphite felt as a pseudocapacitive material for application in a microbial fuel cell. *Polymers* 9(6): 220.

79 Liang, Y., Dong, H., Aurbach, D., et al. (2020). Current status and future directions of multivalent metal-ion batteries. *Nat. Energy* 5(9): 646.

80 Zhang, C.-X., Mei, S.-L., Chen, X.-H., et al. (2021). Electrochemical construction of functional polymers and their application advances in lithium batteries. *J. Mater. Chem. C* 9(48): 17182.

81 Lin, S., Wu, Q., and Lu, Y. (2024). Recent progress of the application of electropolymerization in batteries and supercapacitors: specific design of functions in electrodes. *ChemElectroChem* 11(12): e202300776.

82 Dou, L.-T., Li, B., Nie, H.-L., et al. (2024). Introducing electrolytic electrochemical polymerization for constructing protective layers on Ni-rich cathodes of Li-ion batteries. *Rare Metals* 43(6): 2536.

83 Cíntora-Juárez, D., Pérez-Vicente, C., Ahmad, S., et al. (2014). Improving the cycling performance of $LiFePO_4$ cathode material by poly(3,4-ethylenedioxythiopene) coating. *RSC Advances* 4(50): 26108.

84 Du, Z., Zhang, S., Jiang, T., et al. (2012). Facile synthesis of SnO_2 nanocrystals coated conducting polymer nanowires for enhanced lithium storage. *J. Power Sourc.* 219: 199.

85 Nacimiento, F., González, J.R., Alcántara, R., et al. (2013). Improving the electrochemical properties of self-organized titanium dioxide nanotubes in lithium batteries by surface polyacrylonitrile electropolymerization. *J. Electrochem. Soc.* 160(5): A3026.

86 Zhang, Y., Xiao, R., Liao, X., et al. (2020). Polyaniline/copper composite anode current collectors prepared through electrochemical polymerization for lithium-ion batteries. *ChemElectroChem*, 7(13): 2896.

87 Zhao, C., Chen, Z., Wang, W., et al. (2020). In situ electropolymerization enables ultrafast long cycle life and high-voltage organic cathodes for lithium batteries. *Angew. Chem. Int. Ed.* 59(29): 11992.

88 Liu, T., Kim, K.C., Lee, B., et al. (2020). Enhanced lithium storage of an organic cathode via the bipolar mechanism. *ACS Appl. Energy Mater.* 3(4): 3728.

89 Oka, K., Strietzel, C., Emanuelsson, R., et al. (2019). Characterization of PEDOT-Quinone conducting redox polymers in water-in-salt electrolytes for safe and high-energy Li-ion batteries. *Electrochem. Commun.* 105: 106489.

90 Zhou, D., Shanmukaraj, D., Tkacheva, A., et al. (2019). Polymer electrolytes for lithium-based batteries: advances and prospects. *Chem*, 5(9): 2326.

91 Chen, X., Zhao, C., Yang, K., et al. (2023). Conducting polymers meet lithium–sulfur batteries: progress, challenges, and perspectives. *Energy Environm. Mater.* 6(5): e12483.

92 Huang, S., Guan, R., Wang, S., et al. (2019). Polymers for high performance Li-S batteries: material selection and structure design. *Progr. Polym. Sci.* 89: 19.

93 Cao, D., Shen, X., Wang, Y., et al. (2019). Conductive polymer coated cathodes in Li–O2 batteries. *ACS Appl. Energy Mater.* 3(1): 951.

94 Zhao, Q., Pan, Z., Liu, B., et al. (2023). Electrochromic-induced rechargeable aqueous batteries: an integrated multifunctional system for cross-domain applications. *Nano-Micro Lett.* 15(1): 87.

95 Granqvist, C.G., Lansåker, P.C., Mlyuka, N.R., et al. (2009). Progress in chromogenics: new results for electrochromic and thermochromic materials and devices. *Solar Energy Mater. Solar Cells* 93(12): 2032.

96 Yang, B., Ma, D., Zheng, E., et al. (2019). A self-rechargeable electrochromic battery based on electrodeposited polypyrrole film. *Solar Energy Mater. Solar Cells* 192: 1.

97 Yang, Y., Jeong, S., Hu, L., et al. (2011). Transparent lithium-ion batteries. *Proc. Natl. Acad. Sci. USA* 108(32): 13013.

98 Poh, W.C., Gong, X., Yu, F., et al. (2021). Electropolymerized 1D growth coordination polymer for hybrid electrochromic aqueous zinc battery. *Adv. Sci.* 8(21): 2101944.

99 Wang, Y., Jiang, H., Zheng, R., et al. (2020). A flexible, electrochromic, rechargeable Zn-ion battery based on actiniae-like self-doped polyaniline cathode. *J. Mater. Chem. A* 8(25): 12799.

100 Seshadri, V., Padilla, J., Bircan, H., et al. (2007). Optimization, preparation, and electrical short evaluation for 30 cm^2 active area dual conjugated polymer electrochromic windows. *Organ. Electr.* 8(4): 367.

101 Sabouraud, G., Sadki, S., and Brodie, N. (2000). The mechanisms of pyrrole electropolymerization. *Chem. Soc. Rev.* 29 (5): 283.

16

Summary and Perspective

*Yuanmiao Sun**

Shenzhen Institute of Advanced Technology, Chinese Academy of Sciences, Shenzhen, Guangdong, P. R. China

*Corresponding author: sunym@siat.ac.cn

16.1 Summary

Electrochemical Energy Conversion via Small-Molecule Catalysis provides a comprehensive exploration of the principles, techniques, and applications underpinning the field of electrochemical energy conversion. The book is organized into 16 chapters, covering key aspects from foundational theories to advanced computational methods, which could offer a well-rounded perspective on small-molecule catalysis.

The introductory chapters establish the groundwork by explaining core concepts in electrochemical energy conversion and the fundamental theories behind small-molecule catalysis. A strong emphasis is placed on advanced characterization techniques, such as in situ spectroscopy and X-ray absorption fine structure analysis, enabling readers to understand the dynamic evolution of catalytic systems.

The book then discusses some specific electrochemical reactions that are critical for sustainable energy solutions, including hydrogen and oxygen evolution/conversion, biomass derivative conversion, and CO_2 and nitrogen reduction. Emerging areas such as carbon-nitrogen coupling reactions, electrochemical fluorination, and polymerization are also explored, highlighting their potential to broaden the scope of catalysis applications.

By combining theoretical insights with practical examples, this book equips readers with the knowledge needed to understand and contribute to the development of next-generation energy technologies.

Electrochemical Energy Conversion via Small Molecule Catalysis, First Edition.
Edited by Zhicheng Zhang and Yuanmiao Sun.

16.2 Perspective

The chapters in this book highlight the pivotal role of small-molecule catalysis in shaping the future of electrochemical energy conversion. As the field continues to evolve, several key trends and challenges may emerge in the future:

1. *Catalyst Development:* The design of catalysts with enhanced and dynamic activity, selectivity, and stability remains a primary goal. Efforts must focus on leveraging multi-functional catalysts capable of operating efficiently under varying conditions.
2. *Advanced Characterization:* Cutting-edge techniques, such as in situ and *operando* analyses, will continue to revolutionize the understanding of catalytic mechanisms, particularly in identifying active sites and elucidating reaction pathways.
3. *System Integration:* While individual catalytic processes have seen significant progress, their integration into practical devices and systems requires addressing scalability, cost-effectiveness, and durability under real-world conditions.
4. *Emerging Reactions:* Electrochemical processes like fluorination, polymerization, and carbon–nitrogen coupling offer promising avenues for new applications but require further investigation to unlock their full potential.
5. *Computational Advances:* Machine learning and high-throughput simulations will increasingly guide the discovery of optimal catalysts, speeding up the transition from theoretical designs to experimental validation.

In the future, interdisciplinary collaborations between chemists, engineers, and data scientists will be critical in overcoming technical barriers and accelerating the deployment of sustainable energy technologies. We hope this book not only captures the current state-of-the-art developments of electrochemical energy conversion technologies but also provides a vision for future breakthroughs that could transform how energy is generated, stored, and utilized in a carbon-neutral economy.

Index

Note: Page numbers in *italics* and **bold** refers to figures and tables respectively.

Electrochemical Energy Conversion via Small Molecule Catalysis, First Edition.
Edited by Zhicheng Zhang and Yuanmiao Sun.

d